From Frenet to Cartan: The Method of Moving Frames

GRADUATE STUDIES IN MATHEMATICS 178

From Frenet to Cartan: The Method of Moving Frames

Jeanne N. Clelland

American Mathematical Society
Providence, Rhode Island

2010 *Mathematics Subject Classification.* Primary 22F30, 53A04, 53A05, 53A15, 53A20, 53A55, 53B25, 53B30, 58A10, 58A15.

For additional information and updates on this book, visit
www.ams.org/bookpages/gsm-178

Library of Congress Cataloging-in-Publication Data

Names: Clelland, Jeanne N., 1970-
Title: From Frenet to Cartan : the method of moving frames / Jeanne N. Clelland.
Description: Providence, Rhode Island : American Mathematical Society, [2017] | Series: Graduate studies in mathematics ; volume 178 | Includes bibliographical references and index.
Identifiers: LCCN 2016041073 | ISBN 9781470429522 (alk. paper)
Subjects: LCSH: Frames (Vector analysis) | Vector analysis. | Exterior differential systems. | Geometry, Differential. | Mathematical physics. | AMS: Topological groups, Lie groups – Noncompact transformation groups – Homogeneous spaces. msc | Differential geometry – Classical differential geometry – Curves in Euclidean space. msc | Differential geometry – Classical differential geometry – Surfaces in Euclidean space. msc | Differential geometry – Classical differential geometry – Affine differential geometry. msc | Differential geometry – Classical differential geometry – Projective differential geometry. msc | Differential geometry – Classical differential geometry – Differential invariants (local theory), geometric objects. msc | Differential geometry – Local differential geometry – Local submanifolds. msc | Differential geometry – Local differential geometry – Lorentz metrics, indefinite metrics. msc | Global analysis, analysis on manifolds – General theory of differentiable manifolds – Differential forms. msc | Global analysis, analysis on manifolds – General theory of differentiable manifolds – Exterior differential systems (Cartan theory). msc
Classification: LCC QA433 .C564 2017 | DDC 515/.63–dc23 LC record available at https://lccn.loc.gov/2016041073

♾ The paper used in this book is acid-free and falls within the guidelines
established to ensure permanence and durability.
Visit the AMS home page at `http://www.ams.org/`

10 9 8 7 6 5 4 3 2 1   22 21 20 19 18 17

To Rick, Kevin, and Valerie, who make everything worthwhile

Contents

Preface xi

Acknowledgments xv

Part 1. Background material

Chapter 1. Assorted notions from differential geometry 3

§1.1. Manifolds 3

§1.2. Tensors, indices, and the Einstein summation convention 9

§1.3. Differentiable maps, tangent spaces, and vector fields 15

§1.4. Lie groups and matrix groups 26

§1.5. Vector bundles and principal bundles 32

Chapter 2. Differential forms 35

§2.1. Introduction 35

§2.2. Dual spaces, the cotangent bundle, and tensor products 35

§2.3. 1-forms on $\mathbb{R}^n$ 40

§2.4. p-forms on $\mathbb{R}^n$ 41

§2.5. The exterior derivative 43

§2.6. Closed and exact forms and the Poincaré lemma 46

§2.7. Differential forms on manifolds 47

§2.8. Pullbacks 49

§2.9. Integration and Stokes's theorem 53

§2.10. Cartan's lemma 55

§2.11. The Lie derivative 56
§2.12. Introduction to the `Cartan` package for Maple 59

Part 2. Curves and surfaces in homogeneous spaces via the method of moving frames

Chapter 3. Homogeneous spaces 69
§3.1. Introduction 69
§3.2. Euclidean space 70
§3.3. Orthonormal frames on Euclidean space 75
§3.4. Homogeneous spaces 84
§3.5. Minkowski space 85
§3.6. Equi-affine space 92
§3.7. Projective space 96
§3.8. Maple computations 103

Chapter 4. Curves and surfaces in Euclidean space 107
§4.1. Introduction 107
§4.2. Equivalence of submanifolds of a homogeneous space 108
§4.3. Moving frames for curves in $\mathbb{E}^3$ 111
§4.4. Compatibility conditions and existence of submanifolds with prescribed invariants 115
§4.5. Moving frames for surfaces in $\mathbb{E}^3$ 117
§4.6. Maple computations 134

Chapter 5. Curves and surfaces in Minkowski space 143
§5.1. Introduction 143
§5.2. Moving frames for timelike curves in $\mathbb{M}^{1,2}$ 144
§5.3. Moving frames for timelike surfaces in $\mathbb{M}^{1,2}$ 149
§5.4. An alternate construction for timelike surfaces 161
§5.5. Maple computations 166

Chapter 6. Curves and surfaces in equi-affine space 171
§6.1. Introduction 171
§6.2. Moving frames for curves in $\mathbb{A}^3$ 172
§6.3. Moving frames for surfaces in $\mathbb{A}^3$ 178
§6.4. Maple computations 191

Chapter 7. Curves and surfaces in projective space 203
§7.1. Introduction 203
§7.2. Moving frames for curves in $\mathbb{P}^2$ 204
§7.3. Moving frames for curves in $\mathbb{P}^3$ 214
§7.4. Moving frames for surfaces in $\mathbb{P}^3$ 220
§7.5. MAPLE computations 235

Part 3. Applications of moving frames

Chapter 8. Minimal surfaces in $\mathbb{E}^3$ and $\mathbb{A}^3$ 251
§8.1. Introduction 251
§8.2. Minimal surfaces in $\mathbb{E}^3$ 251
§8.3. Minimal surfaces in $\mathbb{A}^3$ 268
§8.4. MAPLE computations 280

Chapter 9. Pseudospherical surfaces and Bäcklund's theorem 287
§9.1. Introduction 287
§9.2. Line congruences 288
§9.3. Bäcklund's theorem 289
§9.4. Pseudospherical surfaces and the sine-Gordon equation 293
§9.5. The Bäcklund transformation for the sine-Gordon equation 297
§9.6. MAPLE computations 303

Chapter 10. Two classical theorems 311
§10.1. Doubly ruled surfaces in $\mathbb{R}^3$ 311
§10.2. The Cauchy-Crofton formula 324
§10.3. MAPLE computations 329

Part 4. Beyond the flat case: Moving frames on Riemannian manifolds

Chapter 11. Curves and surfaces in elliptic and hyperbolic spaces 339
§11.1. Introduction 339
§11.2. The homogeneous spaces $\mathbb{S}^n$ and $\mathbb{H}^n$ 340
§11.3. A more intrinsic view of $\mathbb{S}^n$ and $\mathbb{H}^n$ 345
§11.4. Moving frames for curves in $\mathbb{S}^3$ and $\mathbb{H}^3$ 348
§11.5. Moving frames for surfaces in $\mathbb{S}^3$ and $\mathbb{H}^3$ 351
§11.6. MAPLE computations 357

Chapter 12. The nonhomogeneous case: Moving frames on Riemannian manifolds 361
§12.1. Introduction 361
§12.2. Orthonormal frames and connections on Riemannian manifolds 362
§12.3. The Levi-Civita connection 370
§12.4. The structure equations 373
§12.5. Moving frames for curves in 3-dimensional Riemannian manifolds 379
§12.6. Moving frames for surfaces in 3-dimensional Riemannian manifolds 381
§12.7. MAPLE computations 388
Bibliography 397
Index 403

Preface

Perhaps the earliest example of a moving frame is the Frenet frame along a nondegenerate curve in the Euclidean space $\mathbb{R}^3$, consisting of a triple of orthonormal vectors (T, N, B) based at each point of the curve. First introduced by Bartels in the early nineteenth century [**Sen31**] and later described by Frenet in his thesis [**Fre47**] and Serret in [**Ser51**], the frame at each point is chosen based on properties of the geometry of the curve near that point, and the fundamental geometric invariants of the curve—curvature and torsion—appear when the derivatives of the frame vectors are expressed in terms of the frame vectors themselves.

In the late nineteenth century, Darboux studied the problem of constructing moving frames on surfaces in Euclidean space [**Dar72a**], [**Dar72b**], [**Dar72c**], [**Dar72d**]. In the early twentieth century, Élie Cartan generalized the notion of moving frames to other geometries (for example, affine and projective geometry) and developed the theory of moving frames extensively. A very nice introduction to Cartan's ideas may be found in Guggenheimer's text [**Gug77**].

More recently, Fels and Olver [**FO98**], [**FO99**] have introduced the notion of an "equivariant moving frame", which expands on Cartan's construction and provides new algorithmic tools for computing invariants. This approach has generated substantial interest and spawned a wide variety of applications in the last several years. This material will not be treated here, but several surveys of recent results are available; for example, see [**Man10**], [**Olv10**], and [**Olv11a**].

The goal of this book is to provide an introduction to Cartan's theory of moving frames at a level suitable for beginning graduate students, with an emphasis on curves and surfaces in various 3-dimensional homogeneous spaces. This book assumes a standard undergraduate mathematics background, including courses in linear algebra, abstract algebra, real analysis, and topology, as well as a course on the differential geometry of curves and surfaces. (An appropriate differential geometry course might be based on a text such as [**dC76**], [**O'N06**], or [**Opr07**].) There are occasional references to additional topics such as differential equations, but these are less crucial.

The first two chapters contain background material that might typically be taught in a graduate differential geometry course; Chapter 1 contains general material from differential geometry, while Chapter 2 focuses more specifically on differential forms. Students who have taken such a course might safely skip these chapters, although it might be wise to skim them to get accustomed to the notation that will be used throughout the book.

Chapters 3–7 are the heart of the book. Chapter 3 introduces the main ingredients for the method of moving frames: homogeneous spaces, frame bundles, and Maurer-Cartan forms. Chapters 4–7 show how to apply the method of moving frames to compute local geometric invariants for curves and surfaces in 3-dimensional Euclidean, Minkowski, affine, and projective spaces. These chapters should be read in order (with the possible exception of Chapter 5), as they build on each other.

Chapters 8–10 show how the method of moving frames may be applied to several classical problems in differential geometry. The first half of Chapter 8, all of Chapter 9, and the last half of Chapter 10 may be read anytime after Chapter 4; the remainder of these chapters may be read anytime after Chapter 6.

Chapters 11 and 12 give a brief introduction to the method of moving frames on non-flat Riemannian manifolds and the additional issues that arise when the underlying space has nonzero curvature. These chapters may be read anytime after Chapter 4.

Exercises are embedded in the text rather than being presented at the end of each chapter. Readers are strongly encouraged to pause and attempt the exercises as they occur, as they are intended to engage the reader and to enhance the understanding of the text. Many of the exercises contain results which are important for understanding the remainder of the text; these exercises are marked with a star and should be given particular attention. (Even if you don't do them, you should at least read them!)

A special feature of this book is that it includes guidance on how to use the mathematical software package MAPLE to perform many of the computations involved in the exercises. (If you do not have access to MAPLE, rest assured that, with very few exceptions, the exercises can be done perfectly well by hand.) The computations here make use of the custom MAPLE package `Cartan`, which was written by myself and Yunliang Yu of Duke University. The `Cartan` package can be downloaded either from the AMS webpage

`www.ams.org/bookpages/gsm-178`

or from my webpage at

`http://euclid.colorado.edu/~jnc/Maple.html.`

(Installation instructions are included with the package.) The last section of Chapter 2 contains an introduction to the `Cartan` package, and beginning with Chapter 3, each chapter includes a section at the end describing how to use MAPLE and the `Cartan` package for some of the exercises in that chapter. Additional exercises are worked out in MAPLE worksheets for each chapter that are available on the AMS webpage.

Remark. As of MAPLE 16 and above, much of `Cartan`'s functionality is now available as part of the `DifferentialGeometry` package, which is included in the standard MAPLE installation and covers a wide range of applications. The two packages have very different syntax, and no attempt will be made here to translate—but interested readers are encouraged to do so!

Acknowledgments

First and foremost, my deepest thanks go to Robert Bryant—my teacher, mentor, and friend—for inviting me to teach alongside him at the Mathematical Sciences Research Institute in the summer of 1999, when I was a mere three years post-Ph.D.; for not laughing out loud when I naively mentioned the idea of turning the lecture notes into a book (although he probably should have); and for unflagging support in more ways than I can count over the years.

Thanks also to Edward Dunne and Sergei Gelfand at the American Mathematical Society for expressing interest in the project early on and for extreme patience and not losing faith in me as it dragged on for many more years than I ever imagined. I am also grateful to the anonymous reviewers for the AMS who read initial drafts of the manuscript, pointed out significant errors, and made valuable suggestions for improvements.

I am forever grateful to Bryan Kaufman and Nathaniel Bushek, who in 2009 asked if I would supervise an independent study course for them. I suggested that they work through my nascent manuscript, and they eagerly agreed, struggling through a version that consisted of little more than the original lecture notes. Their questions and suggestions were invaluable and had a major impact on the tone, content, and structure of the book. This project might have stayed forever on my to-do list if not for them. Thanks especially to Bryan for suggesting that I add the material on curves and surfaces in Minkowski space and to Sunita Vatuk for recommending the book [**Cal00**] on this material.

Thanks to all the other students who have worked through subsequent versions of the manuscript over the last several years: Brian Carlsen, Michael Schmidt, Edward Estrada, Molly May, Jonah Miller, Sean Peneyra, Duff Baker-Jarvis, Akaxia Cruz, Rachel Helm, Peter Joeris, Joshua Karpel, Andrew Jensen, and Michael Mahoney. These independent study courses—and the research projects that followed—have been, hands down, the most rewarding experiences of my teaching career. I hope you all enjoyed them half as much as I did! And thanks to Sunita Vatuk and George Wilkens for sitting in on some of these courses, contributing many valuable insights to our discussions, and making great suggestions for the manuscript.

I am grateful to the Mathematical Sciences Research Institute for sponsoring the 1999 Summer Graduate Workshop where I gave the lectures that were the genesis for this book; videos of the original lectures are available on MSRI's webpage at [**Cle99**]. I am also grateful to the National Science Foundation for research support; portions of this book were written while I was supported by NSF grants DMS-0908456 and DMS-1206272.

Finally, profound thanks to my husband, Rick; his love and support have been constant and unwavering, and I count myself fortunate beyond all measure to have him as my best friend and partner in life.

Part 1

Background material

Chapter 1

Assorted notions from differential geometry

This chapter contains some useful background material from differential geometry. Much of this material would typically be covered in courses on multivariable analysis, algebra, and graduate-level differential geometry and linear algebra. This chapter is *not* intended to be a comprehensive introduction to any of these topics; the focus will be on ideas that will be used in the remainder of the book, and details will be kept to a minimum.

1.1. Manifolds

Just about all the objects that we discuss in this book will be defined on a *manifold* of some sort. The simplest manifolds are things that you already know about: regular curves (1-dimensional manifolds) and regular surfaces (2-dimensional manifolds). Before making a more general definition, let's remind ourselves of some of the fundamental properties of regular surfaces.

1.1.1. Regular surfaces.

First, recall the definition of a regular surface in $\mathbb{R}^3$ (see, e.g., [**dC76**]):

Definition 1.1. A subset $\Sigma \subset \mathbb{R}^3$ is a *regular surface* if for each point $\mathbf{q} \in \Sigma$, there exists a neighborhood $V \subset \mathbb{R}^3$ of $\mathbf{q}$ and a map $\mathbf{x} : U \to V \cap \Sigma$ from an open set $U \subset \mathbb{R}^2$ onto $V \cap \Sigma$ with the following properties:

(1) $\mathbf{x}$ is differentiable.

(2) $\mathbf{x}$ is a homeomorphism.

(3) For each $\mathbf{u} \in U$, the differential $d\mathbf{x}_{\mathbf{u}} : \mathbb{R}^2 \to \mathbb{R}^2$ is one-to-one.

The mapping $\mathbf{x}$ is called a *parametrization* of Σ, or a *system of local coordinates* on Σ, in a neighborhood of $\mathbf{q}$.

Remark 1.2. Throughout this book, the word "differentiable" will be taken to mean "infinitely differentiable" (i.e., C^∞). We will also use the word "smooth" as a synonym.

Any regular surface $\Sigma \subset \mathbb{R}^3$ can be covered by a family of parametrizations $\mathbf{x}_i : U_i \to \Sigma$, where each U_i is an open set in $\mathbb{R}^2$ and each $\mathbf{x}_i$ is an injective, differentiable map from U_i to $\mathbb{R}^3$. The image $\mathbf{x}_i(U_i)$ of each parametrization is an open set in Σ (in the subspace topology inherited from $\mathbb{R}^3$); moreover, if the images of two parametrizations $\mathbf{x}_i : U_i \to \Sigma$ and $\mathbf{x}_j : U_j \to \Sigma$ have nonempty intersection $V \subset \Sigma$, then the composite maps

$$\mathbf{x}_j^{-1} \circ \mathbf{x}_i : \mathbf{x}_i^{-1}(V) \to \mathbf{x}_j^{-1}(V), \qquad \mathbf{x}_i^{-1} \circ \mathbf{x}_j : \mathbf{x}_j^{-1}(V) \to \mathbf{x}_i^{-1}(V) \tag{1.1}$$

are differentiable maps between open sets in $\mathbb{R}^2$. (See Figure 1.1.)

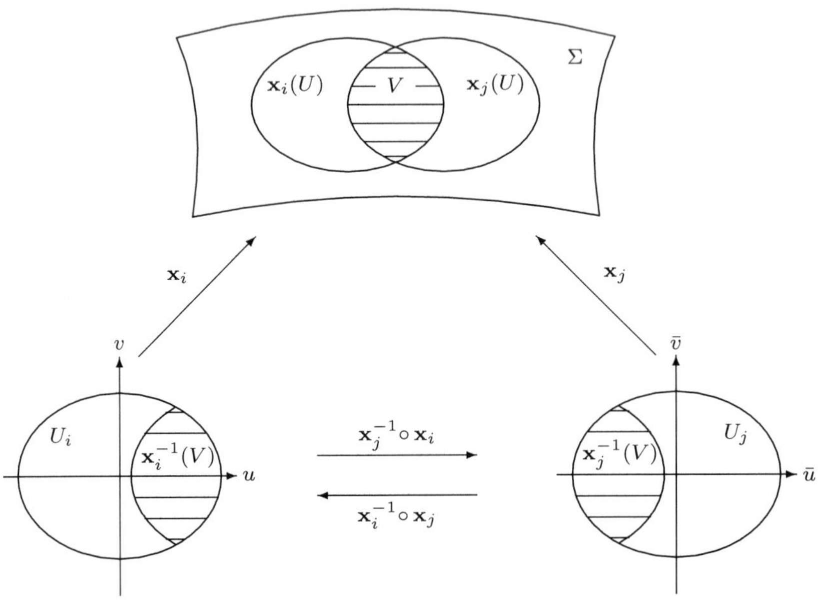

Figure 1.1. Overlapping parametrizations

When Σ is a regular surface in $\mathbb{R}^3$, the differentiability of the maps (1.1) is a theorem. But regular surfaces can also be defined as abstract, *intrin-*

sic objects, not living in any particular Euclidean space. In this case, the differentiability of (1.1) becomes part of the *definition* of a regular surface:

Definition 1.3. A *regular surface* is a set Σ and a family of injective mappings $\mathbf{x}_i : U_i \to \Sigma$, where each U_i is an open set in $\mathbb{R}^2$, with the following properties:

(1) $\displaystyle\bigcup_i \mathbf{x}_i(U_i) = \Sigma$.

(2) For each pair i, j with $V = \mathbf{x}_i(U_i) \cap \mathbf{x}_j(U_j) \neq \emptyset$, the sets $\mathbf{x}_i^{-1}(V)$ and $\mathbf{x}_j^{-1}(V)$ are open sets in $\mathbb{R}^2$, and the mappings $\mathbf{x}_j^{-1} \circ \mathbf{x}_i$ and $\mathbf{x}_i^{-1} \circ \mathbf{x}_j$ are differentiable.

The mappings $\mathbf{x}_i$ are called *parametrizations* of Σ, or *systems of local coordinates* on Σ.

Incorporating condition (2) as an axiom allows us to define objects such as differentiable functions on surfaces and differentiable maps between surfaces in a way that is independent of a choice of parametrization. All the usual notions of differential calculus on $\mathbb{R}^2$ can then be extended to analogous notions on regular surfaces.

For an abstract surface, the maps $\mathbf{x}_i$ are rarely defined explicitly. (This is *very* different from how we view surfaces in $\mathbb{R}^3$, where the parametrizations are often used to define the surface!) Rather, we think of an abstract surface as a collection of open sets U_i in $\mathbb{R}^2$, "glued together" via the transition maps $\mathbf{x}_j^{-1} \circ \mathbf{x}_i$. It is these transition maps between open sets in $\mathbb{R}^2$ that may (or may not) be defined explicitly; the $\mathbf{x}_i$ should just be thought of as a means of identifying a part of the surface with a system of local coordinates (u, v). If (u, v) are local coordinates on U_i, $(\bar{u}, \bar{v})$ are local coordinates on U_j, and the set $V = \mathbf{x}_i(U_i) \cap \mathbf{x}_j(U_j) \subset \Sigma$ is nonempty, then the transition map

$$\mathbf{x}_j^{-1} \circ \mathbf{x}_i : \mathbf{x}_i^{-1}(V) \to \mathbf{x}_j^{-1}(V)$$

is a local coordinate transformation of the form

$$\mathbf{x}_j^{-1} \circ \mathbf{x}_i(u, v) = (\bar{u}(u, v), \bar{v}(u, v)).$$

(See Figure 1.1.)

1.1.2. Manifolds: from 2 to n. A *manifold* (or, more precisely, a *differentiable manifold*) of dimension $n \geq 1$ is simply what we get by replacing the number 2 in Definition 1.3 with n:

Definition 1.4. A *differentiable manifold* of dimension $n \geq 1$ is a set M and a family of injective mappings $\mathbf{x}_i : U_i \to M$, where each U_i is an open set in $\mathbb{R}^n$, with the following properties:

(1) $\bigcup_i \mathbf{x}_i(U_i) = M$.

(2) For any pair i, j with $V = \mathbf{x}_i(U_i) \cap \mathbf{x}_j(U_j) \neq \emptyset$, the sets $\mathbf{x}_i^{-1}(V)$ and $\mathbf{x}_j^{-1}(V)$ are open sets in $\mathbb{R}^n$, and the mappings $\mathbf{x}_j^{-1} \circ \mathbf{x}_i$ and $\mathbf{x}_i^{-1} \circ \mathbf{x}_j$ are differentiable. The mappings $\mathbf{x}_i$ are called *parametrizations* of M, or *systems of local coordinates* on M.

Remark 1.5. When studying regular surfaces in $\mathbb{R}^3$, it is traditional to use (u, v) as local coordinates on $\mathbb{R}^2$ and to write parametrizations $\mathbf{x} : U \to \mathbb{R}^3$ as

$$\mathbf{x}(u, v) = \begin{bmatrix} x(u, v) \\ y(u, v) \\ z(u, v) \end{bmatrix}.$$

When we graduate to manifolds of arbitrary dimension, we need to use variables with indices so that we don't run out of letters. So, depending on the context, we will generally use either $\mathbf{u} = (u^1, \ldots, u^n)$ or $\mathbf{x} = (x^1, \ldots, x^n)$ as local coordinates on an open set $U \subset \mathbb{R}^n$.

Just as for regular surfaces, a manifold M of dimension n can be a subset of some Euclidean space $\mathbb{R}^k$ with $k \geq n$, or it can be an intrinsic object not living in any ambient Euclidean space. When M is a subset of $\mathbb{R}^k$, it may be defined by explicit parametrizations $\mathbf{x}_i : U_i \to \mathbb{R}^k$, where the U_i are open sets in $\mathbb{R}^n$. When M is an intrinsic manifold, the maps $\mathbf{x}_i$ are generally not defined explicitly; we simply think of M as a collection of open sets $U_i \subset \mathbb{R}^n$ that are glued together via the transition maps $\mathbf{x}_j^{-1} \circ \mathbf{x}_i$. As for surfaces, these transition maps are local coordinate transformations of the form

$$\mathbf{x}_j^{-1} \circ \mathbf{x}_i(u^1, \ldots, u^n) = (\bar{u}^1(u^1, \ldots, u^n), \ldots, \bar{u}^n(u^1, \ldots, u^n)). \tag{1.2}$$

1.1.3. Examples.

Example 1.6 (The unit sphere). Let $\mathbb{S}^n \subset \mathbb{R}^{n+1}$ be the set

$$\mathbb{S}^n = \{{}^t[x^1, \ldots, x^{n+1}] \in \mathbb{R}^{n+1} \mid (x^1)^2 + \cdots + (x^{n+1})^2 = 1\};$$

i.e., $\mathbb{S}^n$ is the set of all vectors in $\mathbb{R}^{n+1}$ of Euclidean length 1.

Remark 1.7. Vectors in $\mathbb{R}^k$ are assumed to be column vectors unless otherwise specified. The notation ${}^t[x^1, \ldots, x^{n+1}]$ in Example 1.6 denotes the transpose of the row vector $[x^1, \ldots, x^{n+1}]$, which is the column vector

$\begin{bmatrix} x^1 \\ \vdots \\ x^{n+1} \end{bmatrix}$. We will often write column vectors in this way in order to save space, using the transpose notation in order to maintain the distinction between column vectors and row vectors.

Exercise 1.8. $\mathbb{S}^n$ can be covered by parametrizations in the same fashion as the 2-dimensional sphere $\mathbb{S}^2 \subset \mathbb{R}^3$; it just takes a bit more bookkeeping to keep up with all the indices. For $i = 1, \ldots, n+1$, let

$$V_i^+ = \{{}^t[x^1, \ldots, x^{n+1}] \in \mathbb{S}^n \mid x^i > 0\},$$
$$V_i^- = \{{}^t[x^1, \ldots, x^{n+1}] \in \mathbb{S}^n \mid x^i < 0\}.$$

Let

$$U = \{{}^t[u^1, \ldots, u^n] \in \mathbb{R}^n \mid (u^1)^2 + \cdots + (u^n)^2 < 1\};$$

i.e., U is the open unit ball in $\mathbb{R}^n$. Define maps

$$\mathbf{x}_i^+ : U \to V_i^+, \qquad \mathbf{x}_i^- : U \to V_i^-$$

by

$$\mathbf{x}_i^+(u^1, \ldots, u^n) = {}^t\left[u^1, \ldots, u^{i-1}, \sqrt{1 - (u^1)^2 - \cdots - (u^n)^2}, u^i, \ldots, u^n\right],$$

$$\mathbf{x}_i^-(u^1, \ldots, u^n) = {}^t\left[u^1, \ldots, u^{i-1}, -\sqrt{1 - (u^1)^2 - \cdots - (u^n)^2}, u^i, \ldots, u^n\right].$$

(a) What portion of $\mathbb{S}^n$ is covered by the images of $\mathbf{x}_i^+$ and $\mathbf{x}_i^-$? Show that every point in $\mathbb{S}^n$ is contained in the image of at least one of these parametrizations.

(b) For $i \neq j$, identify the open set $V_{ij}^+ = \mathbf{x}_i^+(U) \cap \mathbf{x}_j^+(U)$ in $\mathbb{S}^n$ and the open sets $(\mathbf{x}_j^+)^{-1}(V_{ij}^+)$ and $(\mathbf{x}_i^+)^{-1}(V_{ij}^+)$ in U, and compute the local coordinate transformation $(\mathbf{x}_j^+)^{-1} \circ \mathbf{x}_i^+$ between the latter two open sets.

Example 1.9 (Real projective space of dimension n). Let $\mathbb{P}^n$ denote the set of lines through the origin in the vector space $\mathbb{R}^{n+1}$. Since any such line is determined by any point on the line other than the origin, we can think of $\mathbb{P}^n$ as the quotient space

$$\mathbb{P}^n = (\mathbb{R}^{n+1} \setminus \{0\}) / \sim,$$

where $\sim$ represents the equivalence relation defined by the condition that two points $\mathbf{x}, \mathbf{y} \in \mathbb{R}^{n+1} \setminus \{0\}$ satisfy $\mathbf{x} \sim \mathbf{y}$ if and only if $\mathbf{x}$ and $\mathbf{y}$ lie on the same line through the origin. It is customary to use local coordinates $(x^0, x^1, \ldots, x^n)$ on $\mathbb{R}^{n+1}$; then for any nonzero real number λ, we have

$$^t[x^0, \ldots, x^n] \sim {}^t[\lambda x^0, \ldots, \lambda x^n].$$

The equivalence class of the point $\mathbf{x} = {}^t[x^0, \ldots, x^n]$ is denoted by

$$[\mathbf{x}] = [x^0 : \cdots : x^n];$$

these are called the *homogeneous coordinates* for a point in $\mathbb{P}^n$.

Note that each line through the origin in $\mathbb{R}^{n+1}$ intersects the unit sphere $\mathbb{S}^n$ in exactly two points, which form an antipodal pair $\{\mathbf{x}, -\mathbf{x}\}$. So, an alternative (but equivalent!) way to think of $\mathbb{P}^n$ is as the manifold obtained from $\mathbb{S}^n$ by identifying every point $\mathbf{x} \in \mathbb{S}^n$ with its polar opposite $-\mathbf{x}$.

Even better, we can identify $\mathbb{P}^n$ with the upper half of $\mathbb{S}^n$ (which is topologically an n-dimensional disk), with its boundary glued together so as to identify opposite points on the boundary. Note that gluing the boundary—which is a copy of $\mathbb{S}^{n-1}$—together in this fashion produces a copy of $\mathbb{P}^{n-1}$. So, yet *another* way to think of $\mathbb{P}^n$ is as an open n-dimensional disk (which is topologically the same as $\mathbb{R}^n$) with a copy of $\mathbb{P}^{n-1}$ attached along its boundary. (See Figure 1.2 for an idea of how this looks for $\mathbb{P}^1$ and $\mathbb{P}^2$.)

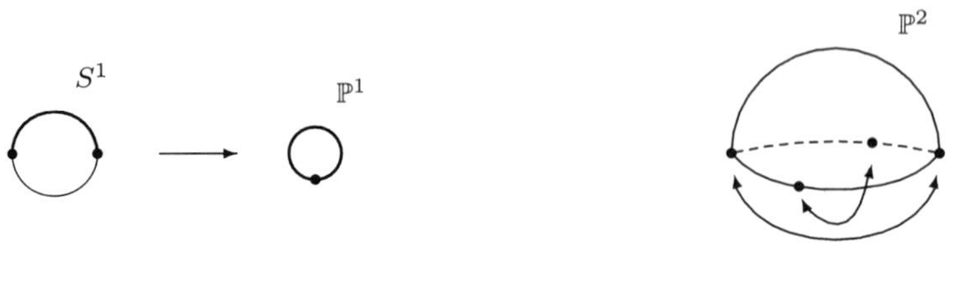

Figure 1.2. Construction of $\mathbb{P}^1$ and $\mathbb{P}^2$

Unlike the unit sphere $\mathbb{S}^n$, $\mathbb{P}^n$ is not obviously defined as a subset of any Euclidean space $\mathbb{R}^k$. The following exercise shows how $\mathbb{P}^n$ can be covered by parametrizations.

Exercise 1.10. For $i = 0, \ldots, n$, let

$$V_i = \{[x^0 : \cdots : x^n] \in \mathbb{P}^n \mid x^i \neq 0\}.$$

Let $U = \mathbb{R}^n$, with coordinates $\mathbf{u} = (u^1, \ldots, u^n)$. Define maps

$$[\mathbf{x}_i] : U \to \mathbb{P}^n$$

by

$$[\mathbf{x}_i(u^1, \ldots, u^n)] = [u^1 : \cdots : u^i : 1 : u^{i+1} : \cdots : u^n].$$

(a) What portion of $\mathbb{P}^n$ is covered by the image of $[\mathbf{x}_i]$? Show that every point in $\mathbb{P}^n$ is contained in the image of at least one of these parametrizations.

(b) For $i \neq j$, identify the set $V_{ij} = [\mathbf{x}_i(U)] \cap [\mathbf{x}_j(U)]$ in $\mathbb{P}^n$ and the open sets $[\mathbf{x}_j]^{-1}(V_{ij})$ and $[\mathbf{x}_i]^{-1}(V_{ij})$ in U, and compute the local coordinate transformation $[\mathbf{x}_j]^{-1} \circ [\mathbf{x}_i]$ between the latter two open sets.

1.2. Tensors, indices, and the Einstein summation convention

In equation (1.2), the variables are indexed with superscripts rather than subscripts. You may be wondering—why on earth would we do *that?* So, before we go any farther, let's discuss how (and why!) indices are used throughout this book. Some objects will be indexed with superscripts ("upper indices" or, less formally, "up indices"), some by subscripts ("lower indices" or "down indices"), and some by a combination of both. Most of these objects will be *tensors* or *tensor fields.* Without getting too specific, a tensor is just an element of a certain type of vector space, and a tensor field on a manifold is defined by assigning a tensor to each point of the manifold. (A precise definition will be given in Chapter 2.) For example, tensors include objects such as vectors and linear transformations, while tensor fields include objects such as vector fields, metrics, and differential forms on manifolds.

A key feature of tensors is the way in which their coordinate expressions change when the basis for the underlying vector space is changed. The following examples demonstrate some typical behavior.

Example 1.11. Consider an n-dimensional vector space V, and let $(\mathbf{e}_1, \ldots, \mathbf{e}_n)$ be a basis for V. Any vector $\mathbf{v} \in V$ can be expressed uniquely as

$$\mathbf{v} = a^1\mathbf{e}_1 + \cdots + a^n\mathbf{e}_n$$

for some real numbers $a^1, \ldots, a^n \in \mathbb{R}$.

Now, what happens if we express $\mathbf{v}$ in terms of a different basis $(\bar{\mathbf{e}}_1, \ldots, \bar{\mathbf{e}}_n)$? Suppose that the new basis can be written in terms of the old basis as

$$\bar{\mathbf{e}}_i = \sum_{k=1}^{n} r_i^k \mathbf{e}_k, \qquad i = 1, \ldots, n,$$

or, in more compact matrix notation,

$$\begin{bmatrix}\bar{\mathbf{e}}_1 & \ldots & \bar{\mathbf{e}}_n\end{bmatrix} = \begin{bmatrix}\mathbf{e}_1 & \ldots & \mathbf{e}_n\end{bmatrix} R,$$

where

$$R = \begin{bmatrix} r_1^1 & \dots & r_n^1 \\ \vdots & & \vdots \\ r_1^n & \dots & r_n^n \end{bmatrix}.$$

Then we have

$$\begin{bmatrix} \mathbf{e}_1 & \dots & \mathbf{e}_n \end{bmatrix} = \begin{bmatrix} \bar{\mathbf{e}}_1 & \dots & \bar{\mathbf{e}}_n \end{bmatrix} R^{-1},$$

and so

$$\begin{aligned} \mathbf{v} &= a^1\mathbf{e}_1 + \dots + a^n\mathbf{e}_n \\ &= \begin{bmatrix} \mathbf{e}_1 & \dots & \mathbf{e}_n \end{bmatrix} \begin{bmatrix} a^1 \\ \vdots \\ a^n \end{bmatrix} \\ &= \begin{bmatrix} \bar{\mathbf{e}}_1 & \dots & \bar{\mathbf{e}}_n \end{bmatrix} R^{-1} \begin{bmatrix} a^1 \\ \vdots \\ a^n \end{bmatrix} \\ &= \bar{a}^1\bar{\mathbf{e}}_1 + \dots + \bar{a}^n\bar{\mathbf{e}}_n, \end{aligned}$$

where

$$\begin{bmatrix} \bar{a}^1 \\ \vdots \\ \bar{a}^n \end{bmatrix} = R^{-1} \begin{bmatrix} a^1 \\ \vdots \\ a^n \end{bmatrix}.$$

While this is a simple (and hopefully familiar!) example, it illustrates some important points:

(1) A vector is an example of a *rank* 1 *tensor*. "Rank 1" means that, when the vector is expressed in terms of a given basis, the *components*—in this case, $(a^1, \dots, a^n)$—each have one index.

(2) The matrix $\mathbf{e} = \begin{bmatrix} \mathbf{e}_1 & \dots & \mathbf{e}_n \end{bmatrix}$ (Careful: It *looks* like a row vector, but each *entry* is really a column vector!) and the column vector $\mathbf{a}$ are basis-dependent, but their product $\mathbf{v} = \mathbf{ea}$ is well-defined independently of a choice of basis: When the basis vectors $(\mathbf{e}_1, \dots, \mathbf{e}_n)$ are changed, the coefficients $(a^1, \dots, a^n)$ change in such a way that the changes cancel each other out in the product.

(3) When the matrix $\mathbf{e}$, whose entries are indexed by lower indices, is transformed by R, the vector $\mathbf{a}$, whose entries are indexed by upper indices, is multiplied by R^{-1}. This is not an accident! In general, the placement—up or down—of indices for the components of a tensor is dictated by how these components transform under a change of basis for the underlying vector space.

Example 1.12. Let V be an m-dimensional vector space with basis $(\mathbf{e}_1, \ldots, \mathbf{e}_m)$ and W an n-dimensional vector space with basis $(\mathbf{f}_1, \ldots, \mathbf{f}_n)$. Let $T : V \to W$ be a linear transformation. In terms of the given bases for V and W, T can be represented as an $n \times m$ matrix

$$A_T = \begin{bmatrix} c_1^1 & \cdots & c_m^1 \\ \vdots & & \vdots \\ c_1^n & \cdots & c_m^n \end{bmatrix}.$$

Let's think carefully about what this means. We usually write a vector $\mathbf{v} \in V$ as a column vector

$$\mathbf{a} = \begin{bmatrix} a^1 \\ \vdots \\ a^m \end{bmatrix}$$

(by which we really mean that $\mathbf{v} = [\mathbf{e}_1 \ \ldots \ \mathbf{e}_m] \cdot \mathbf{a}$), and we write $\mathbf{w} = T(\mathbf{v})$ as the matrix product

$$\mathbf{b} = \begin{bmatrix} b^1 \\ \vdots \\ b^n \end{bmatrix} = \begin{bmatrix} c_1^1 & \cdots & c_m^1 \\ \vdots & & \vdots \\ c_1^n & \cdots & c_m^n \end{bmatrix} \begin{bmatrix} a^1 \\ \vdots \\ a^m \end{bmatrix}$$

(by which we really mean that $\mathbf{w} = [\mathbf{f}_1 \ \ldots \ \mathbf{f}_n] \cdot \mathbf{b}$). The use of different notations here for the vectors $\mathbf{v}$ and $\mathbf{w}$, which are well-defined independently of a choice of basis, and their expressions $\mathbf{a}$ and $\mathbf{b}$ in terms of specific bases, is deliberate: The equation

$$\mathbf{b} = A_T\mathbf{a}$$

is a basis-dependent expression of the basis-independent equation

$$\mathbf{w} = T(\mathbf{v}).$$

What it really means is that

$$\begin{aligned} \mathbf{w} = T(\mathbf{v}) &= T\left(\sum_{i=1}^{m} a^i \mathbf{e}_i\right) \\ &= \sum_{j=1}^{n}\left(\sum_{i=1}^{m} c_i^j a^i\right)\mathbf{f}_j. \end{aligned}$$

In more compact matrix notation,

$$T\left([\mathbf{e}_1 \ \ldots \ \mathbf{e}_m]\begin{bmatrix} a^1 \\ \vdots \\ a^m \end{bmatrix}\right) = [\mathbf{f}_1 \ \ldots \ \mathbf{f}_n]\, A_T \begin{bmatrix} a^1 \\ \vdots \\ a^m \end{bmatrix}. \tag{1.3}$$

Again, we ask what happens if we change bases for V and W. Suppose that we set

$$\begin{aligned} \begin{bmatrix}\bar{\mathbf{e}}_1 & \dots & \bar{\mathbf{e}}_m\end{bmatrix} &= \begin{bmatrix}\mathbf{e}_1 & \dots & \mathbf{e}_m\end{bmatrix} R, \\ \begin{bmatrix}\bar{\mathbf{f}}_1 & \dots & \bar{\mathbf{f}}_n\end{bmatrix} &= \begin{bmatrix}\mathbf{f}_1 & \dots & \mathbf{f}_n\end{bmatrix} S, \end{aligned} \tag{1.4}$$

where R is an invertible $m \times m$ matrix and S is an invertible $n \times n$ matrix. We want to find the matrix $\overline{A}_T$ that satisfies

$$T\left(\begin{bmatrix}\bar{\mathbf{e}}_1 & \dots & \bar{\mathbf{e}}_m\end{bmatrix}\begin{bmatrix}\bar{a}^1\\ \vdots \\ \bar{a}^m\end{bmatrix}\right) = \begin{bmatrix}\bar{\mathbf{f}}_1 & \dots & \bar{\mathbf{f}}_n\end{bmatrix}\overline{A}_T\begin{bmatrix}\bar{a}^1\\ \vdots \\ \bar{a}^m\end{bmatrix}.$$

Using (1.3) and (1.4), we can compute:

$$\begin{aligned} T\left(\begin{bmatrix}\bar{\mathbf{e}}_1 & \dots & \bar{\mathbf{e}}_m\end{bmatrix}\begin{bmatrix}\bar{a}^1\\ \vdots \\ \bar{a}^m\end{bmatrix}\right) &= T\left(\begin{bmatrix}\mathbf{e}_1 & \dots & \mathbf{e}_m\end{bmatrix} R\begin{bmatrix}\bar{a}^1\\ \vdots \\ \bar{a}^m\end{bmatrix}\right) \\ &= \begin{bmatrix}\mathbf{f}_1 & \dots & \mathbf{f}_n\end{bmatrix} A_T R\begin{bmatrix}\bar{a}^1\\ \vdots \\ \bar{a}^m\end{bmatrix} \\ &= \begin{bmatrix}\bar{\mathbf{f}}_1 & \dots & \bar{\mathbf{f}}_n\end{bmatrix} (S^{-1}A_T R)\begin{bmatrix}\bar{a}^1\\ \vdots \\ \bar{a}^m\end{bmatrix}. \end{aligned}$$

So, in terms of the new bases, the new matrix representation for T is

$$\overline{A}_T = S^{-1}A_T R.$$

Some observations about this example:

(1) The linear transformation T is an example of a *rank* 2 *tensor*. The components (c^i_j) of its representation A_T in terms of specific bases for V and W have an upper index i with range $1 \le i \le n$, corresponding to the vector space W, and a lower index j with range $1 \le j \le m$, corresponding to the vector space V.

(2) When the basis for V is transformed by a matrix R and the basis for W is transformed by a matrix S, the matrix representation A_T is multiplied on the left by S^{-1} and on the right by R. This illustrates the general phenomenon that, under a change of basis:
   - Down indices indicate that components will transform by right multiplication by the matrix for the basis transformation.

- Up indices indicate that components will transform by left multiplication by the inverse of the matrix for the basis transformation.

WARNING: When applying this guideline, one must think *very* carefully about which direction the transformation goes in order to get the inverses in the right places!

Remark 1.13. These were fairly simple examples of tensors, dealing with a single vector and a single linear transformation. We will mostly be interested in *tensor fields*, which consist of an underlying manifold with a tensor defined at each point. It may sound complicated, but you already know several examples—e.g., vector fields and metrics on surfaces. The only real differences are that:

(1) The components are *functions* on the base manifold rather than constants.

(2) The vector spaces in question are usually the tangent or cotangent spaces to the manifold at each point.

(3) Basis changes to the vector spaces usually arise as derivatives (i.e., Jacobian matrices) of local coordinate transformations on the manifold.

Remark 1.14. When working with partial derivatives, up indices in the denominator count as down indices, and vice versa. For example, when working on a manifold with local coordinates $(x^1, \ldots, x^n)$, the partial derivative operators $\left(\frac{\partial}{\partial x^1}, \ldots, \frac{\partial}{\partial x^n}\right)$ will often be used as a local basis for the tangent space at each point (the reason for this will be explained in §1.3), and the indices on these operators should be regarded as down indices.

***Exercise 1.15.** A *metric* g on a manifold M is defined by specifying for each point $\mathbf{q} \in M$ a symmetric, positive definite bilinear form $g_{\mathbf{q}}$ on the tangent space $T_{\mathbf{q}}M$. (Tangent spaces will be defined in §1.3; all you need to know here is that $T_{\mathbf{q}}M$ is an n-dimensional vector space.) This means that $g_{\mathbf{q}}$ is a bilinear function

$$g_{\mathbf{q}} : T_{\mathbf{q}}M \times T_{\mathbf{q}}M \to \mathbb{R}$$

such that for all $\mathbf{v}, \mathbf{w} \in T_{\mathbf{q}}M$, we have $g_{\mathbf{q}}(\mathbf{v}, \mathbf{w}) = g_{\mathbf{q}}(\mathbf{w}, \mathbf{v})$ and $g_{\mathbf{q}}(\mathbf{v}, \mathbf{v}) \geq 0$, with $g_{\mathbf{q}}(\mathbf{v}, \mathbf{v}) = 0$ if and only if $\mathbf{v} = \mathbf{0}$. If we have a basis $(\mathbf{e}_1, \ldots, \mathbf{e}_n)$ for the tangent space $T_{\mathbf{q}}M$, then we define the *components* of $g_{\mathbf{q}}$ with respect to this basis to be the real numbers

$$\{g_{ij}(\mathbf{q}) = g_{\mathbf{q}}(\mathbf{e}_i, \mathbf{e}_j) \mid 1 \leq i, j \leq n\}.$$

The assumption that g is symmetric implies that $g_{ji}(\mathbf{q}) = g_{ij}(\mathbf{q})$, and the bilinearity of g implies that for any vectors

$$\mathbf{v} = \sum_{i=1}^{n} a^i \mathbf{e}_i, \qquad \mathbf{w} = \sum_{j=1}^{n} b^j \mathbf{e}_j$$

in $T_{\mathbf{q}}M$, we have

$$g_{\mathbf{q}}(\mathbf{v}, \mathbf{w}) = \sum_{i=1}^{n} \sum_{j=1}^{n} g_{ij}(\mathbf{q}) a^i b^j.$$

For instance, when M is a regular surface $\Sigma \subset \mathbb{R}^3$ with a local parametrization $\mathbf{x} : U \to \Sigma$, we typically use the basis

$$\mathbf{e}_1 = \mathbf{x}_u, \qquad \mathbf{e}_2 = \mathbf{x}_v$$

for $T_{\mathbf{q}}M$, and the bilinear form is given by the dot product in $\mathbb{R}^3$:

$$g_{\mathbf{q}}(\mathbf{v}, \mathbf{w}) = \mathbf{v} \cdot \mathbf{w}.$$

Then we have

$$g_{11} = E = \mathbf{x}_u \cdot \mathbf{x}_u, \qquad g_{12} = g_{21} = F = \mathbf{x}_u \cdot \mathbf{x}_v, \qquad g_{22} = G = \mathbf{x}_v \cdot \mathbf{x}_v.$$

A metric g is often represented by the symmetric matrix

$$A_g = \begin{bmatrix} g_{11} & \cdots & g_{1n} \\ \vdots & & \vdots \\ g_{n1} & \cdots & g_{nn} \end{bmatrix}.$$

Suppose that we make a change of basis

$$\begin{bmatrix} \bar{\mathbf{e}}_1 & \dots & \bar{\mathbf{e}}_n \end{bmatrix} = \begin{bmatrix} \mathbf{e}_1 & \dots & \mathbf{e}_n \end{bmatrix} R$$

for $T_{\mathbf{q}}M$. Compute the components $(\bar{g}_{ij}(\mathbf{q}))$ of $g_{\mathbf{q}}$ with respect to the new basis $(\bar{\mathbf{e}}_1, \dots, \bar{\mathbf{e}}_n)$. How does the matrix A_g transform? Compare the transformation rule for A_g to that given in Example 1.12 for the matrix representation A_T for a linear transformation T from a vector space V to itself.

We close this section by introducing the *Einstein summation convention.* In working with tensors and tensor fields, there are a lot of sums involved: A vector is expressed as

$$\mathbf{v} = \sum_{i=1}^{n} a^i \mathbf{e}_i;$$

a metric g acting on a pair of vectors $\mathbf{v} = \sum_{i=1}^{n} a^i \mathbf{e}_i, \mathbf{w} = \sum_{j=1}^{n} b^j \mathbf{e}_j$ is expressed as

$$g(\mathbf{v}, \mathbf{w}) = \sum_{i=1}^{n} \sum_{j=1}^{n} g_{ij} a^i b^j,$$

etc. So, in order to reduce notational clutter, we omit the explicit sum notation and simply write

$$\mathbf{v} = a^i \mathbf{e}_i,$$
$$g(\mathbf{v}, \mathbf{w}) = g_{ij} a^i b^j.$$

The Einstein summation convention says that *whenever the same index appears twice in an expression, once up and once down, it indicates a sum over the entire range of that index.* This takes a bit of getting used to, but it's actually extremely handy. For instance, it makes the multivariable chain rule look exactly like the single-variable version, where you can "cancel" the intermediate variables:

$$\frac{\partial f}{\partial x^i} = \frac{\partial f}{\partial y^j} \frac{\partial y^j}{\partial x^i}.$$

It also gives a quick method for error checking, much like dimensional analysis in chemistry or physics: Once repeated indices are "canceled", the indices on both sides of an equation should match.

1.3. Differentiable maps, tangent spaces, and vector fields

Let $U \subset \mathbb{R}^m$ be an open set, and consider a function $F : U \to \mathbb{R}^n$. (Such a function is often called a *mapping*, or simply a *map*, from U to $\mathbb{R}^n$.) We can write

$$F(x^1, \ldots, x^m) = \begin{bmatrix} f^1(x^1, \ldots, x^m) \\ \vdots \\ f^n(x^1, \ldots, x^m) \end{bmatrix}$$

for some real-valued functions $f^1, \ldots, f^n : U \to \mathbb{R}$. We say that F is *continuous* if each of the functions $f^1, \ldots, f^n$ is continuous, and F is *differentiable* if each of the functions $f^1, \ldots, f^n$ is differentiable.

Now, if a function is differentiable, then it really ought to have a derivative. So, what is the appropriate notion for the derivative of such a map?

First, consider the case $m = 1$. In this case, $U \subset \mathbb{R}$ is an open interval I, and the image of F is a curve in $\mathbb{R}^n$. (In this case, we usually denote the function by a lowercase Greek letter rather than by a capital Roman letter.) Given a curve $\alpha : I \to \mathbb{R}^n$ defined by

$$\alpha(t) = \begin{bmatrix} y^1(t) \\ \vdots \\ y^n(t) \end{bmatrix},$$

the derivative of α at any point $t \in I$ is often simply defined to be the column vector

$$\alpha'(t) = \begin{bmatrix} \frac{dy^1}{dt}(t) \\ \vdots \\ \frac{dy^n}{dt}(t) \end{bmatrix}, \tag{1.5}$$

also called the *tangent vector* to α at the point $\alpha(t)$. But the expression (1.5) is actually an incomplete description of the tangent vector because the base point of a vector is a crucial part of its definition. A more accurate definition would be

$$\alpha'(t) = \left(\begin{bmatrix} y^1(t) \\ \vdots \\ y^n(t) \end{bmatrix}, \begin{bmatrix} \frac{dy^1}{dt}(t) \\ \vdots \\ \frac{dy^n}{dt}(t) \end{bmatrix} \right), \tag{1.6}$$

including both the base point and its derivative. While the abbreviated notation (1.5) is common, it is important to remember that it represents a vector based at the specific point $\alpha(t)$ in $\mathbb{R}^n$.

Given a point $\mathbf{y} = {}^t[y^1, \ldots, y^n] \in \mathbb{R}^n$, the *tangent space* to $\mathbb{R}^n$ at $\mathbf{y}$, denoted $T_{\mathbf{y}}\mathbb{R}^n$, is the set of all tangent vectors to all curves in $\mathbb{R}^n$ passing through $\mathbf{y}$; i.e.,

$$T_{\mathbf{y}}\mathbb{R}^n = \{\alpha'(0) \mid \alpha : I \to \mathbb{R}^n \text{ is a smooth curve with } \alpha(0) = \mathbf{y}\}.$$

$T_{\mathbf{y}}\mathbb{R}^n$ is an n-dimensional vector space, and it is *canonically* isomorphic to the base space $\mathbb{R}^n$. This means not only that $T_{\mathbf{y}}\mathbb{R}^n$ is isomorphic to $\mathbb{R}^n$ (which is obvious, since both are n-dimensional vector spaces), but also that there is one *particular* isomorphism between them that is somehow the "right" one. Specifically, this isomorphism is defined by (1.5), which identifies the tangent vector $\alpha'(0) \in T_{\mathbf{y}}\mathbb{R}^n$ with the vector ${}^t\left[\frac{dy^1}{dt}(0), \ldots, \frac{dy^n}{dt}(0)\right] \in \mathbb{R}^n$.

Now let m be arbitrary, and let $F : U \subset \mathbb{R}^m \to \mathbb{R}^n$ be a differentiable map. Fix a point $\mathbf{x} \in U$. The *derivative* or *differential* of F (denoted by dF or F_*, depending on the context) at $\mathbf{x}$ is a linear map from $T_{\mathbf{x}}\mathbb{R}^m$ to $T_{F(\mathbf{x})}\mathbb{R}^n$ defined as follows: Given a tangent vector $\mathbf{v} \in T_{\mathbf{x}}\mathbb{R}^m$, let $\alpha : I \to \mathbb{R}^m$ be a smooth curve with $\alpha(0) = \mathbf{x}$ and $\alpha'(0) = \mathbf{v}$. Then

$$dF_{\mathbf{x}}(\mathbf{v}) = (F \circ \alpha)'(0) \in T_{F(\mathbf{x})}\mathbb{R}^n.$$

(See Figure 1.3.)

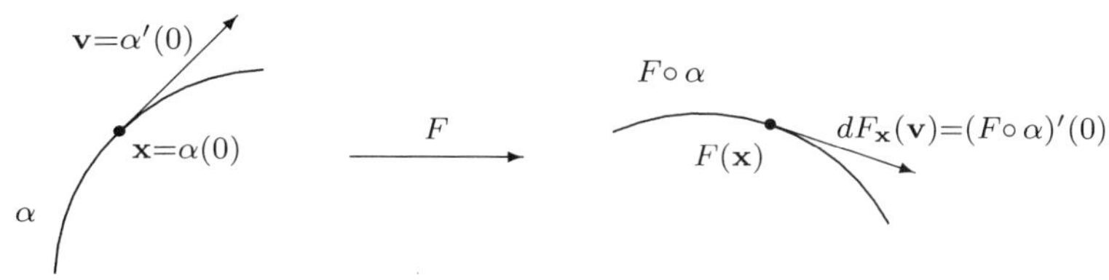

Figure 1.3. Construction of $dF_{\mathbf{x}}(\mathbf{v})$

While it may not be obvious, it is true that:

(1) $dF_{\mathbf{x}}(\mathbf{v})$ is well-defined, independent of the choice of the curve α.

(2) $dF_{\mathbf{x}} : T_{\mathbf{x}}\mathbb{R}^m \to T_{F(\mathbf{x})}\mathbb{R}^n$ is a linear map.

(3) The matrix representation for $dF_{\mathbf{x}}$ in terms of the canonical bases for $T_{\mathbf{x}}\mathbb{R}^m$ and $T_{F(\mathbf{x})}\mathbb{R}^n$ is the Jacobian matrix of F at $\mathbf{x}$.

***Exercise 1.16.** Prove the statements above: Let $F : U \subset \mathbb{R}^m \to \mathbb{R}^n$ be a differentiable map.

(a) Let $\mathbf{x} \in U$, and let $\mathbf{v} \in T_{\mathbf{x}}\mathbb{R}^m$ be a tangent vector. Let $\alpha : I \to \mathbb{R}^m$, $\beta : I \to \mathbb{R}^m$ be two curves with

$$\alpha(0) = \beta(0) = \mathbf{x}, \qquad \alpha'(0) = \beta'(0) = \mathbf{v}.$$

Show by direct computation that

$$(F \circ \alpha)'(0) = (F \circ \beta)'(0) \in T_{F(\mathbf{x})}\mathbb{R}^n.$$

This shows that the differential $dF_{\mathbf{x}}(\mathbf{v})$ is well-defined.

(b) Let $\mathbf{x} \in U$, and let $\mathbf{v}, \mathbf{w} \in T_{\mathbf{x}}\mathbb{R}^m$ be tangent vectors. Show that

$$dF_{\mathbf{x}}(\mathbf{v} + \mathbf{w}) = dF_{\mathbf{x}}(\mathbf{v}) + dF_{\mathbf{x}}(\mathbf{w}),$$

and for any real number c,

$$dF_{\mathbf{x}}(c\mathbf{v}) = c\, dF_{\mathbf{x}}(\mathbf{v}).$$

This shows that $dF_x(\mathbf{v})$ is a linear map. (Hint: Let $\alpha : I \to \mathbb{R}^m$, $\beta : I \to \mathbb{R}^m$ be two curves with

$$\alpha(0) = \beta(0) = \mathbf{x}, \qquad \alpha'(0) = \mathbf{v}, \qquad \beta'(0) = \mathbf{w}.$$

Write

$$\alpha(t) = \begin{bmatrix} x_1^1(t) \\ \vdots \\ x_1^m(t) \end{bmatrix}, \qquad \beta(t) = \begin{bmatrix} x_2^1(t) \\ \vdots \\ x_2^m(t) \end{bmatrix}.$$

Let γ be the curve

$$\gamma(t) = \begin{bmatrix} \frac{1}{2}\left(x_1^1(2t) + x_2^1(2t)\right) \\ \vdots \\ \frac{1}{2}\left(x_1^m(2t) + x_2^m(2t)\right) \end{bmatrix}.$$

Then γ satisfies $\gamma(0) = \mathbf{x}$ and $\gamma'(0) = \mathbf{v} + \mathbf{w}$.)

(c) Let

$$F(x^1, \ldots, x^m) = \begin{bmatrix} y^1(x^1, \ldots, x^m) \\ \vdots \\ y^n(x^1, \ldots, x^m) \end{bmatrix}.$$

Show that the matrix representation for $dF_{\mathbf{x}}$ in terms of the canonical bases for $T_{\mathbf{x}}\mathbb{R}^m$ and $T_{F(\mathbf{x})}\mathbb{R}^n$ is the Jacobian matrix

$$J = \left[\frac{\partial y^i}{\partial x^j}\right].$$

This means that if $\mathbf{v}$ has canonical representation $\mathbf{a} = {}^t[a^1, \ldots, a^m]$, then the canonical representation for $dF_{\mathbf{x}}(\mathbf{v})$ is

$$J\mathbf{a} = \begin{bmatrix} \frac{\partial y^1}{\partial x^j} a^j \\ \vdots \\ \frac{\partial y^n}{\partial x^j} a^j \end{bmatrix}.$$

(Hint: What is the simplest curve $\alpha(t)$ that you can think of with $\alpha(0) = \mathbf{x}$ and $\alpha'(0) = \mathbf{v}$?)

Now let's get a little more abstract and define the analogous notions for manifolds.

Definition 1.17. Suppose that M and N are m- and n-dimensional manifolds, respectively. A map $F : M \to N$ is called *differentiable* if for any parametrizations

$$\mathbf{x} : U \subset \mathbb{R}^m \to M, \qquad \mathbf{y} : V \subset \mathbb{R}^n \to N$$

on M and N such that $F(\mathbf{x}(U)) \subset \mathbf{y}(V)$, the composite map

$$\mathbf{y}^{-1} \circ F \circ \mathbf{x} : U \to V$$

is differentiable. (Definition 1.4 implies that this condition is independent of the choice of parametrizations $\mathbf{x}$ and $\mathbf{y}$.)

Remark 1.18. Keeping track of all these different maps can be challenging, and so they are often described using diagrams. For example, the maps above might be depicted as follows:

$$\begin{array}{ccc} M & \xrightarrow{\quad F \quad} & N \\ {\scriptstyle \mathbf{x}}\big\uparrow & & \big\uparrow{\scriptstyle \mathbf{y}} \\ \mathbb{R}^m \supseteq U & \xrightarrow{\mathbf{y}^{-1}\circ F \circ \mathbf{x}} & V \subset \mathbb{R}^n. \end{array}$$

This is called a *commutative diagram.* (Sometimes we simply say that "the diagram commutes".) This means that if we start with a point in U and map it to N, then we will arrive at the same point of N regardless of which path we take. Specifically, if $\mathbf{u} \in U$, then

$$(F \circ \mathbf{x})(\mathbf{u}) = (\mathbf{y} \circ (\mathbf{y}^{-1} \circ F \circ \mathbf{x}))(\mathbf{u}) \in N.$$

Of course, the commutativity of the diagram is obvious in this case, but sometimes it can indicate something more substantial.

In order to define the derivative of such a map F, we first need to define the tangent space to a manifold at a point. When M is a surface in $\mathbb{R}^3$, we naturally think of the tangent plane at each point $\mathbf{q} \in M$ as a 2-dimensional subspace of the ambient space $\mathbb{R}^3$. But if M is an abstract manifold and not a subset of some larger Euclidean space, then there may not be any natural choice of Euclidean space available for the tangent space to live in. In this case, we need a more self-contained notion for the tangent space.

It turns out that the answer lies in the observation that tangent vectors act on functions via directional derivative: Suppose that $\mathbf{x} \in \mathbb{R}^m$ and $\mathbf{v} \in T_{\mathbf{x}}\mathbb{R}^m$. If $\alpha : I \to \mathbb{R}^m$ is any smooth curve in $\mathbb{R}^m$ with $\alpha(0) = \mathbf{x}$, $\alpha'(0) = \mathbf{v}$ and if $f : \mathbb{R}^m \to \mathbb{R}$ is any differentiable function, then $\mathbf{v}$ acts on f as follows:

$$\mathbf{v}[f] = \left.\frac{d}{dt}\right|_{t=0} f(\alpha(t)).$$

In other words, $\mathbf{v}[f]$ is the directional derivative of f at $\mathbf{x}$ in the direction of $\mathbf{v}$. For an abstract manifold M, we simply turn this observation around and use it as a *definition* for tangent vectors.

Definition 1.19. Let $\alpha : I \to M$ be a smooth curve in M. The *tangent vector* $\alpha'(t_0)$ to α at the point $\alpha(t_0) \in M$ is the differential operator

$$\alpha'(t_0) : C^\infty(M) \to \mathbb{R}$$

(where $C^\infty(M)$ denotes the space of all smooth, real-valued functions on M) defined by

$$\alpha'(t_0)[f] = \left.\frac{d}{dt}\right|_{t=t_0} f(\alpha(t))$$

for $f \in C^\infty(M)$. The *tangent space* $T_\mathbf{q}M$ to M at a point $\mathbf{q} \in M$ is the set of all tangent vectors to all curves in M passing through $\mathbf{q}$; i.e.,

$$T_\mathbf{q}M = \{\alpha'(0) \mid \alpha : I \to M \text{ is a smooth curve with } \alpha(0) = \mathbf{q}\}.$$

Example 1.20. Let Σ be a regular surface in $\mathbb{R}^3$ and $\mathbf{x} : U \to \Sigma$ a parametrization of Σ defined on some open set $U \subset \mathbb{R}^2$, with local coordinates (u, v) on U. The tangent vectors $(\mathbf{x}_u, \mathbf{x}_v)$ form a basis for the tangent plane $T_\mathbf{q}\Sigma$ at each point $\mathbf{q} \in \mathbf{x}(U)$. In order to see how these vectors act as differential operators, let $f : \Sigma \to \mathbb{R}$ be a differentiable, real-valued function on Σ. Let $\mathbf{q} = \mathbf{x}(u_0, v_0) \in \mathbf{x}(U)$, and let $\alpha(t)$ be the curve

$$\alpha(t) = \mathbf{x}(u_0 + t, v_0)$$

in Σ. Then $\alpha'(0) = \mathbf{x}_u$, and so

$$\begin{aligned} \mathbf{x}_u[f] &= \left.\frac{d}{dt}\right|_{t=0} f(\alpha(t)) \\ &= \left.\frac{d}{dt}\right|_{t=0} f(\mathbf{x}(u_0 + t, v_0)) \\ &= \left.\frac{\partial(f \circ \mathbf{x})}{\partial u}\right|_{(u,v)=(u_0,v_0)}. \end{aligned}$$

Similarly, $\mathbf{x}_v[f] = \left.\frac{\partial(f\circ\mathbf{x})}{\partial v}\right|_{(u,v)=(u_0,v_0)}$. So, if we think of the parametrization $\mathbf{x}$ as identifying its image $\mathbf{x}(U) \subset \Sigma$ with the open set $U \subset \mathbb{R}^2$ and identifying a function $f \in C^\infty(M)$ with the composition $f\circ\mathbf{x} \in C^\infty(U)$, then the partial derivative operators $\left(\frac{\partial}{\partial u}, \frac{\partial}{\partial v}\right)$ form a natural basis for the tangent space $T_\mathbf{q}\Sigma$ at each point $\mathbf{q} \in \mathbf{x}(U)$.

***Exercise 1.21.** Let M be a manifold with local coordinates $\mathbf{x} = (x^1, \ldots, x^m)$ on some open set $V \subset M$. (The parametrization $\mathbf{x} : U \subset \mathbb{R}^m \to V \subset M$ is implicit in this statement.) Let $\mathbf{q} \in V$, and write $\mathbf{q} = \mathbf{x}(x_0^1, \ldots, x_0^m)$. The tangent space $T_\mathbf{q}M$ can be given the structure of an m-dimensional vector space, with the differential operators $\left(\frac{\partial}{\partial x^1}, \ldots, \frac{\partial}{\partial x^m}\right)$ as a basis, almost exactly as in Example 1.20

(a) Let $\mathbf{v} = \alpha'(0) \in T_\mathbf{q}M$, where

$$\alpha(t) = \mathbf{x}(x^1(t), \ldots, x^m(t)).$$

Show that for any $f \in C^\infty(M)$, we have

$$\mathbf{v}[f] = (x^i)'(0) \left.\frac{\partial(f \circ \mathbf{x})}{\partial x^i}\right|_{(x^1,\ldots,x^m)=(x_0^1,\ldots,x_0^m)}.$$

Therefore, any tangent vector $\mathbf{v} \in T_\mathbf{q}M$ may be regarded as a linear combination of the differential operators $\left(\frac{\partial}{\partial x^1}, \ldots, \frac{\partial}{\partial x^m}\right)$.

(b) Show that if $c^i \frac{\partial}{\partial x^i} = 0$ for some real numbers $c^1, \ldots, c^m \in \mathbb{R}$, then $c^1 = \cdots = c^m = 0$. This shows that the differential operators $\left(\frac{\partial}{\partial x^1}, \ldots, \frac{\partial}{\partial x^m}\right)$ are linearly independent. (Hint: The statement "$c^i \frac{\partial}{\partial x^i} = 0$" means that $c^i \frac{\partial (f \circ \mathbf{x})}{\partial x^i} = 0$ for every differentiable function $f : M \to \mathbb{R}$. Try to concoct real-valued functions f_i on V (which can then be extended to all of M via standard techniques) with the property that

$$\frac{\partial (f_i \circ \mathbf{x})}{\partial x^j} = \begin{cases} 1, & i = j, \\ 0, & i \neq j. \end{cases} \quad)$$

***Exercise 1.22.** (a) Suppose that we have a local coordinate transformation of the form

$$x^i = x^i(\bar{x}^1, \ldots, \bar{x}^m), \qquad i = 1, \ldots, m,$$

on M. Use the definition of tangent vectors and the multivariable chain rule to show that

$$\left[\frac{\partial}{\partial \bar{x}^1} \cdots \frac{\partial}{\partial \bar{x}^m}\right] = \left[\frac{\partial}{\partial x^1} \cdots \frac{\partial}{\partial x^m}\right] J,$$

where J is the Jacobian matrix of the coordinate transformation, i.e., the matrix whose (i, j)th entry is $\frac{\partial x^i}{\partial \bar{x}^j}$.

(b) Write out the ith component of the vector equation in part (a) using the Einstein summation convention, and make sure that the indices on both sides of the equation match up correctly.

The union of all the tangent spaces $T_{\mathbf{q}}M$ to a manifold M forms an object called the *tangent bundle* TM of M:

$$TM = \bigcup_{\mathbf{q} \in M} T_{\mathbf{q}}M.$$

If M has dimension m, then TM can be given the structure of a smooth manifold of dimension $2m$. Before we consider the general case, recall how this works for surfaces in $\mathbb{R}^3$:

Example 1.23 (The tangent bundle of a surface)**.** Let $\Sigma \subset \mathbb{R}^3$ be a regular surface. At each point $\mathbf{q} \in \Sigma$, the *tangent plane* $T_{\mathbf{q}}\Sigma$ at $\mathbf{q}$ consists of all tangent vectors to Σ at $\mathbf{q}$. The *tangent bundle* $T\Sigma$ of Σ is simply the union of all these tangent planes:

$$T\Sigma = \bigcup_{\mathbf{q} \in \Sigma} T_{\mathbf{q}}\Sigma.$$

Note that the tangent bundle is characterized by

(1) the *base space* Σ and

(2) for each point $\mathbf{q} \in \Sigma$, a vector space $T_{\mathbf{q}}\Sigma$ associated to $\mathbf{q}$, called the *fiber* at $\mathbf{q}$.

Moreover, the vector spaces associated to each point all have the same dimension—2, in this case. $T\Sigma$ is called the *total space* of the tangent bundle, and it can be given the structure of a manifold of dimension 4, as follows. If $\mathbf{x} = (x^1, x^2)$ is a system of local coordinates on an open set $V \subset \Sigma$, it can be canonically extended to a system of local coordinates $(\mathbf{x}, \mathbf{y}) = (x^1, x^2, y^1, y^2)$ on $TV = \bigcup_{\mathbf{q}\in V} T_{\mathbf{q}}V \subset T\Sigma$ by associating to any tangent vector $\mathbf{v} \in T_{\mathbf{q}}V$ its unique representation as

$$\mathbf{v} = y^1 \frac{\partial}{\partial x^1} + y^2 \frac{\partial}{\partial x^2}. \tag{1.7}$$

***Exercise 1.24.** Let $T\Sigma$ be the tangent bundle of a regular surface, and let $\mathbf{x} = (x^1, x^2)$ and $\bar{\mathbf{x}} = (\bar{x}^1, \bar{x}^2)$ be two overlapping systems of local coordinates on Σ, related by a coordinate transformation of the form

$$x^1 = x^1(\bar{x}^1, \bar{x}^2), \qquad x^2 = x^2(\bar{x}^1, \bar{x}^2). \tag{1.8}$$

Show that the local coordinate transformation between the canonically associated local coordinates $(\mathbf{x}, \mathbf{y}) = (x^1, x^2, y^1, y^2)$ and $(\bar{\mathbf{x}}, \bar{\mathbf{y}}) = (\bar{x}^1, \bar{x}^2, \bar{y}^1, \bar{y}^2)$ on $T\Sigma$ is given by equations (1.8) together with the equations

$$\begin{bmatrix} y^1 \\ y^2 \end{bmatrix} = J \begin{bmatrix} \bar{y}^1 \\ \bar{y}^2 \end{bmatrix},$$

where J is the Jacobian matrix of the transformation (1.8). (Hint: The result of Exercise 1.22 should be helpful here.)

Tangent bundles for manifolds of arbitrary dimension work exactly the same way. If M is a manifold of dimension m and $\mathbf{x} : U \subset \mathbb{R}^m \to M$ is a parametrization of M, then the associated *canonical parametrization* of TM is the map $(\mathbf{x}, \mathbf{y}) : U \times \mathbb{R}^m \to TM$ given by

$$(\mathbf{x}, \mathbf{y})(x^1, \ldots, x^m, y^1, \ldots, y^m) = y^i \frac{\partial}{\partial x^i} \in T_{\mathbf{x}(x^1,\ldots,x^m)}M.$$

In other words, for each point $\mathbf{q} \in \mathbf{x}(U)$, the tangent space $T_{\mathbf{q}}M$ is identified with the vector space $\mathbb{R}^m$ by identifying the basis $\left(\frac{\partial}{\partial x^1}, \ldots, \frac{\partial}{\partial x^m}\right)$ for $T_{\mathbf{q}}M$ with the standard basis for $\mathbb{R}^m$. The same calculation as in Exercise 1.24 shows that the transition map between any two canonical parametrizations $(\mathbf{x}, \mathbf{y})$ and $(\bar{\mathbf{x}}, \bar{\mathbf{y}})$ is differentiable and, in fact, has the form

$$x^i = x^i(\bar{x}^1, \ldots, \bar{x}^m), \qquad y^i = \frac{\partial x^i}{\partial \bar{x}^j} \bar{y}^j;$$

therefore, TM is a smooth manifold of dimension $2m$. TM also comes equipped with the *base-point projection map* $\pi : TM \to M$ defined by the condition that for any $\mathbf{v} \in T_{\mathbf{q}}M$, $\pi(\mathbf{v}) = \mathbf{q}$.

Here's the good news: Now that we know how to define tangent spaces for manifolds, the definition of the derivative of a differentiable map $F : M \to N$ is exactly the same as before.

Definition 1.25. Let M be a manifold of dimension m, N a manifold of dimension n, and let $F : M \to N$ be a differentiable map. Fix a point $\mathbf{q} \in M$. The *derivative* or *differential* of F at $\mathbf{q}$ is the linear map $dF_{\mathbf{q}} : T_{\mathbf{q}}M \to T_{F(\mathbf{q})}N$ defined as follows: Given a tangent vector $\mathbf{v} \in T_{\mathbf{q}}M$, let $\alpha : I \to M$ be a smooth curve with $\alpha(0) = \mathbf{q}$ and $\alpha'(0) = \mathbf{v}$. Then

$$dF_{\mathbf{q}}(\mathbf{v}) = (F \circ \alpha)'(0) \in T_{F(\mathbf{q})}N.$$

If we happen to have local coordinates $\mathbf{x} = (x^1, \ldots, x^m)$ on M and $\mathbf{y} = (y^1, \ldots, y^n)$ on N, then we can compute exactly as if M and N were $\mathbb{R}^m$ and $\mathbb{R}^n$, respectively. The matrix representation for $dF_{\mathbf{q}}$ will still look like the Jacobian matrix of F at $\mathbf{q}$, provided that the natural coordinate bases $\left(\frac{\partial}{\partial x^1}, \ldots, \frac{\partial}{\partial x^m}\right)$ and $\left(\frac{\partial}{\partial y^1}, \ldots, \frac{\partial}{\partial y^n}\right)$ are used for the tangent spaces $T_{\mathbf{q}}M$ and $T_{F(\mathbf{q})}N$, respectively. The main difference is that the isomorphisms between $T_{\mathbf{q}}M, T_{F(\mathbf{q})}N$ and $\mathbb{R}^m, \mathbb{R}^n$ defined by these coordinate bases are no longer canonical: Changing local coordinates changes the natural bases for the tangent spaces as well. This is the main reason why we need to know how tensors transform under changes of basis for the underlying vector spaces.

***Exercise 1.26.** Let $F : M \to N$ be a differentiable map. Let

$$\mathbf{x} : U \subset \mathbb{R}^m \to M, \qquad \mathbf{y} : V \subset \mathbb{R}^n \to N$$

be parametrizations, and let

$$H = \mathbf{y}^{-1} \circ F \circ \mathbf{x} : U \to V.$$

We can write

$$H(x^1, \ldots, x^m) = (y^1(x^1, \ldots, x^m), \ldots, y^n(x^1, \ldots, x^m)).$$

(a) Suppose that $\mathbf{q} \in \mathbf{x}(U)$, and let $\mathbf{v} = a^j \frac{\partial}{\partial x^j} \in T_{\mathbf{q}}M$. Show that

$$dF_{\mathbf{q}}(\mathbf{v}) = \left(a^j \frac{\partial y^i}{\partial x^j}\right) \frac{\partial}{\partial y^i} \in T_{F(\mathbf{q})}N;$$

in other words,

$$dF_{\mathbf{q}}\left(\begin{bmatrix}\frac{\partial}{\partial x^1} & \cdots & \frac{\partial}{\partial x^m}\end{bmatrix}\begin{bmatrix}a^1\\ \vdots \\ a^m\end{bmatrix}\right) = \begin{bmatrix}\frac{\partial}{\partial y^1} & \cdots & \frac{\partial}{\partial y^n}\end{bmatrix} J \begin{bmatrix}a^1\\ \vdots \\ a^m\end{bmatrix},$$

where J is the $n \times m$ Jacobian matrix of H.

BONUS: What happens to this formula if you perform local coordinate transformations on M and N?

(b) The differential of F can be extended to a map

$$dF : TM \to TN$$

in the obvious way: For $\mathbf{v} \in T_{\mathbf{q}}M$, define

$$dF(\mathbf{v}) = dF_{\mathbf{q}}(\mathbf{v}).$$

Show that dF is a differentiable map from TM to TN.

***Exercise 1.27.** Let M, N, and P be manifolds, and suppose that $F : M \to N$ and $G : N \to P$ are differentiable maps.

(a) Prove the *chain rule:* For $\mathbf{q} \in M$,

$$d(G \circ F)_{\mathbf{q}} = dG_{F(\mathbf{q})} \circ dF_{\mathbf{q}}.$$

(b) Give an explicit interpretation of the chain rule in terms of local coordinates and Jacobian matrices. (What are the sizes of each of the matrices involved, in terms of the dimensions m, n, p of M, N, and P?) Write out the (i, j)th component of this matrix equation using the Einstein summation convention, and make sure that the indices on both sides of the equation match up correctly.

We will often need to consider *vector fields* on manifolds. Intuitively, a vector field on a manifold M is simply a choice of a single tangent vector in each tangent space $T_{\mathbf{q}}M$, but we also need to know what it means for a vector field to be differentiable. Fortunately, we now have all the tools that we need in order to make this idea precise:

Definition 1.28. A *(smooth) vector field* $\mathbf{v}$ on a manifold M is a differentiable map

$$\mathbf{v} : M \to TM$$

with the property that for any $\mathbf{q} \in M$, we have $\mathbf{v}(\mathbf{q}) \in T_{\mathbf{q}}M$. (Equivalently, the composition $\pi \circ \mathbf{v} : M \to M$ is the identity map.)

A vector field $\mathbf{v}$ on a manifold M may be regarded as a first-order, linear, homogeneous differential operator $\mathbf{v} : C^\infty(M) \to C^\infty(M)$ defined as follows: For $f \in C^\infty(M)$, the function $\mathbf{v}[f] : M \to \mathbb{R}$ is given by

$$\mathbf{v}[f](\mathbf{q}) = \mathbf{v}(\mathbf{q})[f].$$

This means that the value of $\mathbf{v}[f]$ at $\mathbf{q} \in M$ is just the value of the tangent vector $\mathbf{v}(\mathbf{q}) \in T_{\mathbf{q}}M$ acting on f at $\mathbf{q}$.

In terms of local coordinates $(x^1, \ldots, x^m)$ on M, a vector field $\mathbf{v}$ is generally expressed as

$$\mathbf{v}(x^1, \ldots, x^m) = a^i(x^1, \ldots, x^m)\frac{\partial}{\partial x^i}. \tag{1.9}$$

Then for $f \in C^\infty(M)$, the function $\mathbf{v}[f]$ is expressed in terms of these local coordinates as

$$\mathbf{v}[f](x^1, \ldots, x^m) = a^i(x^1, \ldots, x^m)\frac{\partial f}{\partial x^i}(x^1, \ldots, x^m). \tag{1.10}$$

Conversely, any (smooth) first-order, linear, homogeneous differential operator $L : C^\infty(M) \to C^\infty(M)$ given in local coordinates by (1.10) defines a vector field $\mathbf{v}$ on M described in local coordinates by (1.9).

We close this section by defining four important types of differentiable maps between manifolds.

Definition 1.29. Let M be a manifold of dimension m, N a manifold of dimension n, and let $F : M \to N$ be a differentiable map. F is called

(1) a *diffeomorphism* if F is bijective and the inverse map $F^{-1} : N \to M$ is also differentiable; if such a map F exists, then we say that M and N are *diffeomorphic*;

(2) an *immersion* if $dF_{\mathbf{q}} : T_{\mathbf{q}}M \to T_{F(\mathbf{q})}N$ is injective at every point $\mathbf{q} \in M$ (note that this requires $m \leq n$);

(3) an *embedding* if F is an injective immersion that is also a homeomorphism from M onto its image $F(M) \subset N$, where $F(M)$ is given the subspace topology inherited from the topology of N (note that this requires $m \leq n$);

(4) a *submersion* if $dF_{\mathbf{q}} : T_{\mathbf{q}}M \to T_{F(\mathbf{q})}N$ is surjective at every point $\mathbf{q} \in M$ (note that this requires $m \geq n$).

Remark 1.30. The definition of an embedding looks rather technical, but in practice, an embedding is usually just an injective immersion. The image of an immersion may have self-intersections, but the restriction of an immersion to a sufficiently small neighborhood of any point in its domain is always an embedding.

1.4. Lie groups and matrix groups

We begin with the following definition.

Definition 1.31. A *Lie group* is a set G that is both a group and a differentiable manifold, with the additional property that the map $\mu : G \times G \to G$ given by

$$\mu(g, h) = gh^{-1}$$

is differentiable.

This is a concise way of stating that group multiplication and group inverse are both differentiable operations on G, as the following exercise shows.

Exercise 1.32. Let G be a Lie group. Show that the multiplication map $G \times G \to G$ defined by

$$(g, h) \to gh$$

and the inverse map $G \to G$ defined by

$$g \to g^{-1}$$

are both differentiable. Conversely, show that differentiability of both of these maps implies that the map μ in Definition 1.31 is differentiable. (Hint: Consider the inverse map first.)

To any Lie group G is associated a *Lie algebra* $\mathfrak{g}$. The Lie algebra $\mathfrak{g}$ is simply the tangent space to G at the identity element $e \in G$; thus $\mathfrak{g}$ is a vector space of the same dimension as the manifold G. There is also a product structure on $\mathfrak{g}$ known as the *Lie bracket.* Defining this structure will require a bit of effort.

For each element $g \in G$, the *left translation* or *left multiplication* map $L_g : G \to G$ is defined by

$$L_g(h) = gh.$$

(Similarly, the *right translation* map $R_g : G \to G$ is defined by

$$R_g(h) = hg.)$$

For any element $h \in G$, the differential $(dL_g)_h$ of L_g at h is a linear map from T_hG to $T_{gh}G$. In particular, the differential of L_g at the identity element e is a linear map from the Lie algebra $\mathfrak{g}$ to T_gG. We can use this map to associate to any element $\mathbf{v} \in \mathfrak{g}$ a *left-invariant vector field* $\tilde{\mathbf{v}}$ on G defined at each point $g \in G$ by

$$\tilde{\mathbf{v}}(g) = (dL_g)_e(\mathbf{v}). \tag{1.11}$$

This terminology is explained in the following exercise:

***Exercise 1.33.** Let G be a Lie group with Lie algebra $\mathfrak{g}$, let $\mathbf{v} \in \mathfrak{g}$, and let $\tilde{\mathbf{v}}$ be the vector field on G defined by (1.11).

(a) Show that for any elements $g, h \in G$, the value of $\tilde{\mathbf{v}}$ at $gh \in G$ is given by

$$\tilde{\mathbf{v}}(gh) = dL_g(\tilde{\mathbf{v}}(h)),$$

where

$$dL_g : T_hG \to T_{gh}G$$

is the differential of the left multiplication map L_g. This is why the vector field $\tilde{\mathbf{v}}$ is called *left-invariant:* It is invariant under the action of left multiplication by any element of G. (It is also true that any left-invariant vector field $\tilde{\mathbf{v}}$ on a Lie group G is smooth; see, e.g., [**Lee13**].)

(b) Conversely, suppose that a smooth vector field $\mathbf{V}$ on G has the property that for any $g, h \in G$,

$$\mathbf{V}(gh) = dL_g(\mathbf{V}(h)). \tag{1.12}$$

Show that $\mathbf{V} = \tilde{\mathbf{v}}$, where $\mathbf{v} = \mathbf{V}(e)$.

Recall that a vector field on G is really a first-order, linear, homogeneous differential operator on smooth functions $f : G \to \mathbb{R}$. Given any two smooth vector fields $\mathbf{V}, \mathbf{W}$ on G (or indeed, any two smooth vector fields on any manifold), the *Lie bracket* $[\mathbf{V}, \mathbf{W}]$ is the smooth vector field defined by the property that for any smooth function $f : G \to \mathbb{R}$,

$$[\mathbf{V}, \mathbf{W}][f] = \mathbf{V}[\mathbf{W}[f]] - \mathbf{W}[\mathbf{V}[f]].$$

Wait a minute—a vector field is a first-order differential operator, but that looks like a second-order differential operator! The following exercise should allay your concerns:

***Exercise 1.34.** Let $\mathbf{V}, \mathbf{W}$ be two vector fields on a manifold M. There is no harm in working within a parametrization, so without loss of generality we may assume that

$$\mathbf{V}(\mathbf{x}) = a^i(\mathbf{x})\frac{\partial}{\partial x^i}, \qquad \mathbf{W}(\mathbf{x}) = b^j(\mathbf{x})\frac{\partial}{\partial x^j}$$

for some functions $(a^i(\mathbf{x}), b^j(\mathbf{x}))$. Show by direct computation that the Lie bracket $[\mathbf{V}, \mathbf{W}]$ is a vector field—i.e., a first-order, linear, homogeneous differential operator—on M. How are its coefficients related to $(a^i(\mathbf{x}), b^j(\mathbf{x}))$? (Hint: Compute the action of the operator $[\mathbf{V}, \mathbf{W}]$ on a smooth function $f : M \to \mathbb{R}$. And since "smooth" means "infinitely differentiable", you can safely assume that mixed partial derivatives commute.)

Finally, we use the Lie bracket of the left-invariant vector fields $\tilde{\mathbf{v}}, \tilde{\mathbf{w}}$ to define the Lie bracket operation on the Lie algebra $\mathfrak{g}$:

Definition 1.35. Let G be a Lie group with associated Lie algebra $\mathfrak{g}$. For any two elements $\mathbf{v}, \mathbf{w} \in \mathfrak{g}$, the *Lie bracket* $[\mathbf{v}, \mathbf{w}]$ of $\mathbf{v}$ and $\mathbf{w}$ is the unique element $\mathbf{z} \in \mathfrak{g}$ such that

$$[\tilde{\mathbf{v}}, \tilde{\mathbf{w}}] = \tilde{\mathbf{z}},$$

where tildes denote the extensions to left-invariant vector fields on G, as defined above.

In order for this definition to make sense, we need to know that the Lie bracket of two left-invariant vector fields on G is also left-invariant.

***Exercise 1.36.** Let $\tilde{\mathbf{v}}, \tilde{\mathbf{w}}$ be left-invariant vector fields on a Lie group G. Show that the Lie bracket $[\tilde{\mathbf{v}}, \tilde{\mathbf{w}}]$ satisfies the condition (1.12) and that it is therefore equal to $\tilde{\mathbf{z}}$ for some $\mathbf{z} \in \mathfrak{g}$. Conclude that the Lie bracket is a well-defined operation on $\mathfrak{g}$.

While this all sounds fairly abstract, the most common examples of Lie groups—and the only ones that we will encounter in this book—are subgroups of the group $GL(n)$ of invertible $n \times n$ matrices. The differentiable structure on $GL(n)$ is simply that inherited by regarding $GL(n)$ as an open set in $\mathbb{R}^{n^2}$. The Lie algebra of any such Lie group is a subspace of the vector space of $n \times n$ matrices.

Remark 1.37. If it seems a little strange to think of a matrix as a vector, let G be any subgroup of $GL(n)$, and let $\alpha(t)$ be a curve in G with $\alpha(0) = I$. Then we can write

$$\alpha(t) = g(t),$$

where $g(t)$ is an $n \times n$ matrix. The tangent vector to this curve at $t = 0$ is, of course,

$$\alpha'(0) = g'(0),$$

which, while it *could* be thought of as a vector in $\mathbb{R}^{n^2}$, is most naturally regarded as an $n \times n$ matrix.

The following exercise shows that the Lie bracket on the Lie algebra of any subgroup of $GL(n)$ is simply the matrix commutator:

$$[A, B] = AB - BA.$$

***Exercise 1.38.** Let G be a subgroup of $GL(n)$.

(a) Let $g \in G$, and let $L_g : G \to G$ be the left multiplication map. Show that the differential $dL_g : \mathfrak{g} \to T_gG$ is given explicitly by

$$dL_g(A) = g \cdot A, \qquad A \in \mathfrak{g},$$

where the right-hand side is just the matrix product of the two $n \times n$ matrices g and A. Therefore, the vector field $\tilde{A}$ on G defined by

$$\tilde{A}(g) = g \cdot A$$

is left-invariant. (Hint: Consider a curve $h(t)$ in G with $h(0) = I, h'(0) = A$.)

(b) Let's get explicit about exactly what differential operator the vector field $\tilde{A}$ represents on G: If we use local coordinates (x^i_j) on the manifold of $n \times n$ matrices, then a general element $g \in G$ is represented by a matrix $[x^i_j]$. If we write $A = [a^i_j]$, then the tangent vector $A \in \mathfrak{g}$ represents the differential operator

$$A = a^i_j \frac{\partial}{\partial x^i_j},$$

and the matrix product $\tilde{A} = g \cdot A$, whose entries are

$$\tilde{A}^i_j = x^i_r a^r_j,$$

represents the differential operator

$$\tilde{A} = x^i_r a^r_j \frac{\partial}{\partial x^i_j}.$$

Now, let $A = [a^i_j]$, $B = [b^k_l] \in \mathfrak{g}$. Then we can write

$$\tilde{A} = x^i_r a^r_j \frac{\partial}{\partial x^i_j}, \qquad \tilde{B} = x^k_s b^s_l \frac{\partial}{\partial x^k_l}.$$

Use the formula that you computed in Exercise 1.34 (and some very careful index manipulation!) to show that

$$[\tilde{A}, \tilde{B}] = x^i_r \left(a^r_l b^l_j - b^r_l a^l_j \right) \frac{\partial}{\partial x^i_j}.$$

(c) Conclude from part (b) that

$$[\tilde{A}, \tilde{B}] = g \cdot (AB - BA) = (\widetilde{AB - BA})$$

and that the Lie bracket on $\mathfrak{g}$ is therefore just the matrix commutator:

$$[A, B] = AB - BA.$$

The following examples describe some of the most commonly encountered Lie groups; we will be seeing all of these groups later on in the book.

Example 1.39. The general linear group $GL(n)$, consisting of all invertible $n \times n$ matrices. As a manifold, $GL(n)$ is an open subset of the space $M_{n\times n}$ of all $n \times n$ matrices; specifically,

$$GL(n) = \{g \in M_{n\times n} \mid \det(g) \neq 0\}.$$

Thus, its Lie algebra $\mathfrak{gl}(n)$ is the tangent space to the space of all $n \times n$ matrices at the identity I_n. This tangent space is naturally isomorphic to $M_{n\times n}$.

Example 1.40. The special linear group $SL(n)$. $SL(n)$ is defined as

$$SL(n) = \{A \in GL(n) \mid \det(A) = 1\}.$$

A standard computation using the implicit function theorem (which we omit here) shows that $SL(n)$ is a manifold of dimension $n^2 - 1$. The Lie algebra $\mathfrak{sl}(n)$ consists of all tangent vectors to curves in $SL(n)$ passing through the identity I_n at $t = 0$. So, suppose that

$$A(t) = \left[a^i_j(t)\right]$$

is a curve of matrices in $SL(n)$ with $A(0) = I_n$. Since $A(t) \in SL(n)$ for all t, we have

$$\det(A(t)) = \sum_{\sigma\in S_n} \operatorname{sgn}(\sigma) a^1_{\sigma(1)}(t) \cdots a^n_{\sigma(n)}(t) = 1,$$

where the sum is over all permutations σ in the symmetric group S_n and $\operatorname{sgn}(\sigma) = \pm 1$ is the sign of σ. In order to determine what conditions this imposes on the tangent vector $A'(0)$, we need to differentiate this equation and evaluate the result at $t = 0$. Fortunately, this computation is not as bad as it looks: The fact that $A(0) = I_n$ means that $a^i_j(0) = 0$ unless $i = j$, in which case $a^i_j(0) = 1$. So most of the terms in the derivative will drop out at $t = 0$, leaving the equation

$$(a^i_i)'(0) = 0.$$

(Don't forget the summation convention!) In other words, $A'(0)$ must have trace equal to zero.

Exercise 1.41. Do this computation explicitly for $SL(2)$: Let

$$A(t) = \begin{bmatrix} a^1_1(t) & a^1_2(t) \\ a^2_1(t) & a^2_2(t) \end{bmatrix},$$

with $A(0) = I$ and

$$\det(A(t)) = a^1_1(t)a^2_2(t) - a^1_2(t)a^2_1(t) = 1.$$

Differentiate this equation and set $t = 0$, and show that

$$(a^1_1)'(0) + (a^2_2)'(0) = 0;$$

i.e., $\operatorname{tr}(A'(0)) = 0$.

Aside from the trace condition, there are no other restrictions on $A'(0)$. This is most easily verified by comparing dimensions: $SL(n)$ has dimension $n^2 - 1$, and the set of $n \times n$ matrices with trace equal to zero is a vector space of dimension $n^2 - 1$. Since $\mathfrak{sl}(n)$ is contained in this vector space and has the same dimension as the entire space, it must be equal to the entire space. Therefore,

$$\mathfrak{sl}(n) = \{B \in M_{n\times n} \mid \operatorname{tr}(B) = 0\}.$$

Example 1.42. The orthogonal group $O(n)$. $O(n)$ is defined as

$$O(n) = \{A \in GL(n) \mid {}^tA\,A = I_n\},$$

where ${}^tA$ denotes the transpose of the matrix A. A standard computation using the implicit function theorem (which we omit here) shows that $O(n)$ is a manifold of dimension $\frac{1}{2}n(n-1)$. In order to compute its Lie algebra $\mathfrak{o}(n)$, suppose that $A(t)$ is a curve of matrices in $O(n)$ with $A(0) = I_n$. Since $A(t) \in O(n)$ for all t, we have

$${}^tA(t)\,A(t) = I_n.$$

Differentiating this equation and setting $t = 0$ yields

$${}^tA'(0) + A'(0) = 0;$$

in other words, $A'(0)$ must be a skew-symmetric matrix. Comparing dimensions as in the previous example shows that there are no other restrictions on $A'(0)$, and so

$$\mathfrak{o}(n) = \{B \in M_{n\times n} \mid {}^tB + B = 0\}.$$

Example 1.43. The special orthogonal group $SO(n)$. $SO(n)$ is defined as

$$SO(n) = \{A \in GL(n) \mid {}^tA\,A = I_n \text{ and } \det(A) = 1\}.$$

Clearly, $SO(n)$ is a subgroup of $O(n)$; it consists of those matrices in $O(n)$ having determinant equal to 1. Now, it follows from the defining equation for $O(n)$ that every element of $O(n)$ has determinant equal to either 1 or -1. Moreover, the function

$$\det : O(n) \to \mathbb{R}$$

is continuous, with image equal to the set $\{1, -1\}$. Therefore, the subsets

$$SO(n) = \{A \in O(n) \mid \det(A) = 1\},$$
$$O(n) \setminus SO(n) = \{A \in O(n) \mid \det(A) = -1\}$$

are disconnected. In fact, $O(n)$ consists of two connected components, one of which is $SO(n)$ and the other of which is the coset of $SO(n)$ consisting of those elements with determinant -1. It follows that the Lie algebra $\mathfrak{so}(n)$ is, in fact, *equal* to the Lie algebra $\mathfrak{o}(n)$. This Lie algebra is usually denoted by $\mathfrak{so}(n)$.

For a more thorough discussion of Lie groups and Lie algebras, see, e.g., [**Lee13**].

1.5. Vector bundles and principal bundles

The tangent bundle TM of a manifold M is an example of a *vector bundle*. The formal definition of a vector bundle is a bit complicated, but the general idea is much like this example.

Definition 1.44. A *rank k vector bundle* consists of a manifold B called the *base space*, a manifold E called the *total space*, and a differentiable *projection map* $\pi : E \to B$ such that for each point $\mathbf{q} \in B$, the inverse image $\pi^{-1}(\mathbf{q}) \subset E$ is a vector space of dimension k. This vector space is called the *fiber* of E at $\mathbf{q}$. Moreover, every point $\mathbf{q} \in B$ must have a neighborhood $U \subset B$ such that the set $\pi^{-1}(U) \subset E$ is diffeomorphic to $U \times \mathbb{R}^k$ via a diffeomorphism that, for every $\tilde{\mathbf{q}} \in U$, maps the fiber $\pi^{-1}(\tilde{\mathbf{q}})$ linearly onto $\{\tilde{\mathbf{q}}\} \times \mathbb{R}^k$. Such a diffeomorphism is called a *local trivialization* of the vector bundle.

It is important to note that, while vector bundles always have local trivializations, they may not have *global* trivializations. That is to say, the local trivializations may patch together in such a way that their union becomes "twisted" in a way that cannot be straightened out. So, a vector bundle $\pi : E \to B$ is *not* necessarily diffeomorphic to $B \times \mathbb{R}^k$. For example, the tangent bundle of a compact surface has no global trivialization unless the surface has Euler characteristic equal to zero.

Next, we introduce the notion of a *section* of a vector bundle.

Definition 1.45. A *(smooth) section* of a vector bundle $\pi : E \to B$ is a differentiable map $\sigma : B \to E$ with the property that the composition $\pi \circ \sigma : B \to B$ is the identity map on B.

A section $\sigma : B \to E$ is kind of like a vector-valued function on the base space B, but not quite. A function has a domain space and a range space; in particular, the function takes values in the same range space at each point of the domain. A section σ has a domain space B, but at each point $\mathbf{q} \in B$, the value of $\sigma(\mathbf{q})$ must lie in the fiber of E at $\mathbf{q}$. Intuitively, this means that *the range space is different at each point of the domain space.* We may consider both *local sections*, whose domain consists of some open set in B, and *global sections*, whose domain is the entire space B. A section is often identified with its image in E, which is a submanifold of E consisting of exactly one point in each fiber. This may sound complicated, but again, you already know an example: A vector field on a surface Σ is a section of the tangent bundle $T\Sigma$.

Any vector bundle has a distinguished section called the *zero section*; this is the section that assigns to each point $\mathbf{q} \in B$ the zero vector in the fiber of E at $\mathbf{q}$. Other than the zero section, there is no canonical way of choosing "constant" sections unless an additional structure known as a *connection* is introduced. (And even then, it may only be possible to choose a section which is "constant" along a curve, but not on any open set.) A connection allows sections of vector bundles to be differentiated through a process called *covariant differentiation.* The tangent bundle of a regular surface in $\mathbb{R}^3$ carries a canonical connection called the *Levi-Civita connection*, which is used to define parallelisms, geodesics, and so forth. (These concepts will be discussed in more detail in Chapters 11 and 12.)

A section is called *nonvanishing* if its image does not intersect the image of the zero section. If a vector bundle does not have a global trivialization, then it may be that there do not exist *any* nonvanishing global sections. If this sounds surprising, then recall the *Poincaré-Hopf theorem:* Any smooth vector field on a compact surface with nonzero Euler characteristic must vanish at some point on the surface. This is equivalent to the statement that the tangent bundle has no nonvanishing global sections.

As if vector bundles weren't enough, we will also frequently encounter *principal bundles.* A principal bundle is similar to a vector bundle, in that it consists of a manifold B called the *base space*, a manifold P called the *total space*, and a differentiable projection map $\pi : P \to B$. The main difference is that the fiber $\pi^{-1}(\mathbf{q})$ at any point $\mathbf{q} \in B$ is a manifold diffeomorphic to some Lie group G rather than to a vector space. However, this diffeomorphism is generally *not* a group isomorphism in any canonical way. In particular, there is usually no well-defined "identity section" of P, and a principal bundle generally has no distinguished section similar to the zero section of a vector bundle. (In fact, it is possible for a principal bundle to have *no global sections whatsoever.*) Rather, each fiber $\pi^{-1}(\mathbf{q})$ is a manifold on which G acts freely and transitively.

A principal bundle $\pi : P \to B$ whose fiber at each point is diffeomorphic to a Lie group G is often represented by the diagram

$$\begin{array}{ccc} G & \longrightarrow & P \\ & & \downarrow{\scriptstyle \pi} \\ & & B. \end{array}$$

The arrow $G \to P$ doesn't really represent a map here; it just indicates that each fiber of P is diffeomorphic to G and that G acts freely and transitively on the fibers of P.

We will see some examples of principal bundles beginning in Chapter 3. In many cases, the total space P will itself be a Lie group, and the fibers will all be diffeomorphic to some subgroup G of P.

Exercise 1.46. Let Σ be the unit sphere $\mathbb{S}^2 \subset \mathbb{R}^3$. Let $\mathcal{F}(\mathbb{S}^2)$ denote the *orthonormal frame bundle* of $\mathbb{S}^2$; this is the set of all orthonormal bases (a.k.a. "frames") $\mathbf{e} = (\mathbf{e}_1, \mathbf{e}_2)$ for the tangent space $T_{\mathbf{q}}\mathbb{S}^2$ at each point $\mathbf{q} \in \mathbb{S}^2$. Given any orthonormal frame $\mathbf{e}$ at a point $\mathbf{q} \in \mathbb{S}^2$, any other frame at $\mathbf{q}$ can be obtained from $\mathbf{e}$ by composition of a rotation and (possibly) a reflection, i.e., an element of the Lie group $O(2)$. Conversely, every element of $O(2)$ acts on $\mathbf{e}$ to produce a new orthonormal frame at $\mathbf{q}$, and distinct elements of $O(2)$ will produce different frames. Thus, the fiber of $\mathcal{F}(\mathbb{S}^2)$ over any point $\mathbf{q} \in \mathbb{S}^2$ is diffeomorphic to $O(2)$, and $\mathcal{F}(\mathbb{S}^2)$ is a principal bundle over $\mathbb{S}^2$ with fiber group $O(2)$.

Prove that $\mathcal{F}(\mathbb{S}^2)$ has no global sections. (Hint: Use the Poincaré-Hopf theorem.)

Chapter 2

Differential forms

2.1. Introduction

In calculus, you may have seen the *differential* or *exterior derivative* df of a function $f : \mathbb{R}^3 \to \mathbb{R}$ defined to be

$$df = \frac{\partial f}{\partial x} dx + \frac{\partial f}{\partial y} dy + \frac{\partial f}{\partial z} dz.$$

The expression df is called a 1-*form* and you probably learned various ways of manipulating such things—for instance, how to integrate a 1-form along a parametrized curve in $\mathbb{R}^3$. But what *is* a 1-form, really?

In this chapter, we will give formal definitions for 1-forms and then more generally for p-forms with $p \geq 0$ on $\mathbb{R}^n$, and we will explore a bit about how they work. Once we understand how differential forms behave in $\mathbb{R}^n$, we will see how to define them more generally on manifolds. But before we launch into all that, it will be helpful to go into a bit more detail about certain types of tensors.

2.2. Dual spaces, the cotangent bundle, and tensor products

Definition 2.1. Let V be an n-dimensional vector space. The *dual space* V^* is the set of all linear mappings $\alpha : V \to \mathbb{R}$.

It should be clear that V^* is closed under addition and scalar multiplication, and so V^* is a vector space. The following exercise shows that V^* has dimension n:

***Exercise 2.2.** Let $(\mathbf{e}_1, \ldots, \mathbf{e}_n)$ be a basis for V. Define linear functions $\mathbf{e}^{*i} : V \to \mathbb{R}$, $1 \leq i \leq n$, by the property that

$$\mathbf{e}^{*i}(\mathbf{e}_j) = \delta^i_j = \begin{cases} 1, & i = j, \\ 0, & i \neq j. \end{cases}$$

(Note that requiring $\mathbf{e}^{*i}$ to be linear and specifying the value of $\mathbf{e}^{*i}$ on all of the basis vectors $\mathbf{e}_j$ completely determines the function $\mathbf{e}^{*i}$ on V.)

(a) Show that the functions $(\mathbf{e}^{*1}, \ldots, \mathbf{e}^{*n})$ are linearly independent. (Hint: The hypothesis "$c_i\mathbf{e}^{*i} = 0$" means that for every $\mathbf{v} \in V$, $c_i\mathbf{e}^{*i}(\mathbf{v}) = 0$.)

(b) Show that the functions $(\mathbf{e}^{*1}, \ldots, \mathbf{e}^{*n})$ span V^*. (Hint: Let $\alpha : V \to \mathbb{R}$ be a linear mapping, and let $c_i = \alpha(\mathbf{e}_i)$. Consider the function $c_i\mathbf{e}^{*i} : V \to \mathbb{R}$.)

(c) Show that $(V^*)^* \cong V$. (Hint: Associate to any vector $\mathbf{v} \in V$ the function $\hat{\mathbf{v}} : V^* \to \mathbb{R}$ defined by

$$\hat{\mathbf{v}}(\alpha) = \alpha(\mathbf{v}).)$$

The basis $(\mathbf{e}^{*1}, \ldots, \mathbf{e}^{*n})$ for V^* constructed in Exercise 2.2 is called the *dual basis* to the basis $(\mathbf{e}_1, \ldots, \mathbf{e}_n)$ for V. Also, note that the isomorphism $(V^*)^* \cong V$ is *canonical*: It is completely independent of any choice of basis for V. Thus it is accurate—and common—to regard $(V^*)^*$ as being *equal* to V.

The most common example of a dual space that we will encounter is the following:

Example 2.3. Let M be a manifold of dimension m, let $\mathbf{q} \in M$, and let $V = T_\mathbf{q}M$. The dual space $V^* = T^*_\mathbf{q}M$ is called the *cotangent space* to M at $\mathbf{q}$. The *cotangent bundle* of M is the union of all these cotangent spaces:

$$T^*M = \bigcup_{\mathbf{q} \in M} T^*_\mathbf{q}M.$$

T^*M has the structure of a smooth manifold of dimension $2m$: Given any parametrization $\mathbf{x} : U \subset \mathbb{R}^m \to M$ of M, there is an associated canonical parametrization $(\mathbf{x}, \mathbf{p}) : U \times \mathbb{R}^m \to T^*M$ of T^*M, defined in a manner analogous to the canonical parametrizations of the tangent bundle TM. Specifically, the element $\alpha = (\mathbf{x}, \mathbf{p})(x^1, \ldots, x^m, p_1, \ldots, p_m) \in T^*_{\mathbf{x}(x^1,\ldots,x^m)}M$ is defined by the condition that for any $\mathbf{v} = y^i \frac{\partial}{\partial x^i} \in T_{\mathbf{x}(x^1,\ldots,x^m)}M$, we have

$$\alpha(\mathbf{v}) = p_i y^i \in \mathbb{R}.$$

Remark 2.4. More generally, given a vector space V, the dual space V^* is sometimes referred to as the *covector space* of V, and the elements of V^* as *covectors.* If the elements of V are represented by column vectors, then the elements of V^* are represented by row vectors, and vice versa.

Most of the tensor fields that commonly appear in geometry are sections of vector bundles that are constructed from the tangent and cotangent bundles of a manifold. The construction used to build more complicated bundles from these two is the *tensor product.* The official definition starts by defining tensor products for dual spaces:

Definition 2.5. Let V be a vector space of dimension m and W a vector space of dimension n. Let $\alpha \in V^*, \beta \in W^*$. The *tensor product* of α and β, denoted $\alpha \otimes \beta$, is the bilinear function

$$\alpha \otimes \beta : V \times W \to \mathbb{R}$$

defined by

$$(\alpha \otimes \beta)(\mathbf{v}, \mathbf{w}) = \alpha(\mathbf{v})\beta(\mathbf{w}).$$

The *tensor product* of V^* and W^*, denoted $V^* \otimes W^*$, is the vector space consisting of all linear combinations of such tensor products; i.e.,

$$V^* \otimes W^* = \operatorname{span}\{\alpha \otimes \beta \mid \alpha \in V^*, \beta \in W^*\}.$$

***Exercise 2.6.** (a) Show that $\alpha \otimes \beta$ is indeed a bilinear function on $V \times W$, i.e., that for any vectors $\mathbf{v}, \mathbf{v}_1, \mathbf{v}_2 \in V$, $\mathbf{w}, \mathbf{w}_1, \mathbf{w}_2 \in W$, and real numbers a, b,

$$\begin{aligned}(\alpha \otimes \beta)(a\mathbf{v}_1 + b\mathbf{v}_2, \mathbf{w}) &= a(\alpha \otimes \beta)(\mathbf{v}_1, \mathbf{w}) + b(\alpha \otimes \beta)(\mathbf{v}_2, \mathbf{w}),\\ (\alpha \otimes \beta)(\mathbf{v}, a\mathbf{w}_1 + b\mathbf{w}_2) &= a(\alpha \otimes \beta)(\mathbf{v}, \mathbf{w}_1) + b(\alpha \otimes \beta)(\mathbf{v}, \mathbf{w}_2).\end{aligned}$$

(b) Show that $V^* \otimes W^*$ is a vector space of dimension mn. (Hint: Let $(\mathbf{e}_1, \ldots, \mathbf{e}_m)$ be a basis for V, with dual basis $(\mathbf{e}^{*1}, \ldots, \mathbf{e}^{*m})$ for V^*, and let $(\mathbf{f}_1, \ldots, \mathbf{f}_n)$ be a basis for W, with dual basis $(\mathbf{f}^{*1}, \ldots, \mathbf{f}^{*n})$ for W^*. Show that

$$\{\mathbf{e}^{*i} \otimes \mathbf{f}^{*j} \mid 1 \le i \le m,\ 1 \le j \le n\}$$

is a basis for $V^* \otimes W^*$.)

So, how do we define the tensor product $V \otimes W$? We simply take advantage of the canonical identification $V = (V^*)^*$ that we proved in Exercise 2.2:

Definition 2.7. Let V be a vector space of dimension m and W a vector space of dimension n. For any vectors $\mathbf{v} \in V, \mathbf{w} \in W$, define $\hat{\mathbf{v}} \in (V^*)^*$, $\hat{\mathbf{w}} \in (W^*)^*$ by

$$\begin{aligned}\hat{\mathbf{v}}(\alpha) &= \alpha(\mathbf{v}), \qquad \alpha \in V^*,\\ \hat{\mathbf{w}}(\beta) &= \beta(\mathbf{w}), \qquad \beta \in W^*.\end{aligned}$$

The *tensor product* of $\mathbf{v}$ and $\mathbf{w}$, denoted $\mathbf{v} \otimes \mathbf{w}$, is given by

$$\mathbf{v} \otimes \mathbf{w} = \hat{\mathbf{v}} \otimes \hat{\mathbf{w}} \in (V^*)^* \otimes (W^*)^*.$$

The *tensor product* of V and W, denoted $V \otimes W$, is the vector space

$$V \otimes W = (V^*)^* \otimes (W^*)^*.$$

Of course, this definition is much too convoluted to use in practice! For practical purposes, if $(\mathbf{e}_1, \ldots, \mathbf{e}_m)$ is a basis for V and $(\mathbf{f}_1, \ldots, \mathbf{f}_n)$ is a basis for W, then $V \otimes W$ is simply a vector space of dimension mn, with basis given by the formal symbols

$$\{\mathbf{e}_i \otimes \mathbf{f}_j \mid 1 \leq i \leq m, 1 \leq j \leq n\}.$$

Remark 2.8. The tensor product is associative; i.e.,

$$(\mathbf{v} \otimes \mathbf{w}) \otimes \mathbf{x} = \mathbf{v} \otimes (\mathbf{w} \otimes \mathbf{x}),$$

and so it makes sense to write $\mathbf{v} \otimes \mathbf{w} \otimes \mathbf{x}$. But it is *not* commutative; even when $V = W$, $\mathbf{v} \otimes \mathbf{w}$ is, in general, not equal to $\mathbf{w} \otimes \mathbf{v}$.

Definition 2.9. A *rank k tensor* is an element of a tensor product of the form

$$V_1 \otimes \cdots \otimes V_k,$$

where $V_1, \ldots, V_k$ are vector spaces, some or all of which may be dual spaces. In particular, a rank 1 tensor is simply an element of a vector space V (or V^*).

Example 2.10 (Cf. Example 1.12)**.** Let V be an m-dimensional vector space with basis $(\mathbf{e}_1, \ldots, \mathbf{e}_m)$ and dual basis $(\mathbf{e}^{*1}, \ldots, \mathbf{e}^{*m})$ for V^*, and let W be an n-dimensional vector space with basis $(\mathbf{f}_1, \ldots, \mathbf{f}_n)$ and dual basis $(\mathbf{f}^{*1}, \ldots, \mathbf{f}^{*n})$ for W^*. Let $T : V \to W$ be a linear transformation, represented in terms of the given bases for V and W by the $n \times m$ matrix

$$A_T = \begin{bmatrix} c_1^1 & \ldots & c_m^1 \\ \vdots & & \vdots \\ c_1^n & \ldots & c_m^n \end{bmatrix}.$$

Then T is a rank 2 tensor; it is an element of the vector space $W \otimes V^*$, and it can be written in terms of the given bases as

$$T = c_j^i \mathbf{f}_i \otimes \mathbf{e}^{*j}.$$

Two important categories of tensors are the *symmetric* and *skew-symmetric* tensors of a vector space V.

Definition 2.11. Let V be an n-dimensional vector space, and let $V^{\otimes 2} = V \otimes V$. (Similarly, let $V^{\otimes k}$ denote the tensor product of k copies of V.) The space of *symmetric 2-tensors* of V is the subspace S^2V of $V^{\otimes 2}$ defined by

$$S^2V = \operatorname{span}\{\mathbf{v} \otimes \mathbf{v} \mid \mathbf{v} \in V\}.$$

The space of *skew-symmetric 2-tensors* of V is the subspace $\Lambda^2 V$ of $V^{\otimes 2}$ defined by

$$\Lambda^2 V = \operatorname{span}\{\mathbf{v} \otimes \mathbf{w} - \mathbf{w} \otimes \mathbf{v} \mid \mathbf{v}, \mathbf{w} \in V\}.$$

Exercise 2.12. Show that for any vectors $\mathbf{v}, \mathbf{w} \in V$,

$$\mathbf{v} \otimes \mathbf{w} + \mathbf{w} \otimes \mathbf{v} \in S^2V.$$

(Hint: Consider $(\mathbf{v} + \mathbf{w}) \otimes (\mathbf{v} + \mathbf{w})$.)

Definition 2.13. The *symmetric product* $\mathbf{v} \circ \mathbf{w}$ and *wedge product* $\mathbf{v} \wedge \mathbf{w}$ are defined by

$$\mathbf{v} \circ \mathbf{w} = \tfrac{1}{2}(\mathbf{v} \otimes \mathbf{w} + \mathbf{w} \otimes \mathbf{v}),$$
$$\mathbf{v} \wedge \mathbf{w} = \mathbf{v} \otimes \mathbf{w} - \mathbf{w} \otimes \mathbf{v}.$$

(Note that some authors insert a factor of $\frac{1}{2}$ into the definition of the wedge product as well.) More generally, if $\mathbf{v}_1, \dots, \mathbf{v}_k \in V$, then

$$\mathbf{v}_1 \circ \cdots \circ \mathbf{v}_k = \frac{1}{k!} \sum_{\sigma} \mathbf{v}_{\sigma(1)} \otimes \cdots \otimes \mathbf{v}_{\sigma(k)},$$
$$\mathbf{v}_1 \wedge \cdots \wedge \mathbf{v}_k = \sum_{\sigma} \operatorname{sgn}(\sigma) \mathbf{v}_{\sigma(1)} \otimes \cdots \otimes \mathbf{v}_{\sigma(k)},$$

where the sum is over all permutations σ in the symmetric group S_k and $\operatorname{sgn}(\sigma) = \pm 1$ is the sign of σ. The spaces of *symmetric* and *skew-symmetric k-tensors* of V are the subspaces S^kV, $\Lambda^k V$ of $V^{\otimes k}$ defined by

$$S^kV = \operatorname{span}\{\mathbf{v}_1 \circ \cdots \circ \mathbf{v}_k \mid \mathbf{v}_1, \dots, \mathbf{v}_k \in V\},$$
$$\Lambda^k V = \operatorname{span}\{\mathbf{v}_1 \wedge \cdots \wedge \mathbf{v}_k \mid \mathbf{v}_1, \dots, \mathbf{v}_k \in V\}.$$

Exercise 2.14. (a) Show that $V \otimes V = S^2V \oplus \Lambda^2 V$; i.e., every element of $V \otimes V$ can be uniquely expressed as the sum of a symmetric product and a skew-symmetric product.

(b) Show that the analogous decomposition does not hold for $V^{\otimes 3}$. (Hint: Compute the dimensions of $V^{\otimes 3}, S^3V$, and $\Lambda^3 V$ in terms of the dimension n of V.)

All these tensor products can be used to form *tensor bundles* on a manifold M. A tensor bundle on M is simply a vector bundle on M where each fiber is isomorphic to a fixed tensor product of vector spaces. In the most common examples, the fiber over each point $\mathbf{q} \in M$ is a tensor product of some numbers of copies of the tangent space $T_{\mathbf{q}}M$ and the cotangent space $T^*_{\mathbf{q}}M$, or some natural subspace of such a tensor product. Finally, this allows us to give a formal definition of tensor fields:

Definition 2.15. A *rank k tensor field* on a manifold M is a section of a vector bundle $\pi : E \to M$, where the fiber $E_{\mathbf{q}}$ of E over each point $\mathbf{q} \in M$ has the form

$$E_{\mathbf{q}} = V_1 \otimes \cdots \otimes V_k,$$

where $V_1, \ldots, V_k$ are vector spaces. (In most cases, each of the vector spaces $V_1, \ldots, V_k$ is equal to either $T_{\mathbf{q}}M$ or $T^*_{\mathbf{q}}M$.)

Remark 2.16. In practice, tensor fields on a manifold M are almost always referred to as "tensors on M". For example, a Riemannian manifold M has a "metric tensor" g and a "Riemann curvature tensor" R. Strictly speaking, these are both tensor fields on M, but you will probably never encounter the phrase "Riemann curvature tensor field".

Example 2.17 (Cf. Exercise 1.15)**.** Let M be a manifold of dimension n, and let

$$S^2(T^*M) = \bigcup_{\mathbf{q} \in M} S^2(T^*_{\mathbf{q}}M).$$

$S^2(T^*M)$ is a subbundle of the tensor bundle $T^*M \otimes T^*M$, and it is a bundle of symmetric 2-tensors on M. (Note that I didn't say "the" bundle of symmetric 2-tensors on M, because there are other bundles that would satisfy the same description, e.g., $S^2(TM)$.) A metric g on M is a section of $S^2(T^*M)$ because, at every point $\mathbf{q} \in M$, g defines a symmetric, bilinear function

$$g_{\mathbf{q}} : T_{\mathbf{q}}M \times T_{\mathbf{q}}M \to \mathbb{R}.$$

Now we're ready to start talking about differential forms!

2.3. 1-forms on $\mathbb{R}^n$

Definition 2.18. A *(smooth)* 1*-form* ϕ on $\mathbb{R}^n$ is a (smooth) section of the cotangent bundle $T^*\mathbb{R}^n$. Equivalently,

$$\phi : T\mathbb{R}^n \to \mathbb{R}$$

is a real-valued function on the set of all tangent vectors to $\mathbb{R}^n$, with the properties that:

(1) For each $\mathbf{x} \in \mathbb{R}^n$, the restriction

$$\phi_{\mathbf{x}} : T_{\mathbf{x}}\mathbb{R}^n \to \mathbb{R}$$

of ϕ to $T_{\mathbf{x}}\mathbb{R}^n$ is a linear map.

(2) For any smooth vector field $\mathbf{v}$ on $\mathbb{R}^n$, the function $\phi(\mathbf{v}) : \mathbb{R}^n \to \mathbb{R}$ is smooth.

In particular, the 1-forms $(dx^1, \ldots, dx^n)$ are defined by the property that for any vector $\mathbf{v} = {}^t[v^1, \ldots, v^n] \in T_{\mathbf{x}}\mathbb{R}^n$,

$$dx^i(\mathbf{v}) = v^i. \tag{2.1}$$

***Exercise 2.19** (Cf. Exercise 2.2)**.** In this exercise, we will show directly that for any point $\mathbf{x} \in \mathbb{R}^n$, the 1-forms $(dx^1, \ldots, dx^n)$ defined by (2.1) form a basis for the cotangent space $T^*_{\mathbf{x}}\mathbb{R}^n$.

(a) In order to show that the 1-forms $(dx^1, \ldots, dx^n)$ are linearly independent, suppose that

$$c_i\, dx^i = 0.$$

This means that $c_i\, dx^i(\mathbf{v}) = 0$ for every vector $\mathbf{v} \in T_{\mathbf{x}}\mathbb{R}^n$. Show that this implies that $c_1 = \cdots = c_n = 0$.

(b) In order to show that the 1-forms $(dx^1, \ldots, dx^n)$ span $T^*_{\mathbf{x}}\mathbb{R}^n$, let $\phi_{\mathbf{x}} \in T^*_{\mathbf{x}}\mathbb{R}^n$ be arbitrary. Let $(\underline{\mathbf{e}}_1, \ldots, \underline{\mathbf{e}}_n)$ be the standard basis for $T_{\mathbf{x}}\mathbb{R}^n$, and let

$$c_i = \phi_{\mathbf{x}}(\underline{\mathbf{e}}_i), \qquad i = 1, \ldots, n.$$

Show that $\phi_{\mathbf{x}} = c_i\, dx^i$. (It suffices to show that $\phi_{\mathbf{x}}(\mathbf{v}) = c_i\, dx^i(\mathbf{v})$ for every $\mathbf{v} \in T_{\mathbf{x}}\mathbb{R}^n$.)

It follows from Exercise 2.19 that the 1-forms $(dx^1, \ldots, dx^n)$ form a basis for the 1-forms on $\mathbb{R}^n$ as a *module* over $C^\infty(\mathbb{R}^n)$; this means that any 1-form ϕ on $\mathbb{R}^n$ can be expressed uniquely as

$$\phi = f_i(\mathbf{x})\, dx^i \tag{2.2}$$

for some smooth, real-valued functions $f_1, \ldots, f_n : \mathbb{R}^n \to \mathbb{R}$. If a vector field $\mathbf{v}$ on $\mathbb{R}^n$ has the form

$$\mathbf{v}(\mathbf{x}) = {}^t[v^1(\mathbf{x}), \ldots, v^n(\mathbf{x})],$$

then we have

$$\phi(\mathbf{v}(\mathbf{x})) = f_i(\mathbf{x})\, v^i(\mathbf{x}).$$

2.4. p-forms on $\mathbb{R}^n$

The 1-forms on $\mathbb{R}^n$ are the building blocks of an algebra, called the *algebra of differential forms* on $\mathbb{R}^n$. The multiplication in this algebra is the wedge product of Definition 2.13. The wedge product is multilinear and skew-symmetric; i.e., for any two 1-forms ϕ, ψ, we have

$$\phi \wedge \psi = -\psi \wedge \phi.$$

In particular, $\phi \wedge \phi = 0$ for any 1-form ϕ.

If each summand of a differential form Φ is a wedge product of p 1-forms, then the form is called a *p-form*. Scalar-valued functions are considered to be 0-forms, and any form on $\mathbb{R}^n$ of degree $p > n$ must be zero due to the skew-symmetry.

Exercise 2.20. Prove this last statement. (Hint: Let $\phi_1, \dots, \phi_p$ be 1-forms. According to (2.2), we can write

$$\phi_j = f_{ji}(\mathbf{x})\, dx^i, \qquad j = 1, \dots, p.$$

Show by direct computation that if $p > n$, then $\phi_1 \wedge \cdots \wedge \phi_p = 0$.)

More formally, we have the following definition:

Definition 2.21. A *p-form* on $\mathbb{R}^n$ is a section of the tensor bundle $\Lambda^p(T^*\mathbb{R}^n)$. The *algebra of differential forms* on $\mathbb{R}^n$ consists of all sections of the bundle

$$\bigoplus_{p \geq 0} \Lambda^p(T^*\mathbb{R}^n) = \bigcup_{\mathbf{x} \in \mathbb{R}^n} \bigoplus_{p \geq 0} \Lambda^p(T^*_{\mathbf{x}}\mathbb{R}^n),$$

with multiplication given by the wedge product

$$\wedge : \Lambda^p(T^*\mathbb{R}^n) \times \Lambda^q(T^*\mathbb{R}^n) \to \Lambda^{p+q}(T^*\mathbb{R}^n).$$

The space of p-forms on $\mathbb{R}^n$ is generally denoted by $\Omega^p(\mathbb{R}^n)$, and the algebra of all differential forms on $\mathbb{R}^n$ is denoted by

$$\Omega^*(\mathbb{R}^n) = \bigoplus_{p \geq 0} \Omega^p(\mathbb{R}^n).$$

A basis for the p-forms on $\mathbb{R}^n$ (as a module over $C^\infty(\mathbb{R}^n)$) is given by the set

$$\{dx^{i_1} \wedge \cdots \wedge dx^{i_p} \mid 1 \leq i_1 < i_2 < \cdots < i_p \leq n\};$$

this simply means that any p-form Φ on $\mathbb{R}^n$ can be expressed uniquely as

$$\Phi = \sum_{|I| = p} f_I(\mathbf{x})\, dx^{i_1} \wedge \cdots \wedge dx^{i_p},$$

where I ranges over all increasing multi-indices $I = (i_1, \dots, i_p)$ of length p and the coefficients $f_I(\mathbf{x})$ represent smooth functions $f_I : \mathbb{R}^n \to \mathbb{R}$.

Just as 1-forms act on vector fields to give real-valued functions, so p-forms act on p-tuples of vector fields to give real-valued functions. For instance, if ϕ, ψ are 1-forms and $\mathbf{v}, \mathbf{w}$ are vector fields, then

$$(\phi \wedge \psi)(\mathbf{v}, \mathbf{w}) = \phi(\mathbf{v})\psi(\mathbf{w}) - \phi(\mathbf{w})\psi(\mathbf{v}).$$

More generally, if $\phi_1, \dots, \phi_p$ are 1-forms and $\mathbf{v}_1, \dots, \mathbf{v}_p$ are vector fields, then

$$(\phi_1 \wedge \cdots \wedge \phi_p)(\mathbf{v}_1, \dots, \mathbf{v}_p) = \sum_{\sigma \in S_p} \operatorname{sgn}(\sigma)\, \phi_1(\mathbf{v}_{\sigma(1)})\, \phi_2(\mathbf{v}_{\sigma(2)}) \cdots \phi_n(\mathbf{v}_{\sigma(n)}),$$

where the sum is over all permutations σ in the symmetric group S_p and $\operatorname{sgn}(\sigma) = \pm 1$ is the sign of σ.

Exercise 2.22. Show that the wedge product is related to the determinant as follows: If $\phi_1, \dots, \phi_p$ are 1-forms and $\mathbf{v}_1, \dots, \mathbf{v}_p$ are vector fields, then

$$(\phi_1 \wedge \cdots \wedge \phi_p)(\mathbf{v}_1, \dots, \mathbf{v}_p) = \det \begin{bmatrix} \phi_1(\mathbf{v}_1) & \cdots & \phi_1(\mathbf{v}_p) \\ \vdots & & \vdots \\ \phi_p(\mathbf{v}_1) & \cdots & \phi_p(\mathbf{v}_p) \end{bmatrix}.$$

2.5. The exterior derivative

The *exterior derivative* is an operator that takes p-forms to $(p+1)$-forms. We will define it first for functions and then extend this definition to higher degree forms.

Definition 2.23. If $f : \mathbb{R}^n \to \mathbb{R}$ is differentiable, then the *exterior derivative* of f is the 1-form df with the property that for any $\mathbf{x} \in \mathbb{R}^n$, $\mathbf{v} \in T_{\mathbf{x}}\mathbb{R}^n$,

$$df_{\mathbf{x}}(\mathbf{v}) = \mathbf{v}[f];$$

i.e., $df_{\mathbf{x}}(\mathbf{v})$ is the directional derivative of f at $\mathbf{x}$ in the direction of $\mathbf{v}$.

***Exercise 2.24.** Show that dx^i as defined by equation (2.1) really is the exterior derivative of the ith coordinate function x^i on $\mathbb{R}^n$. (So the notation is consistent!)

It is not difficult to show that, as one might expect,

$$df = \frac{\partial f}{\partial x^i} dx^i. \tag{2.3}$$

The exterior derivative also obeys the Leibniz rule

$$d(fg) = g\, df + f\, dg \tag{2.4}$$

for functions $f, g : \mathbb{R}^n \to \mathbb{R}$ and the chain rule

$$d(h \circ f) = (h' \circ f)\, df \tag{2.5}$$

for functions $f : \mathbb{R}^n \to \mathbb{R}$, $h : \mathbb{R} \to \mathbb{R}$.

***Exercise 2.25.** Verify equations (2.3), (2.4), and (2.5).

We extend the definition of the exterior derivative to p-forms on $\mathbb{R}^n$ as follows:

Definition 2.26. Given a p-form $\Phi = \sum_{|I|=p} f_I(\mathbf{x})\, dx^{i_1} \wedge \cdots \wedge dx^{i_p}$ on $\mathbb{R}^n$, the *exterior derivative* $d\Phi$ of Φ is the $(p+1)$-form

$$d\Phi = \sum_{|I|=p} df_I \wedge dx^{i_1} \wedge \cdots \wedge dx^{i_p}. \tag{2.6}$$

If Φ is a p-form and Ψ is a q-form, then the Leibniz rule takes the form

$$d(\Phi \wedge \Psi) = d\Phi \wedge \Psi + (-1)^p \Phi \wedge d\Psi. \tag{2.7}$$

***Exercise 2.27.** Prove the Leibniz rule (2.7) in the case $p = q = 1$: If ϕ, ψ are 1-forms on $\mathbb{R}^n$, then

$$d(\phi \wedge \psi) = d\phi \wedge \psi - \phi \wedge d\psi.$$

The following theorem is possibly the most important one in this entire book!

Theorem 2.28. *$d \circ d = 0$; i.e., for any differential form Φ on $\mathbb{R}^n$,*

$$d(d\Phi) = 0.$$

Proof. First, suppose that f is a function (i.e., a 0-form). Then

$$\begin{aligned}
d(df) &= d\left(\frac{\partial f}{\partial x^i} dx^i\right) \\
&= \sum_{i,j=1}^{n} \frac{\partial^2 f}{\partial x^i\, \partial x^j} dx^j \wedge dx^i \\
&= \sum_{i<j} \left(\frac{\partial^2 f}{\partial x^j\, \partial x^i} - \frac{\partial^2 f}{\partial x^i\, \partial x^j}\right) dx^i \wedge dx^j \\
&= 0
\end{aligned}$$

because mixed partials commute.

Next, note that by Exercise 2.24 and the argument above, $d(dx^i) = 0$. Now suppose that

$$\Phi = \sum_{|I|=p} f_I(\mathbf{x})\, dx^{i_1} \wedge \cdots \wedge dx^{i_p}.$$

Then by the Leibniz rule,

$$\begin{aligned} d(d\Phi) &= d\left(\sum_{|I|=p} df_I \wedge dx^{i_1} \wedge \cdots \wedge dx^{i_p}\right) \\ &= \sum_{|I|=p} \left(d(df_I) \wedge dx^{i_1} \wedge \cdots \wedge dx^{i_p} - df_I \wedge d(dx^{i_1}) \wedge \cdots \wedge dx^{i_p} + \cdots\right) \\ &= 0. \end{aligned}$$

□

***Exercise 2.29.** In multivariable calculus, there are three basic operations based on partial differentiation involving functions and vector fields on $\mathbb{R}^3$:

(1) the *gradient* of a function $f : \mathbb{R}^3 \to \mathbb{R}$:

$$\operatorname{grad}(f) = \sum_{i=1}^{3} \frac{\partial f}{\partial x^i} \frac{\partial}{\partial x^i};$$

(2) the *curl* of a vector field $\mathbf{v}(\mathbf{x}) = v^i(\mathbf{x}) \dfrac{\partial}{\partial x^i}$:

$$\operatorname{curl}(\mathbf{v}) = \left(\frac{\partial v^3}{\partial x^2} - \frac{\partial v^2}{\partial x^3}\right) \frac{\partial}{\partial x^1} + \left(\frac{\partial v^1}{\partial x^3} - \frac{\partial v^3}{\partial x^1}\right) \frac{\partial}{\partial x^2} + \left(\frac{\partial v^2}{\partial x^1} - \frac{\partial v^1}{\partial x^2}\right) \frac{\partial}{\partial x^3};$$

(3) the *divergence* of a vector field $\mathbf{v}(\mathbf{x}) = v^i(\mathbf{x}) \dfrac{\partial}{\partial x^i}$:

$$\operatorname{div}(\mathbf{v}) = \sum_{i=1}^{3} \frac{\partial v^i}{\partial x^i}.$$

These operations can all be expressed in terms of differential forms and their exterior derivatives via the following correspondences between forms on the one hand and vector fields and functions on the other:

$$f_1(\mathbf{x})dx^1 + f_2(\mathbf{x})dx^2 + f_3(\mathbf{x})dx^3 \longleftrightarrow \sum_{i=1}^{3} f_i(\mathbf{x}) \frac{\partial}{\partial x^i},$$

$$f_1(\mathbf{x})dx^2 \wedge dx^3 + f_2(\mathbf{x})dx^3 \wedge dx^1 + f_3(\mathbf{x})dx^1 \wedge dx^2 \longleftrightarrow \sum_{i=1}^{3} f_i(\mathbf{x}) \frac{\partial}{\partial x^i},$$

$$f(\mathbf{x})\, dx^1 \wedge dx^2 \wedge dx^3 \longleftrightarrow f(\mathbf{x}).$$

Prove that under this correspondence:

(1) If f is a function on $\mathbb{R}^3$, then $df \longleftrightarrow \operatorname{grad}(f)$.

(2) If ϕ is a 1-form on $\mathbb{R}^3$ and $\phi \longleftrightarrow \mathbf{v}$, then $d\phi \longleftrightarrow \operatorname{curl}(\mathbf{v})$.

(3) If η is a 2-form on $\mathbb{R}^3$ and $\eta \longleftrightarrow \mathbf{v}$, then $d\eta \longleftrightarrow \operatorname{div}(\mathbf{v})$.

Show that the identities

$$\operatorname{curl}(\operatorname{grad}(f)) = 0,$$
$$\operatorname{div}(\operatorname{curl}(\mathbf{v})) = 0$$

follow from the fact that $d \circ d = 0$.

2.6. Closed and exact forms and the Poincaré lemma

In light of Theorem 2.28, we introduce the following important classes of differential forms:

Definition 2.30. A p-form Φ on $\mathbb{R}^n$ is called *closed* if $d\Phi = 0$. Φ is called *exact* if there exists a $(p-1)$-form Ψ on $\mathbb{R}^n$ such that $\Phi = d\Psi$.

By Theorem 2.28, every exact form is closed. The converse is only partially true: Every closed form is *locally* exact. Precisely, we have

Theorem 2.31 (The Poincaré lemma). *Given a closed p-form Φ on an open set $U \subset \mathbb{R}^n$ that is homeomorphic to an open ball, there exists a $(p-1)$-form Ψ on U with $d\Psi = \Phi$.*

(For proofs of the results in this section, see Volume 1 of [**Spi79**].) In particular, when $p = 1$ we have

Corollary 2.32. *Given a closed 1-form ϕ on an open set $U \subset \mathbb{R}^n$ that is homeomorphic to an open ball, there exists a differentiable function $f : U \to \mathbb{R}$ such that $\phi = df$.*

Closely related to Corollary 2.32 is the following special case of the *Frobenius theorem*:

Theorem 2.33 (Frobenius). *Let ϕ be a nonvanishing 1-form on an open set $U \subset \mathbb{R}^n$, and suppose that $\phi \wedge d\phi = 0$. Then for any point $\mathbf{x} \in U$, there exist a neighborhood $V \subset U$ of $\mathbf{x}$ and differentiable functions $f, g : V \to \mathbb{R}$ such that the restriction of ϕ to V is given by $\phi = g\,df$.*

2.7. Differential forms on manifolds

Now that we know how differential forms behave on $\mathbb{R}^n$, we might expect them to behave similarly in terms of any local coordinate system on an arbitrary manifold M. But if we want differential forms to be well-defined objects on M, then they need to be defined independently of any particular choice of local coordinate system. So, we start with the following coordinate-free definition:

Definition 2.34. Given a smooth manifold M, a *smooth* 1-*form* ϕ on M is a smooth section of the cotangent bundle T^*M. Equivalently, ϕ is a real-valued function $\phi : TM \to \mathbb{R}$ on the tangent bundle of M such that:

(1) For each $\mathbf{q} \in M$, the restriction
$$\phi : T_{\mathbf{q}}M \to \mathbb{R}$$
of ϕ to $T_{\mathbf{q}}M$ is a linear map.

(2) For any smooth vector field $\mathbf{v}$ on M, the function $\phi(\mathbf{v}) : M \to \mathbb{R}$ is smooth.

Now, suppose that we have a system of local coordinates $\mathbf{x} = (x^1, \ldots, x^n)$ on an open set $U \subset M$. Associated to these local coordinates, we have the basis $\left(\frac{\partial}{\partial x^1}, \ldots, \frac{\partial}{\partial x^n}\right)$ for the local sections of the tangent bundle TM over U. We define the 1-forms $(dx^1, \ldots, dx^n)$ on U to be the dual basis for the local sections of the cotangent bundle T^*M over U; in other words, the 1-forms (dx^i) are defined by the property that
$$dx^i\left(\frac{\partial}{\partial x^j}\right) = \begin{cases} 0, & i \neq j, \\ 1, & i = j. \end{cases}$$
Any p-form Φ on U then has a unique expression as
$$\Phi = \sum_{|I|=p} f_I(\mathbf{x})\, dx^{i_1} \wedge \cdots \wedge dx^{i_p}$$
for some functions f_I on U.

Wedge products, p-forms, and exterior derivatives of functions are defined just as for $\mathbb{R}^n$. (Go back and convince yourself that these definitions are coordinate-independent, and so they make sense on an arbitrary manifold without having to specify a particular choice of local coordinates.) However, Definition 2.26 for the exterior derivative of a p-form with $p \geq 1$ *does* depend on a choice of local coordinates, and so extending this definition to an arbitrary manifold requires some care. The most straightforward approach is to go ahead and use Definition 2.26 *and* to prove that this definition is consistent under local coordinate transformations.

Definition 2.35. Let Φ be a p-form on a manifold M, and suppose that Φ can be expressed as

$$\Phi = \sum_{|I|=p} f_I(\mathbf{x})\, dx^{i_1} \wedge \cdots \wedge dx^{i_p}$$

in terms of some system of local coordinates $\mathbf{x} = (x^1, \ldots, x^n)$ on M. Then the *exterior derivative* $d\Phi$ of Φ is the $(p+1)$-form whose local coordinate expression is

$$d\Phi = \sum_{|I|=p} df_I \wedge dx^{i_1} \wedge \cdots \wedge dx^{i_p}.$$

***Exercise 2.36.** Show that Definition 2.35 is well-defined for 1-forms, as follows: Suppose that $\mathbf{x} = (x^1, \ldots, x^n)$ and $\bar{\mathbf{x}} = (\bar{x}^1, \ldots, \bar{x}^n)$ are two systems of local coordinates on an open set $U \subset M$, with transition functions

$$(\bar{x}^1, \ldots, \bar{x}^n) \to \left(x^1(\bar{x}^1, \ldots, \bar{x}^n), \ldots, x^n(\bar{x}^1, \ldots, \bar{x}^n)\right).$$

Then according to (2.3),

$$dx^i = \frac{\partial x^i}{\partial \bar{x}^j}\, d\bar{x}^j.$$

(Note that this equation may be expressed as

$$\begin{bmatrix} dx^1 \\ \vdots \\ dx^n \end{bmatrix} = J \begin{bmatrix} d\bar{x}^1 \\ \vdots \\ d\bar{x}^n \end{bmatrix},$$

where J is the Jacobian matrix of the coordinate transformation. Compare with Exercise 1.22.)

(a) Let ϕ be a 1-form on U, defined by

$$\phi = f_i(\mathbf{x})\, dx^i$$

with respect to the local coordinates $\mathbf{x} = (x^1, \ldots, x^n)$. Show that

$$\phi = f_i(\mathbf{x}(\bar{\mathbf{x}}))\, \frac{\partial x^i}{\partial \bar{x}^k}\, d\bar{x}^k$$

with respect to the local coordinates $\bar{\mathbf{x}} = (\bar{x}^1, \ldots, \bar{x}^n)$.

(b) Compute $d\phi$ for both of the local coordinate expressions in part (a), and use the chain rule to show that both expressions for $d\phi$ represent the same 2-form on U.

Alternatively, the exterior derivative may be defined independently of any choice of local coordinates: According to *Cartan's formula*, if Φ is a p-form, then $d\Phi$ is the $(p+1)$-form defined by the property that for any $p+1$ vector fields $\mathbf{v}_1, \ldots, \mathbf{v}_{p+1}$ on M,

$$(2.8)\quad d\Phi(\mathbf{v}_1, \ldots, \mathbf{v}_{p+1}) = \sum_{i=1}^{p+1} (-1)^{i+1} \mathbf{v}_i[\Phi(\mathbf{v}_1, \ldots, \hat{\mathbf{v}}_i, \ldots, \mathbf{v}_{p+1})] + \sum_{i<j} (-1)^{i+j} \Phi([\mathbf{v}_i, \mathbf{v}_j], \mathbf{v}_1, \ldots, \hat{\mathbf{v}}_i, \ldots, \hat{\mathbf{v}}_j, \ldots, \mathbf{v}_{p+1}).$$

Here a hat over a symbol indicates that that element is omitted from the list; $\mathbf{v}_i[\Phi(\mathbf{v}_1, \ldots, \hat{\mathbf{v}}_i, \ldots, \mathbf{v}_{p+1})]$ denotes the directional derivative of the scalar-valued function $\Phi(\mathbf{v}_1, \ldots, \hat{\mathbf{v}}_i, \ldots, \mathbf{v}_{p+1})$ in the direction of $\mathbf{v}_i$, and $[\mathbf{v}_i, \mathbf{v}_j]$ denotes the Lie bracket of the vector fields $\mathbf{v}_i$ and $\mathbf{v}_j$.

Remark 2.37. This definition is obviously somewhat cumbersome, and it is rarely used in practice. Its importance lies in its theoretical significance: Since the exterior derivative is well-defined independently of a choice of local coordinates, it really should have a coordinate-free definition, and Cartan's formula provides one.

Exercise 2.38. Show by direct computation that Cartan's formula agrees with Definition 2.35 in the case when Φ is a 1-form: Specifically, let

$$\phi = f_i(\mathbf{x})dx^i, \qquad \mathbf{v} = a^j(\mathbf{x})\frac{\partial}{\partial x^j}, \qquad \mathbf{w} = b^k(\mathbf{x})\frac{\partial}{\partial x^k},$$

and compute $d\phi(\mathbf{v}, \mathbf{w})$ both in local coordinates as in (2.6) and according to Cartan's formula (2.8). This gives another proof (at least for 1-forms) that the exterior derivative in Definition 2.35 is, in fact, well-defined.

2.8. Pullbacks

Now, suppose that M, N are smooth manifolds and that $F : M \to N$ is a differentiable map. Recall from Definition 1.25 that for any $\mathbf{q} \in M$, the *differential* of F at $\mathbf{q}$ is the linear map

$$dF_{\mathbf{q}} : T_{\mathbf{q}}M \to T_{F(\mathbf{q})}N$$

defined by the property that for any curve $\alpha : \mathbb{R} \to M$ with $\alpha(0) = \mathbf{q}$, $\alpha'(0) = \mathbf{v}$,

$$dF_{\mathbf{q}}(\mathbf{v}) = (F \circ \alpha)'(0).$$

The differential dF may be thought of as a *vector-valued* (more specifically, TN-valued) 1-form on M because for each $\mathbf{q} \in M$, it defines a linear map on $T_{\mathbf{q}}M$ that takes values in the vector space $T_{F(\mathbf{q})}N$.

Remark 2.39. This is why the same notation is used both for the differential dF of a map $F : M \to N$ and for the exterior derivative df of a function $f : M \to \mathbb{R}$: The exterior derivative of a function really is just the differential of the function regarded as a map between manifolds. (This equivalence relies on the canonical isomorphism $T_{f(\mathbf{q})}\mathbb{R} \cong \mathbb{R}$.)

***Exercise 2.40.** Let M be a manifold of dimension m, N a manifold of dimension n, and $F : M \to N$ a differentiable map with differential $dF : TM \to TN$. Let $(\mathbf{f}_1, \ldots, \mathbf{f}_n)$ be vector fields on N that form a basis for $T_{\mathbf{q}}N$ at each point $\mathbf{q} \in N$. Show that there exist ordinary, scalar-valued 1-forms $(\phi^1, \ldots, \phi^n)$ on M defined by the condition that

$$dF = \mathbf{f}_i \phi^i.$$

(Recall that this means that for any $\mathbf{v} \in T_{\mathbf{q}}M$,

$$dF_{\mathbf{q}}(\mathbf{v}) = \mathbf{f}_i \phi^i(\mathbf{v}) \in T_{F(\mathbf{q})}N.)$$

If the basis $(\mathbf{f}_1, \ldots, \mathbf{f}_n)$ is understood, then we may identify dF with the $\mathbb{R}^n$-valued 1-form

$$\phi = \begin{bmatrix} \phi^1 \\ \vdots \\ \phi^n \end{bmatrix}$$

on M.

Because the differential takes tangent vectors on M to tangent vectors on N, it is sometimes called the "push-forward" map from TM to TN, and it is often denoted by F_* rather than dF. (So, for $\mathbf{v} \in T_{\mathbf{q}}M$, the vector $dF(\mathbf{v}) \in T_{F(\mathbf{q})}N$ is often written as $F_*(\mathbf{v})$.) There is an analogous map in the opposite direction for differential forms, called the *pullback* and denoted F^*. It is defined as follows.

Definition 2.41. Let $F : M \to N$ be a differentiable map.

(1) If $f : N \to \mathbb{R}$ is a differentiable function, then $F^*f : M \to \mathbb{R}$ is the composite function defined by

$$(F^*f)(\mathbf{q}) = (f \circ F)(\mathbf{q})$$

for $\mathbf{q} \in M$. This may be represented by the following commutative diagram:

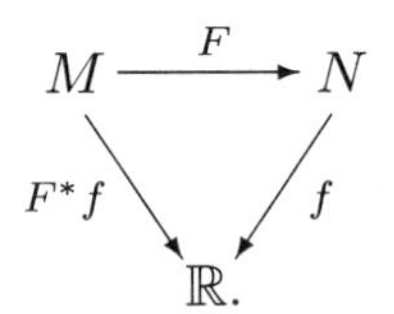

(2) If ϕ is a p-form on N, then $F^*\phi$ is the p-form on M defined by its action on tangent vectors to M as follows: If $\mathbf{q} \in M$ and $\mathbf{v}_1, \ldots, \mathbf{v}_p \in T_{\mathbf{q}}M$, then

$$(F^*\phi)(\mathbf{v}_1, \ldots, \mathbf{v}_p) = \phi(F_*(\mathbf{v}_1), \ldots, F_*(\mathbf{v}_p)).$$

(Note that $F_*(\mathbf{v}_1), \ldots, F_*(\mathbf{v}_p)$ are elements of $T_{F(\mathbf{q})}N$, so the right-hand side makes sense.) This may be represented by the following commutative diagram:

$$\begin{array}{ccc} \Lambda^p(T_{\mathbf{q}}M) & \xrightarrow{F_*} & \Lambda^p(T_{F(\mathbf{q})}N) \\ & {\scriptstyle F^*\phi} \searrow \quad \swarrow {\scriptstyle \phi} & \\ & \mathbb{R}. & \end{array}$$

In this diagram, the map $F_* : \Lambda^p(T_{\mathbf{q}}M) \to \Lambda^p(T_{F(\mathbf{q})}N)$ represents the obvious extension of the map $F_* : T_{\mathbf{q}}M \to T_{F(\mathbf{q})}N$; namely,

$$F_*(\mathbf{v}_1 \wedge \cdots \wedge \mathbf{v}_p) = F_*(\mathbf{v}_1) \wedge \cdots \wedge F_*(\mathbf{v}_p).$$

In the special case where $F : M \to N$ is a diffeomorphism, the pullback F^* defines a differentiable map from T^*N to T^*M; in particular, for $\tilde{\mathbf{q}} \in N$, F^* takes the cotangent space $T^*_{\tilde{\mathbf{q}}}N$ to $T^*_{F^{-1}(\tilde{\mathbf{q}})}M$. Note that this map goes in the opposite direction from the differential F_*, which maps TM to TN; the following commutative diagrams show how these maps operate:

$$\begin{array}{ccc} \mathbf{v} \in TM & \xrightarrow{F_*} & TN \ni F_*\mathbf{v} \\ \big\downarrow & & \big\downarrow \\ M & \xrightarrow{F} & N \end{array}$$

and

$$\begin{array}{ccc} F^*\phi \in T^*M & \xleftarrow{F^*} & T^*N \ni \phi \\ \big\downarrow & & \big\downarrow \\ M & \xrightarrow{F} & N. \end{array}$$

Just as the differential $F_* : TM \to TN$ can be extended to a differentiable map $F_* : \Lambda^p(TM) \to \Lambda^p(TN)$, so the pullback $F^* : T^*N \to T^*M$ can be extended to a differentiable map $F^* : \Lambda^p(T^*N) \to \Lambda^p(T^*M)$ as follows:

$$F^*(\phi_1 \wedge \cdots \wedge \phi_p) = F^*(\phi_1) \wedge \cdots \wedge F^*(\phi_p).$$

For more general differentiable maps $F : M \to N$, the picture is slightly more complicated: According to Definition 2.41, the pullback F^* maps p-forms on N (i.e., *sections* of the tensor bundle $\Lambda^p(T^*N)$) to p-forms on M (i.e., sections of the tensor bundle $\Lambda^p(T^*M)$). But in general, there may be no well-defined map $F^* : \Lambda^p(T^*N) \to \Lambda^p(T^*M)$, because for any given point $\tilde{\mathbf{q}} \in N$, there may not exist a unique point $\mathbf{q} \in M$ (or indeed, any point at all) with $F(\mathbf{q}) = \tilde{\mathbf{q}}$.

In terms of local coordinates $\mathbf{x} = (x^1, \ldots, x^m)$ on M and $\mathbf{y} = (y^1, \ldots, y^n)$ on N, suppose that the map F is described by

$$F(x^1, \ldots, x^m) = \left(y^1(x^1, \ldots, x^m), \ldots, y^n(x^1, \ldots, x^m)\right).$$

Recall from Chapter 1 that the differential F_* at each point $\mathbf{q} \in M$ may be represented in terms of the bases $\left(\frac{\partial}{\partial x^1}, \ldots, \frac{\partial}{\partial x^m}\right)$ and $\left(\frac{\partial}{\partial y^1}, \ldots, \frac{\partial}{\partial y^n}\right)$ for $T_{\mathbf{q}}M$ and $T_{F(\mathbf{q})}N$, respectively, by the matrix

$$\left[\frac{\partial y^i}{\partial x^j}\right];$$

specifically,

$$F_*\left(\frac{\partial}{\partial x^j}\right) = \frac{\partial y^i}{\partial x^j}\frac{\partial}{\partial y^i}.$$

What about F^*? The pullback map takes forms on N to forms on M, and it acts on the basis 1-forms dy^i by

$$F^*(dy^i) = \frac{\partial y^i}{\partial x^j}dx^j. \tag{2.9}$$

***Exercise 2.42.** Use Definition 2.41 to prove the formula (2.9).

Exercise 2.43. Let (r, θ) be coordinates on $\mathbb{R}^2$ and (x, y, z) coordinates on $\mathbb{R}^3$. Let $F : \mathbb{R}^2 \to \mathbb{R}^3$ be defined by

$$F(r, \theta) = (\cos(\theta), \sin(\theta), r).$$

Describe the differential F_* in terms of these coordinates, and compute the pullbacks $F^*(dx), F^*(dy), F^*(dz)$.

The pullback map behaves as nicely as one could hope with respect to the various operations on differential forms, as described in the following theorem.

Theorem 2.44. *Let $F : M \to N$ be a differentiable map, and let Φ, Ψ be differential forms on N. Then:*

(1) $F^*(\Phi + \Psi) = F^*\Phi + F^*\Psi$.

(2) $F^*(\Phi \wedge \Psi) = (F^*\Phi) \wedge (F^*\Psi)$.

(3) $F^*(d\Phi) = d(F^*\Phi)$.

***Exercise 2.45.** Prove Theorem 2.44 in the case where Φ, Ψ are 1-forms.

***Exercise 2.46.** Suppose that $m = n = 2$ and that $F : M \to N$ is described in local coordinates by

$$F(x^1, x^2) = \left(y^1(x^1, x^2), y^2(x^1, x^2)\right).$$

Show that

$$F^*\left(dy^1 \wedge dy^2\right) = \begin{vmatrix} \frac{\partial y^1}{\partial x^1} & \frac{\partial y^1}{\partial x^2} \\ \frac{\partial y^2}{\partial x^1} & \frac{\partial y^2}{\partial x^2} \end{vmatrix} dx^1 \wedge dx^2.$$

(You may recognize this expression from the change of variables formula for double integrals.)

Exercise 2.46 is a special case of the following generalization of equation (2.9), which follows from Theorem 2.44:

$$\text{(2.10)} \quad F^*\left(\sum_{|I|=p} f_I(\mathbf{y})\, dy^{i_1} \wedge \cdots \wedge dy^{i_p}\right)$$
$$= \left(\sum_{|I|=p} \left(\sum_{j_1,\ldots,j_p=1}^{m} f_I(\mathbf{y}(\mathbf{x})) \frac{\partial y^{i_1}}{\partial x^{j_1}} \cdots \frac{\partial y^{i_p}}{\partial x^{j_p}}\right)\right) dx^{j_1} \wedge \cdots dx^{j_p}.$$

2.9. Integration and Stokes's theorem

While we won't need it for the topics in this book, no discussion of differential forms is complete without a mention of Stokes's theorem, which has to do with the integration of differential forms over manifolds. Generally speaking, a p-form Φ on a manifold M can be integrated over a p-dimensional submanifold P of M, and pullbacks allow us to define such integrals in terms of ordinary integrals over regions in $\mathbb{R}^p$. This construction requires the notion of a "manifold with boundary", which means more or less exactly what it sounds like; common examples might include a curve segment in $\mathbb{R}^n$ (including its endpoints) or a hemisphere in $\mathbb{R}^3$ (including the boundary circle). It also requires that the submanifold P be *orientable*, which means that P can be covered by a collection of parametrizations $\mathbf{x}_i : U \subset \mathbb{R}^p \to P$ with the property that all the transition functions $\mathbf{x}_j^{-1} \circ \mathbf{x}_i$ have Jacobian matrices with positive determinant. The choice of such a collection of parametrizations is called an *orientation* on P, and P must be given an orientation before integrals over P can be defined.

In order to define the integral $\int_P \Phi$, let $D \subset \mathbb{R}^p$ be a closed and bounded region in $\mathbb{R}^p$, and let $\mathbf{x} : D \to P$ be a parametrization of P that is compatible with the orientation of P. (Differentiability at boundary points of D is defined via one-sided limits, and "compatible with the orientation" means that the Jacobian of the transition function between $\mathbf{x}$ and any of the parametrizations $\mathbf{x}_i$ in the orientation has positive determinant.) For any p-form Φ on M, the integral of Φ over the region $\mathbf{x}(D) \subset P$ is defined to be

$$\int_{\mathbf{x}(D)} \Phi = \int_D \mathbf{x}^*(\Phi).$$

(Integrals over regions in $\mathbb{R}^p$ are defined exactly as you would expect.) In order to compute the integral $\int_P \Phi$, simply divide P into regions that can each be covered by a single parametrization and add the integrals of Φ over each region. The formula (2.10) can be used to show that $\int_P \Phi$ is well-defined, independent of the choice of parametrizations used to compute it.

If P is a manifold with boundary ∂P, then an orientation on P induces an orientation on the boundary ∂P. The precise definition for the induced orientation is a bit complicated (see, e.g., [**Lee13**]), but intuitively it involves constructing parametrizations for ∂P by choosing oriented parametrizations $\mathbf{x}_i$ for neighborhoods $V \subset P$ containing points of ∂P so that $V \cap \partial P$ is defined by the equation $x^p = 0$, with the remaining coordinates chosen so that the tangent vector $\frac{\partial}{\partial x^p}$ points "outward" from P.

Theorem 2.47 (Stokes). *Let M be a closed, oriented, $(p+1)$-dimensional manifold, possibly with boundary, and let ∂M denote the p-dimensional boundary (possibly empty) of M, with orientation induced from the orientation of M. Let Φ be a p-form on M. Then*

$$\int_{\partial M} \Phi = \int_M d\Phi. \tag{2.11}$$

***Exercise 2.48.** Show that the following theorems are all special cases of Stokes's theorem:

(a) The Fundamental Theorem of Calculus: If $f : [a,b] \to \mathbb{R}$ is a differentiable function, then

$$\int_a^b f'(x)\, dx = f(b) - f(a).$$

(Hint: Let $M \subset \mathbb{R}$ be the closed interval $[a,b]$. The oriented boundary of M is the formal sum $(b)-(a)$, and the "integral" of a function $f:[a,b]\to\mathbb{R}$ over ∂M is $f(b)-f(a)$.)

(b) Green's theorem: Let $D \subset \mathbb{R}^2$ be a region bounded by a simple closed curve C (oriented counterclockwise), and let $P, Q : D \to \mathbb{R}$ be differentiable functions. Then

$$\oint_C (P\,dx + Q\,dy) = \iint_D (Q_x - P_y)\,dx \wedge dy.$$

(c) Stokes's theorem (multivariable calculus version): Let S be an oriented surface in $\mathbb{R}^3$ bounded by a simple closed curve C with the induced orientation, and let $P, Q, R : U \to \mathbb{R}$ be differentiable functions on an open set $U \subset \mathbb{R}^3$ containing S. Then

$$\oint_C (P\,dx + Q\,dy + R\,dz) \\ = \iint_S ((R_y - Q_z)\,dy \wedge dz + (P_z - R_x)\,dz \wedge dx + (Q_x - P_y)\,dx \wedge dy).$$

(Bonus points if you can express this in the traditional vector form as

$$\oint_C \mathbf{F} \cdot d\mathbf{r} = \iint_S (\nabla \times \mathbf{F}) \cdot \mathbf{n}\,dA\,!)$$

(d) The Divergence theorem: Let $V \subset \mathbb{R}^3$ be a closed and bounded region with piecewise smooth boundary S (oriented with outward-pointing normal vector), and let $P, Q, R : V \to \mathbb{R}$ be differentiable functions. Then

$$\iint_S (P\,dy \wedge dz + Q\,dz \wedge dx + R\,dx \wedge dy) = \iiint_V (P_x + Q_y + R_z)\,dx \wedge dy \wedge dz.$$

(Bonus points if you can express this in the traditional vector form as

$$\iint_S (\mathbf{F} \cdot \mathbf{n})\,dA = \iiint_V (\nabla \cdot \mathbf{F})\,dV\,!)$$

2.10. Cartan's lemma

In this section, we present a technical result known as *Cartan's lemma.* We will make frequent use of this result beginning in Chapter 4.

Lemma 2.49 (Cartan). *Suppose that $\eta^1, \ldots, \eta^n$ are linearly independent, smooth 1-forms on a manifold M and that $\phi_1, \ldots, \phi_n$ are smooth 1-forms on M such that*

$$\phi_i \wedge \eta^i = 0. \tag{2.12}$$

Then there exist smooth functions $\{h_{ij} = h_{ji} : M \to \mathbb{R} \mid 1 \leq i, j \leq n\}$, *symmetric in their lower indices, such that*

$$\phi_i = h_{ij}\eta^j.$$

PROOF. Suppose that M has dimension $s \geq n$, and let $\eta^{n+1}, \ldots, \eta^s$ be any smooth 1-forms on M with the property that $(\eta^1, \ldots, \eta^n, \eta^{n+1}, \ldots, \eta^s)$ form a basis for the cotangent space $T^*_{\mathbf{q}}M$ at each point $\mathbf{q} \in M$. Then we can write

$$\phi_i = h_{i\alpha}\eta^\alpha, \qquad 1 \leq i \leq n, \tag{2.13}$$

for some smooth functions $(h_{i\alpha})$ on M, where the sum is over $1 \leq \alpha \leq s$.

***Exercise 2.50.** Show that substituting these expressions into (2.12) and taking the skew-symmetry of the wedge product into account yields

$$h_{ir} = 0, \qquad n+1 \leq r \leq s,$$
$$h_{ji} = h_{ij}, \qquad 1 \leq i, j \leq n.$$

Therefore, $\phi_i = h_{ij}\eta^j$, where the sum is over $1 \leq j \leq n$.

□

Remark 2.51. Technically, this proof only works if the tangent bundle TM has a global trivialization, because otherwise there might not exist complementary 1-forms $\eta^{n+1}, \ldots, \eta^s$ that are globally defined on M. But such 1-forms always exist on some neighborhood of any given point in M, so this argument can be used to prove the lemma in general by showing that equation (2.13) holds in a neighborhood of any point in M and hence on all of M.

2.11. The Lie derivative

The final operation that we will define on differential forms is the Lie derivative. This is a generalization of the notion of the directional derivative of a function.

Suppose that $\mathbf{v}$ is a vector field on a manifold M, and let φ be the *flow* of $\mathbf{v}$. This is, by definition, the unique map $\varphi : V \to M$, defined on some open neighborhood V of the set $M \times \{0\}$ in $M \times \mathbb{R}$, that satisfies the conditions

$$\frac{\partial \varphi}{\partial t}(\mathbf{q}, t) = \mathbf{v}(\varphi(\mathbf{q}, t)), \tag{2.14}$$
$$\varphi(\mathbf{q}, 0) = \mathbf{q}.$$

(The existence and uniqueness of the flow on some neighborhood V of $M \times \{0\}$ is equivalent to the existence and uniqueness of local solutions to first-order systems of ordinary differential equations; see, e.g., [**Lee13**].) In other words, $\varphi(\mathbf{q}, t)$ is the point reached at time t by flowing along the vector field $\mathbf{v}$ starting from the point $\mathbf{q}$ at time 0. The expression $\varphi(\mathbf{q}, t)$ is sometimes denoted $\varphi_t(\mathbf{q})$ to emphasize that, for each fixed t, the flow defines a differentiable map $\varphi_t : U \to M$, where $U = \{\mathbf{q} \in M \mid (\mathbf{q}, t) \in V\}$.

Recall that if $f : M \to \mathbb{R}$ is a smooth function, then the directional derivative of f at $\mathbf{q}$ in the direction of $\mathbf{v}$ may be defined as

$$\begin{aligned}\mathbf{v}[f] &= \lim_{t\to 0} \frac{f(\varphi_t(\mathbf{q})) - f(\mathbf{q})}{t} \\ &= \lim_{t\to 0} \frac{(\varphi_t^*(f) - f)(\mathbf{q})}{t}.\end{aligned}$$

This motivates the following definition:

Definition 2.52. Given a p-form Φ and a vector field $\mathbf{v}$ on M, the *Lie derivative* of Φ along the vector field $\mathbf{v}$ is the p-form

$$\mathcal{L}_{\mathbf{v}}\Phi = \lim_{t\to 0} \frac{\varphi_t^*\Phi - \Phi}{t}, \tag{2.15}$$

where $\varphi_t^*\Phi$ denotes the pullback of the p-form Φ via the map φ_t.

It is rarely possible to compute the limit in (2.15) directly; among other things, finding an explicit formula for the map $\varphi_t : U \to M$ requires solving the system of differential equations (2.14). But fortunately, there is a practical way to compute the Lie derivative. First, we need the notion of the *interior product* of a differential form with a vector field.

Definition 2.53. Given a p-form Φ and a vector field $\mathbf{v}$ on M, the *interior product* of Φ with $\mathbf{v}$ (also called the *left-hook* of $\mathbf{v}$ with Φ) is the $(p-1)$-form $\mathbf{v} \lrcorner \Phi$ on M defined by the property that for any vector fields $\mathbf{w}_1, \ldots, \mathbf{w}_{p-1}$ on M,

$$(\mathbf{v} \lrcorner \Phi)(\mathbf{w}_1, \ldots, \mathbf{w}_{p-1}) = \Phi(\mathbf{v}, \mathbf{w}_1, \ldots, \mathbf{w}_{p-1}).$$

Example 2.54. Let Φ be the 2-form $\Phi = dx \wedge dy + dz \wedge dx$ on $\mathbb{R}^3$, and let $\mathbf{v} = \frac{\partial}{\partial x}$. Then $\mathbf{v} \lrcorner \Phi$ is a 1-form, and we can compute it by computing its action on an arbitrary vector field $\mathbf{w}$ on $\mathbb{R}^3$:

$$\begin{aligned}\left(\frac{\partial}{\partial x} \lrcorner \Phi\right)(\mathbf{w}) &= \Phi\left(\frac{\partial}{\partial x}, \mathbf{w}\right) \\ &= (dx \wedge dy + dz \wedge dx)\left(\frac{\partial}{\partial x}, \mathbf{w}\right) \\ &= dx\left(\frac{\partial}{\partial x}\right) dy(\mathbf{w}) - dx(\mathbf{w})\, dy\left(\frac{\partial}{\partial x}\right) \\ &\quad + dz\left(\frac{\partial}{\partial x}\right) dx(\mathbf{w}) - dz(\mathbf{w})\, dx\left(\frac{\partial}{\partial x}\right) \\ &= 1 \cdot dy(\mathbf{w}) - 0 \cdot dx(\mathbf{w}) + 0 \cdot dx(\mathbf{w}) - 1 \cdot dz(\mathbf{w}) \\ &= (dy - dz)(\mathbf{w}).\end{aligned}$$

Therefore,

$$\frac{\partial}{\partial x} \lrcorner (dx \wedge dy + dz \wedge dx) = dy - dz.$$

With the interior product in hand, the following theorem tells us how to compute the Lie derivative:

Theorem 2.55. *Given a p-form Φ and a vector field $\mathbf{v}$ on a manifold M,*

$$\mathcal{L}_{\mathbf{v}}\phi = \mathbf{v} \lrcorner d\phi + d(\mathbf{v} \lrcorner \phi). \tag{2.16}$$

The formula (2.16) is called *Cartan's formula* for the Lie derivative.

Exercise 2.56. Use the following steps to prove Cartan's formula for the Lie derivative:

(a) Prove that the Lie derivative (Definition 2.52) is a *derivation* of order zero; i.e., it takes p-forms to p-forms and satisfies the Leibniz rule

$$\mathcal{L}_{\mathbf{v}}(\Phi \wedge \Psi) = (\mathcal{L}_{\mathbf{v}}\Phi) \wedge \Psi + \Phi \wedge (\mathcal{L}_{\mathbf{v}}\Psi).$$

(b) Prove that for any vector field $\mathbf{v}$, the operator

$$(\mathbf{v} \lrcorner) \circ d + d \circ (\mathbf{v} \lrcorner)$$

is also a derivation of order zero; i.e., it takes p-forms to p-forms, and for any forms Φ, Ψ,

$$\begin{aligned}&[(\mathbf{v} \lrcorner) \circ d + d \circ (\mathbf{v} \lrcorner)](\Phi \wedge \Psi) \\ &\quad = [(\mathbf{v} \lrcorner) \circ d + d \circ (\mathbf{v} \lrcorner)](\Phi) \wedge \Psi + \Phi \wedge [(\mathbf{v} \lrcorner) \circ d + d \circ (\mathbf{v} \lrcorner)](\Psi).\end{aligned}$$

(c) Show that the operators $\mathcal{L}_{\mathbf{v}}$ and $[(\mathbf{v} \lrcorner) \circ d + d \circ (\mathbf{v} \lrcorner)]$ agree on 0-forms and 1-forms.

(d) It is shown in [**Wil61**] that any two derivations of the same degree that agree on 0-forms and 1-forms must be equal. Use this fact to conclude that Cartan's formula holds.

2.12. Introduction to the `Cartan` package for MAPLE

The `Cartan` package is a descendant (and an enhancement) of the `Forms` package, originally written by Yungliang Yu of Duke University for performing computations involving differential forms in MAPLE. The purpose of this section is to introduce some of the commands in this package and to illustrate how they may be used in practice. First, you need to download and install the `Cartan` package; see the Preface for details on how to do this, and be sure that you install the appropriate version for your preferred MAPLE interface ("Worksheet" or "Document" mode). This section assumes that you are using the Worksheet mode, with the input mode set to "Maple Input", so that your input appears verbatim in red print; if you prefer the Document mode, you must replace the `&^` command for wedge product with the alternative command `&w`.

Once you have `Cartan` installed, start up MAPLE and type

```
> with(Cartan);
```

This command loads the package; if you have it properly installed, MAPLE should return something like the following:

*The symbol for wedge product is &^*

[CartanKahler, Forder, Form, L, LieDeriv, ScalarForm, Simf, Wdegree, WedgeProduct, d, hook, makebacksub, mixpar, pick, prolong]

(From now on, we will always assume that this is the first thing that you do when you start up MAPLE.)

The `Cartan` commands that will be most useful for the purposes of this book are the following:

The `Form` command. Before you can do any computations with differential forms, you have to declare them. For instance, the command

```
> Form(omega=1);
```

declares the object `omega` to be a 1-form. You can declare multiple forms with a single command; e.g., the command

```
> Form(omega=1, theta=1, Omega=2);
```

declares `omega` and `theta` to be 1-forms and `Omega` to be a 2-form. You can declare functions (i.e., 0-forms); e.g.,

```
> Form(f=0);
```

but it isn't necessary; anything which hasn't been declared with a `Form` command is automatically assumed to be a 0-form. You can also declare constants by assigning them to have degree -1 (this is purely a MAPLE convention and has nothing to do with their degree as differential forms!); e.g., the command

```
> Form(a=-1, b=-1);
```

declares `a` and `b` to be constants. Finally, if you want to declare an object to be a 1-form (certainly the most common use of this command), the "`=1`" can be omitted; e.g., the command

```
> Form(omega, theta, eta);
```

declares the objects `omega`, `theta`, and `eta` to be 1-forms.

The &^ command. This is the command for wedge product. For instance, the commands

```
> omega:= x*d(y);
> theta:= y*d(z);
> omega &^ theta;
```

can be used to compute the wedge product of the 1-forms $x\,dy$ and $y\,dz$.

MAPLE will choose a default ordering of forms for wedge products; e.g., the command

```
> d(y) &^ d(x);
```

returns

$$-(d(x)\ \&\hat{}\ d(y))$$

If you don't like the order that MAPLE chooses, you can change it with the `Forder` command; see MAPLE's help page for the `Cartan` package for more details. Furthermore, if you don't like this symbol for the wedge product, you can change it with the `WedgeProduct` command.

The `d` command. This is the all-purpose exterior derivative command. (WARNING: When you're using the `Cartan` package, don't use the letter `d` as a variable or MAPLE will get hopelessly confused!) It can be used both for computation and for assignment, and it knows how to use the chain rule for functions. For example, the command

```
> d(f(x,y,z));
```

returns

$$\left(\frac{\partial}{\partial x}f(x,y,z)\right)d(x)+\left(\frac{\partial}{\partial y}f(x,y,z)\right)d(y)+\left(\frac{\partial}{\partial z}f(x,y,z)\right)d(z)$$

If you like, you can clean this up a little bit by using the `PDETools[declare]` command to tell MAPLE that f is a function of the variables (x, y, z):

```
> PDETools[declare](f(x,y,z));
> d(f(x,y,z));
```

returns

$$f_x\, d(x) + f_y\, d(y) + f_z\, d(z)$$

(Unfortunately, you still have to type out `f(x,y,z)` wherever it appears in the input.) This works on forms of any degree; e.g.,

```
> d(x^2*d(y) + y*d(z));
```

returns

$$2x(d(x)\ \&\hat{}\ d(y)) + (d(y)\ \&\hat{}\ d(z))$$

Note that for coordinate 1-forms, you must type, e.g., `d(x)` rather than `dx`; MAPLE will regard `dx` as a completely different variable having nothing to do with the variable `x` or its exterior derivative.

An important feature of the `Cartan` package is that exterior derivatives can be *assigned* as well as computed, and no explicit local coordinates are required. (The value of this feature for the method of moving frames will become apparent in later chapters!) For instance, the commands

```
> Form(omega, theta, eta);
> d(omega):= theta &^ eta;
```

declare that `omega, theta`, and `eta` are 1-forms and that the exterior derivative of `omega` is the 2-form `(theta &^ eta)`.

The `Simf` command. This is the all-purpose simplification command for differential forms, and you should use it liberally. (MAPLE fails to make some fairly obvious simplifications without it.) For instance, the command

```
> d(x^2*d(y) + y^2*d(x));
```

returns

$$-2y\ (d(x)\ \&\hat{}\ d(y)) + 2x\ (d(x)\ \&\hat{}\ d(y))$$

and then the command

```
> Simf(%);
```

returns

$$(-2y + 2x)(d(x) \ \&\hat{}\ d(y))$$

(The % operator refers to the output of the previous command.) Of course, these two commands can be combined into the single command

```
> Simf(d(x^2*d(y) + y^2*d(x)));
```

The pick command. The command

```
> pick(bigform, omega);
```

where `omega` is a 1-form, finds all the terms in the expression `bigform` that involve wedge products with `omega`. It then writes `bigform` as

```
bigform = form1 &^ omega + form2
```

where `form2` contains no terms involving `omega` and returns `form1`. For instance, the command

```
> pick(x*d(y) &^ d(z) + y*d(z) &^ d(x) + z*d(x) &^ d(y),
  d(x));
```

returns

$$-z\ dy + y\ dz$$

This command can also be invoked with additional 1-forms as arguments in order to find all the terms in `bigform` that involve specific wedge products of higher degree. So, for example, the command

```
> pick(bigform, omega1, omega2);
```

where `omega1, omega2` are 1-forms, finds all the terms in the expression `bigform` that involve wedge products with `omega1 &^ omega2`. (Note that this command is equivalent to

```
> pick(pick(bigform, omega2), omega1);
```

and pay attention to the order of the 1-forms in both versions!) For instance, the command

```
> pick(x*d(y) &^ d(z) + y*d(z) &^ d(x) + z*d(x) &^ d(y), d(x),
  d(y));
```

simply returns the coefficient z.

The `ScalarForm` command. This command takes the scalar coefficient of each summand in a differential form and returns them all as a list. For instance, the command

```
> ScalarForm(x*d(y) &^ d(z) + y*d(z) &^ d(x)
  + z*d(x) &^ d(y));
```

returns

$$[z, y, -x]$$

(The signs and the order of the elements in this list may vary, depending on what order MAPLE has decided to assign to all the 1-forms involved.) This command is particularly useful when you have a differential form that you want to set equal to zero (often something that you computed as $d \circ d$ of something else), and you therefore want to set all its coefficients equal to zero.

The `ScalarForm` command has an optional second argument, which must be a string. If this argument is present, then the command assigns to it a list of the decomposable forms in each summand, in order corresponding to the order of the coefficients of these forms produced by the main command. For instance, the command

```
> ScalarForm(x*d(y) &^ d(z) + y*d(z) &^ d(x) + z*d(x) &^ d(y),
  'terms');
```

returns

$$[z, y, -x]$$

as before, but now if you type

```
> terms;
```

it returns

$$[(d(x)\ \&\hat{}\ d(y)),\ (d(z)\ \&\hat{}\ d(x)),\ (d(z)\ \&\hat{}\ d(y))]$$

Again, the order may vary, but the order of the coefficients in the first line will match the order of the forms in the second line.

The `makebacksub` command. It often happens that we have two different bases for the 1-forms on a manifold, and it's handy to be able to go back and forth between them. We usually do this using the `subs` command to make substitutions. (And you should pretty much always follow up a `subs` command with a `Simf` command.) For instance, suppose that we are working on $\mathbb{R}^5$ with coordinates (x, y, z, p, q) and we want to use the basis

$$\theta = dz - p\,dx - q\,dy, \qquad \omega^1 = dx, \qquad \omega^2 = dy,$$
$$\pi_1 = dp - e^z dy, \qquad \pi_2 = dq - e^z dx$$

for the 1-forms on $\mathbb{R}^5$. (Don't worry about why you might want to do such a thing; it has to do with an exterior differential system representing the partial differential equation

$$z_{xy} = e^z,$$

but that's well beyond the scope of this book!) You could simply make assignments such as

```
> theta:= d(z) - p*d(x) - q*d(y);
```

etc., but then everything that follows would be expressed in terms of the coordinate basis. If you want to do something like computing $d\theta$ and expressing it in terms of the basis $(\theta, \omega^1, \omega^2, \pi_1, \pi_2)$, it's more effective to set up substitutions to go back and forth between the two bases. You can do this as follows. First, you need to declare the 1-forms in your new basis:

```
> Form(theta, omega1, omega2, pi1, pi2);
```

(You don't have to declare the coordinate basis; the functions x, y, z, p, q are automatically assumed to be 0-forms, so their exterior derivatives are 1-forms.) Then define the substitution:

```
> sub1:= [theta = d(z) - p*d(x) - q*d(y), omega1 = d(x),
  omega2 = d(y), pi1 = d(p) - exp(z)*d(y),
  pi2 = d(q) - exp(z)*d(x)];
```

You can now use the substitution `sub1` to go from the basis $(\theta, \omega^1, \omega^2, \pi_1, \pi_2)$ to the coordinate basis; e.g., the command

```
> Simf(subs(sub1, theta));
```

yields

$$d(z) - p\; d(x) - q\; d(y)$$

Where the `makebacksub` command comes in is when you want to go the other way. The command

```
> backsub1:= makebacksub(sub1);
```

produces a substitution `backsub1` that is the inverse of the substitution `sub1`; thus, it will go from the coordinate basis to the basis $(\theta, \omega^1, \omega^2, \pi_1, \pi_2)$. So, e.g., the command

```
> Simf(subs(backsub1, d(z)));
```

yields

$$\theta + p\;\omega 1 + q\;\omega 2$$

One warning about the `makebacksub` command: It only works properly when the substitution `sub1` is a *complete* list of the elements of one basis of

1-forms expressed in terms of another basis. So, for instance, if you define

```
> sub2:= [theta = d(z) - p*d(x) - q*d(y),
  pi1 = d(p) - exp(z)*d(y), pi2 = d(q) - exp(z)*d(x)];
```

without including omega1 and omega2 in the list, the command

```
> makebacksub(sub2);
```

returns a less than helpful backwards substitution, and it might not even do it consistently if you execute it more than once. (Try it and see what happens!)

You can use these two substitutions to express $d\theta$ in terms of the basis $(\theta, \omega^1, \omega^2, \pi_1, \pi_2)$ via the following sequence of commands:

```
> Simf(subs(sub1, theta));
```

$$d(z) - p\ d(x) - q\ d(y)$$

```
> Simf(d(%));
```

$$(d(x))\ \&\hat{}\ (d(p)) + (d(y))\ \&\hat{}\ (d(q))$$

```
> Simf(subs(backsub1, %));
```

$$\omega 1\ \&\hat{}\ \pi 1 + \omega 2\ \&\hat{}\ \pi 2$$

And, of course, these commands can be combined into a single command:

```
> Simf(subs(backsub1, Simf(d(Simf(subs(sub1, theta))))));
```

(It's probably not really necessary to put *all* of those Simf commands in there, but it doesn't hurt, and sometimes it prevents problems.)

Having done all this, you can compute—and then assign—the exterior derivatives of the new basis in terms of this basis, via the commands

```
> d(theta):= Simf(subs(backsub1, Simf(d(Simf(subs(sub1,
  theta))))));
> d(omega1):= Simf(subs(backsub1, Simf(d(Simf(subs(sub1,
  omega1))))));
> d(omega2):= Simf(subs(backsub1, Simf(d(Simf(subs(sub1,
  omega2))))));
> d(pi1):= Simf(subs(backsub1, Simf(d(Simf(subs(sub1,
  pi1))))));
> d(pi2):= Simf(subs(backsub1, Simf(d(Simf(subs(sub1,
  pi2))))));
```

Exercise 2.57. Use commands such as

```
> Simf(d(d(theta)));
```

to check that all the 1-forms in this basis, with their assigned exterior derivatives, satisfy the identity $d \circ d = 0$. What goes wrong with $d(d\pi_1)$ and $d(d\pi_2)$? Why does this happen, and what can you do to fix it?

Part 2

Curves and surfaces in homogeneous spaces via the method of moving frames

Chapter 3

Homogeneous spaces

3.1. Introduction

In the late nineteenth century, there were several different types of geometry under investigation: There was the classical Euclidean geometry with its standard notions of lengths and angles, various non-Euclidean geometries in which the parallel postulate was replaced by alternative versions and lengths were measured differently than in Euclidean geometry, and even affine and projective geometries, where lengths and angles weren't well-defined notions.

In 1872, Felix Klein published a revolutionary treatise on geometry ([**Kle93b**]; an English translation is available in [**Kle93a**]), in which he proposed that the most useful way to study a geometric structure is to study its group of *symmetries*, i.e., the group of transformations that preserve the key features of the structure. This approach revolutionized the study of geometry, and it continues to influence the development of the subject today. For instance, when studying curves in $\mathbb{R}^3$ (with the standard Euclidean metric on $\mathbb{R}^3$), you probably learned the following theorem:

Theorem 3.1 (Fundamental Theorem of Space Curves). *Given any smooth functions $\kappa(s), \tau(s)$ on an interval $I \subset \mathbb{R}$ with $\kappa(s) > 0$, there exists a smooth, unit-speed curve $\alpha : I \to \mathbb{R}^3$ with curvature $\kappa(s)$ and torsion $\tau(s)$. Moreover, α is* unique up to rigid motion*: Any other such curve β differs from α by a translation and rotation in $\mathbb{R}^3$.*

The rigid motions—translations and rotations—are the *symmetries* of the Euclidean space $\mathbb{R}^3$: They are exactly the transformations of $\mathbb{R}^3$ that preserve the Euclidean metric. Curvature and torsion are the *invariants* of smooth curves in $\mathbb{R}^3$: They are exactly the properties of smooth curves that remain unchanged when a curve is transformed by a rigid motion.

Remark 3.2. Technically, translations and rotations are the *orientation-preserving* symmetries of Euclidean space. Reflections are also symmetries of Euclidean space, but they reverse orientation. (They also reverse certain other quantities, such as the sign of the torsion of a curve.) It is generally advantageous to restrict consideration to orientation-preserving symmetries, mainly because doing so typically allows us to work with connected Lie groups. (Think $SO(n)$ vs. $O(n)$.)

Fundamental to Klein's approach—and to the remainder of this book—is the notion of a *homogeneous space.* We will look at curves, surfaces, etc., as submanifolds of homogeneous spaces, and our primary tool for studying such submanifolds will be the *method of moving frames*, which was introduced by Élie Cartan in 1935 ([**Car35**]). In this chapter, we will start with a detailed discussion of Euclidean space as a homogeneous space; we will then give some general definitions and explore several other homogeneous spaces (Minkowski space, equi-affine space, and projective space) that are commonly studied in geometry. In subsequent chapters, we will develop the theory of curves and surfaces in each of these spaces.

3.2. Euclidean space

3.2.1. Inner products.

Definition 3.3. An *inner product* on the vector space $\mathbb{R}^n$ is a function

$$\langle \cdot, \cdot \rangle : \mathbb{R}^n \times \mathbb{R}^n \to \mathbb{R}$$

with the following properties:

(1) *Symmetry:* For any vectors $\mathbf{v}, \mathbf{w} \in \mathbb{R}^n$, $\langle \mathbf{v}, \mathbf{w} \rangle = \langle \mathbf{w}, \mathbf{v} \rangle$.

(2) *Bilinearity:* For any vectors $\mathbf{v}, \mathbf{w}, \mathbf{z} \in \mathbb{R}^n$ and any scalars $a, b \in \mathbb{R}$,

$$\langle a\mathbf{v} + b\mathbf{w}, \mathbf{z} \rangle = a\langle \mathbf{v}, \mathbf{z} \rangle + b\langle \mathbf{w}, \mathbf{z} \rangle$$

and

$$\langle \mathbf{z}, a\mathbf{v} + b\mathbf{w} \rangle = a\langle \mathbf{z}, \mathbf{v} \rangle + b\langle \mathbf{z}, \mathbf{w} \rangle.$$

(3) *Positive definiteness:* For any vector $\mathbf{v} \in \mathbb{R}^n$, $\langle \mathbf{v}, \mathbf{v} \rangle \geq 0$, with equality if and only if $\mathbf{v} = \mathbf{0}$.

Definition 3.4. The vector space $\mathbb{R}^n$ endowed with an inner product $\langle \cdot, \cdot \rangle$ is called *Euclidean space* and is denoted $\mathbb{E}^n$.

An inner product provides a means for measuring lengths of vectors and angles between vectors in Euclidean space: Given vectors $\mathbf{v}, \mathbf{w} \in \mathbb{E}^n$, the length of $\mathbf{v}$ is

$$|\mathbf{v}| = \sqrt{\langle \mathbf{v}, \mathbf{v} \rangle},$$

and the angle between $\mathbf{v}$ and $\mathbf{w}$ is the angle θ, $0 \leq \theta \leq \pi$, satisfying

$$\cos(\theta) = \frac{\langle \mathbf{v}, \mathbf{w} \rangle}{|\mathbf{v}||\mathbf{w}|}.$$

Exercise 3.5. Let $(\mathbf{e}_1, \dots, \mathbf{e}_n)$ be any basis for $\mathbb{R}^n$, and let $\langle \cdot, \cdot \rangle$ be any inner product on $\mathbb{R}^n$. Define constants $\{g_{ij}, 1 \leq i, j, \leq n\}$ by

$$g_{ij} = \langle \mathbf{e}_i, \mathbf{e}_j \rangle.$$

(a) Show that for any vectors $\mathbf{v} = a^i \mathbf{e}_i, \mathbf{w} = b^j \mathbf{e}_j$,

$$\langle \mathbf{v}, \mathbf{w} \rangle = g_{ij} a^i b^j.$$

(b) (Cf. Exercise 1.15) Let A_g be the matrix

$$A_g = \begin{bmatrix} g_{11} & \cdots & g_{1n} \\ \vdots & & \vdots \\ g_{n1} & \cdots & g_{nn} \end{bmatrix},$$

and let $\mathbf{v}, \mathbf{w}$ be represented by the column vectors

$$\mathbf{a} = \begin{bmatrix} a^1 \\ \vdots \\ a^n \end{bmatrix}, \qquad \mathbf{b} = \begin{bmatrix} b^1 \\ \vdots \\ b^n \end{bmatrix},$$

respectively. Show that

$$\langle \mathbf{v}, \mathbf{w} \rangle = {}^t\mathbf{a} A_g \mathbf{b},$$

where we identify the 1×1 matrix on the right-hand side with its single real-valued entry.

Definition 3.4 makes it sound as though there might be many different Euclidean spaces corresponding to different inner products, but in fact they are all equivalent, as the following exercise shows.

Exercise 3.6. Let $\langle \cdot, \cdot \rangle$ be any inner product on $\mathbb{R}^n$, and construct a basis $(\mathbf{e}_1, \dots, \mathbf{e}_n)$ for $\mathbb{R}^n$ inductively as follows:

(Initial step) Choose any nonzero vector $\mathbf{v}_1 \in \mathbb{R}^n$, and set

$$\mathbf{e}_1 = \frac{\mathbf{v}_1}{\sqrt{\langle \mathbf{v}_1, \mathbf{v}_1 \rangle}}.$$

Let $(\mathbf{e}_1)^\perp$ denote the *orthogonal complement* of $\mathbf{e}_1$ with respect to $\langle \cdot, \cdot \rangle$; i.e.,

$$(\mathbf{e}_1)^\perp = \{\mathbf{v} \in \mathbb{R}^n \mid \langle \mathbf{v}, \mathbf{e}_1 \rangle = 0\}.$$

(a) Show that $(\mathbf{e}_1)^\perp \subset \mathbb{R}^n$ is a vector space of dimension $n-1$. (Hint: Consider the linear map $f : \mathbb{R}^n \to \mathbb{R}$ defined by $f(\mathbf{v}) = \langle \mathbf{v}, \mathbf{e}_1 \rangle$, and compute the dimension of $\ker(f)$.)

(Inductive step) Suppose that $\mathbf{e}_1, \ldots, \mathbf{e}_k$ have been chosen. Let

$$(\mathbf{e}_1, \ldots, \mathbf{e}_k)^\perp = \{\mathbf{v} \in \mathbb{R}^n \mid \langle \mathbf{v}, \mathbf{e}_1 \rangle = \cdots = \langle \mathbf{v}, \mathbf{e}_k \rangle = 0\}.$$

Let $\mathbf{v}_{k+1}$ be any nonzero vector in $(\mathbf{e}_1, \ldots, \mathbf{e}_k)^\perp$, and set

$$\mathbf{e}_{k+1} = \frac{\mathbf{v}_{k+1}}{\sqrt{\langle \mathbf{v}_{k+1}, \mathbf{v}_{k+1} \rangle}}.$$

(b) Show that $\mathbf{e}_{k+1}$ is linearly independent from $(\mathbf{e}_1, \ldots, \mathbf{e}_k)$.

(c) Show that $(\mathbf{e}_1, \ldots, \mathbf{e}_k)^\perp \subset \mathbb{R}^n$ is a vector space of dimension $n-k$. (Hint: Consider the linear map $F : \mathbb{R}^n \to \mathbb{R}^k$ defined by

$$F(\mathbf{v}) = {}^t[\langle \mathbf{v}, \mathbf{e}_1 \rangle, \ldots, \langle \mathbf{v}, \mathbf{e}_k \rangle],$$

and compute the dimension of $\ker(F)$. Note that, by part (b), the vectors $(\mathbf{e}_1, \ldots, \mathbf{e}_k)$ are linearly independent.)

(d) By parts (b) and (c), this process results in the construction of a basis $(\mathbf{e}_1, \ldots, \mathbf{e}_n)$ for $\mathbb{R}^n$. Show that with respect to this basis, the (g_{ij}) of Exercise 3.5 are

$$g_{ij} = \langle \mathbf{e}_i, \mathbf{e}_j \rangle = \begin{cases} 1, & i = j, \\ 0, & i \neq j. \end{cases}$$

(We say that the basis $(\mathbf{e}_1, \ldots, \mathbf{e}_n)$ is *orthonormal* with respect to the inner product $\langle \cdot, \cdot \rangle$.) Therefore, the isomorphism $\phi : \mathbb{R}^n \to \mathbb{E}^n$ that identifies $(\mathbf{e}_1, \ldots, \mathbf{e}_n)$ with the standard basis $(\underline{\mathbf{e}}_1, \ldots, \underline{\mathbf{e}}_n)$ for $\mathbb{E}^n$ identifies $\langle \cdot, \cdot \rangle$ with the standard inner product on $\mathbb{E}^n$.

Exercise 3.7. Let $(\mathbf{e}_1, \ldots, \mathbf{e}_n)$ be an orthonormal basis for $\mathbb{E}^n$. Show that the dual basis $(\mathbf{e}^{*1}, \ldots, \mathbf{e}^{*n})$ for the dual space $(\mathbb{E}^n)^*$ consists of the linear mappings $\mathbf{e}^{*i} : \mathbb{E}^n \to \mathbb{R}$ defined by

$$\mathbf{e}^{*i}(\mathbf{v}) = \langle \mathbf{e}_i, \mathbf{v} \rangle$$

for $\mathbf{v} \in \mathbb{E}^n$.

3.2.2. Symmetries and isotropy groups. Now we consider the issue of orientation-preserving symmetries (cf. Remark 3.2) of Euclidean space: What kinds of orientation-preserving transformations $\varphi : \mathbb{E}^n \to \mathbb{E}^n$ preserve the fundamental properties of lengths of vectors and angles between vectors—specifically, lengths and angles of vectors *tangent* to $\mathbb{R}^n$ and based at the same point of $\mathbb{R}^n$? The answer (which hopefully you already know) is *translations and rotations*, collectively known as "rigid motions". Any such transformation has the form

$$\varphi(\mathbf{x}) = A\mathbf{x} + \mathbf{b},$$

where A is an element of the special orthogonal group $SO(n)$ and $\mathbf{b} \in \mathbb{E}^n$. The set of such transformations forms a Lie group, called the *Euclidean group* $E(n)$. This group can be represented as a group of $(n+1) \times (n+1)$ matrices as follows: Let

$$E(n) = \left\{ \begin{bmatrix} 1 & {}^t\mathbf{0} \\ \mathbf{b} & A \end{bmatrix} : A \in SO(n),\ \mathbf{b} \in \mathbb{E}^n \right\}.$$

Here A is an $n \times n$ matrix, $\mathbf{b}$ is an $n \times 1$ column vector, and ${}^t\mathbf{0}$ represents a $1 \times n$ row of 0's, i.e., the transpose of the $n \times 1$ column vector $\mathbf{0}$. If we represent a vector $\mathbf{x} \in \mathbb{E}^n$ by the $(n+1)$-dimensional column vector $\begin{bmatrix} 1 \\ \mathbf{x} \end{bmatrix}$, then elements of $E(n)$ act by matrix multiplication:

$$\begin{bmatrix} 1 & {}^t\mathbf{0} \\ \mathbf{b} & A \end{bmatrix} \begin{bmatrix} 1 \\ \mathbf{x} \end{bmatrix} = \begin{bmatrix} 1 \\ A\mathbf{x} + \mathbf{b} \end{bmatrix}.$$

Remark 3.8. In the literature on Lie groups, it is common to write the matrix as $\begin{bmatrix} A & \mathbf{b} \\ {}^t\mathbf{0} & 1 \end{bmatrix}$ and the vector as $\begin{bmatrix} \mathbf{x} \\ 1 \end{bmatrix}$. We have chosen to write them in this way so that, once we get around to defining moving frames, the order of the columns in the matrix will correspond to the order of the vectors in the associated moving frame.

Given a point $\mathbf{x} \in \mathbb{E}^n$, it is natural to ask: Which elements of $E(n)$ leave $\mathbf{x}$ fixed? In other words, which transformations $\varphi \in E(n)$ have the property that $\varphi(\mathbf{x}) = \mathbf{x}$? The set of such transformations is a subgroup of $E(n)$, called the *isotropy group* of the point $\mathbf{x} \in \mathbb{E}^n$ and denoted $H_{\mathbf{x}}$.

This question is easiest to answer when $\mathbf{x} = \mathbf{0}$: An element $\begin{bmatrix} 1 & {}^t\mathbf{0} \\ \mathbf{b} & A \end{bmatrix}$ fixes the point $\mathbf{x} = \mathbf{0}$ if and only if $\mathbf{b} = \mathbf{0}$. Thus, the isotropy group $H_{\mathbf{0}}$ of the point $\mathbf{0} \in \mathbb{E}^n$ consists of all elements of $E(n)$ of the form $\begin{bmatrix} 1 & {}^t\mathbf{0} \\ \mathbf{0} & A \end{bmatrix}$. This subgroup is clearly isomorphic to $SO(n)$.

What about other points $\mathbf{x} \in \mathbb{E}^n$? It seems reasonable to expect that there shouldn't be anything special about $\mathbf{0}$ because we can move any point to any other point via a translation. And, in fact, the isotropy group $H_{\mathbf{x}}$ of any point $\mathbf{x} \in \mathbb{E}^n$ is also isomorphic to $SO(n)$. We can define an explicit isomorphism $\phi : H_{\mathbf{0}} \to H_{\mathbf{x}}$ as follows. Let $t_{\mathbf{x}} = \begin{bmatrix} 1 & {}^t\mathbf{0} \\ \mathbf{x} & I \end{bmatrix}$ (where I represents the $n\times n$ identity matrix), so that $t_{\mathbf{x}}$ represents the translation $t_{\mathbf{x}}(\mathbf{y}) = \mathbf{y}+\mathbf{x}$. Then for any element $h \in H_{\mathbf{0}}$, define

$$\phi(h) = t_{\mathbf{x}} h t_{\mathbf{x}}^{-1} \in E(n).$$

The following exercise shows that ϕ is indeed an isomorphism from $H_{\mathbf{0}}$ to $H_{\mathbf{x}}$:

Exercise 3.9. (a) Show that for any $h \in H_{\mathbf{0}}$, we have $\phi(h) \in H_{\mathbf{x}}$; i.e., $\phi(h)(\mathbf{x}) = \mathbf{x}$. Therefore, ϕ is a map from $H_{\mathbf{0}}$ to $H_{\mathbf{x}}$.

(b) Show that $\phi : H_{\mathbf{0}} \to H_{\mathbf{x}}$ is a *homomorphism*; i.e., show that for any two elements $h_1, h_2 \in H_{\mathbf{0}}$, we have

$$\phi(h_1 h_2) = \phi(h_1)\phi(h_2).$$

(c) Show that $\phi : H_{\mathbf{0}} \to H_{\mathbf{x}}$ is *injective*; i.e., show that if $\phi(h)$ is the identity element in $H_{\mathbf{x}}$, then h must be the identity element in $H_{\mathbf{0}}$.

(d) Show that $\phi : H_{\mathbf{0}} \to H_{\mathbf{x}}$ is *surjective*; i.e., show that for any $\tilde{h} \in H_{\mathbf{x}}$, there exists $h \in H_{\mathbf{0}}$ with $\phi(h) = \tilde{h}$.

The isomorphism defined by ϕ is called *conjugation*; we describe it using the notation

$$H_{\mathbf{x}} = t_{\mathbf{x}} H_{\mathbf{0}}\, t_{\mathbf{x}}^{-1},$$

and we say that all the isotropy groups $H_{\mathbf{x}} \subset E(n)$ are *conjugate* in $E(n)$. Moreover, they are all isomorphic to $SO(n)$.

Definition 3.10. The *left coset* $t_{\mathbf{x}} H_{\mathbf{0}}$ is the subset of $E(n)$ defined by

$$t_{\mathbf{x}} H_{\mathbf{0}} = \{t_{\mathbf{x}} h \mid h \in H_{\mathbf{0}}\}.$$

Note that this set is *not* a subgroup of $E(n)$ unless $\mathbf{x} = \mathbf{0}$.

The following exercise shows that $E(n)$ is the *disjoint union* of all the left cosets $t_{\mathbf{x}} H_{\mathbf{0}}$, $\mathbf{x} \in \mathbb{E}^n$:

***Exercise 3.11.** (a) Show that

$$t_{\mathbf{x}} H_{\mathbf{0}} = \left\{ \begin{bmatrix} 1 & {}^t\mathbf{0} \\ \mathbf{x} & A \end{bmatrix} : A \in SO(n) \right\}.$$

(b) Conclude from part (a) that:

(1) If $\mathbf{x} \neq \mathbf{y}$, then the left cosets $t_{\mathbf{x}} H_{\mathbf{0}}$ and $t_{\mathbf{y}} H_{\mathbf{0}}$ are disjoint.

(2) Every transformation $\varphi \in E(n)$ belongs to some left coset $t_{\mathbf{x}} H_{\mathbf{0}}$.

In general, if G is a group and H is a subgroup of G, then the set of left cosets of H in G is denoted by G/H. Since H_0 is isomorphic to $SO(n)$, Exercise 3.11 shows that we have a natural correspondence

$$\mathbb{E}^n \cong E(n)/SO(n).$$

The set $E(n)/SO(n)$ can be given a manifold structure so that this correspondence becomes a diffeomorphism.

3.3. Orthonormal frames on Euclidean space

3.3.1. The orthonormal frame bundle. Another way to look at all this is in terms of *orthonormal frames* on $\mathbb{E}^n$.

Definition 3.12. An (oriented) *orthonormal frame* $\mathbf{f}$ on $\mathbb{E}^n$ is a list of vectors $\mathbf{f} = (\mathbf{x}; \mathbf{e}_1, \ldots, \mathbf{e}_n)$, where $\mathbf{x} \in \mathbb{E}^n$ and $(\mathbf{e}_1, \ldots, \mathbf{e}_n)$ is an oriented, orthonormal basis for the tangent space $T_{\mathbf{x}}\mathbb{E}^n$. Alternatively, we may say that $(\mathbf{e}_1, \ldots, \mathbf{e}_n)$ is a *orthonormal frame based at* $\mathbf{x}$.

If we regard the vectors $(\mathbf{e}_1, \ldots, \mathbf{e}_n)$ as the columns of a matrix $A \in SO(n)$, then we see that there is a one-to-one correspondence between the set of frames on $\mathbb{E}^n$ and the Euclidean group $E(n)$: The vector $\mathbf{x}$ represents the translation component, and the matrix A represents the rotation component. Regarded in this way, we can define a projection map $\pi : E(n) \to \mathbb{E}^n$ by

$$\pi(\mathbf{x}; \mathbf{e}_1, \ldots, \mathbf{e}_n) = \mathbf{x}.$$

This map is differentiable, and the fiber over any point $\mathbf{x} \in \mathbb{E}^n$ is the set of all oriented, orthonormal frames based at $\mathbf{x}$. $SO(n)$ acts freely and transitively on each fiber, and so this map gives an explicit description of $E(n)$ as a principal bundle over $\mathbb{E}^n$ with fiber group $SO(n)$:

$$\begin{array}{ccc} SO(n) & \longrightarrow & E(n) \\ & & \downarrow{\scriptstyle \pi} \\ & & \mathbb{E}^n \cong E(n)/SO(n). \end{array}$$

In this context, $E(n)$ is also called the (oriented) *orthonormal frame bundle* of $\mathbb{E}^n$, and it is denoted $\mathcal{F}(\mathbb{E}^n)$.

3.3.2. Dual forms, connection forms, and structure equations. A guiding principle as we proceed is that *the power of the method of moving frames lies in expressing the derivatives of a frame in terms of the frame itself.* (Remember how well this worked when you studied Frenet frames for curves?) So, our next step is to consider the derivatives—specifically, the exterior derivatives—of the components $(\mathbf{x}, \mathbf{e}_1, \ldots, \mathbf{e}_n)$ of a frame. These components may all be thought of as $\mathbb{E}^n$-valued functions on $\mathcal{F}(\mathbb{E}^n)$. (For example, the function $\mathbf{x} : \mathcal{F}(\mathbb{E}^n) \to \mathbb{E}^n$ is just the projection map π.)

Remark 3.13. In fact, this requires some abuse of notation. The function $\mathbf{x}$ is legitimately $\mathbb{E}^n$-valued, but the functions $(\mathbf{e}_1, \ldots, \mathbf{e}_n)$ actually take values in the tangent bundle $T\mathbb{E}^n$, and their derivatives—which we will get to shortly—take values in the bundle $T(T\mathbb{E}^n)$. This distinction is sometimes important, but for the most part we will ignore it by making use of the canonical isomorphisms $T_{\mathbf{x}}\mathbb{E}^n \cong \mathbb{E}^n$, which allow us to regard $(\mathbf{e}_1, \ldots, \mathbf{e}_n)$ as $\mathbb{E}^n$-valued functions.

Now, consider the exterior derivatives $d\mathbf{x}$, $d\mathbf{e}_i$ of the functions $\mathbf{x}, \mathbf{e}_i$ on $\mathcal{F}(\mathbb{E}^n)$. (Recall from Remark 2.39 that these are simply the differentials of the maps $\mathbf{x}, \mathbf{e}_i : \mathcal{F}(\mathbb{E}^n) \to \mathbb{E}^n$.) For any point $\mathbf{f} = (\mathbf{x}; \mathbf{e}_1, \ldots, \mathbf{e}_n) \in \mathcal{F}(\mathbb{E}^n)$, these are maps

$$\begin{aligned} d\mathbf{x} &: T_{\mathbf{f}}\mathcal{F}(\mathbb{E}^n) \to T_{\mathbf{x}}\mathbb{E}^n, \\ d\mathbf{e}_i &: T_{\mathbf{f}}\mathcal{F}(\mathbb{E}^n) \to T_{\mathbf{e}_i}\mathbb{E}^n \cong T_{\mathbf{e}_i}(T_{\mathbf{x}}\mathbb{E}^n) \cong T_{\mathbf{x}}\mathbb{E}^n. \end{aligned}$$

Moreover, since for any frame $\mathbf{f} = (\mathbf{x}; \mathbf{e}_1, \ldots, \mathbf{e}_n)$ the vectors $(\mathbf{e}_1, \ldots, \mathbf{e}_n)$ form a basis for the tangent space $T_{\mathbf{x}}\mathbb{E}^n$, the vector-valued 1-forms $d\mathbf{x}$ and $d\mathbf{e}_i$ can be expressed as linear combinations of $(\mathbf{e}_1, \ldots, \mathbf{e}_n)$ whose coefficients are ordinary scalar-valued 1-forms (cf. Exercise 2.40).

These considerations lead us to define scalar-valued 1-forms (ω^i, ω^i_j) on $\mathcal{F}(\mathbb{E}^n)$ by the equations

$$\begin{aligned} d\mathbf{x} &= \mathbf{e}_i \omega^i, \\ d\mathbf{e}_i &= \mathbf{e}_j \omega^j_i, \end{aligned} \tag{3.1}$$

where $1 \le i, j, \le n$.

Remark 3.14. While it may look strange to write the scalar-valued 1-forms after the vectors in these equations, order is important here: For example, the first equation may be written as the matrix product

$$d\mathbf{x} = \begin{bmatrix} \mathbf{e}_1 & \cdots & \mathbf{e}_n \end{bmatrix} \begin{bmatrix} \omega^1 \\ \vdots \\ \omega^n \end{bmatrix}, \tag{3.2}$$

and it wouldn't make sense in the other order without interchanging the roles of row and column vectors.

In other words, the 1-forms (ω^i, ω^i_j) on $\mathcal{F}(\mathbb{E}^n)$ are defined by the property that for any $\mathbf{f} = (\mathbf{x}; \mathbf{e}_1, \ldots, \mathbf{e}_n) \in \mathcal{F}(\mathbb{E}^n)$, $\mathbf{v} \in T_{\mathbf{f}}\mathcal{F}(\mathbb{E}^n)$,

$$\begin{aligned} d\mathbf{x}(\mathbf{v}) &= \mathbf{e}_i\omega^i(\mathbf{v}) \in T_{\mathbf{x}}\mathbb{E}^n, \\ d\mathbf{e}_i(\mathbf{v}) &= \mathbf{e}_j\omega^j_i(\mathbf{v}) \in T_{\mathbf{e}_i}(T_{\mathbf{x}}\mathbb{E}^n) \cong T_{\mathbf{x}}\mathbb{E}^n. \end{aligned}$$

Remark 3.15. Note that the 1-forms (ω^i, ω^i_j) are *not* well-defined on the base space $\mathbb{E}^n$ since they are defined relative to a particular choice of frame $(\mathbf{e}_1, \ldots, \mathbf{e}_n)$ for the tangent space $T_{\mathbf{x}}\mathbb{E}^n$.

We can actually describe these 1-forms fairly explicitly. Given a point $\mathbf{f} = (\mathbf{x}; \mathbf{e}_1, \ldots, \mathbf{e}_n) \in \mathcal{F}(\mathbb{E}^n)$, let $A \in SO(n)$ denote the matrix

$$A = \begin{bmatrix} \mathbf{e}_1 & \cdots & \mathbf{e}_n \end{bmatrix}.$$

Then equation (3.2) can be written as

$$d\mathbf{x} = A \begin{bmatrix} \omega^1 \\ \vdots \\ \omega^n \end{bmatrix}.$$

Therefore,

$$\begin{bmatrix} \omega^1 \\ \vdots \\ \omega^n \end{bmatrix} = A^{-1} d\mathbf{x} = A^{-1} \begin{bmatrix} dx^1 \\ \vdots \\ dx^n \end{bmatrix}. \tag{3.3}$$

From this expression, it is clear that the 1-forms $(\omega^1, \ldots, \omega^n)$ are linearly independent and form a basis for the 1-forms on $\mathbb{E}^n$.

Remark 3.16. This, too, requires some abuse of notation: While $(\omega^1, \ldots, \omega^n)$ are not well-defined on $\mathbb{E}^n$, they *are* linearly independent, linear combinations of the 1-forms $(dx^1, \ldots, dx^n)$—or, more precisely, of the pullbacks $(\pi^*(dx^1), \ldots, \pi^*(dx^n))$ of $(dx^1, \ldots, dx^n)$ to $\mathcal{F}(\mathbb{E}^n)$. Therefore, $(dx^1, \ldots, dx^n)$ can be expressed as linear combinations of $(\omega^1, \ldots, \omega^n)$, with coefficients that are functions on $\mathcal{F}(\mathbb{E}^n)$—and indeed, the first equation in (3.1) does just that.

Similarly, the equations for $d\mathbf{e}_i$ in (3.1) can be combined into the matrix equation

$$dA = \begin{bmatrix} d\mathbf{e}_1 & \cdots & d\mathbf{e}_n \end{bmatrix} = \begin{bmatrix} \mathbf{e}_1 & \cdots & \mathbf{e}_n \end{bmatrix} \begin{bmatrix} \omega_1^1 & \cdots & \omega_n^1 \\ \vdots & & \vdots \\ \omega_1^n & \cdots & \omega_n^n \end{bmatrix}$$
$$= A \begin{bmatrix} \omega_1^1 & \cdots & \omega_n^1 \\ \vdots & & \vdots \\ \omega_1^n & \cdots & \omega_n^n \end{bmatrix}.$$

Therefore,

$$\begin{bmatrix} \omega_1^1 & \cdots & \omega_n^1 \\ \vdots & & \vdots \\ \omega_1^n & \cdots & \omega_n^n \end{bmatrix} = A^{-1}dA. \tag{3.4}$$

Example 3.17. Consider the case $n = 2$. Let $\mathbf{x} = (x^1, x^2)$ denote the coordinates of an arbitrary point in $\mathbb{E}^2$. Any orthonormal frame $(\mathbf{e}_1, \mathbf{e}_2)$ for the tangent space $T_{\mathbf{x}}\mathbb{E}^2$ can be written as

$$\mathbf{e}_1 = \begin{bmatrix} \cos(\theta) \\ \sin(\theta) \end{bmatrix}, \qquad \mathbf{e}_2 = \begin{bmatrix} -\sin(\theta) \\ \cos(\theta) \end{bmatrix},$$

where θ is the angle between $\mathbf{e}_1$ and the standard basis vector $\underline{\mathbf{e}}_1 = \frac{\partial}{\partial x^1}$. Thus, θ may be regarded as a local coordinate on the 1-dimensional fibers of the orthonormal frame bundle $\mathcal{F}(\mathbb{E}^2)$. Now, write

$$A = \begin{bmatrix} \mathbf{e}_1 & \mathbf{e}_2 \end{bmatrix} = \begin{bmatrix} \cos(\theta) & -\sin(\theta) \\ \sin(\theta) & \cos(\theta) \end{bmatrix}.$$

Then (ω^1, ω^2) are given by

$$\begin{bmatrix} \omega^1 \\ \omega^2 \end{bmatrix} = A^{-1}d\mathbf{x}$$
$$= \begin{bmatrix} \cos(\theta) & \sin(\theta) \\ -\sin(\theta) & \cos(\theta) \end{bmatrix} \begin{bmatrix} dx^1 \\ dx^2 \end{bmatrix} \tag{3.5}$$
$$= \begin{bmatrix} \cos(\theta)\, dx^1 + \sin(\theta)\, dx^2 \\ -\sin(\theta)\, dx^1 + \cos(\theta)\, dx^2 \end{bmatrix},$$

and the 1-forms (ω^i_j) are given by

$$\begin{aligned}
\begin{bmatrix} \omega^1_1 & \omega^1_2 \\ \omega^2_1 & \omega^2_2 \end{bmatrix} &= A^{-1}dA \\
&= \begin{bmatrix} \cos(\theta) & \sin(\theta) \\ -\sin(\theta) & \cos(\theta) \end{bmatrix} \begin{bmatrix} -\sin(\theta)\,d\theta & -\cos(\theta)\,d\theta \\ \cos(\theta)\,d\theta & -\sin(\theta)\,d\theta \end{bmatrix} \\
&= \begin{bmatrix} 0 & -d\theta \\ d\theta & 0 \end{bmatrix}.
\end{aligned} \tag{3.6}$$

The 1-forms $(\omega^1, \ldots, \omega^n)$ are often called the *dual forms* on the orthonormal frame bundle $\mathcal{F}(\mathbb{E}^n)$ because they have the property that at any point $\mathbf{f} \in \mathcal{F}(\mathbb{E}^n)$,

$$\omega^i(\mathbf{e}_j) = \delta^i_j = \begin{cases} 1, & i = j, \\ 0, & i \neq j. \end{cases} \tag{3.7}$$

Exercise 3.18. Verify equations (3.7) by direct computation.

The dual forms also have the property that $\omega^i(\mathbf{v}) = 0$ for any vector $\mathbf{v} \in T\mathcal{F}(\mathbb{E}^n)$ that is tangent to the fibers of the projection $\pi : \mathcal{F}(\mathbb{E}^n) \to \mathbb{E}^n$. The technical way to say this is that *the pullback of ω^i to any fiber $\pi^{-1}(\mathbf{x}_0)$ of π via the inclusion map $\iota : \pi^{-1}(\mathbf{x}_0) \hookrightarrow \mathcal{F}(\mathbb{E}^n)$ vanishes.*

Exercise 3.19. Prove this statement. (Hint: Let $\alpha : I \to \mathcal{F}(\mathbb{E}^n)$ be a curve tangent to $\mathbf{v}$. Since $\mathbf{v}$ is tangent to some fiber $\pi^{-1}(\mathbf{x}_0) \subset \mathcal{F}(\mathbb{E}^n)$, you can assume that $\alpha(I) \subset \pi^{-1}(\mathbf{x}_0)$. So, what is the value of $\mathbf{x}$ at any point $\alpha(t)$? Now compute $d\mathbf{x}(\mathbf{v}) = \frac{d}{dt}\big|_{t=0}\mathbf{x}(\alpha(t))$.)

Forms with this property (i.e., their pullbacks to each fiber of π vanish) are called *semi-basic* for the projection π; for this reason, the dual forms are also sometimes called the *semi-basic forms* on $\mathcal{F}(\mathbb{E}^n)$. The pullbacks of (ω^i_j), on the other hand, form a *basis* for the 1-forms on each fiber of π. They are called the *connection forms* on $\mathcal{F}(\mathbb{E}^n)$.

Now for the fun part: Start differentiating! In order to compute the exterior derivatives of the dual forms and connection forms, we need to differentiate equations (3.1).

***Exercise 3.20.** Differentiate equations (3.1) (taking the second equation into account!) and do some careful index-juggling to obtain

$$\begin{aligned}
\mathbf{0} &= \mathbf{e}_i(\omega^i_j \wedge \omega^j + d\omega^i), \\
\mathbf{0} &= \mathbf{e}_i(\omega^i_k \wedge \omega^k_j + d\omega^i_j).
\end{aligned}$$

Conclude that these 1-forms satisfy the *Cartan structure equations*

$$
\begin{aligned}
d\omega^i &= -\omega^i_j \wedge \omega^j, \\
d\omega^i_j &= -\omega^i_k \wedge \omega^k_j.
\end{aligned}
\tag{3.8}
$$

(Hint: Recall the Leibniz rule for differentiating p-forms, and observe that the vector field $\mathbf{e}_i$ is a vector-valued 0-form. Note that if we wrote the terms in (3.1) in the other order—and some authors do!—the Leibniz rule would lead to different signs in the structure equations.)

Exercise 3.21. Verify by direct computation that the dual forms (3.5) and connection forms (3.6) on $\mathcal{F}(\mathbb{E}^2)$ satisfy the Cartan structure equations (3.8).

Up to this point, we haven't taken advantage of the fact that we have a Euclidean structure on $\mathbb{E}^n$. Since $(\mathbf{e}_1, \dots, \mathbf{e}_n)$ are orthonormal vectors, we have

$$
\langle \mathbf{e}_i, \mathbf{e}_j \rangle = \delta_{ij} = \begin{cases} 1, & i = j, \\ 0, & i \neq j. \end{cases}
\tag{3.9}
$$

***Exercise 3.22.** Differentiate equations (3.9) and conclude that the connection forms (ω^i_j) are skew-symmetric in their indices; that is, they have the property that $\omega^j_i = -\omega^i_j$.

Exercise 3.23. Write out equations (3.1) and the structure equations (3.8) explicitly (i.e., without the summation convention) in the case $n = 3$. What is the dimension of $\mathcal{F}(\mathbb{E}^3)$? How many linearly independent connection forms are there?

3.3.3. The Maurer-Cartan form.

This all fits into a larger structure, which is easier to see if we go back to regarding $\mathcal{F}(\mathbb{E}^n)$ as the Lie group $E(n)$.

Exercise 3.24. Prove that the Lie algebra of $E(n)$ is

$$
\mathfrak{e}(n) = \left\{ \begin{bmatrix} 0 & {}^t\mathbf{0} \\ \mathbf{b} & B \end{bmatrix} : B \in \mathfrak{so}(n),\ \mathbf{b} \in \mathbb{E}^n \right\}.
$$

(Recall that $B \in \mathfrak{so}(n)$ if and only if B is a skew-symmetric $n \times n$ matrix.)

We can define an $\mathfrak{e}(n)$-valued 1-form ω on $E(n)$ as follows: Recall that for $h \in E(n)$, the left multiplication map $L_h : E(n) \to E(n)$ is defined by

$$
L_h(g) = hg,
$$

and for any $g \in E(n)$, we have the differential

$$
(L_h)_* : T_g E(n) \to T_{hg} E(n).
$$

Now, for any $g \in E(n), \mathbf{v} \in T_g E(n)$, define

$$\omega(\mathbf{v}) = (L_{g^{-1}})_*(\mathbf{v}).$$

Since $(L_{g^{-1}})_*$ maps $T_g E(n)$ to $T_I E(n) = \mathfrak{e}(n)$, we have $\omega(\mathbf{v}) \in \mathfrak{e}(n)$ for all $\mathbf{v} \in TE(n)$. The form ω is called the *Maurer-Cartan* form on $E(n)$.

***Exercise 3.25.** Prove that the Maurer-Cartan form ω on $E(n)$ is left-invariant; i.e., show that for any $h \in E(n)$, $L_h^* \omega = \omega$. (Hint: Let $g \in E(n)$, $\mathbf{v} \in T_g E(n)$, and compute

$$(L_h^* \omega)(\mathbf{v}) = \omega((L_h)_*(\mathbf{v})).$$

Try to keep track of which tangent space each object lives in!)

While this definition is fairly simple in theory, it is not so easy to work with computationally. The Maurer-Cartan form is more commonly written as

$$\omega = g^{-1}\, dg. \tag{3.10}$$

This notation requires some explanation. The variable g here essentially denotes the identity map $g : E(n) \to E(n)$, but it really should be thought of as a coordinate function on $E(n)$, which realizes any element of the abstract Lie group $E(n)$ as its $(n+1) \times (n+1)$ matrix representation. Specifically, for any $\mathbf{f} = (\mathbf{x}; \mathbf{e}_1, \ldots, \mathbf{e}_n) \in E(n)$,

$$g(\mathbf{f}) = \begin{bmatrix} 1 & 0 & \cdots & 0 \\ \mathbf{x} & \mathbf{e}_1 & \cdots & \mathbf{e}_n \end{bmatrix}. \tag{3.11}$$

Now, let $\mathbf{f}_0$ be any element of $E(n)$, and let $g_0 = g(\mathbf{f}_0)$. (So $\mathbf{f}_0$ and g_0 represent the same element of $E(n)$; the difference in notation is meant to suggest that $\mathbf{f}_0$ is an element of the abstract group and g_0 is its matrix representation.) Differentiating (3.11) shows that the mapping

$$dg : T_{\mathbf{f}_0} E(n) \to T_{g_0} E(n)$$

can be written as the matrix-valued 1-form

$$dg = \begin{bmatrix} 0 & 0 & \cdots & 0 \\ d\mathbf{x} & d\mathbf{e}_1 & \cdots & d\mathbf{e}_n \end{bmatrix}. \tag{3.12}$$

Since g is essentially the identity map, the same is true for dg: It simply identifies any tangent vector $\mathbf{v} \in T_{\mathbf{f}_0} E(n)$ with its matrix representation as a tangent vector to the matrix group $E(n)$ at g_0. So

$$dg(\mathbf{v}) = \mathbf{v} \in T_{g_0} E(n).$$

The left multiplication by g^{-1}, applied to a vector $\mathbf{v} \in T_{g_0} E(n)$, then means to multiply the matrix representation for $\mathbf{v}$ by the matrix representation for g_0^{-1}.

Remark 3.26. The realization of $E(n)$ as a matrix group is crucial in order for this notation to make any sense at all: In the abstract setting, there is no way to multiply $\mathbf{v}$ by an element of $E(n)$ because $\mathbf{v}$ is a tangent vector and not a group element. The only sensible interpretation of "$g^{-1}dg(\mathbf{v})$" is to act on the vector $dg(\mathbf{v}) = \mathbf{v} \in T_{g_0}E(n)$ by the *differential* of the left-multiplication map $L_{g_0^{-1}}$. (Note that this is our original definition for ω!) But because everything here is matrix-valued, $(L_{g_0^{-1}})_*(\mathbf{v})$ is, in fact, given by the product of the matrices g_0^{-1} and $\mathbf{v}$.

As you might suspect, it is no accident that we have used the letter ω both for the scalar-valued dual and connection forms and for the matrix-valued Maurer-Cartan form. The following exercise shows how they are related:

***Exercise 3.27.** (a) Show that the Maurer-Cartan form on $E(n)$ is given by

$$\omega = \begin{bmatrix} 0 & 0 & \cdots & 0 \\ \omega^1 & \omega^1_1 & \cdots & \omega^1_n \\ \vdots & \vdots & & \vdots \\ \omega^n & \omega^n_1 & \cdots & \omega^n_n \end{bmatrix}.$$

(Hint: Use equations (3.1), (3.11), (3.12) and the fact that $\omega = g^{-1}dg$ is equivalent to $dg = g\omega$.) Write out this matrix explicitly in the case $n = 3$.

(b) Use the skew-symmetry of the forms (ω^i_j) to confirm that for any $\mathbf{v} \in TE(n)$, we have $\omega(\mathbf{v}) \in \mathfrak{e}(n)$. (This is what it means to say that ω is "$\mathfrak{e}(n)$-valued".)

Because the matrix-valued Maurer-Cartan form ω contains the scalar-valued dual forms and connection forms as its entries, the dual forms and connection forms are collectively referred to as "the Maurer-Cartan forms". The fact that ω is left-invariant implies that the Maurer-Cartan forms are left-invariant as well; in fact, they form a *basis* for the left-invariant 1-forms on $E(n)$.

***Exercise 3.28.** Show that the structure equations (3.8) are equivalent to the *Maurer-Cartan* equation

$$d\omega = -\omega \wedge \omega.$$

(The wedge product of two matrices of 1-forms is computed just like the ordinary matrix product, substituting wedge product for ordinary multiplication of the appropriate entries.) Note that, despite the skew-symmetry of the wedge product for scalar-valued 1-forms, the wedge product of a matrix-valued 1-form with itself does not necessarily vanish!

***Exercise 3.29.** Suppose that we choose a *particular* orthonormal frame $(\mathbf{e}_1(\mathbf{x}), \ldots, \mathbf{e}_n(\mathbf{x}))$ for $T_{\mathbf{x}}\mathbb{E}^n$ at each point $\mathbf{x} \in \mathbb{E}^n$. This amounts to choosing a section $\sigma : \mathbb{E}^n \to \mathcal{F}(\mathbb{E}^n)$ of the orthonormal frame bundle $\pi : \mathcal{F}(\mathbb{E}^n) \to \mathbb{E}^n$, also known as an *orthonormal frame field* on $\mathbb{E}^n$. The pullbacks $(\sigma^*(\omega^i), \sigma^*(\omega^i_j))$ are then 1-forms on $\mathbb{E}^n$. In order to reduce notational clutter, we will denote these pullbacks by $(\bar{\omega}^i, \bar{\omega}^i_j)$. Show that if we set

$$A(\mathbf{x}) = \begin{bmatrix}\mathbf{e}_1(\mathbf{x}) & \ldots & \mathbf{e}_n(\mathbf{x})\end{bmatrix},$$

then the dual forms and connection forms associated to this frame field are given by

$$\begin{bmatrix}\bar{\omega}^1 \\ \vdots \\ \bar{\omega}^n\end{bmatrix} = A(\mathbf{x})^{-1}\begin{bmatrix}dx^1 \\ \vdots \\ dx^n\end{bmatrix}, \qquad \begin{bmatrix}\bar{\omega}^1_1 & \ldots & \bar{\omega}^1_n \\ \vdots & & \vdots \\ \bar{\omega}^n_1 & \ldots & \bar{\omega}^n_n\end{bmatrix} = A(\mathbf{x})^{-1}\, dA(\mathbf{x}).$$

(Compare with equations (3.3), (3.4).) Show that if we write

$$g(\mathbf{x}) = \begin{bmatrix}1 & {}^t\mathbf{0} \\ \mathbf{x} & A(\mathbf{x})\end{bmatrix}, \qquad \bar{\omega} = \begin{bmatrix}0 & 0 & \cdots & 0 \\ \bar{\omega}^1 & \bar{\omega}^1_1 & \cdots & \bar{\omega}^1_n \\ \vdots & \vdots & & \vdots \\ \bar{\omega}^n & \bar{\omega}^n_1 & \cdots & \bar{\omega}^n_n\end{bmatrix},$$

then these equations imply that

$$\bar{\omega} = g(\mathbf{x})^{-1} dg(\mathbf{x}).$$

Remark 3.30. The dual forms (ω^i) and connection forms (ω^i_j) are all linearly independent "upstairs" on the frame bundle $\mathcal{F}(\mathbb{E}^n)$, but this is no longer the case when these forms are pulled back "downstairs" to $\mathbb{E}^n$ via a section, as in Exercise 3.29. The pulled-back dual forms $(\bar{\omega}^i)$ are linearly independent, linear combinations of $(dx^1, \ldots, dx^n)$, but the pulled-back connection forms $(\bar{\omega}^i_j)$ are no longer linearly independent from the dual forms; indeed, they may be expressed as linear combinations of $(\bar{\omega}^1, \ldots, \bar{\omega}^n)$, which form a basis for the 1-forms on $\mathbb{E}^n$.

Exercise 3.31. Consider $\mathbb{E}^3$ (minus the z-axis) with cylindrical coordinates (r, θ, z), related to the usual Euclidean coordinates (x, y, z) by

$$x = r\cos(\theta), \qquad y = r\sin(\theta), \qquad z = z.$$

Apply the result of Exercise 3.29 to the cylindrical frame field

$$\mathbf{e}_1 = \begin{bmatrix}\cos(\theta) \\ \sin(\theta) \\ 0\end{bmatrix}, \qquad \mathbf{e}_2 = \begin{bmatrix}-\sin(\theta) \\ \cos(\theta) \\ 0\end{bmatrix}, \qquad \mathbf{e}_3 = \begin{bmatrix}0 \\ 0 \\ 1\end{bmatrix}$$

to compute the dual forms $(\bar{\omega}^i)$ and the connection forms $(\bar{\omega}^i_j)$ for this frame field. Show by direct computation that these forms satisfy the structure equations (3.8).

Exercise 3.32. Repeat Exercise 3.31 for the spherical frame field

$$\mathbf{e}_1 = \begin{bmatrix} \cos(\varphi)\cos(\theta) \\ \cos(\varphi)\sin(\theta) \\ \sin(\varphi) \end{bmatrix}, \qquad \mathbf{e}_2 = \begin{bmatrix} -\sin(\theta) \\ \cos(\theta) \\ 0 \end{bmatrix}, \qquad \mathbf{e}_3 = \begin{bmatrix} -\sin(\varphi)\cos(\theta) \\ -\sin(\varphi)\sin(\theta) \\ \cos(\varphi) \end{bmatrix},$$

where (ρ, φ, θ) are spherical coordinates on $\mathbb{E}^3$ (minus the z-axis), related to the usual Euclidean coordinates (x, y, z) by

$$x = \rho\cos(\varphi)\cos(\theta), \qquad y = \rho\cos(\varphi)\sin(\theta), \qquad z = \rho\sin(\varphi).$$

3.4. Homogeneous spaces

Let's take a moment to summarize our work so far: We started with Euclidean space $\mathbb{E}^n$, i.e., the vector space $\mathbb{R}^n$ endowed with an inner product $\langle\cdot,\cdot\rangle$. By considering the group of symmetries of this structure, we arrived at a description of $\mathbb{E}^n$ as the set of left cosets of the closed subgroup $SO(n)$ of the Lie group $E(n)$. These particular groups arose as the symmetry group of the structure $(E(n))$ and the isotropy group of a particular point $(SO(n))$. This description of $\mathbb{E}^n$ as the group quotient $E(n)/SO(n)$ realizes $\mathbb{E}^n$ as a *homogeneous space*; it is a special case of the following general construction:

Definition 3.33. A *homogeneous space* is the set G/H of left cosets of a closed subgroup H of a Lie group G, endowed with the unique manifold structure with respect to which the natural action of G on G/H is smooth. (The existence and uniqueness of this manifold structure on G/H is a consequence of a theorem of Cartan [**Car30**] which states that every closed subgroup of a Lie group is a Lie subgroup.) The Lie group G is called the *symmetry group* of the homogeneous space G/H.

In most commonly encountered examples, G is a subgroup of some matrix group $GL(m)$. G acts on G/H by left multiplication in the obvious way, and we are generally interested in those properties (such as the inner product on Euclidean space) that are preserved under this action. For any point $\mathbf{x} = gH \in G/H$, the isotropy group $H_{\mathbf{x}}$ of $\mathbf{x}$ is the group

$$H_{\mathbf{x}} = gHg^{-1},$$

where g is any element of the coset gH. All the isotropy groups $H_{\mathbf{x}}$ are clearly conjugate to H in G. Moreover, the projection map $\pi : G \to G/H$

defined by

$$\pi(g) = gH$$

describes G as a principal bundle over G/H with fiber H:

$$\begin{array}{ccc} H & \longrightarrow & G \\ & & \Big\downarrow \pi \\ & & G/H. \end{array}$$

Just as we identified the group $E(n)$ with the set of orthonormal frames on Euclidean space, there is often some natural way to think of the group G as the set of "frames" on the space G/H. The Maurer-Cartan form on G is defined by

$$\omega(\mathbf{v}) = (L_{g^{-1}})_*(\mathbf{v})$$

for any $\mathbf{v} \in T_g G$, and if G is a matrix group, then we can write

$$\omega = g^{-1}\,dg.$$

Either way, ω takes values in the Lie algebra $\mathfrak{g}$ of G, and it satisfies the Maurer-Cartan equation:

$$d\omega = -\omega \wedge \omega. \tag{3.13}$$

This equation will turn out to play a crucial role in the geometry of submanifolds of the space G/H.

***Exercise 3.34.** Suppose that G is a matrix Lie group, so that $\omega = g^{-1}\,dg$. Prove directly that ω satisfies the Maurer-Cartan equation (3.13). (Hint: Show by differentiating the equation $gg^{-1} = I$ that $d(g^{-1}) = -g^{-1}\,dg\,g^{-1}$.)

In the remainder of this chapter, we will explore some other examples of homogeneous spaces: Minkowski space, equi-affine space, and projective space.

3.5. Minkowski space

3.5.1. The Minkowski inner product. Minkowski space was introduced in 1907 by Hermann Minkowski [**Min78**] as a geometric setting for Einstein's theory of special relativity. In Minkowski's original formulation, the three dimensions of Euclidean space are combined with a single time dimension to create a 4-dimensional *spacetime*. More generally, we can consider Minkowski spaces consisting of n space dimensions and one time dimension.

What distinguishes Minkowski space from Euclidean space is its inner product structure: Instead of a positive definite inner product, Minkowski space is equipped with an *indefinite* inner product.

Given any symmetric bilinear form Q on the vector space $\mathbb{R}^n$, there exists a basis $(\mathbf{e}_1, \dots, \mathbf{e}_n)$ for $\mathbb{R}^n$ and integers $p, q \geq 0$ with the property that for any $\mathbf{v} = v^i \mathbf{e}_i \in \mathbb{R}^n$,

$$Q(\mathbf{v}, \mathbf{v}) = (v^1)^2 + \cdots + (v^p)^2 - (v^{p+1})^2 - \cdots - (v^{p+q})^2.$$

The integers p, q are invariants of Q, and the ordered pair (p, q) is called the *signature* of Q. The form Q is nondegenerate if and only if $p + q = n$. A positive definite form Q has signature $(n, 0)$, whereas Q is called *indefinite* if p and q are both greater than zero.

Definition 3.35. The $(1 + n)$-dimensional *Minkowski space* $\mathbb{M}^{1,n}$ is the vector space $\mathbb{R}^{n+1}$ endowed with a nondegenerate, symmetric bilinear form $\langle \cdot, \cdot \rangle$ with signature $(1, n)$. This bilinear form is called the *Minkowski inner product* on $\mathbb{M}^{1,n}$.

Equivalently, for any basis $(\mathbf{e}_0, \dots, \mathbf{e}_n)$ for $\mathbb{R}^{n+1}$, the symmetric matrix

$$g = [g_{\alpha\beta}] = [\langle \mathbf{e}_\alpha, \mathbf{e}_\beta \rangle]$$

that describes the Minkowski inner product $\langle \cdot, \cdot \rangle$ in terms of this basis has one positive eigenvalue and n negative eigenvalues.

Remark 3.36. Even when n has a fixed value, say $n = 3$, the Minkowski space $\mathbb{M}^{1,3}$ is still often referred to as "(1+3)-dimensional Minkowski space" to emphasize the distinction between the time and space dimensions.

Remark 3.37. The Minkowski inner product is sometimes taken to have signature $(n, 1)$ rather than $(1, n)$. Our convention is the same as that used in [**Cal00**]; it is chosen so that tangent vectors $\mathbf{v}$ to curves corresponding to the world lines of particles in spacetime will have $\langle \mathbf{v}, \mathbf{v} \rangle > 0$.

Remark 3.38. For clarity, we will generally use Roman letters (i, j, etc.) for indices that range from 1 to n and Greek letters (α, β, etc.) for indices that range from 0 to n.

Just as in the Euclidean case, all Minkowski inner products on $\mathbb{R}^{n+1}$ are equivalent, as the following exercise shows. (Cf. Exercise 3.6.)

Exercise 3.39. Let $\langle \cdot, \cdot \rangle$ be any Minkowski inner product on $\mathbb{R}^{n+1}$. Construct a basis $(\mathbf{e}_0, \dots, \mathbf{e}_n)$ for $\mathbb{R}^{n+1}$ as follows: Choose any nonzero vector $\mathbf{v}_0 \in \mathbb{R}^{n+1}$ with $\langle \mathbf{v}_0, \mathbf{v}_0 \rangle > 0$, and set

$$\mathbf{e}_0 = \frac{\mathbf{v}_0}{\sqrt{\langle \mathbf{v}_0, \mathbf{v}_0 \rangle}}.$$

Let $(\mathbf{e}_0)^\perp$ denote the *orthogonal complement* of $\mathbf{e}_0$ with respect to $\langle\cdot,\cdot\rangle$; i.e.,

$$(\mathbf{e}_0)^\perp = \{\mathbf{v} \in \mathbb{R}^{n+1} \mid \langle \mathbf{v}, \mathbf{e}_0 \rangle = 0\}.$$

(a) Show that for any nonzero vector $\mathbf{v} \in (\mathbf{e}_0)^\perp$,

$$\langle \mathbf{v}, \mathbf{v} \rangle < 0.$$

(Hint: Let $(\mathbf{v}_1, \dots, \mathbf{v}_n)$ be any basis for $(\mathbf{e}_0)^\perp$. Show that the matrix of g with respect to the basis $(\mathbf{e}_0, \mathbf{v}_1, \dots, \mathbf{v}_n)$ has the form

$$g = \begin{bmatrix} 1 & {}^t\mathbf{0} \\ \mathbf{0} & A \end{bmatrix},$$

where A is a symmetric $n \times n$ matrix with all negative eigenvalues. Use this to prove the result.)

(b) By part (a), the restriction of $\langle\cdot,\cdot\rangle$ to $(\mathbf{e}_0)^\perp$ is *negative* definite. Construct an orthonormal basis $(\mathbf{e}_1, \dots, \mathbf{e}_n)$ for $(\mathbf{e}_0)^\perp$ as in Exercise 3.6. (For purposes of this construction, there is essentially no difference between negative definite and positive definite inner products.)

(c) Show that with respect to the basis $(\mathbf{e}_0, \dots, \mathbf{e}_n)$, the $(g_{\alpha\beta})$ are given by

$$g_{\alpha\beta} = \langle \mathbf{e}_\alpha, \mathbf{e}_\beta \rangle = \begin{cases} 1, & \alpha = \beta = 0, \\ -1, & \alpha = \beta = 1, \dots, n, \\ 0, & \alpha \neq \beta. \end{cases} \tag{3.14}$$

We say that the basis $(\mathbf{e}_0, \dots, \mathbf{e}_n)$ is *orthonormal* with respect to the inner product $\langle\cdot,\cdot\rangle$, and (3.14) is the *standard* Minkowski inner product on $\mathbb{M}^{1,n}$.

Henceforth, we will let $(\underline{\mathbf{e}}_0, \dots, \underline{\mathbf{e}}_n)$ denote the standard basis for $\mathbb{M}^{1,n}$.

Nonzero vectors in $\mathbb{M}^{1,n}$ are divided into three types:

Definition 3.40. A nonzero vector $\mathbf{v} \in \mathbb{M}^{1,n}$ is called

- *timelike* if $\langle \mathbf{v}, \mathbf{v} \rangle > 0$;
- *spacelike* if $\langle \mathbf{v}, \mathbf{v} \rangle < 0$;
- *lightlike* or *null* if $\langle \mathbf{v}, \mathbf{v} \rangle = 0$.

A timelike vector $\mathbf{v} \in \mathbb{M}^{1,n}$ is called *future-pointing* if $\langle \mathbf{v}, \underline{\mathbf{e}}_0 \rangle > 0$ and *past-pointing* if $\langle \mathbf{v}, \underline{\mathbf{e}}_0 \rangle < 0$. The lightlike vectors form a cone, called the *light cone* or *null cone*, opening in both directions with axis parallel to $\underline{\mathbf{e}}_0$. The interior of this cone consists of timelike vectors, and the exterior consists of spacelike vectors.

Remark 3.41. This terminology comes from special relativity, where $\underline{\mathbf{e}}_0$ is regarded as the time direction and $(\underline{\mathbf{e}}_1, \dots, \underline{\mathbf{e}}_n)$ as the spatial directions. The path of a particle traces out a curve in spacetime, called its *world line*. Since nothing can travel faster than the speed of light (which is normalized to $c = 1$ for the standard Minkowski inner product), the world line of any particle traveling at speed $v < c$ must be a timelike curve (i.e., a curve whose tangent vectors are all timelike). The world line of a photon traveling at the speed of light is a lightlike curve (i.e., a curve whose tangent vectors are all lightlike). Any two points $\mathbf{q}_1, \mathbf{q}_2 \in \mathbb{M}^{1,n}$ with spacelike separation (i.e., for which the vector $\mathbf{q}_2 - \mathbf{q}_1$ is spacelike) represent events that cannot have any causal impact on each other because they cannot be connected by a smooth timelike curve.

Definition 3.42. The *Minkowski norm* $\|\mathbf{v}\|$ of a nonzero vector $\mathbf{v} \in \mathbb{M}^{1,n}$ is defined to be

(1) $\|\mathbf{v}\| = \sqrt{\langle \mathbf{v}, \mathbf{v} \rangle}$ if $\mathbf{v}$ is timelike;

(2) $\|\mathbf{v}\| = \sqrt{-\langle \mathbf{v}, \mathbf{v} \rangle}$ if $\mathbf{v}$ is spacelike;

(3) $\|\mathbf{v}\| = 0$ if $\mathbf{v}$ is lightlike.

(Note that $\|\mathbf{v}\| \geq 0$ for all $\mathbf{v} \in \mathbb{M}^{1,n}$.)

Exercise 3.43. For $n \geq 2$ and any $\lambda \in \mathbb{R}$, show that the "sphere" $\mathbb{S}_\lambda$ consisting of all vectors $\mathbf{v} \in \mathbb{M}^{1,n}$ satisfying $\langle \mathbf{v}, \mathbf{v} \rangle = \lambda$ is

(1) a hyperboloid of two sheets when $\lambda > 0$;

(2) a cone when $\lambda = 0$;

(3) a hyperboloid of one sheet when $\lambda < 0$.

(See Figure 3.1; note that the $\underline{\mathbf{e}}_0$-axis is drawn as the vertial axis.)

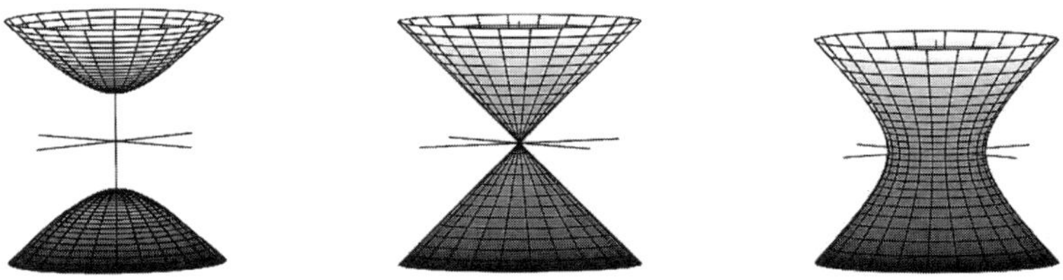

Figure 3.1. Minkowski "spheres" $\mathbb{S}_\lambda$ with $\lambda = 1$, $\lambda = 0$, $\lambda = -1$

3.5.2. Symmetries and isotropy groups. Let $\mathbb{M}^{1,n}$ denote the Minkowski space of dimension $1 + n$ with the standard inner product (3.14). Just as we did for Euclidean space, we ask: What are the symmetries of $\mathbb{M}^{1,n}$?

***Exercise 3.44.** Let $A \in GL(n+1)$. Show that

$$\langle A\mathbf{v}, A\mathbf{v}\rangle = \langle \mathbf{v}, \mathbf{v}\rangle$$

for every $\mathbf{v} \in \mathbb{M}^{1,n}$ if and only if A is an element of the Lie group

$$O(1,n) = \left\{ A \in GL(n+1) : {}^tA \begin{bmatrix} 1 & 0 & \cdots & 0 \\ 0 & -1 & \cdots & 0 \\ \vdots & \vdots & & \vdots \\ 0 & 0 & \cdots & -1 \end{bmatrix} A = \begin{bmatrix} 1 & 0 & \cdots & 0 \\ 0 & -1 & \cdots & 0 \\ \vdots & \vdots & & \vdots \\ 0 & 0 & \cdots & -1 \end{bmatrix} \right\}.$$

Furthermore, show that $O(1,n)$ consists of all matrices of the form

$$A = \begin{bmatrix} \mathbf{e}_0 & \cdots & \mathbf{e}_n \end{bmatrix},$$

where $(\mathbf{e}_0, \ldots, \mathbf{e}_n)$ is an orthonormal basis for $\mathbb{M}^{1,n}$. $O(1,n)$ is called the *Lorentz group*, and elements of $O(1,n)$ are called *Lorentz transformations.*

Remark 3.45. Unlike $O(n)$, which has two connected components, $O(1,n)$ has four connected components. The connected component that contains the identity matrix is denoted $SO^+(1,n)$, and it consists of those transformations in $O(1,n)$ that have determinant equal to 1 and map the vector $\underline{\mathbf{e}}_0$ to a future-pointing vector. These transformations are called *proper* (because they preserve the orientation of the spatial dimensions) and *orthochronous* (because they preserve the orientation of the time direction). $SO^+(1,n)$ is therefore referred to as the "proper, orthochronous Lorentz group" or, more succinctly, the "restricted Lorentz group". We will restrict our attention to the proper, orthochronous symmetries of $\mathbb{M}^{1,n}$, and for simplicity we will refer to elements of this group as *Lorentz transformations.*

***Exercise 3.46.** (a) Show that

$$SO^+(1,1) = \left\{ \begin{bmatrix} \cosh(\theta) & \sinh(\theta) \\ \sinh(\theta) & \cosh(\theta) \end{bmatrix} : \theta \in \mathbb{R} \right\}.$$

What are the other components of $O(1,1)$?

(b) Show that the light cone

$$\left\{ {}^t[x^0, x^1] \in \mathbb{M}^{1,1} \mid (x^0)^2 - (x^1)^2 = 0 \right\}$$

and the hyperbolas

$$\left\{ {}^t[x^0, x^1] \in \mathbb{M}^{1,1} \mid (x^0)^2 - (x^1)^2 = \pm r^2 \right\}$$

are each preserved under the action of $SO^+(1,1)$.

(c) Experiment with the action of $SO^+(1,1)$ on various shapes in the plane. For instance, consider the unit circle, parametrized by

$$\mathbf{x}(t) = {}^t[x^0(t), x^1(t)] = {}^t[\cos(t), \sin(t)].$$

Use MAPLE to plot the curve $A\mathbf{x}(t)$, where $A \in SO^+(1,1)$, for various values of the group parameter θ.

Exercise 3.47. Show that the Lie algebra $\mathfrak{so}(1,n)$ of $O(1,n)$ is given by

$$\mathfrak{so}(1,n) = \left\{ B \in M_{(n+1)\times(n+1)} : {}^tB \begin{bmatrix} 1 & 0 & \cdots & 0 \\ 0 & -1 & \cdots & 0 \\ \vdots & \vdots & & \vdots \\ 0 & 0 & \cdots & -1 \end{bmatrix} + \begin{bmatrix} 1 & 0 & \cdots & 0 \\ 0 & -1 & \cdots & 0 \\ \vdots & \vdots & & \vdots \\ 0 & 0 & \cdots & -1 \end{bmatrix} B = 0 \right\}.$$

In particular, this implies that the entries (b^α_β) of B satisfy the relations

$$b^0_0 = 0,$$
$$b^i_0 = b^0_i, \quad i = 1, \ldots, n,$$
$$b^j_i = -b^i_j, \quad i, j = 1, \ldots, n.$$

Since the Minkowski inner product is also preserved by translation, the full symmetry group of $\mathbb{M}^{1,n}$ consists of all transformations of the form

$$\varphi(\mathbf{x}) = A\mathbf{x} + \mathbf{b},$$

where $A \in SO^+(1,n)$ and $\mathbf{b} \in \mathbb{M}^{1,n}$. These transformations form a Lie group, called the *Poincaré group* $M(1,n)$, which can be represented as

$$M(1,n) = \left\{ \begin{bmatrix} 1 & {}^t\mathbf{0} \\ \mathbf{b} & A \end{bmatrix} : A \in SO^+(1,n),\ \mathbf{b} \in \mathbb{M}^{1,n} \right\}.$$

As in the Euclidean case, if we represent a vector $\mathbf{x} \in \mathbb{M}^{1,n}$ by the $(n+2)$-dimensional vector $\begin{bmatrix} 1 \\ \mathbf{x} \end{bmatrix}$, then elements of $M(1,n)$ act by matrix multiplication:

$$\begin{bmatrix} 1 & {}^t\mathbf{0} \\ \mathbf{b} & A \end{bmatrix} \begin{bmatrix} 1 \\ \mathbf{x} \end{bmatrix} = \begin{bmatrix} 1 \\ A\mathbf{x} + \mathbf{b} \end{bmatrix}.$$

Once again, we ask: Given a vector $\mathbf{x} \in \mathbb{M}^{1,n}$, what is the isotropy group of $\mathbf{x}$?

***Exercise 3.48.** Show that:

(a) The isotropy group $H_\mathbf{0}$ of $\mathbf{0} \in \mathbb{M}^{1,n}$ is

$$H_\mathbf{0} = \left\{ \begin{bmatrix} 1 & {}^t\mathbf{0} \\ \mathbf{0} & A \end{bmatrix} : A \in SO^+(1,n) \right\} \cong SO^+(1,n).$$

(b) The isotropy group $H_{\mathbf{x}}$ of any other point $\mathbf{x} \in \mathbb{M}^{1,n}$ is

$$H_{\mathbf{x}} = t_{\mathbf{x}} H_{\mathbf{0}} t_{\mathbf{x}}^{-1},$$

where $t_{\mathbf{x}} = \begin{bmatrix} 1 & {}^t\mathbf{0} \\ \mathbf{x} & I \end{bmatrix}$ represents the translation $t_{\mathbf{x}}(\mathbf{y}) = \mathbf{y} + \mathbf{x}$.

(c) There is a natural correspondence between $\mathbb{M}^{1,n}$ and $M(1,n)/SO^+(1,n)$, the set of left cosets of $SO^+(1,n)$ in $M(1,n)$.

3.5.3. Orthonormal frames and Maurer-Cartan forms. Orthonormal frames on $\mathbb{M}^{1,n}$ are defined as follows.

Definition 3.49. An *orthonormal frame* $\mathbf{f}$ on $\mathbb{M}^{1,n}$ is a list of vectors $\mathbf{f} = (\mathbf{x}; \mathbf{e}_0, \ldots, \mathbf{e}_n)$, where $\mathbf{x} \in \mathbb{M}^{1,n}$ and $(\mathbf{e}_0, \ldots, \mathbf{e}_n)$ is an orthonormal basis for the tangent space $T_{\mathbf{x}}\mathbb{M}^{1,n}$. (We may also say that $(\mathbf{e}_0, \ldots, \mathbf{e}_n)$ is an *orthonormal frame based at* $\mathbf{x}$.)

The same reasoning as in the Euclidean case shows that the set of orthonormal frames may be regarded as the Poincaré group $M(1,n)$ via the one-to-one correspondence

$$g(\mathbf{x}; \mathbf{e}_0, \ldots, \mathbf{e}_n) = \begin{bmatrix} 1 & 0 & \cdots & 0 \\ \mathbf{x} & \mathbf{e}_0 & \cdots & \mathbf{e}_n \end{bmatrix}.$$

We can define a projection map $\pi : M(1,n) \to \mathbb{M}^{1,n}$ by

$$\pi(\mathbf{x}; \mathbf{e}_0, \ldots, \mathbf{e}_n) = \mathbf{x};$$

the fiber of this map is the set of all orthonormal frames based at $\mathbf{x}$. $SO^+(1,n)$ acts freely and transitively on each fiber, and so this map gives an explicit description of $M(1,n)$ as a principal bundle over $\mathbb{M}^{1,n}$ with fiber group $SO^+(1,n)$:

$$\begin{array}{ccc} SO^+(1,n) & \longrightarrow & M(1,n) \\ & & \big\downarrow \pi \\ & & \mathbb{M}^{1,n} \cong M(1,n)/SO^+(1,n). \end{array}$$

In this context, $M(1,n)$ is also called the *orthonormal frame bundle* of $\mathbb{M}^{1,n}$, and it is denoted $\mathcal{F}(\mathbb{M}^{1,n})$.

The Maurer-Cartan forms on $M(1,n)$ are defined exactly as in the Euclidean case by equations (3.1). The structure equations (3.8) are also the same as in the Euclidean case; the only difference is that with the indefinite inner product, the connection forms (ω^α_β) are no longer *quite* skew-symmetric.

***Exercise 3.50.** Differentiate equations (3.14) and conclude that

(1) $\omega_0^0 = 0$;

(2) $\omega_0^i = \omega_i^0$ for $i = 1, \ldots, n$;

(3) $\omega_i^j = -\omega_j^i$ for $i, j = 1, \ldots, n$.

***Exercise 3.51.** Prove that the Lie algebra of $M(1, n)$ is

$$\mathfrak{m}(1, n) = \left\{ \begin{bmatrix} 0 & {}^t\mathbf{0} \\ \mathbf{b} & B \end{bmatrix} : B \in \mathfrak{so}(1, n), \ \mathbf{b} \in \mathbb{M}^{1,n} \right\}.$$

Exercise 3.52. Repeat Exercise 3.23 for $\mathbb{M}^{1,2}$.

3.6. Equi-affine space

3.6.1. Volume forms. Affine geometry is the study of geometric properties that are preserved by the action of invertible linear transformations and translations on $\mathbb{R}^n$. There is less structure in affine geometry than in Euclidean or Minkowski geometry; there are, for instance, no invariant notions of lengths of vectors or angles between vectors. Affine transformations do, however, preserve collinearity of points, and the notion of parallel lines is still well-defined in affine geometry.

In this book, we will study *equi-affine geometry* (also known as *special affine geometry*), which has slightly more structure than general affine geometry, in that we endow $\mathbb{R}^n$ with a *volume form*.

Definition 3.53. A *volume form* on $\mathbb{R}^n$ is a nonzero element dV of the 1-dimensional vector space $\Lambda^n(\mathbb{R}^n)^*$. Equivalently, dV is a skew-symmetric, multilinear function

$$dV : \overbrace{\mathbb{R}^n \times \cdots \times \mathbb{R}^n}^{n \text{ copies}} \to \mathbb{R}.$$

A volume form provides a way to measure the volume of the parallelepiped spanned by any n vectors $(\mathbf{v}_1, \ldots, \mathbf{v}_n)$. It does *not* provide a way to measure lengths of individual vectors or angles between vectors.

The following exercise shows that all volume forms on $\mathbb{R}^n$ are equivalent.

***Exercise 3.54.** Let dV_0 be the standard volume form on $\mathbb{R}^n$, defined by the property that

$$dV_0(\underline{\mathbf{e}}_1, \ldots, \underline{\mathbf{e}}_n) = 1,$$

where $(\underline{\mathbf{e}}_1, \ldots, \underline{\mathbf{e}}_n)$ is the standard basis for $\mathbb{R}^n$.

(a) Show that for any n vectors $(\mathbf{v}_1, \dots, \mathbf{v}_n) \in \mathbb{R}^n$,

$$dV_0(\mathbf{v}_1, \dots, \mathbf{v}_n) = \det \begin{bmatrix} \mathbf{v}_1 & \cdots & \mathbf{v}_n \end{bmatrix}.$$

(b) Let dV be any other volume form on $\mathbb{R}^n$. Show that

$$dV = \lambda\, dV_0$$

for some nonzero real number λ. (Hint: Let $\lambda = dV(\underline{\mathbf{e}}_1, \dots, \underline{\mathbf{e}}_n)$.)

(c) Let $(\mathbf{v}_1, \dots, \mathbf{v}_n), (\mathbf{w}_1, \dots, \mathbf{w}_n)$ be two bases for $\mathbb{R}^n$, and let dV be a volume form on $\mathbb{R}^n$. Show that

$$dV(\mathbf{v}_1, \dots, \mathbf{v}_n) = dV(\mathbf{w}_1, \dots, \mathbf{w}_n)$$

if and only if

$$\begin{bmatrix} \mathbf{w}_1 & \cdots & \mathbf{w}_n \end{bmatrix} = A \begin{bmatrix} \mathbf{v}_1 & \cdots & \mathbf{v}_n \end{bmatrix}$$

for some matrix $A \in SL(n)$. (Hint: Consider the matrix

$$\begin{bmatrix} \mathbf{w}_1 & \cdots & \mathbf{w}_n \end{bmatrix} \begin{bmatrix} \mathbf{v}_1 & \cdots & \mathbf{v}_n \end{bmatrix}^{-1}.)$$

(d) Let $dV = \lambda\, dV_0$. Show that the isomorphism $\phi : \mathbb{R}^n \to \mathbb{R}^n$ defined by

$$\phi(\mathbf{v}) = \lambda^{(1/n)}\mathbf{v}$$

identifies dV with the standard volume form dV_0, in the sense that

$$dV(\mathbf{v}_1, \dots, \mathbf{v}_n) = dV_0(\phi(\mathbf{v}_1), \dots, \phi(\mathbf{v}_n)).$$

(If $\lambda < 0$ and n is even, first perform a linear transformation that interchanges $\underline{\mathbf{e}}_1$ and $\underline{\mathbf{e}}_2$; this will have the effect of reversing the sign of λ.) Therefore, all volume forms on $\mathbb{R}^n$ are equivalent.

Definition 3.55. The vector space $\mathbb{R}^n$ endowed with a volume form dV is called *equi-affine space* and is denoted $\mathbb{A}^n$.

Remark 3.56. The volume form is not really part of the definition of equi-affine space; where it comes into play is when we decide what transformations to allow in our symmetry group. *Affine transformations* are those that preserve the linear structure of $\mathbb{A}^n$; *equi-affine* (or *special affine*) *transformations* are those that preserve the volume form as well. It is most common—although not universal—to study geometric properties that are preserved under the group of equi-affine transformations, and this is the approach that we will take. It is fairly common to use the terms "affine space" and "affine transformations" to refer to the equi-affine versions, but we will attempt to resist the temptation to do so.

By Exercise 3.54, we may assume without loss of generality that dV is the standard volume form on $\mathbb{R}^n$, so that for any n vectors $(\mathbf{v}_1, \dots, \mathbf{v}_n) \in \mathbb{R}^n$,

$$dV(\mathbf{v}_1, \dots, \mathbf{v}_n) = \det \begin{bmatrix} \mathbf{v}_1 & \cdots & \mathbf{v}_n \end{bmatrix}.$$

3.6.2. Symmetries and isotropy groups. Now we ask: What are the symmetries of equi-affine space? (It turns out that "orientation-preserving" is an unneccesary stipulation here, because any volume-preserving transformation necessarily preserves orientation.) Since equi-affine space has less structure that must be preserved by a symmetry than Euclidean space does, we might expect that the symmetry group would be larger.

***Exercise 3.57.** Let $A \in GL(n)$. Show that for any basis $(\mathbf{e}_1, \ldots, \mathbf{e}_n)$ for $\mathbb{A}^n$, we have

$$dV(A\mathbf{e}_1, \ldots, A\mathbf{e}_n) = (\det A)\, dV(\mathbf{e}_1, \ldots, \mathbf{e}_n).$$

Therefore, $dV(A\mathbf{e}_1, \ldots, A\mathbf{e}_n) = dV(\mathbf{e}_1, \ldots, \mathbf{e}_n)$ if and only if $A \in SL(n)$.

Since the volume form is also preserved by translation, the full symmetry group of $\mathbb{A}^n$ consists of all transformations $\varphi : \mathbb{A}^n \to \mathbb{A}^n$ of the form

$$\varphi(\mathbf{x}) = A\mathbf{x} + \mathbf{b},$$

where $A \in SL(n)$ and $\mathbf{b} \in \mathbb{A}^n$. These transformations form a Lie group, called the *equi-affine group* or *special affine group* $A(n)$, which can be represented as

$$A(n) = \left\{ \begin{bmatrix} 1 & {}^t\mathbf{0} \\ \mathbf{b} & A \end{bmatrix} : A \in SL(n),\ \mathbf{b} \in \mathbb{A}^n \right\}.$$

Now, given a vector $\mathbf{x} \in \mathbb{A}^n$, what is the isotropy group of $\mathbf{x}$?

***Exercise 3.58.** Show that:

(a) The isotropy group $H_{\mathbf{0}}$ of $\mathbf{0} \in \mathbb{A}^n$ is

$$H_{\mathbf{0}} = \left\{ \begin{bmatrix} 1 & {}^t\mathbf{0} \\ \mathbf{0} & A \end{bmatrix} : A \in SL(n) \right\} \cong SL(n).$$

(b) The isotropy group $H_{\mathbf{x}}$ of any other point $\mathbf{x} \in \mathbb{A}^n$ is

$$H_{\mathbf{x}} = t_{\mathbf{x}} H_{\mathbf{0}} t_{\mathbf{x}}^{-1},$$

where $t_{\mathbf{x}} = \begin{bmatrix} 1 & {}^t\mathbf{0} \\ \mathbf{x} & I \end{bmatrix}$ represents the translation $t_{\mathbf{x}}(\mathbf{y}) = \mathbf{y} + \mathbf{x}$.

(c) There is a natural correspondence between $\mathbb{A}^n$ and $A(n)/SL(n)$, the set of left cosets of $SL(n)$ in $A(n)$.

3.6.3. Unimodular frames and Maurer-Cartan forms. Unimodular frames on $\mathbb{A}^n$ are defined as follows:

Definition 3.59. A *unimodular frame* $\mathbf{f}$ on $\mathbb{A}^n$ is a list of vectors $\mathbf{f} = (\mathbf{x}; \mathbf{e}_1, \ldots, \mathbf{e}_n)$, where $\mathbf{x} \in \mathbb{A}^n$ and $(\mathbf{e}_1, \ldots, \mathbf{e}_n)$ is a basis for the tangent space $T_{\mathbf{x}}\mathbb{A}^n$ that spans a parallelepiped of volume 1. (We may also say that $(\mathbf{e}_1, \ldots, \mathbf{e}_n)$ is a *unimodular frame based at* $\mathbf{x}$.)

The same reasoning as in the Euclidean case shows that the set of unimodular frames may be regarded as the equi-affine group $A(n)$ via the one-to-one correspondence

$$g(\mathbf{x}; \mathbf{e}_1, \ldots, \mathbf{e}_n) = \begin{bmatrix} 1 & 0 & \cdots & 0 \\ \mathbf{x} & \mathbf{e}_1 & \cdots & \mathbf{e}_n \end{bmatrix}.$$

We can define a projection map $\pi : A(n) \to \mathbb{A}^n$ by

$$\pi(\mathbf{x}; \mathbf{e}_1, \ldots, \mathbf{e}_n) = \mathbf{x};$$

the fiber of this map is the set of all unimodular frames based at $\mathbf{x}$. $SL(n)$ acts freely and transitively on each fiber, and so this map gives an explicit description of $A(n)$ as a principal bundle over $\mathbb{A}^n$ with fiber group $SL(n)$:

$$\begin{array}{ccc} SL(n) & \longrightarrow & A(n) \\ & & \Big\downarrow \pi \\ & & \mathbb{A}^n \cong A(n)/SL(n). \end{array}$$

In this context, $A(n)$ is also called the *unimodular frame bundle* of $\mathbb{A}^n$, and it is denoted $\mathcal{F}(\mathbb{A}^n)$.

Just as for Euclidean and Minkowski spaces, the Maurer-Cartan forms on $A(n)$ are defined by equations (3.1). The structure equations (3.8) are also the same as before, but without any sort of inner product structure, there is no symmetry or skew-symmetry condition on the connection forms (ω^i_j). There is, however, one relation among the (ω^i_j) that comes from the unimodular condition

$$\det \begin{bmatrix} \mathbf{e}_1 & \cdots & \mathbf{e}_n \end{bmatrix} = 1.$$

We can differentiate this condition more easily if we express it as

$$\mathbf{e}_1 \wedge \cdots \wedge \mathbf{e}_n = \underline{\mathbf{e}}_1 \wedge \cdots \wedge \underline{\mathbf{e}}_n, \tag{3.15}$$

where $(\underline{\mathbf{e}}_1, \ldots, \underline{\mathbf{e}}_n)$ denotes the standard basis for $\mathbb{A}^n$. (Recall that wedge product is simply the skew-symmetrization of the tensor product.)

Exercise 3.60. Show that equation (3.15) holds for any unimodular basis by showing that for *any* basis $(\mathbf{e}_1, \ldots, \mathbf{e}_n)$ of $\mathbb{A}^n$, unimodular or not, we have

$$\mathbf{e}_1 \wedge \cdots \wedge \mathbf{e}_n = \left(\det \begin{bmatrix} \mathbf{e}_1 & \cdots & \mathbf{e}_n \end{bmatrix}\right) \underline{\mathbf{e}}_1 \wedge \cdots \wedge \underline{\mathbf{e}}_n.$$

***Exercise 3.61.** Differentiate equation (3.15) and conclude that the connection forms (ω^i_j) satisfy the trace condition $\omega^i_i = 0$. (Hint: Don't let the wedge product confuse you: $(\mathbf{e}_1, \ldots, \mathbf{e}_n)$ are vector-valued 0-forms, so there are no minus signs in the Leibniz rule. Also, note that the right-hand side of (3.15) is constant.)

***Exercise 3.62.** Prove that the Lie algebra of the $A(n)$ is

$$\mathfrak{a}(n) = \left\{ \begin{bmatrix} 0 & {}^t\mathbf{0} \\ \mathbf{b} & B \end{bmatrix} : B \in \mathfrak{sl}(n),\ \mathbf{b} \in \mathbb{A}^n \right\}.$$

(Recall that $B \in \mathfrak{sl}(n)$ if and only if $\operatorname{tr}(B) = 0$.)

Exercise 3.63. Repeat Exercise 3.23 for $\mathbb{A}^3$.

3.7. Projective space

3.7.1. The structure of projective space. Projective space is related to the study of *perspective*—for instance, how a 3-dimensional object appears when projected onto a 2-dimensional image. All points in 3-dimensional space that lie on a line intersecting a common focal point (such as the lens of a camera) are projected onto the same point in the 2-dimensional image. In this geometry, lines that are parallel in 3-dimensional space may appear to intersect "at infinity", but this intersection point may appear at some finite distance in the 2-dimensional projection. *Projective transformations* of 2-dimensional space are the result of changing the focal point in 3-dimensional space. Projective transformations are more general than affine transformations; among other things, they can map points at a finite distance from the origin to points "at infinity", and vice versa. Because of this, n-dimensional projective space must be constructed so as to include not only the space $\mathbb{R}^n$ of finite-distance points, but also a point "at infinity" for every direction in $\mathbb{R}^n$.

Recall the definition from Example 1.9:

Definition 3.64. The n-dimensional (real) *projective space* $\mathbb{P}^n$ is the set of lines through the origin in the vector space $\mathbb{R}^{n+1}$.

In order to see that this captures the idea of "$\mathbb{R}^n$ plus points at infinity", we will use homogeneous coordinates $[x^0 : \cdots : x^n]$ (cf. Example 1.9) to represent a point in $\mathbb{P}^n$. Consider the dense, open subset V_0 of $\mathbb{P}^n$ given by

$$V_0 = \{[x^0 : \cdots : x^n] \in \mathbb{P}^n \mid x^0 \neq 0\}.$$

Any point $[\mathbf{x}] \in V_0$ can be written as

$$[\mathbf{x}] = [x^0 : \cdots : x^n] = \left[1 : \frac{x^1}{x^0} : \cdots : \frac{x^n}{x^0}\right]$$

and so may be represented by the point $\left(\frac{x^1}{x^0}, \ldots, \frac{x^n}{x^0}\right) \in \mathbb{R}^n$. We will denote this point by $\bar{\mathbf{x}} = (\bar{x}^1, \ldots, \bar{x}^n)$; these are referred to as *affine coordinates* on the open set V_0. We can visualize this by observing that any line in $\mathbb{R}^{n+1}$ that passes through the origin and another point $\mathbf{x}$ with $x^0 \neq 0$ intersects the plane $x^0 = 1$ in exactly one point. (See Figure 3.2.)

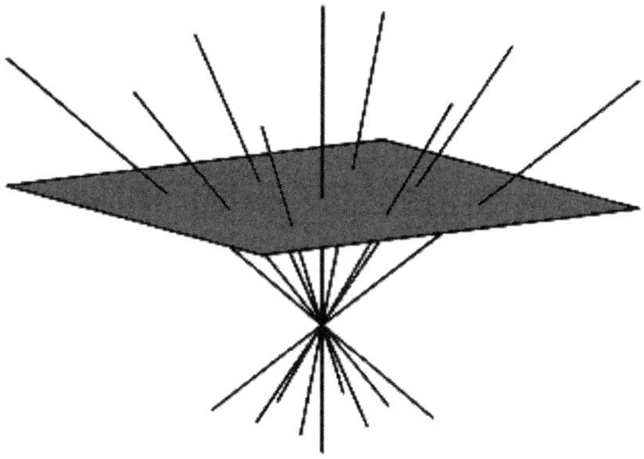

Figure 3.2. $V_0 \subset \mathbb{P}^n$

On the other hand, any point $[\mathbf{x}] \in \mathbb{P}^n \setminus V_0$ has the form

$$[\mathbf{x}] = [0 : \bar{x}^1 : \cdots : \bar{x}^n].$$

This point occurs as a limit point of the line

$$[\mathbf{x}(t)] = [1 : t\bar{x}^1 : \cdots : t\bar{x}^n]$$

in V_0 as $t \to \pm\infty$ and so may be thought of as a "point at infinity" corresponding to the direction of the vector $\bar{\mathbf{x}} = {}^t[\bar{x}^1, \ldots, \bar{x}^n] \in \mathbb{R}^n$.

***Exercise 3.65.** Show that the set $\mathbb{P}^n \setminus V_0$ may be naturally identified with $\mathbb{P}^{n-1}$. So inductively, we can think of $\mathbb{P}^n$ as

$$\mathbb{P}^n = \mathbb{R}^n \cup \mathbb{R}^{n-1} \cup \cdots \cup \mathbb{R} \cup \{0\},$$

where $\mathbb{P}^0 = \{0\}$ represents a single point.

3.7.2. Symmetries and isotropy groups.

The symmetries of projective space are called *projective transformations*. Based on Definition 3.64 of $\mathbb{P}^n$ as the set of lines through the origin in $\mathbb{R}^{n+1}$, the symmetry group of $\mathbb{P}^n$ is simply the group of linear transformations of $\mathbb{R}^{n+1}$, i.e, $GL(n+1)$. Note that every element of $GL(n+1)$ maps any line through the origin to another line through the origin; thus the action of $GL(n+1)$ on $\mathbb{R}^{n+1}$ gives a well-defined action on $\mathbb{P}^n$.

However, there is a slight catch. While it is true that all elements of $GL(n+1)$ are symmetries of $\mathbb{P}^n$, some of them act *trivially* on $\mathbb{P}^n$. A matrix

$A \in GL(n+1)$ fixes every line in $\mathbb{R}^{n+1}$—and therefore acts as the identity transformation on $\mathbb{P}^n$—if and only if $A = \lambda I$ for some $\lambda \neq 0$. Moreover, any two matrices $A, B \in GL(n+1)$ induce the same transformation on $\mathbb{P}^n$ if and only if $A = \lambda B$ for some $\lambda \neq 0$. Therefore, the most natural choice for the symmetry group of $\mathbb{P}^n$ is the quotient group $GL(n+1)/\sim$, where $A \sim B$ if and only if $A = \lambda B$ for some $\lambda \neq 0$.

***Exercise 3.66.** Show that $GL(m)/\sim$ is isomorphic to

(1) $SL(m)$ if m is odd;

(2) a semidirect product $(SL(m)/\{\pm I\}) \rtimes \{\pm 1\}$ if m is even. (The quotient of $SL(m)$ by $\{\pm I\}$ reflects the fact that $-I$ is an element of $SL(m)$, and it acts trivially on $\mathbb{P}^n$. The semidirect product with $\{\pm 1\}$ reflects the fact that the sign of the determinant is fixed under scaling, and so this group has two components: one whose elements have determinant 1 and one whose elements have determinant -1.)

Remark 3.67. The group $GL(m)/\sim$ is called the *projective general linear group*, denoted $PGL(m)$; similarly, the group $SL(m)/\sim$ is called the *projective special linear group*, denoted $PSL(m)$. When m is odd, these two groups are identical and isomorphic to $SL(m)$; when m is even, $PGL(m)$ is the group $(SL(m)/\{\pm I\}) \rtimes \{\pm 1\}$ of Exercise 3.66, while $PSL(m)$ is the proper subgroup $SL(m)/\{\pm I\}$ of $PGL(m)$.

In order to keep things as simple as possible, we will generally take the symmetry group of $\mathbb{P}^n$ to be $SL(n+1)$ regardless of whether n is odd or even. In the odd case, this means that we will restrict to the identity component of the symmetry group (as we did with orientation-preserving transformations in the Euclidean case) and that our choices of frames on $\mathbb{P}^n$ will be determined only up to sign.

Now, it is not at all obvious that this symmetry group has much to do with the idea of projective transformations as representing a change in perspective. The following exercise will illustrate how this works in the case of $\mathbb{P}^1$.

***Exercise 3.68.** Consider the plane $\mathbb{R}^2$ with coordinates $(\bar{x}, \bar{y})$, and let the $\bar{x}$-axis represent the open set

$$V_0 = \{[x^0 : x^1] \in \mathbb{P}^1 \mid x^0 \neq 0\}$$

via the identification $\bar{x} = \frac{x^1}{x^0}$. Suppose that an object lies along the line L with equation $\bar{y} = m\bar{x} + b$ and that two viewers located at points $\mathbf{p} = (p^1, p^2)$

and $\mathbf{q} = (q^1, q^2)$ see the object as if it were projected from their respective viewpoints onto the $\bar{x}$-axis. (See Figure 3.3.) Observer $\mathbf{p}$ sees the point $\mathbf{r}$ as

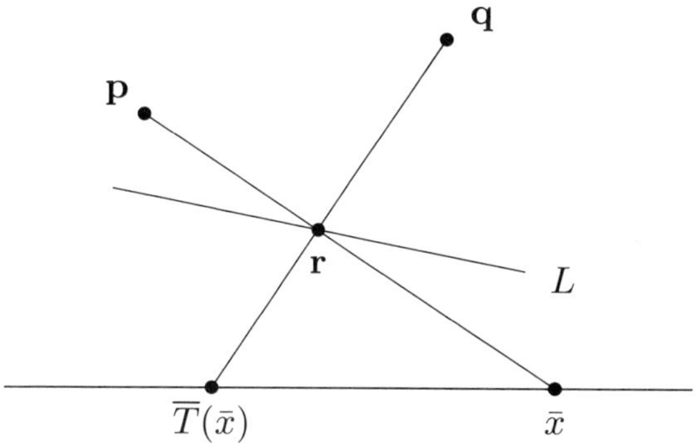

Figure 3.3. Projective transformation of $\mathbb{P}^1$

lying at $(\bar{x}, 0)$, and observer $\mathbf{q}$ sees $\mathbf{r}$ as lying at $(\overline{T}(\bar{x}), 0)$. Use the following steps to compute the transformation $\overline{T}(\bar{x})$:

(a) Show that the point $\mathbf{r}$ of intersection between L and the line joining the point $(\bar{x}, 0)$ to $\mathbf{p}$ has coordinates

$$\mathbf{r} = \left(\frac{(p^2 - b)\bar{x} + bp^1}{m\bar{x} + p^2 - mp^1}, \frac{mp^2\bar{x} + bp^2}{m\bar{x} + p^2 - mp^1} \right).$$

(b) Show that the line joining $\mathbf{q}$ to $\mathbf{r}$ intersects the $\bar{x}$-axis at the point $(\overline{T}(\bar{x}), 0)$, where

$$\overline{T}(\bar{x}) = \frac{\alpha\bar{x} + \beta}{\gamma\bar{x} + \delta} \tag{3.16}$$

for some constants $\alpha, \beta, \gamma, \delta$ with $\alpha\delta - \beta\gamma \neq 0$. (You can keep up with the precise constants if you want, but they're kind of a mess!) The map (3.16) is called a *linear fractional transformation.*

(c) For what value(s) of $\bar{x}$ would it make sense to define $\overline{T}(\bar{x}) = \infty$? What value would you assign to $\overline{T}(\infty)$? Can you see how to interpret these assignments in terms of the observers in Figure 3.3?

(d) Recall that $\bar{x} = \frac{x^1}{x^0}$, where $[x^0 : x^1]$ are homogeneous coordinates on $\mathbb{P}^1$. Show that $\overline{T}$ corresponds to the map

$$T([x^0 : x^1]) = [\delta x^0 + \gamma x^1 : \beta x^0 + \alpha x^1]$$

on $\mathbb{P}^1$, which in turn corresponds to the linear transformation

$$\widetilde{T}\left(\begin{bmatrix} x^0 \\ x^1 \end{bmatrix}\right) = \begin{bmatrix} \delta & \gamma \\ \beta & \alpha \end{bmatrix} \begin{bmatrix} x^0 \\ x^1 \end{bmatrix}$$

on $\mathbb{R}^2$, where $\begin{bmatrix} \delta & \gamma \\ \beta & \alpha \end{bmatrix} \in GL(2)$.

(e) Show that the matrix $\begin{bmatrix} \delta & \gamma \\ \beta & \alpha \end{bmatrix}$ can be modified so as to have determinant equal to ± 1 without changing the transformation T. Thus, T can be regarded as an element of the group $PGL(2)$.

***Exercise 3.69.** Let $V_0 \subset \mathbb{P}^2$ be the open set

$$V_0 = \{[x^0 : x^1 : x^2] \in \mathbb{P}^2 \mid x^0 \neq 0\},$$

with affine coordinates

$$(\bar{x}^1, \bar{x}^2) = \left(\frac{x^1}{x^0}, \frac{x^2}{x^0}\right).$$

(a) Let

$$A = \begin{bmatrix} a_0^0 & a_1^0 & a_2^0 \\ a_0^1 & a_1^1 & a_2^1 \\ a_0^2 & a_1^2 & a_2^2 \end{bmatrix} \in SL(3).$$

Show that the transformation

$$T([\mathbf{x}]) = [A\mathbf{x}]$$

corresponds to the map

$$\overline{T}(\bar{x}^1, \bar{x}^2) = \left(\frac{a_0^1 + a_1^1\bar{x}^1 + a_2^1\bar{x}^2}{a_0^0 + a_1^0\bar{x}^1 + a_2^0\bar{x}^2}, \frac{a_0^2 + a_1^2\bar{x}^1 + a_2^2\bar{x}^2}{a_0^0 + a_1^0\bar{x}^1 + a_2^0\bar{x}^2}\right)$$

on V_0.

(b) Show that if A has the form

$$A = \begin{bmatrix} 1 & 0 & 0 \\ b^1 & a_1^1 & a_2^1 \\ b^2 & a_1^2 & a_2^2 \end{bmatrix}$$

with $a_1^1 a_2^2 - a_2^1 a_1^2 = 1$, then $\overline{T} : V_0 \to V_0$ is the equi-affine transformation

$$\overline{T}(\bar{\mathbf{x}}) = \bar{A}\bar{\mathbf{x}} + \bar{\mathbf{b}},$$

where $\bar{A} = \begin{bmatrix} a_1^1 & a_2^1 \\ a_1^2 & a_2^2 \end{bmatrix}, \bar{\mathbf{b}} = \begin{bmatrix} b^1 \\ b^2 \end{bmatrix}$. Therefore, the equi-affine group $A(2)$ is a proper subgroup of the group of projective transformations.

(c) (MAPLE recommended) Consider the image of the unit circle

$$\{(\bar{x}^1, \bar{x}^2) \in V_0 \mid (\bar{x}^1)^2 + (\bar{x}^2)^2 = 1\}$$

in V_0 under the transformation corresponding to the matrix

$$A = \begin{bmatrix} \cos(\varphi) & -\sin(\varphi) & 0 \\ \sin(\varphi) & \cos(\varphi) & 0 \\ 0 & 0 & 1 \end{bmatrix}.$$

Show that:

(1) If $0 < \varphi < \frac{\pi}{4}$, then the image of the circle is an ellipse in V_0.

(2) If $\varphi = \frac{\pi}{4}$, then the image of the circle is a parabola in V_0.

(3) If $\frac{\pi}{4} < \varphi \leq \frac{\pi}{2}$, then the image of the circle is a hyperbola in V_0.

Thus, projective transformations can transform any nondegenerate conic section into any other nondegenerate conic section. (However, they do preserve the *family* of nondegenerate conic sections.)

Now, given a point $[\mathbf{x}] \in \mathbb{P}^n$, what is the isotropy group of $[\mathbf{x}]$?

***Exercise 3.70.** Let $[\mathbf{x}_0] = [1 : 0 : \cdots : 0] \in \mathbb{P}^n$. Show that:

(a) The isotropy group $H_{[\mathbf{x}_0]}$ of $[\mathbf{x}_0]$ in $SL(n+1)$ is

$$H_{[\mathbf{x}_0]} = \left\{ \begin{bmatrix} (\det A)^{-1} & r_1 & \cdots & r_n \\ 0 & & & \\ \vdots & & A & \\ 0 & & & \end{bmatrix} : A \in GL(n), r_1, \ldots, r_n \in \mathbb{R} \right\}. \tag{3.17}$$

(b) The isotropy group $H_{[\mathbf{x}]}$ of any other point $[\mathbf{x}] \in \mathbb{P}^n$ is

$$H_{[\mathbf{x}]} = t_{[\mathbf{x}]} H_{[\mathbf{x}_0]} t_{[\mathbf{x}]}^{-1},$$

where $t_{[\mathbf{x}]}$ is any matrix in $SL(n+1)$ whose first column is $[\mathbf{x}]$. ($t_{[\mathbf{x}]}$ will then have the property that $t_{[\mathbf{x}]}([\mathbf{x}_0]) = [\mathbf{x}]$.)

(c) There is a natural correspondence between $\mathbb{P}^n$ and $SL(n+1)/H_{[\mathbf{x}_0]}$, the set of left cosets of $H_{[\mathbf{x}_0]}$ in $SL(n+1)$.

3.7.3. Projective frames and Maurer-Cartan forms.

Projective frames on $\mathbb{P}^n$ are defined as follows:

Definition 3.71. A *projective frame* $\mathbf{f}$ on $\mathbb{P}^n$ is a list of vectors $\mathbf{f} = (\mathbf{e}_0, \ldots, \mathbf{e}_n)$, where $\mathbf{e}_\alpha \in \mathbb{R}^{n+1}$ and $\det \begin{bmatrix} \mathbf{e}_0 & \cdots & \mathbf{e}_n \end{bmatrix} = 1$. We identify $[\mathbf{e}_0]$ with the position vector $[\mathbf{x}] \in \mathbb{P}^n$, and $(\mathbf{e}_1, \ldots, \mathbf{e}_n)$ may be regarded as a basis for the tangent space $T_{[\mathbf{e}_0]}\mathbb{P}^n$. (We may also say that $(\mathbf{e}_0, \ldots, \mathbf{e}_n)$ is a *projective frame based at* $[\mathbf{e}_0]$.) Note that $\mathbf{e}_0$ needs to be part of the frame since it is not uniquely determined by its equivalence class $[\mathbf{e}_0]$ and choosing a different representative for $[\mathbf{e}_0]$ would affect the scaling of the vectors $(\mathbf{e}_1, \ldots, \mathbf{e}_n)$.

The same reasoning as in the Euclidean case shows that the set of projective frames may be regarded as the group $SL(n+1)$ via the one-to-one correspondence

$$g(\mathbf{e}_0, \ldots, \mathbf{e}_n) = \begin{bmatrix} \mathbf{e}_0 & \cdots & \mathbf{e}_n \end{bmatrix}.$$

We can define a projection map $\pi : SL(n+1) \to \mathbb{P}^n$ by

$$\pi(\mathbf{e}_0, \ldots, \mathbf{e}_n) = [\mathbf{e}_0];$$

the fiber of this map is the set of all projective frames based at $[\mathbf{e}_0]$. $H_{[\mathbf{x}_0]}$ acts freely and transitively on each fiber, and so this map gives an explicit description of $SL(n+1)$ as a principal bundle over $\mathbb{P}^n$ with fiber group $H_{[\mathbf{x}_0]}$:

$$\begin{array}{ccc} H_{[\mathbf{x}_0]} & \longrightarrow & SL(n+1) \\ & & \big\downarrow \pi \\ & & \mathbb{P}^n \cong SL(n+1)/H_{[\mathbf{x}_0]}. \end{array}$$

In this context, $SL(n+1)$ is also called the *projective frame bundle* of $\mathbb{P}^n$, and it is denoted $\mathcal{F}(\mathbb{P}^n)$.

The Maurer-Cartan forms on $SL(n+1)$ are defined by the equations

$$d\mathbf{e}_\alpha = \mathbf{e}_\beta\, \omega^\beta_\alpha, \tag{3.18}$$

where $0 \leq \alpha, \beta \leq n$.

The structure equations are

$$d\omega^\beta_\alpha = -\omega^\beta_\gamma \wedge \omega^\gamma_\alpha, \tag{3.19}$$

where the sum is over $0 \leq \gamma \leq n$. Since the Maurer-Cartan form $\omega = [\omega^\beta_\alpha]$ takes values in the Lie algebra $\mathfrak{sl}(n+1)$, the only relation among the 1-forms (ω^β_α) is the trace condition

$$\omega^\alpha_\alpha = 0.$$

As for which forms are semi-basic for the projection $\pi : SL(n+1) \to \mathbb{P}^n$ and which should be regarded as connection forms, recall that the semi-basic forms are those that pull back to each fiber of π to be zero, i.e., those with the property that $\omega^\beta_\alpha(\mathbf{v}) = 0$ whenever $d[\mathbf{e}_0](\mathbf{v}) = 0$. But $d[\mathbf{e}_0](\mathbf{v}) = 0$ if and only if $d\mathbf{e}_0(\mathbf{v})$ is a multiple of $\mathbf{e}_0$. Therefore, $d[\mathbf{e}_0](\mathbf{v}) = 0$ if and only if

$$\omega_0^1(\mathbf{v}) = \cdots = \omega_0^n(\mathbf{v}) = 0,$$

and so the semi-basic forms for π are $(\omega_0^1, \ldots, \omega_0^n)$. The remaining (ω_α^β) are the connection forms.

Exercise 3.72. Repeat Exercise 3.23 for $\mathbb{P}^3$.

3.8. MAPLE computations

In this section, we will explore how the **Cartan** package may be used to facilitate some of the computations involved in the exercises in this chapter. While most of these computations are simple enough to be done by hand, seeing how to do them in MAPLE will help you to familiarize yourself with how the package works.

You should begin by loading the **Cartan** package into MAPLE, and the **LinearAlgebra** package will be useful as well:

```
> with(Cartan);
> with(LinearAlgebra);
```

EXERCISE 3.32: Define A to be the matrix whose columns are the frame field vectors:

```
> A:= Matrix([
  [cos(phi)*cos(theta), -sin(theta), -sin(phi)*cos(theta)],
  [cos(phi)*sin(theta), cos(theta), -sin(phi)*sin(theta)],
  [sin(phi),0,cos(phi)]
  ]);
```

Let $d\mathbf{x}$ be the vector of the Cartesian coordinate 1-forms:

```
> dx:= Vector([d(x), d(y), d(z)]);
```

We know that the dual forms $(\bar{\omega}^1, \bar{\omega}^2, \bar{\omega}^3)$ are the components of the following vector:

```
> dualforms:= simplify(MatrixInverse(A).dx);
```

We can assign them as follows:

```
> for i from 1 to 3 do
    omega[i]:= dualforms[i];
  end do;
```

Similarly, the connection forms $(\bar{\omega}^i_j)$ are given by the following matrix:

```
> connectionforms:= simplify(MatrixInverse(A).map(d, A));
```

Assign these as follows:

```
> for i from 1 to 3 do
    for j from 1 to 3 do
      omega[i,j]:= connectionforms[i,j];
    end do;
  end do;
```

In order to verify the structure equations for the $(d\bar{\omega}^i)$, check that the following quantities are all equal to zero:

```
> for i from 1 to 3 do
    Simf(d(omega[i]) + add(omega[i,j] &^ omega[j], j=1..3));
  end do;
```

In order to verify the structure equations for the $(d\bar{\omega}^i_j)$, check that the following quantities are all equal to zero. (The `print` command is to force MAPLE to print the output; normally output is suppressed for computations nested this deeply.)

```
> for i from 1 to 3 do
    for j from 1 to 3 do
      print(Simf(d(omega[i,j]) + add(omega[i,k] &^ omega[k,j],
      k=1..3)));
    end do;
  end do;
```

The following two exercises don't require the `Cartan` package, but they nicely illustrate some of MAPLE's graphic capabilities, so we will include them here anyway. First, we need to load the `plots` package:

```
> with(plots);
```

EXERCISE 3.46, part (c): The general element of $SO^+(1,1)$ is represented by the following matrix:

```
> A:= Matrix([[cosh(theta), sinh(theta)], [sinh(theta),
  cosh(theta)]]);
```

Consider the unit circle, parametrized as follows:

```
> unitcircle:= Vector([cos(t), sin(t)]);
```

The image of this curve under the action of the matrix A (for a fixed value of θ) is given by

```
> newcurve:= A.unitcircle;
```

We can now define plots of this curve for various values of θ; by using the `display` command, we can view them all on one graph:

```
> g1:= plot([subs([theta=0], newcurve[1]), subs([theta=0],
  newcurve[2]), t=0..2*Pi], scaling=constrained):
  g2:= plot([subs([theta=0.5], newcurve[1]),
  subs([theta=0.5], newcurve[2]), t=0..2*Pi],
  scaling=constrained):
  g3:= plot([subs([theta=1], newcurve[1]),
  subs([theta=1], newcurve[2]), t=0..2*Pi],
  scaling=constrained):
  display(g1, g2, g3);
```

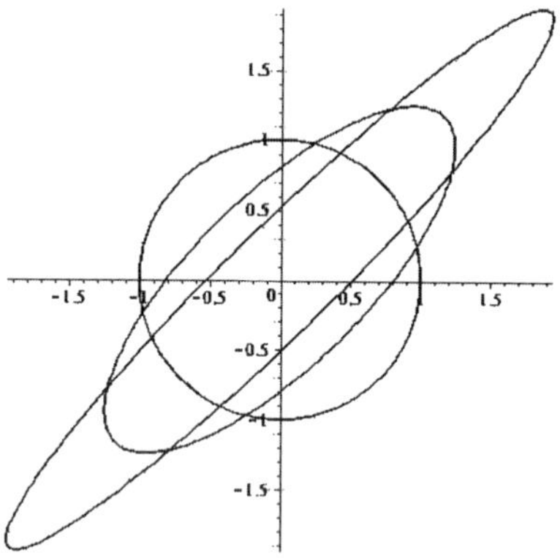

EXERCISE 3.69, part (c): Define A to be the transformation matrix:

```
> A:= Matrix([[cos(phi), -sin(phi), 0],
  [sin(phi), cos(phi), 0], [0,0,1]]);
```

Parametrize the unit circle in terms of homogeneous coordinates with $x^0 = 1$:

```
-> x_homog:= Vector([1, cos(t), sin(t)]);
```

Apply the transformation:

```
> Tx_homog:= A.x_homog;
```

If we write the transformed curve in terms of homogeneous coordinates with $x^0 = 1$, then the other two coordinates are given by

```
> Tx[1]:= Tx_homog[2]/Tx_homog[1];
  Tx[2]:= Tx_homog[3]/Tx_homog[1];
```

We can animate the resulting curve, using φ as the time parameter, to see how the unit circle transforms through a family of quadrics and back again as φ varies from 0 to π:

```
> animate(plot, [ [Tx[1], Tx[2], t=0..2*Pi],
  view = [-5..5, -5..5], scaling=constrained], phi = 0..Pi,
  frames=100);
```

In order to view the animation, click on the plot, and then from the "Plot" menu choose "Animation → Play". (Alternatively, right-click on the plot and select "Animation → Play" from the pop-up menu.)

Chapter 4

Curves and surfaces in Euclidean space

4.1. Introduction

In this chapter, we get to the heart of the matter: Cartan's method of moving frames. This method is used to study the geometry of submanifolds of homogeneous spaces; in this chapter, we will see how it applies to curves and surfaces in $\mathbb{E}^3$. The main idea goes something like this: By associating a frame to each point of a submanifold in some geometrically natural way and then studying how the frame varies along the submanifold, we can construct a complete set of *invariants* for a given class of submanifolds. Invariants are quantities associated to a submanifold (such as curvature and torsion for curves in $\mathbb{E}^3$) that remain unchanged when the submanifold is acted on by an element of the symmetry group of the homogeneous space. A *complete* set of invariants contains enough information to determine a submanifold uniquely up to the group action. This perspective naturally leads to two questions:

(1) How can we tell when we have found a complete set of invariants? This is a question about *uniqueness:* Given two submanifolds of a homogeneous space, when is it possible to transform one into the other via an element of the symmetry group? This is also known as the *equivalence problem:* When are two submanifolds equivalent under the action of the symmetry group?

(2) Once we know what a complete set of invariants should look like, can they be prescribed arbitrarily? This is a question about *existence:* Given prescribed values for the invariants, does there necessarily exist a submanifold whose invariants coincide with the given values?

In §4.2, we will address the theory underlying the first question, and in §4.3 we will show how it applies to curves in $\mathbb{E}^3$. Then in §4.4 and §4.5, we will take up the second question and show how the theory as a whole applies to surfaces in $\mathbb{E}^3$.

4.2. Equivalence of submanifolds of a homogeneous space

We will approach the equivalence problem for submanifolds of a homogeneous space G/H by considering the restriction of certain frames on the underlying space G/H to the submanifold in question.

Remark 4.1. If $M \subset G/H$ and $f : U \to G/H$ is an immersion with $f(U) = M$ (typically U is some open, connected, and simply connected region in $\mathbb{R}^n$ and f is a parametrization of M), then "restriction to M" really means pullback to U. The *pullback bundle* (or *induced bundle*) f^*G of the principal bundle $\pi : G \to G/H$ is the bundle over U whose fiber over a point $\mathbf{u} \in U$ is just the fiber of G over the point $f(\mathbf{u}) \in G/H$:

$$f^*G = \{(\mathbf{u}, \mathbf{f}) \in U \times G \mid f(\mathbf{u}) = \pi(\mathbf{f})\}.$$

The bundle f^*G is a principal bundle over U, with fiber group H. There is a natural map $\hat{f} : f^*G \to G$ defined by

$$\hat{f}(\mathbf{u}, \mathbf{f}) = \mathbf{f}.$$

When $\pi : G \to G/H$ is regarded as a frame bundle, the image $\hat{f}(\mathbf{u}, \mathbf{f})$ of any element $(\mathbf{u}, \mathbf{f}) \in f^*G$ may be thought of as a frame based at the point $f(\mathbf{u}) \in G/H$. These maps may be represented by the following commutative diagram:

$$\begin{array}{ccc} f^*G & \xrightarrow{\;\hat{f}\;} & G \\ {\scriptstyle \bar{\pi}}\downarrow & & \downarrow{\scriptstyle \pi} \\ U & \xrightarrow{\;f\;} & G/H. \end{array}$$

We will generally be interested in choosing a frame at each point of $M \subset G/H$—i.e., a "frame field" on M—according to certain geometric considerations. Technically, this means choosing a section of the bundle f^*G over U, but we will usually regard it as choosing a *lifting* $\tilde{f} : U \to G$, i.e., a function $\tilde{f}$ with the property that for any $\mathbf{u} \in U$,

$$(\pi \circ \tilde{f})(\mathbf{u}) = f(\mathbf{u}) \in M \subset G/H.$$

In other words, we choose $\tilde{f}$ so that the following diagram commutes:

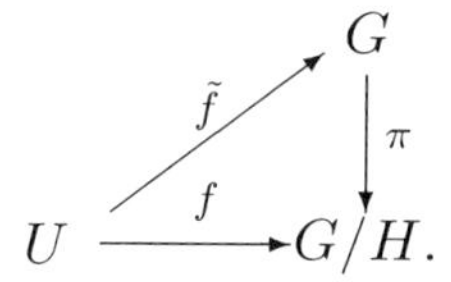

When choosing a lifting $\tilde{f} : U \to G$, we will want to choose frame fields that are *adapted* to M. This means that, instead of just choosing arbitrary frame fields, we will use the geometry of M to choose "nice" frame fields. This is somewhat analogous to choosing "nice" coordinates on a neighborhood of a point on a surface to study the geometry at that point; the beauty of the method of moving frames is that we can do this at all points simultaneously.

Once we have chosen a nice lifting (called an *adapted frame field*, or sometimes simply an *adapted frame*) $\tilde{f} : U \to G$, we can consider the pullback $\tilde{f}^*\omega$ of the Maurer-Cartan form ω of G and its structure equations to U. The pulled-back Maurer-Cartan form $\tilde{f}^*\omega$ will generally contain quantities that are *invariants* of M: If we act on M by a symmetry of the ambient space G/H, then these quantities remain unchanged. (The invariance of $\tilde{f}^*\omega$ under such an action follows from the left-invariance of ω under action by an element of G.) Typical examples of invariants are quantities such as arc length, curvature, etc.

In order for the adapted frame field $\tilde{f} : U \to G$ to contain useful information about the invariants of M, the algorithm for choosing $\tilde{f}$ should be completely determined in some canonical way by the geometry of M. Moreover, the adapted frame field itself should be *equivariant*; this means that

$$\widetilde{(g \cdot f)} = g \cdot \tilde{f}$$

for any $g \in G$. If such an equivariant adapted frame field exists, then the question of equivalence is completely answered by the following important lemma:

Lemma 4.2. *Let $U \subset \mathbb{R}^n$ be a connected, open set, and let $\tilde{f}_1, \tilde{f}_2 : U \to G$ be two immersions. Then there exists an element $g \in G$ such that*

$$\tilde{f}_1(\mathbf{u}) = g \cdot \tilde{f}_2(\mathbf{u})$$

for all $\mathbf{u} \in U$ if and only if $\tilde{f}_1^*\omega = \tilde{f}_2^*\omega$, where ω is the Maurer-Cartan form of G.

PROOF. First, observe that for any map $\tilde{f} : U \to G$, we have

$$\tilde{f}^*\omega = \tilde{f}^{-1}d\tilde{f}. \tag{4.1}$$

Remark 4.3. What does equation (4.1) really mean? Recall that for any $g \in G$, ω is a linear map from T_gG to $T_eG = \mathfrak{g}$ defined by

$$\omega(\mathbf{w}) = (L_{g^{-1}})_*(\mathbf{w})$$

for $\mathbf{w} \in T_gG$. Now, if $\tilde{f} : U \to G$ is a differentiable map, then $\tilde{f}^*\omega$ is a linear map from $T_\mathbf{u}U$ to $\mathfrak{g}$ defined by

$$\tilde{f}^*\omega(\mathbf{v}) = \omega(\tilde{f}_*(\mathbf{v})) = (L_{(\tilde{f}(\mathbf{u}))^{-1}})_*(\tilde{f}_*(\mathbf{v})) = (\tilde{f}(\mathbf{u}))^{-1} \cdot d\tilde{f}(\mathbf{v})$$

for $\mathbf{v} \in T_\mathbf{u}U$. Therefore,

$$\tilde{f}^*\omega = \tilde{f}^{-1}d\tilde{f}.$$

Now, given $\tilde{f}_1, \tilde{f}_2 : U \to G$, there exists a unique function $g : U \to G$ satisfying the condition that

$$\tilde{f}_2(\mathbf{u}) = g(\mathbf{u})\tilde{f}_1(\mathbf{u}) \tag{4.2}$$

for all $\mathbf{u} \in U$—specifically, $g(\mathbf{u}) = \tilde{f}_2(\mathbf{u})(\tilde{f}_1(\mathbf{u}))^{-1}$. Differentiating (4.2) yields

$$d\tilde{f}_2 = dg\,\tilde{f}_1 + g\,d\tilde{f}_1;$$

therefore,

$$\begin{aligned}
\tilde{f}_2^*\omega &= \tilde{f}_2^{-1}d\tilde{f}_2 \\
&= \tilde{f}_2^{-1}dg\,\tilde{f}_1 + \tilde{f}_2^{-1}g\,d\tilde{f}_1 \\
&= \tilde{f}_2^{-1}dg\,\tilde{f}_1 + (g\tilde{f}_1)^{-1}g\,d\tilde{f}_1 \\
&= \tilde{f}_2^{-1}dg\,\tilde{f}_1 + \tilde{f}_1^{-1}g^{-1}g\,d\tilde{f}_1 \\
&= \tilde{f}_2^{-1}dg\,\tilde{f}_1 + \tilde{f}_1^{-1}d\tilde{f}_1 \\
&= \tilde{f}_2^{-1}dg\,\tilde{f}_1 + \tilde{f}_1^*\omega.
\end{aligned}$$

It follows that $\tilde{f}_1^*\omega = \tilde{f}_2^*\omega$ if and only if $dg = 0$, i.e., if and only if $g(\mathbf{u})$ is constant. □

This lemma is more powerful than it looks; it says that:

(1) Whatever geometric information is contained in $\tilde{f}^*\omega$ remains unchanged when M is transformed by a symmetry g of the ambient homogeneous space G/H.

(2) Conversely, $\tilde{f}^*\omega$ contains enough information about the geometry of M to *completely* determine it up to a symmetry of the ambient space.

So, our approach from here on will go something like this: Given an immersion $f : U \to G/H$, we will look for an equivariant method of constructing a canonical adapted frame field $\tilde{f} : U \to G$. Then we will examine the pulled-back Maurer-Cartan form $\tilde{f}^*\omega$, which will contain a complete set of geometric invariants for the original immersion f. This is known as the *method of moving frames*, and we will start by demonstrating how to carry it out for curves in $\mathbb{E}^3$.

4.3. Moving frames for curves in $\mathbb{E}^3$

Consider a smooth, parametrized curve $\alpha : I \to \mathbb{E}^3$ that maps some open interval $I \subset \mathbb{R}$ into Euclidean space. $\mathbb{E}^3$ has the structure of the homogeneous space $E(3)/SO(3)$, so an adapted frame field along α should be a lifting $\tilde{\alpha} : I \to E(3)$. Any such lifting can be written as

$$\tilde{\alpha}(t) = (\alpha(t); \mathbf{e}_1(t), \mathbf{e}_2(t), \mathbf{e}_3(t)),$$

where for each $t \in I$, $(\mathbf{e}_1(t), \mathbf{e}_2(t), \mathbf{e}_3(t))$ is an oriented, orthonormal basis for the tangent space $T_{\alpha(t)}\mathbb{E}^3$. Such an adapted frame field is usually called an *orthonormal frame field* along α. If the curve is "nice enough" (the precise meaning of this will become clear shortly), then we will be able to choose such a frame field in a canonical way, based on the geometry of the curve.

Remark 4.4. While the orthonormal frame field is technically the image of the map $\tilde{\alpha}$ and so includes the position vector $\alpha(t)$ at each point, it is common to refer to the triple of vector fields $(\mathbf{e}_1(t), \mathbf{e}_2(t), \mathbf{e}_3(t))$ as an "orthonormal frame field along α". Hopefully this terminology will not cause any confusion.

Recall that α is *regular* if $\alpha'(t) \neq \mathbf{0}$ for every $t \in I$. The first condition that we will require in order for α to be "nice enough" is that α must be a regular curve. With this assumption, we can make our first frame adaptation by setting

$$\mathbf{e}_1(t) = \frac{\alpha'(t)}{|\alpha'(t)|};$$

i.e., we require that $\mathbf{e}_1(t)$ be the unit tangent vector to the curve at $\alpha(t)$.

Exercise 4.5. Show that this choice of $\mathbf{e}_1(t)$ is equivariant under the action of $E(3)$: If we replace α by $g \cdot \alpha$ for some $g \in E(3)$, then $\mathbf{e}_1(t) \in T_{\alpha(t)}\mathbb{E}^3$ will be replaced by $(L_g)_*(\mathbf{e}_1(t)) \in T_{g\cdot\alpha(t)}\mathbb{E}^3$.

The vector $\mathbf{e}_1(t)$ is now uniquely determined, but we still have the freedom to vary the pair $(\mathbf{e}_2(t), \mathbf{e}_3(t))$ by an arbitrary rotation in $SO(2)$. We will need to delve deeper into the geometry of the curve α in order to determine how to choose the remainder of the adapted frame field.

Here we make an observation that will simplify the remainder of our computations. Fix $t_0 \in I$ and define the *arc length* function along α to be

$$s(t) = \int_{t_0}^{t} |\alpha'(u)|\, du.$$

Exercise 4.6. Show that $s(t)$ is invariant under the action of $E(3)$; that is, for any $g \in E(3)$, the curves α and $g \cdot \alpha$ have the same arc length function.

Since $\alpha'(t) \neq \mathbf{0}$ for all $t \in I$, the inverse function theorem implies that $s(t)$ has a differentiable inverse function $t(s)$. By setting $\alpha(s) = \alpha(t(s))$, we may assume that α is parametrized by arc length, so that $|\alpha'(s)| = 1$ and $\mathbf{e}_1(s) = \alpha'(s)$.

In order to make the next adaptation, we need to make another assumption about the curve. We will say that α is *nondegenerate* if α is regular and, in addition, $\mathbf{e}_1'(s) \neq \mathbf{0}$ for all $s \in I$. In this case, differentiating the equation

$$\langle \mathbf{e}_1(s), \mathbf{e}_1(s) \rangle = 1$$

with respect to s yields

$$\langle \mathbf{e}_1'(s), \mathbf{e}_1(s) \rangle = 0.$$

Thus, $\mathbf{e}_1'(s)$ is orthogonal to $\mathbf{e}_1(s)$, and we can make our second adaptation by setting

$$\mathbf{e}_2(s) = \frac{\mathbf{e}_1'(s)}{|\mathbf{e}_1'(s)|}.$$

This vector is called the *unit normal vector* to the curve at $\alpha(s)$.

Exercise 4.7. Show that $\mathbf{e}_2(s)$ is equivariant under the action of $E(3)$: If we replace α by $g \cdot \alpha$ for some $g \in E(3)$, then $\mathbf{e}_2(s) \in T_{\alpha(s)}\mathbb{E}^3$ will be replaced by $(L_g)_*(\mathbf{e}_2(s)) \in T_{g\cdot\alpha(s)}\mathbb{E}^3$.

The adapted frame field is now uniquely determined: Because the frame must be oriented and orthonormal, $\mathbf{e}_3(s)$ is uniquely determined by the condition that

$$\mathbf{e}_3(s) = \mathbf{e}_1(s) \times \mathbf{e}_2(s).$$

The vector $\mathbf{e}_3(s)$ is called the *binormal vector* to the curve at $\alpha(s)$. The adapted frame field $(\mathbf{e}_1(s), \mathbf{e}_2(s), \mathbf{e}_3(s))$ is called the *Frenet frame* of the curve $\alpha(s)$; it determines a canonical, left-invariant lifting $\tilde{\alpha} : I \to E(3)$ given by

$$\tilde{\alpha}(s) = (\alpha(s); \mathbf{e}_1(s), \mathbf{e}_2(s), \mathbf{e}_3(s))$$

for any nondegenerate curve α parametrized by arc length. (See Figure 4.1.)

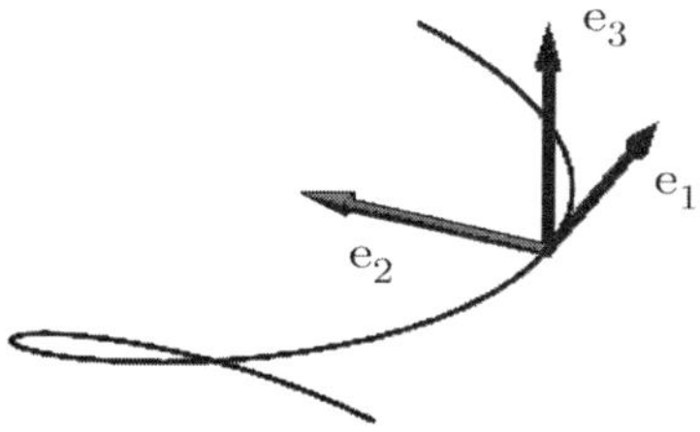

Figure 4.1. Frenet frame at a point of a curve in $\mathbb{E}^3$

Now consider the pullbacks of equations (3.1) to I via $\tilde{\alpha}$. We have

$$\tilde{\alpha}^*(\mathbf{x}) = \alpha(s), \qquad \tilde{\alpha}^*(\mathbf{e}_i) = \mathbf{e}_i(s).$$

Therefore,

$$\begin{aligned} \tilde{\alpha}^*(d\mathbf{x}) &= d(\tilde{\alpha}^*(\mathbf{x})) = \alpha'(s)ds, \\ \tilde{\alpha}^*(d\mathbf{e}_i) &= d(\tilde{\alpha}^*(\mathbf{e}_i)) = \mathbf{e}_i'(s)ds. \end{aligned}$$

As in Chapter 3, write $(\bar{\omega}^i, \bar{\omega}^i_j)$ for the pulled-back forms $(\tilde{\alpha}^*\omega^i, \tilde{\alpha}^*\omega^i_j)$. Note that, since these are all 1-forms on I, they must all be multiples of ds.

We can write the pullbacks of equations (3.1) as

$$\begin{aligned} \alpha'(s)ds &= \mathbf{e}_i(s)\,\bar{\omega}^i, \\ \mathbf{e}_i'(s)ds &= \mathbf{e}_j(s)\,\bar{\omega}^j_i. \end{aligned} \tag{4.3}$$

Now recall how we constructed our adapted frame field. First, we chose $\mathbf{e}_1(s)$ so that $\alpha'(s) = \mathbf{e}_1(s)$; therefore, the first equation in (4.3) implies that

$$\bar{\omega}^1 = ds, \qquad \bar{\omega}^2 = \bar{\omega}^3 = 0.$$

Then we chose $\mathbf{e}_2(s)$ so that $\mathbf{e}_1'(s)$ is a multiple of $\mathbf{e}_2(s)$, say $\mathbf{e}_1'(s) = \kappa(s)\mathbf{e}_2(s)$. The function $\kappa(s)$ is called the *curvature* of α at s; note that α is nondegenerate if and only if $\kappa(s) > 0$ for all $s \in I$. So the equation for $\mathbf{e}_1'(s)$ in (4.3) implies that

$$\bar{\omega}^2_1 = \kappa(s)ds, \qquad \bar{\omega}^3_1 = 0.$$

(Recall that $\bar{\omega}^1_1 = 0$ by the skew-symmetry of the (ω^i_j).) The only remaining Maurer-Cartan form is $\bar{\omega}^2_3$; it must be equal to some multiple of ds, so define a function $\tau(s)$ by the condition that

$$\bar{\omega}^2_3 = -\tau(s)ds.$$

The function $\tau(s)$ is called the *torsion* of α at s.

Remark 4.8. The minus sign in the definition of $\tau(s)$ is a convention preferred by some authors, but it is not universal. This choice of sign has the feature that it results in a positive value of τ for the standard right-handed helix

$$\alpha(s) = {}^t[a\cos(s), a\sin(s), b],$$

where $a, b > 0$ and $a^2 + b^2 = 1$.

Using the skew-symmetry of the $(\bar{\omega}^i_j)$, the remaining two equations in (4.3) become

$$\begin{aligned} \mathbf{e}_2'(s)ds &= \mathbf{e}_1(s)\,\bar{\omega}^1_2 + \mathbf{e}_3(s)\,\bar{\omega}^3_2 = (-\mathbf{e}_1(s)\kappa(s) + \mathbf{e}_3(s)\tau(s))ds, \\ \mathbf{e}_3'(s)ds &= \mathbf{e}_1(s)\,\bar{\omega}^1_3 + \mathbf{e}_2(s)\,\bar{\omega}^2_3 = -\mathbf{e}_2(s)\tau(s)ds. \end{aligned}$$

Thus, we have the familiar *Frenet equations*:

$$\begin{bmatrix}\alpha'(s) & \mathbf{e}_1'(s) & \mathbf{e}_2'(s) & \mathbf{e}_3'(s)\end{bmatrix} = \begin{bmatrix}\alpha(s) & \mathbf{e}_1(s) & \mathbf{e}_2(s) & \mathbf{e}_3(s)\end{bmatrix} \begin{bmatrix} 0 & 0 & 0 & 0 \\ 1 & 0 & -\kappa(s) & 0 \\ 0 & \kappa(s) & 0 & -\tau(s) \\ 0 & 0 & \tau(s) & 0 \end{bmatrix}.$$

Note that if we regard $\tilde{\alpha}(s)$ as the matrix

$$\tilde{\alpha}(s) = \begin{bmatrix}\alpha(s) & \mathbf{e}_1(s) & \mathbf{e}_2(s) & \mathbf{e}_3(s)\end{bmatrix},$$

then the matrix on the right multiplied by the 1-form ds is equal to

$$\tilde{\alpha}(s)^{-1} d(\tilde{\alpha}(s)),$$

and so it is exactly the pullback of the Maurer-Cartan form $\omega = g^{-1}dg$ on $E(3)$ via $\tilde{\alpha}$.

Applying Lemma 4.2 yields the following theorem:

Theorem 4.9. *Two nondegenerate curves $\alpha_1, \alpha_2 : I \to \mathbb{E}^3$ parametrized by arc length differ by a rigid motion if and only if they have the same curvature $\kappa(s)$ and torsion $\tau(s)$.*

This is the uniqueness portion of the fundamental theorem of space curves (cf. Theorem 3.1). We will address the existence portion in §4.4.

Exercise 4.10. Repeat the analysis of this section for curves in $\mathbb{E}^4$. Here are some things to think about along the way:

- Is there a natural choice of parametrization for the curve?
- How should you choose the vectors $(\mathbf{e}_1(s), \mathbf{e}_2(s), \mathbf{e}_3(s), \mathbf{e}_4(s))$ of the frame field? (And how do you ensure that these vectors form an orthonormal frame field?) Prove that your choice is equivariant

under the action of $E(4)$. (Hint: The tricky part is how to choose $\mathbf{e}_3(s)$ so that it is orthogonal to both $\mathbf{e}_1(s)$ and $\mathbf{e}_2(s)$. For guidance, use the Frenet equations to convince yourself that for curves in $\mathbb{E}^3$,

$$\mathbf{e}_3(s) = \frac{\mathbf{e}_2'(s) - \langle \mathbf{e}_2'(s), \mathbf{e}_1(s) \rangle \mathbf{e}_1(s)}{|\mathbf{e}_2'(s) - \langle \mathbf{e}_2'(s), \mathbf{e}_1(s) \rangle \mathbf{e}_1(s)|}.$$

In other words, $\mathbf{e}_3(s)$ is obtained by taking the orthogonal projection of $\mathbf{e}_2'(s)$ onto the orthogonal complement of $\mathbf{e}_1(s)$ and $\mathbf{e}_2(s)$ and then normalizing it to have unit length.)

- What is the right definition of "nondegenerate" for curves in $\mathbb{E}^4$?
- Where do invariants appear in the pullbacks of equations (3.1)? What can you conclude from the skew-symmetry of the connection forms?
- What is the 4-dimensional analog of the Frenet equations?
- How do you think the analysis would go for curves in $\mathbb{E}^n$?

4.4. Compatibility conditions and existence of submanifolds with prescribed invariants

In §4.3, we saw that a curve $\alpha : I \to \mathbb{E}^3$ parametrized by arc length s is completely determined up to rigid motions of $\mathbb{E}^3$ by its curvature $\kappa(s)$ and torsion $\tau(s)$. We may express this by saying that the curvature and torsion form a *complete set of invariants* for curves in $\mathbb{E}^3$.

In general, Lemma 4.2 tells us when we have found a complete set of invariants for a "nice" immersion $f : U \to G/H$: Assuming that we can find a canonical, equivariant way of choosing a lifting $\tilde{f} : U \to G$ (this is what "nice" means), a complete set of invariants is contained in $\tilde{f}^*\omega$, the pullback via $\tilde{f}$ of the Maurer-Cartan form ω of G.

For curves in $\mathbb{E}^3$, it is now natural to ask whether the functions $\kappa(s)$ and $\tau(s)$ may be prescribed arbitrarily. In other words, given arbitrary functions $\kappa(s), \tau(s)$ with $\kappa(s) > 0$, does there necessarily exist a curve $\alpha : I \to \mathbb{E}^3$ that is parametrized by arc length and has curvature $\kappa(s)$ and torsion $\tau(s)$?

Exercise 4.11. Why must we require $\kappa(s) > 0$?

The answer to this existence question is yes, but this result is particular to 1-dimensional submanifolds of homogeneous spaces G/H. It follows from the following lemma:

Lemma 4.12. *Let G be a Lie group with Lie algebra $\mathfrak{g}$, and suppose that $\bar{\omega}$ is a $\mathfrak{g}$-valued 1-form on a connected and simply connected manifold U. Then there exists a smooth map $\tilde{f} : U \to G$ with $\tilde{f}^*\omega = \bar{\omega}$ if and only if $\bar{\omega}$ satisfies the Maurer-Cartan equation*

$$d\bar{\omega} = -\bar{\omega} \wedge \bar{\omega}. \tag{4.4}$$

Outline of Proof. The full proof of this lemma requires the Frobenius theorem and is beyond the scope of this book. (If you're curious, the proof may be found in [**Gri74**].) However, the main idea goes something like this: $\tilde{f}^*\omega$ contains quantities involving derivatives of the unknown function $\tilde{f} : U \to G$, and for any given $\mathfrak{g}$-valued 1-form $\bar{\omega}$ on U, the equation $\tilde{f}^*\omega = \bar{\omega}$ may be regarded as a system of partial differential equations for $\tilde{f}$. In general, this system is overdetermined and may have no solutions. However, equation (4.4) is precisely the compatibility condition that must be satisfied in order to guarantee that solutions exist, at least locally. (In this case, it turns out that a solution is uniquely determined by specifying an initial condition $\tilde{f}(\mathbf{u}_0)$ for any $\mathbf{u}_0 \in U$.) Once we know that local solutions exist, a patching argument can be used to construct a solution $\tilde{f}$ on the entire domain U. □

Remark 4.13. Even without the hypothesis that U is simply connected, the result of Lemma 4.12 holds in some neighborhood of any point $\mathbf{u} \in U$; simple connectivity is only necessary to ensure that these local solutions can be patched together to form a single solution that is globally defined on U. For simplicity of exposition, we will not explicitly state topological hypotheses on the domain U every time we introduce an immersion $f : U \to G/H$. But keep in mind that if U is topologically nontrivial, then many of our constructions may be possible only locally and not globally on U. For example, because a frame bundle over a topologically nontrivial base space may have no global sections, it might not be possible to construct an adapted frame field globally on U. Because of these limitations, the method of moving frames is a tool best suited to the study of the *local* geometry of submanifolds of homogeneous spaces; it has very little to say about global properties.

Assuming that the conditions of Lemma 4.12 are satisfied, composing the map $\tilde{f}$ with the natural projection $\pi : G \to G/H$ gives a smooth map $f : U \to G/H$ that, in most cases of interest, realizes $M = f(U)$ as a submanifold of the homogeneous space G/H. According to Lemma 4.2, specifying a $\mathfrak{g}$-valued 1-form $\bar{\omega}$ on U is equivalent to prescribing the values of a complete set of invariants for an unknown submanifold $M \subset G/H$;

Lemma 4.12 then gives a necessary and sufficient condition for the existence of a smooth map $f : U \to G/H$ whose image has the prescribed invariants. Moreover, Lemma 4.2 implies that any such f is unique up to left action by an element $g \in G$.

***Exercise 4.14.** Let $(\omega^1, \dots, \omega^n)$ be a basis for the left-invariant 1-forms on G that are semi-basic for the projection $\pi : G \to G/H$, and let $(\bar{\omega}^1, \dots, \bar{\omega}^n)$ be the corresponding components of the $\mathfrak{g}$-valued 1-form $\bar{\omega}$ on U. Show that the map $f = \pi \circ \tilde{f}$ of Lemma 4.12 is an immersion if and only if $(\bar{\omega}^1, \dots, \bar{\omega}^n)$ span the cotangent space $T^*_{\mathbf{u}}U$ at every point $\mathbf{u} \in U$. (Note that typically the dimension of U is less than n, so the forms $(\bar{\omega}^1, \dots, \bar{\omega}^n)$ will generally *not* be linearly independent.)

Corollary 4.15. *Let $I \subset \mathbb{R}$ be an open interval, and let $\kappa, \tau : I \to \mathbb{R}$ be any differentiable functions satisfying $\kappa(s) > 0$ for all $s \in I$. Then there exists a nondegenerate curve $\alpha : I \to \mathbb{E}^3$, parametrized by arc length, with curvature $\kappa(s)$ and torsion $\tau(s)$.*

***Exercise 4.16.** Show how Corollary 4.15 follows from Lemma 4.12. (Hint: Observe that both sides of equation (4.4) are 2-forms on I.)

Corollary 4.15 applies more generally to curves in any homogeneous space G/H: Once we know how to construct equivariant frame fields and find a complete set of invariants, Lemma 4.12 implies that these invariants may be prescribed arbitrarily. But for surfaces (and generally for submanifolds of any dimension greater than one), equation (4.4) will give compatibility conditions that a prescribed set of invariants must satisfy in order for an immersed submanifold with the given invariants to exist.

4.5. Moving frames for surfaces in $\mathbb{E}^3$

Let U be an open set in $\mathbb{R}^2$ (assumed here and throughout the remainder of the book to be connected and simply connected; cf. Remark 4.13), and let $\mathbf{x} : U \to \mathbb{E}^3$ be an immersion whose image is a regular surface $\Sigma = \mathbf{x}(U)$. Just as for curves, an adapted frame field along Σ should be a lifting $\tilde{\mathbf{x}} : U \to E(3)$ of the form

$$\tilde{\mathbf{x}}(\mathbf{u}) = (\mathbf{x}(\mathbf{u}); \mathbf{e}_1(\mathbf{u}), \mathbf{e}_2(\mathbf{u}), \mathbf{e}_3(\mathbf{u})),$$

where for each $\mathbf{u} \in U$, $(\mathbf{e}_1(\mathbf{u}), \mathbf{e}_2(\mathbf{u}), \mathbf{e}_3(\mathbf{u}))$ is an oriented, orthonormal basis for the tangent space $T_{\mathbf{x}(\mathbf{u})}\mathbb{E}^3$.

Since $\mathbf{x}$ is an immersion, there is a well-defined tangent plane $T_{\mathbf{x}(\mathbf{u})}\Sigma$ for each point $\mathbf{x}(\mathbf{u}) \in \Sigma$. Thus, we can make our first frame adaptation by requiring that $\mathbf{e}_3(\mathbf{u})$ be orthogonal to $T_{\mathbf{x}(\mathbf{u})}\Sigma$. This determines $\mathbf{e}_3(\mathbf{u})$ uniquely up to sign, and an orthonormal frame field satisfying this condition will be called *adapted.*

Exercise 4.17. Show that this choice of $\mathbf{e}_3(\mathbf{u})$ is equivariant (up to sign) under the action of $E(3)$.

Having chosen $\mathbf{e}_3(\mathbf{u})$ in this way, $\mathbf{e}_1(\mathbf{u})$ and $\mathbf{e}_2(\mathbf{u})$ must form a basis for $T_{\mathbf{x}(\mathbf{u})}\Sigma$ no matter how we choose them. We will explore how we might refine our choices later, but for now, we allow $(\mathbf{e}_1(\mathbf{u}), \mathbf{e}_2(\mathbf{u}))$ to be an arbitrary orthonormal basis of $T_{\mathbf{x}(\mathbf{u})}\Sigma$.

Now consider the pullbacks of equations (3.1) to U via $\tilde{\mathbf{x}}$. (As in Chapter 3, write $(\bar{\omega}^i, \bar{\omega}^i_j)$ for the pulled-back forms $(\tilde{\mathbf{x}}^*\omega^i, \tilde{\mathbf{x}}^*\omega^i_j)$ on U.) Our first observation about these forms is the following:

Proposition 4.18. *Let $U \subset \mathbb{R}^2$ be an open set, and let $\mathbf{x} : U \to \mathbb{E}^3$ be an immersion. For any adapted orthonormal frame field $(\mathbf{e}_1(\mathbf{u}), \mathbf{e}_2(\mathbf{u}), \mathbf{e}_3(\mathbf{u}))$ along $\Sigma = \mathbf{x}(U)$, the associated dual and connection forms $(\bar{\omega}^i, \bar{\omega}^i_j)$ have the property that $\bar{\omega}^3 = 0$.*

Proof. The pullback of the first equation in (3.1) is

$$d\mathbf{x} = \mathbf{e}_i\bar{\omega}^i.$$

Let $\mathbf{u} \in U$. Then $d\mathbf{x}_{\mathbf{u}}$ is a linear map from $T_{\mathbf{u}}U$ to $T_{\mathbf{x}(\mathbf{u})}\Sigma$, and so for any $\mathbf{v} \in T_{\mathbf{u}}U$, we must have

$$d\mathbf{x}_{\mathbf{u}}(\mathbf{v}) = \mathbf{e}_i(\mathbf{u})\bar{\omega}^i(\mathbf{v}) \in T_{\mathbf{x}(\mathbf{u})}\Sigma.$$

Since $T_{\mathbf{x}(\mathbf{u})}\Sigma$ is spanned by $\mathbf{e}_1(\mathbf{u})$ and $\mathbf{e}_2(\mathbf{u})$, the $\mathbf{e}_3(\mathbf{u})$ term in this sum must vanish; therefore, $\bar{\omega}^3(\mathbf{v}) = 0$. And since $\mathbf{v} \in T_{\mathbf{u}}U$ is arbitrary, it follows that $\bar{\omega}^3 = 0$. □

***Exercise 4.19.** Show that $(\bar{\omega}^1, \bar{\omega}^2)$ are linearly independent 1-forms on U. Therefore, they form a basis for the 1-forms on U. (Hint: Evaluate $\bar{\omega}^1$ and $\bar{\omega}^2$ on vectors $\mathbf{v}_1, \mathbf{v}_2 \in T_{\mathbf{u}}U$ with the property that $d\mathbf{x}(\mathbf{v}_i) = \mathbf{e}_i(\mathbf{u})$ for $i = 1, 2$.)

You may recall that the metric properties of a regular surface in $\mathbb{E}^3$ are encapsulated in the *first fundamental form* of the surface.

Definition 4.20. Let $U \subset \mathbb{R}^2$ be an open set, and let $\mathbf{x} : U \to \mathbb{E}^3$ be an immersion. The *first fundamental form* of $\Sigma = \mathbf{x}(U)$ is the quadratic form I on TU defined by

$$\mathrm{I}(\mathbf{v}) = \langle d\mathbf{x}(\mathbf{v}), d\mathbf{x}(\mathbf{v})\rangle$$

for $\mathbf{v} \in T_{\mathbf{u}}U$.

In other words, I is just the restriction of the Euclidean metric on $\mathbb{E}^3$ to vectors which are tangent to Σ. Its primary function is to describe how to compute this metric in terms of the local coordinates $\mathbf{u}$ on Σ that are given by the parametrization $\mathbf{x} : U \to \mathbb{E}^3$.

***Exercise 4.21.** Show that for any $\mathbf{v} \in T_{\mathbf{u}}U$,

$$\mathrm{I}(\mathbf{v}) = \left(\bar{\omega}^1(\mathbf{v})\right)^2 + \left(\bar{\omega}^2(\mathbf{v})\right)^2.$$

This is often written more concisely as

$$\mathrm{I} = (\bar{\omega}^1)^2 + (\bar{\omega}^2)^2,$$

and each term in the sum should be interpreted as the *symmetric* product $\bar{\omega}^i \circ \bar{\omega}^i$.

While the first fundamental form is defined as a function of a single tangent vector, it can be used to define an inner product $\langle \cdot, \cdot \rangle_{\mathbf{u}}$ on each tangent space $T_{\mathbf{u}}U$ through a process called *polarization*.

Definition 4.22. The inner product $\langle \cdot, \cdot \rangle_{\mathbf{u}}$ is defined by

$$\langle \mathbf{v}, \mathbf{w} \rangle_{\mathbf{u}} = \tfrac{1}{4}\left(\mathrm{I}(\mathbf{v} + \mathbf{w}) - \mathrm{I}(\mathbf{v} - \mathbf{w})\right)$$

for $\mathbf{v}, \mathbf{w} \in T_{\mathbf{u}}U$.

***Exercise 4.23.** (a) Show that

$$\langle \mathbf{v}, \mathbf{w} \rangle_{\mathbf{u}} = \bar{\omega}^1(\mathbf{v})\,\bar{\omega}^1(\mathbf{w}) + \bar{\omega}^2(\mathbf{v})\,\bar{\omega}^2(\mathbf{w}).$$

(b) Convince yourself that $\langle \cdot, \cdot \rangle_{\mathbf{u}}$ is a section of the symmetric tensor bundle $S^2(T^*U)$. Any section of this bundle defines a symmetric bilinear form

$$B : TU \times TU \to \mathbb{R},$$

which in turn defines a quadratic form

$$Q : TU \to \mathbb{R}$$

by setting $Q(\mathbf{v}) = B(\mathbf{v}, \mathbf{v})$.

***Exercise 4.24.** If you've seen the first fundamental form before, you probably saw it written as

$$\mathrm{I} = E\, du^2 + 2F\, du\, dv + G\, dv^2,$$

where (u, v) are local coordinates on U and

$$E = \langle \mathbf{x}_u, \mathbf{x}_u \rangle, \qquad F = \langle \mathbf{x}_u, \mathbf{x}_v \rangle, \qquad G = \langle \mathbf{x}_v, \mathbf{x}_v \rangle.$$

Suppose that $\mathbf{x} : U \to \mathbb{E}^3$ is an immersion with $F = 0$. (Such a parametrization for a given surface Σ always exists, at least locally; the proof is beyond the scope of this book but can be found in [**dC76**]. This assumption isn't necessary, but it keeps the calculations simpler.)

(a) Show that the frame field

$$\mathbf{e}_1(\mathbf{u}) = \frac{1}{\sqrt{E}}\mathbf{x}_u, \qquad \mathbf{e}_2(\mathbf{u}) = \frac{1}{\sqrt{G}}\mathbf{x}_v, \qquad \mathbf{e}_3(\mathbf{u}) = \mathbf{e}_1(\mathbf{u}) \times \mathbf{e}_2(\mathbf{u})$$

is an oriented, orthonormal frame field along $\Sigma = \mathbf{x}(U)$, with $\mathbf{e}_3(\mathbf{u})$ orthogonal to $T_{\mathbf{x}(\mathbf{u})}\Sigma$.

(b) Show that the dual forms of this frame field are

$$\bar{\omega}^1 = \sqrt{E}\,du, \qquad \bar{\omega}^2 = \sqrt{G}\,dv, \qquad \bar{\omega}^3 = 0$$

and that

$$\mathrm{I} = (\bar{\omega}^1)^2 + (\bar{\omega}^2)^2 = E\,du^2 + G\,dv^2.$$

(c) Show that if $(\mathbf{e}_1(\mathbf{u}), \mathbf{e}_2(\mathbf{u}), \mathbf{e}_3(\mathbf{u}))$ is replaced by another adapted frame field $(\tilde{\mathbf{e}}_1(\mathbf{u}), \tilde{\mathbf{e}}_2(\mathbf{u}), \tilde{\mathbf{e}}_3(\mathbf{u}))$ of the form

$$\begin{aligned} \tilde{\mathbf{e}}_1 &= \cos(\theta)\,\mathbf{e}_1 + \sin(\theta)\,\mathbf{e}_2, \\ \tilde{\mathbf{e}}_2 &= -\sin(\theta)\,\mathbf{e}_1 + \cos(\theta)\,\mathbf{e}_2, \\ \tilde{\mathbf{e}}_3 &= \mathbf{e}_3, \end{aligned}$$

then the dual forms $(\tilde{\bar{\omega}}^1, \tilde{\bar{\omega}}^2)$ of the new adapted frame field are

$$\begin{aligned} \tilde{\bar{\omega}}^1 &= \cos(\theta)\,\bar{\omega}^1 + \sin(\theta)\,\bar{\omega}^2, \\ \tilde{\bar{\omega}}^2 &= -\sin(\theta)\,\bar{\omega}^1 + \cos(\theta)\,\bar{\omega}^2. \end{aligned}$$

Moreover,

$$\mathrm{I} = (\bar{\omega}^1)^2 + (\bar{\omega}^2)^2 = (\tilde{\bar{\omega}}^1)^2 + (\tilde{\bar{\omega}}^2)^2.$$

Now let's see what we can learn by differentiating! Since $\bar{\omega}^3 = 0$, we must have $d\bar{\omega}^3 = 0$ as well. According to the Cartan structure equations (3.8), this implies that

$$d\bar{\omega}^3 = -\bar{\omega}^3_1 \wedge \bar{\omega}^1 - \bar{\omega}^3_2 \wedge \bar{\omega}^2 = 0.$$

Since $(\bar{\omega}^1, \bar{\omega}^2)$ are linearly independent 1-forms, Cartan's lemma (cf. Lemma 2.49) implies that there exist real-valued functions h_{11}, h_{12}, h_{22} on U such that

$$\begin{bmatrix} \bar{\omega}^3_1 \\ \bar{\omega}^3_2 \end{bmatrix} = \begin{bmatrix} h_{11} & h_{12} \\ h_{12} & h_{22} \end{bmatrix} \begin{bmatrix} \bar{\omega}^1 \\ \bar{\omega}^2 \end{bmatrix}.$$

How should we interpret the functions (h_{ij})? Recall that

$$d\mathbf{e}_3 = \mathbf{e}_1\omega^1_3 + \mathbf{e}_2\omega^2_3 = -(\mathbf{e}_1\omega^3_1 + \mathbf{e}_2\omega^3_2).$$

For any tangent vector $\mathbf{w} \in T_{\mathbf{x}}\Sigma$, $d\mathbf{e}_3(\mathbf{w})$ measures the directional derivative of the normal vector field $\mathbf{e}_3$ in the direction of $\mathbf{w}$. So, up to sign, $\omega^3_1(\mathbf{w})$ measures the $\mathbf{e}_1$ component of this directional derivative, and $\omega^3_2(\mathbf{w})$ measures its $\mathbf{e}_2$ component. In other words, $\omega^3_i(\mathbf{w})$ measures how rapidly $\mathbf{e}_3$

rotates towards $\mathbf{e}_i$ if we move in the direction $\mathbf{w}$. When we pull everything back to U via the parametrization $\mathbf{x}$ and express $\bar{\omega}_i^3$ as a linear combination of $\bar{\omega}^1$ and $\bar{\omega}^2$, we see that h_{ij} measures how rapidly $\mathbf{e}_3$ rotates towards $\mathbf{e}_i$ if we move in the direction $\mathbf{e}_j$.

Recall that, in addition to the metric properties of a regular surface, there are various types of curvature that arise from the geometry of the *Gauss map* of the surface. This is the map from the surface to the unit sphere $\mathbb{S}^2 \subset \mathbb{E}^3$ that sends any point of the surface to the unit normal vector of the surface at that point. In our context, it can be defined as follows:

Definition 4.25. Let $U \subset \mathbb{R}^2$ be an open set, and let $\mathbf{x} : U \to \mathbb{E}^3$ be an immersion with image $\Sigma = \mathbf{x}(U)$. The *Gauss map* of $\Sigma = \mathbf{x}(U)$ is the map $N : \Sigma \to \mathbb{S}^2$ defined by

$$N(\mathbf{x}(\mathbf{u})) = \mathbf{e}_3(\mathbf{u}),$$

where $(\mathbf{e}_1(\mathbf{u}), \mathbf{e}_2(\mathbf{u}), \mathbf{e}_3(\mathbf{u}))$ is any adapted frame field on $\Sigma = \mathbf{x}(U)$. (Note that N is well-defined up to sign.)

Notions of curvature typically associated with surfaces (Gauss curvature, mean curvature, etc.) arise as linear-algebraic properties of the differential dN of the Gauss map, also known as the *shape operator* of the surface. The relevant information is contained in the *second fundamental form* of the surface.

Definition 4.26. Let $U \subset \mathbb{R}^2$ be an open set, and let $\mathbf{x} : U \to \mathbb{E}^3$ be an immersion. The *second fundamental form* of $\Sigma = \mathbf{x}(U)$ is the quadratic form II on TU defined by

$$\mathrm{II}(\mathbf{v}) = -\langle d\mathbf{e}_3(\mathbf{v}), d\mathbf{x}(\mathbf{v})\rangle$$

for $\mathbf{v} \in T_{\mathbf{u}}U$, where $(\mathbf{e}_1(\mathbf{u}), \mathbf{e}_2(\mathbf{u}), \mathbf{e}_3(\mathbf{u}))$ is any adapted frame field on $\Sigma = \mathbf{x}(U)$.

Since curvature is related to how rapidly the normal vector varies as we move around the surface, we might expect the functions (h_{ij}) to show up in the second fundamental form.

***Exercise 4.27.** (a) Show that for any $\mathbf{v} \in T_{\mathbf{u}}U$,

$$\begin{aligned}\mathrm{II}(\mathbf{v}) &= \bar{\omega}_1^3(\mathbf{v})\,\bar{\omega}^1(\mathbf{v}) + \bar{\omega}_2^3(\mathbf{v})\,\bar{\omega}^2(\mathbf{v}) \\ &= h_{11}\,(\bar{\omega}^1(\mathbf{v}))^2 + 2h_{12}\,\bar{\omega}^1(\mathbf{v})\,\bar{\omega}^2(\mathbf{v}) + h_{22}\,(\bar{\omega}^2(\mathbf{v}))^2.\end{aligned}$$

This is often written more concisely as

$$\mathrm{II} = \bar{\omega}_1^3\,\bar{\omega}^1 + \bar{\omega}_2^3\,\bar{\omega}^2 = h_{11}\,(\bar{\omega}^1)^2 + 2h_{12}\,\bar{\omega}^1\,\bar{\omega}^2 + h_{22}\,(\bar{\omega}^2)^2.$$

(b) Suppose that $\mathbf{x} : U \to \mathbb{E}^3$ is a parametrization with $F = 0$, and let $(\mathbf{e}_1(\mathbf{u}), \mathbf{e}_2(\mathbf{u}), \mathbf{e}_3(\mathbf{u}))$ be the frame field in part (a) of Exercise 4.24. Show that

$$\mathrm{II} = Eh_{11}\,du^2 + 2\sqrt{EG}h_{12}\,du\,dv + Gh_{22}\,dv^2.$$

(c) The second fundamental form is more commonly written as

$$\mathrm{II} = e\,du^2 + 2f\,du\,dv + g\,dv^2,$$

where

$$\begin{aligned} e &= \langle \mathbf{e}_3, \mathbf{x}_{uu}\rangle = -\langle (\mathbf{e}_3)_u, \mathbf{x}_u\rangle = -\langle d\mathbf{e}_3\left(\tfrac{\partial}{\partial u}\right), d\mathbf{x}\left(\tfrac{\partial}{\partial u}\right)\rangle, \\ f &= \langle \mathbf{e}_3, \mathbf{x}_{uv}\rangle = -\langle (\mathbf{e}_3)_v, \mathbf{x}_u\rangle = -\langle d\mathbf{e}_3\left(\tfrac{\partial}{\partial v}\right), d\mathbf{x}\left(\tfrac{\partial}{\partial u}\right)\rangle \\ &= -\langle (\mathbf{e}_3)_u, \mathbf{x}_v\rangle = -\langle d\mathbf{e}_3\left(\tfrac{\partial}{\partial u}\right), d\mathbf{x}\left(\tfrac{\partial}{\partial v}\right)\rangle, \\ g &= \langle \mathbf{e}_3, \mathbf{x}_{vv}\rangle = -\langle (\mathbf{e}_3)_v, \mathbf{x}_v\rangle = -\langle d\mathbf{e}_3\left(\tfrac{\partial}{\partial v}\right), d\mathbf{x}\left(\tfrac{\partial}{\partial v}\right)\rangle. \end{aligned}$$

(Some authors use ℓ, m, n or L, M, N in place of e, f, g.) Show that this agrees with Definition 4.26, and conclude that

$$h_{11} = \frac{e}{E}, \qquad h_{12} = \frac{f}{\sqrt{EG}}, \qquad h_{22} = \frac{g}{G}.$$

Now, we still haven't figured out how we should choose the vectors $(\mathbf{e}_1(\mathbf{u}), \mathbf{e}_2(\mathbf{u}))$, except that they should form an orthonormal basis for $T_{\mathbf{x}(\mathbf{u})}\Sigma$ at each point. In order to refine our adapted frame field further, we will examine how the matrix $[h_{ij}]$ changes if we vary the frame. So, let $(\mathbf{e}_1(\mathbf{u}), \mathbf{e}_2(\mathbf{u}), \mathbf{e}_3(\mathbf{u}))$ be any adapted frame field, with associated Maurer-Cartan forms $(\bar{\omega}^i, \bar{\omega}^i_j)$. Any other adapted frame field $(\tilde{\mathbf{e}}_1(\mathbf{u}), \tilde{\mathbf{e}}_2(\mathbf{u}), \tilde{\mathbf{e}}_3(\mathbf{u}))$ has the form (up to sign)

$$\begin{bmatrix} \tilde{\mathbf{e}}_1(\mathbf{u}) & \tilde{\mathbf{e}}_2(\mathbf{u}) & \tilde{\mathbf{e}}_3(\mathbf{u}) \end{bmatrix} = \begin{bmatrix} \mathbf{e}_1(\mathbf{u}) & \mathbf{e}_2(\mathbf{u}) & \mathbf{e}_3(\mathbf{u}) \end{bmatrix} \begin{bmatrix} \cos(\theta) & -\sin(\theta) & 0 \\ \sin(\theta) & \cos(\theta) & 0 \\ 0 & 0 & 1 \end{bmatrix}$$

for some function θ on U. Let $(\tilde{\bar{\omega}}^i, \tilde{\bar{\omega}}^i_j)$ be the Maurer-Cartan forms associated to the new frame field, and set $B = \begin{bmatrix} \cos(\theta) & -\sin(\theta) \\ \sin(\theta) & \cos(\theta) \end{bmatrix}$.

***Exercise 4.28.** (a) Show that the result in part (c) of Exercise 4.24 can be expressed as

$$\begin{bmatrix} \tilde{\bar{\omega}}^1 \\ \tilde{\bar{\omega}}^2 \end{bmatrix} = B^{-1} \begin{bmatrix} \bar{\omega}^1 \\ \bar{\omega}^2 \end{bmatrix}. \tag{4.5}$$

(b) Show that

$$\begin{bmatrix} \tilde{\bar{\omega}}^3_1 \\ \tilde{\bar{\omega}}^3_2 \end{bmatrix} = B^{-1} \begin{bmatrix} \bar{\omega}^3_1 \\ \bar{\omega}^3_2 \end{bmatrix} = {}^tB \begin{bmatrix} \bar{\omega}^3_1 \\ \bar{\omega}^3_2 \end{bmatrix}. \tag{4.6}$$

(Hint: Use the equation for $d\mathbf{e}_3$ in (3.1).)

(c) Cartan's lemma implies that there exist functions $\tilde{h}_{11}, \tilde{h}_{12}, \tilde{h}_{22}$ on U such that

$$\begin{bmatrix} \tilde{\bar{\omega}}^3_1 \\ \tilde{\bar{\omega}}^3_2 \end{bmatrix} = \begin{bmatrix} \tilde{h}_{11} & \tilde{h}_{12} \\ \tilde{h}_{12} & \tilde{h}_{22} \end{bmatrix} \begin{bmatrix} \tilde{\bar{\omega}}^1 \\ \tilde{\bar{\omega}}^2 \end{bmatrix}.$$

Show that

$$\begin{bmatrix} \tilde{h}_{11} & \tilde{h}_{12} \\ \tilde{h}_{12} & \tilde{h}_{22} \end{bmatrix} = B^{-1} \begin{bmatrix} h_{11} & h_{12} \\ h_{12} & h_{22} \end{bmatrix} B = {}^tB \begin{bmatrix} h_{11} & h_{12} \\ h_{12} & h_{22} \end{bmatrix} B. \tag{4.7}$$

Recall from linear algebra that any symmetric matrix can be transformed to a *diagonal* matrix by just such an orthogonal change of basis. Therefore, for each $\mathbf{u} \in U$ there exists an adapted frame $(\mathbf{e}_1(\mathbf{u}), \mathbf{e}_2(\mathbf{u}), \mathbf{e}_3(\mathbf{u}))$ at the point $\mathbf{x}(\mathbf{u}) \in \Sigma$ with the property that

$$\begin{bmatrix} h_{11}(\mathbf{u}) & h_{12}(\mathbf{u}) \\ h_{12}(\mathbf{u}) & h_{22}(\mathbf{u}) \end{bmatrix} = \begin{bmatrix} \kappa_1(\mathbf{u}) & 0 \\ 0 & \kappa_2(\mathbf{u}) \end{bmatrix} \tag{4.8}$$

for some real numbers $\kappa_1(\mathbf{u}), \kappa_2(\mathbf{u})$.

***Exercise 4.29.** Let $(\mathbf{e}_1(\mathbf{u}), \mathbf{e}_2(\mathbf{u}), \mathbf{e}_3(\mathbf{u}))$ be an adapted frame satisfying (4.8), and let $\mathbf{v}_1, \mathbf{v}_2 \in T_{\mathbf{u}}U$ be vectors with the property that $d\mathbf{x}(\mathbf{v}_i) = \mathbf{e}_i(\mathbf{u})$ for $i = 1, 2$. Show that

$$d(\mathbf{e}_3)_{\mathbf{u}}(\mathbf{v}_i) = dN_{\mathbf{x}(\mathbf{u})}(\mathbf{e}_i(\mathbf{u})) = -\kappa_i(\mathbf{u})\mathbf{e}_i(\mathbf{u}), \quad i = 1, 2.$$

This implies that $\mathbf{e}_1(\mathbf{u})$ and $\mathbf{e}_2(\mathbf{u})$ are eigenvectors for the linear transformation $dN_{\mathbf{x}(\mathbf{u})}$, the differential of the Gauss map $N : \Sigma \to \mathbb{S}^2$ at the point $\mathbf{x}(\mathbf{u}) \in \Sigma$, with eigenvalues $-\kappa_1(\mathbf{u}), -\kappa_2(\mathbf{u})$, respectively.

You may recall the following definition:

Definition 4.30. The eigenvectors for $-dN_{\mathbf{x}(\mathbf{u})}$ are called *principal vectors* or *principal directions* at the point $\mathbf{x}(\mathbf{u}) \in \Sigma$. The associated eigenvalues $\kappa_1(\mathbf{u})$, $\kappa_2(\mathbf{u})$ are called the *principal curvatures* of Σ at $\mathbf{x}(\mathbf{u})$.

Therefore, there exists an adapted frame $(\mathbf{e}_1(\mathbf{u}), \mathbf{e}_2(\mathbf{u}), \mathbf{e}_3(\mathbf{u}))$ at each point $\mathbf{x}(\mathbf{u}) \in \Sigma$ with the property that $\mathbf{e}_1(\mathbf{u})$ and $\mathbf{e}_2(\mathbf{u})$ are principal vectors at $\mathbf{x}(\mathbf{u})$. Such a frame will be called a *principal adapted frame* at the point $\mathbf{x}(\mathbf{u}) \in \Sigma$, and an adapted frame field on Σ which has this property at every point $\mathbf{x}(\mathbf{u}) \in \Sigma$ will be called a *principal adapted frame field* on Σ.

Definition 4.31. If $\kappa_1(\mathbf{u}) = \kappa_2(\mathbf{u})$ for some point $\mathbf{u} \in U$, then the corresponding point $\mathbf{x}(\mathbf{u})$ of Σ is called an *umbilic point* of Σ.

If Σ has no umbilic points, then a principal adapted frame field can be determined uniquely (up to sign) by requiring that $\kappa_1 > \kappa_2$; moreover, this frame field determines a smooth map $\tilde{\mathbf{x}} : U \to \mathbb{E}(3)$. However, it can happen that a principal adapted frame field cannot be chosen smoothly in a neighborhood of an umbilic point; for this reason, umbilic points can be somewhat problematic.

***Exercise 4.32.** (a) Show that if Σ has no umbilic points, then the choice of a principal adapted frame field $(\mathbf{e}_1(\mathbf{u}), \mathbf{e}_2(\mathbf{u}), \mathbf{e}_3(\mathbf{u}))$ is equivariant (up to sign) under the action of $E(3)$.

(b) Show that for a principal adapted frame field, the second fundamental form is given by

$$\mathrm{II} = \kappa_1(\bar{\omega}^1)^2 + \kappa_2(\bar{\omega}^2)^2.$$

Remark 4.33. Exactly how much freedom does the phrase "up to sign" represent? Given any principal adapted frame field $(\mathbf{e}_1(\mathbf{u}), \mathbf{e}_2(\mathbf{u}), \mathbf{e}_3(\mathbf{u}))$, we can

(1) replace $\mathbf{e}_3(\mathbf{u})$ by $-\mathbf{e}_3(\mathbf{u})$;

(2) depending on whether or not we changed the sign of $\mathbf{e}_3(\mathbf{u})$, replace one or both of $\mathbf{e}_1(\mathbf{u})$ and $\mathbf{e}_2(\mathbf{u})$ by their opposites so as to preserve the orientation of the basis $(\mathbf{e}_1(\mathbf{u}), \mathbf{e}_2(\mathbf{u}), \mathbf{e}_3(\mathbf{u}))$. (We might also exchange $\mathbf{e}_1(\mathbf{u})$ and $\mathbf{e}_2(\mathbf{u})$ with appropriately chosen signs, but for the most part, we will ignore this option.)

So the other choices for a principal adapted frame field are

$$\begin{gathered}(-\mathbf{e}_1(\mathbf{u}), \mathbf{e}_2(\mathbf{u}), -\mathbf{e}_3(\mathbf{u})),\\ (\mathbf{e}_1(\mathbf{u}), -\mathbf{e}_2(\mathbf{u}), -\mathbf{e}_3(\mathbf{u})),\\ (-\mathbf{e}_1(\mathbf{u}), -\mathbf{e}_2(\mathbf{u}), \mathbf{e}_3(\mathbf{u})).\end{gathered}$$

***Exercise 4.34.** For each of the principal adapted frame fields in Remark 4.33, how do the sign changes to the frame vectors affect the Maurer-Cartan forms $(\bar{\omega}^i, \bar{\omega}^i_j)$?

Suppose that $\Sigma = \mathbf{x}(U)$ has no umbilic points. Now that we (finally!) have a way of defining a canonical adapted frame field along Σ, we can apply Lemma 4.2 to find a complete set of invariants for the surface.

Theorem 4.35 (Bonnet). *Let $U \subset \mathbb{R}^2$ be an open set. Two immersions $\mathbf{x}_1, \mathbf{x}_2 : U \to \mathbb{E}^3$ without umbilic points differ by a rigid motion if and only if they have the same first and second fundamental forms.*

Remark 4.36. This theorem is true even for surfaces with umbilic points, but the proof is slightly more involved due to the issue of how to choose a canonical adapted frame field near umbilic points.

PROOF. One direction is clear: Since all our constructions are equivariant under the action of $E(3)$, any two surfaces that differ by a rigid motion must have the same first and second fundamental forms.

Conversely, suppose that $\mathbf{x}_1, \mathbf{x}_2$ have the same first and second fundamental forms. Let $\tilde{\mathbf{x}}_1, \tilde{\mathbf{x}}_2 : U \to E(3)$ be principal adapted frame fields for $\mathbf{x}_1, \mathbf{x}_2$, respectively; let $(\bar{\omega}^i, \bar{\omega}^i_j)$ denote the pulled-back dual and connection forms for $\mathbf{x}_1$ and let $(\bar{\Omega}^i, \bar{\Omega}^i_j)$ denote those for $\mathbf{x}_2$. By hypothesis,

$$\mathrm{I}_{\mathbf{x}_1} = (\bar{\omega}^1)^2 + (\bar{\omega}^2)^2 = (\bar{\Omega}^1)^2 + (\bar{\Omega}^2)^2 = \mathrm{I}_{\mathbf{x}_2},$$
$$\mathrm{II}_{\mathbf{x}_1} = (\kappa_1)_{\mathbf{x}_1}(\bar{\omega}^1)^2 + (\kappa_2)_{\mathbf{x}_1}(\bar{\omega}^2)^2 = (\kappa_1)_{\mathbf{x}_2}(\bar{\Omega}^1)^2 + (\kappa_2)_{\mathbf{x}_2}(\bar{\Omega}^2)^2 = \mathrm{II}_{\mathbf{x}_2}.$$

Equality of the first fundamental forms implies that

$$\begin{bmatrix} \bar{\Omega}^1 \\ \bar{\Omega}^2 \end{bmatrix} = \begin{bmatrix} \cos(\theta) & -\sin(\theta) \\ \pm\sin(\theta) & \pm\cos(\theta) \end{bmatrix} \begin{bmatrix} \bar{\omega}^1 \\ \bar{\omega}^2 \end{bmatrix}$$

for some function θ on U, where the signs on the bottom row of the matrix are the same. Substituting this relation into the equation for the second fundamental forms yields

$$\begin{aligned}(\kappa_1)_{\mathbf{x}_1}(\bar{\omega}^1)^2 + (\kappa_2)_{\mathbf{x}_1}(\bar{\omega}^2)^2 &= \left((\kappa_1)_{\mathbf{x}_2}\cos^2(\theta) + (\kappa_2)_{\mathbf{x}_2}\sin^2(\theta)\right)(\bar{\omega}^1)^2 \\ &\quad + 2\left(((\kappa_2)_{\mathbf{x}_2} - (\kappa_1)_{\mathbf{x}_2})\sin(\theta)\cos(\theta)\right)\bar{\omega}^1\bar{\omega}^2 \\ &\quad + \left((\kappa_1)_{\mathbf{x}_2}\sin^2(\theta) + (\kappa_2)_{\mathbf{x}_2}\cos^2(\theta)\right)(\bar{\omega}^2)^2.\end{aligned}$$

Since $(\kappa_1)_{\mathbf{x}_2} > (\kappa_2)_{\mathbf{x}_2}$, the vanishing of the middle term on the right-hand side implies that θ is a multiple of $\frac{\pi}{2}$. Then the remaining terms, together with the inequality $\kappa_1 > \kappa_2$ on both sides, imply that θ is a multiple of π. Therefore, $(\kappa_1)_{\mathbf{x}_2} = (\kappa_1)_{\mathbf{x}_1}$, $(\kappa_2)_{\mathbf{x}_2} = (\kappa_2)_{\mathbf{x}_1}$, and

$$\bar{\Omega}^1 = \pm\bar{\omega}^1, \qquad \bar{\Omega}^2 = \pm\bar{\omega}^2.$$

By making one of the permissible frame changes described in Remark 4.33 on one side or the other if necessary, we can arrange that both signs above are positive.

Since we now have $\bar{\Omega}^1 = \bar{\omega}^1$, we must have $d\bar{\Omega}^1 = d\bar{\omega}^1$. According to the Cartan structure equations (3.8), this implies that

$$(\bar{\Omega}^1_2 - \bar{\omega}^1_2) \wedge \bar{\omega}^2 = 0.$$

By Cartan's lemma, $\bar{\Omega}^1_2 - \bar{\omega}^1_2$ must be a multiple of $\bar{\omega}^2$. But the same reasoning applied to the equation $d\bar{\Omega}^2 = d\bar{\omega}^2$ implies that $\bar{\Omega}^1_2 - \bar{\omega}^1_2$ must also be a multiple of $\bar{\omega}^1$. Since $(\bar{\omega}^1, \bar{\omega}^2)$ are linearly independent, it follows that

$$\bar{\Omega}^1_2 = \bar{\omega}^1_2.$$

Finally, since $\tilde{\mathbf{x}}_1$ and $\tilde{\mathbf{x}}_2$ are both principal adapted frame fields, we have

$$\bar{\Omega}^3_1 = (\kappa_1)_{\mathbf{x}_2}\bar{\Omega}^1 = (\kappa_1)_{\mathbf{x}_1}\bar{\omega}^1 = \bar{\omega}^3_1,$$
$$\bar{\Omega}^3_2 = (\kappa_2)_{\mathbf{x}_2}\bar{\Omega}^2 = (\kappa_2)_{\mathbf{x}_1}\bar{\omega}^2 = \bar{\omega}^3_2.$$

The theorem now follows from Lemma 4.2. □

Now we consider the question discussed in §4.4; namely, can the first and second fundamental forms be prescribed arbitrarily? We must require that I be a *positive definite* quadratic form (i.e., that $\mathrm{I}(\mathbf{v}) > 0$ for every $\mathbf{v} \neq \mathbf{0} \in TU$) in order to define a metric on the surface. And in order to avoid the issue of umbilic points, we will assume that I and II are prescribed in such a way that $\mathrm{II}_{\mathbf{u}}$ is not a scalar multiple of $\mathrm{I}_{\mathbf{u}}$ at any point $\mathbf{u} \in U$.

Exercise 4.37. Why is this the right condition to impose on I and II in order to avoid umbilic points?

We saw in the proof of Theorem 4.35 that prescribing these fundamental forms determines the 1-forms $(\bar{\omega}^1, \bar{\omega}^2, \bar{\omega}^3_1, \bar{\omega}^3_2)$ associated to a principal adapted frame field up to sign and that these forms will have the properties that $(\bar{\omega}^1, \bar{\omega}^2)$ are linearly independent and that $\bar{\omega}^3_i$ is a multiple of $\bar{\omega}^i$ for $i = 1, 2$. So, suppose that we are given 1-forms $(\bar{\omega}^1, \bar{\omega}^2, \bar{\omega}^3_1, \bar{\omega}^3_2)$ on an open set $U \subset \mathbb{R}^2$ that satisfy these conditions. What additional conditions must these forms satisfy in order that there exist an embedding $\mathbf{x} : U \to \mathbb{E}^3$ whose first and second fundamental forms are

$$\mathrm{I} = (\bar{\omega}^1)^2 + (\bar{\omega}^2)^2,$$
$$\mathrm{II} = \bar{\omega}^3_1\,\bar{\omega}^1 + \bar{\omega}^3_2\,\bar{\omega}^2?$$

Lemma 4.12 gives the answer: The forms $\bar{\omega}^1$, $\bar{\omega}^2$, $\bar{\omega}^3_1 = -\bar{\omega}^1_3$, $\bar{\omega}^3_2 = -\bar{\omega}^2_3$, together with the form $\bar{\omega}^3 = 0$ and some additional form $\bar{\omega}^1_2 = -\bar{\omega}^2_1$, must satisfy the structure equations (3.8) for the Maurer-Cartan forms on $E(3)$. Because $\bar{\omega}^3 = 0$, the first three of these equations may be written as

$$\begin{aligned} d\bar{\omega}^1 &= -\bar{\omega}^1_2 \wedge \bar{\omega}^2, \\ d\bar{\omega}^2 &= \bar{\omega}^1_2 \wedge \bar{\omega}^1, \\ d\bar{\omega}^3 &= 0 = -\bar{\omega}^3_1 \wedge \bar{\omega}^1 - \bar{\omega}^3_2 \wedge \bar{\omega}^2. \end{aligned} \tag{4.9}$$

***Exercise 4.38.** Show that the first two equations in (4.9) uniquely determine the 1-form $\bar{\omega}^1_2$. (Hint: $\bar{\omega}^1_2$ must be equal to some linear combination of $(\bar{\omega}^1, \bar{\omega}^2)$. Show that each of the first two equations determines one of the unknown coefficients.)

The form $\bar{\omega}_2^1$ determined by the first two equations in (4.9) is called the *Levi-Civita connection form* of the metric defined by the first fundamental form $\mathrm{I} = (\bar{\omega}^1)^2 + (\bar{\omega}^2)^2$. The third equation just says that $(\bar{\omega}_1^3, \bar{\omega}_2^3)$ must be symmetric linear combinations of $(\bar{\omega}^1, \bar{\omega}^2)$, which will automatically be true under our assumptions.

The remaining structure equations may be written as

$$\begin{aligned} d\bar{\omega}_2^1 &= \bar{\omega}_1^3 \wedge \bar{\omega}_2^3, \\ d\bar{\omega}_1^3 &= \bar{\omega}_2^3 \wedge \bar{\omega}_2^1, \\ d\bar{\omega}_2^3 &= -\bar{\omega}_1^3 \wedge \bar{\omega}_2^1. \end{aligned} \tag{4.10}$$

The first of these equations is called the *Gauss equation*, and the last two are called the *Codazzi-Mainardi equations*, or simply the *Codazzi equations.* By Lemma 4.12, we have the following theorem:

Theorem 4.39 (Bonnet). *Let $(\bar{\omega}^1, \bar{\omega}^2, \bar{\omega}_1^3, \bar{\omega}_2^3)$ be 1-forms on a connected and simply connected open set $U \subset \mathbb{R}^2$ satisfying the conditions that $(\bar{\omega}^1, \bar{\omega}^2)$ are linearly independent at each point of U and that $\bar{\omega}_i^3$ is a scalar multiple of $\bar{\omega}^i$ for $i = 1, 2$. Suppose that, together with the Levi-Civita connection form $\bar{\omega}_2^1$ determined by $\bar{\omega}^1$ and $\bar{\omega}^2$, these forms satisfy the Gauss and Codazzi equations* (4.10)*. Then there exists an immersed surface $\mathbf{x} : U \to \mathbb{E}^3$, unique up to rigid motion, whose first and second fundamental forms are*

$$\begin{aligned} \mathrm{I} &= (\bar{\omega}^1)^2 + (\bar{\omega}^2)^2, \\ \mathrm{II} &= \bar{\omega}_1^3 \, \bar{\omega}^1 + \bar{\omega}_2^3 \, \bar{\omega}^2. \end{aligned}$$

Because of this result, the Gauss and Codazzi equations are also referred to as the *compatibility equations* of the theory of surfaces in $\mathbb{E}^3$.

Exercise 4.40. Let (u, v) be local coordinates on $\mathbb{R}^2$. Use the following steps to determine whether there exists an immersion $\mathbf{x} : \mathbb{R}^2 \to \mathbb{E}^3$ with first and second fundamental forms

$$\begin{aligned} \mathrm{I} &= \cosh^2(v) \, (du^2 + dv^2), \\ \mathrm{II} &= du^2 - dv^2. \end{aligned}$$

(a) Show that the 1-forms $(\bar{\omega}^1, \bar{\omega}^2, \bar{\omega}_1^3, \bar{\omega}_2^3)$ determined by I and II according to the conditions of Theorem 4.39 are

$$\begin{aligned} \bar{\omega}^1 &= \cosh(v) \, du, & \bar{\omega}^2 &= \cosh(v) \, dv, \\ \bar{\omega}_1^3 &= \frac{1}{\cosh(v)} \, du, & \bar{\omega}_2^3 &= -\frac{1}{\cosh(v)} \, dv. \end{aligned}$$

(b) Show that the Levi-Civita connection form is

$$\bar{\omega}_2^1 = \tanh(v)\, du.$$

(Hint: Set

$$\bar{\omega}_2^1 = a\, du + b\, dv$$

for some unknown functions a, b on $\mathbb{R}^2$. Use the structure equations for $d\bar{\omega}^1$ and $d\bar{\omega}^2$ to determine a and b.)

(c) Check that these forms satisfy the Gauss and Codazzi equations.

Therefore, Theorem 4.39 implies that the desired surface exists. (In fact, it is a catenoid.)

***Exercise 4.41.** This exercise is a continuation of Exercise 4.24. Suppose that $\mathbf{x} : U \to \mathbb{E}^3$ is an immersion whose coordinate curves are all principal curves. (This means that $\mathbf{x}_u, \mathbf{x}_v$ are both principal vectors at each point of $\Sigma = \mathbf{x}(U)$.)

(a) Show that the frame field in part (a) of Exercise 4.24 is a principal adapted frame field along Σ.

(b) Show that the condition that all coordinate curves of $\mathbf{x}$ are principal curves is equivalent to the condition that the first and second fundamental forms

$$\mathrm{I} = E\, du^2 + 2F\, du\, dv + G\, dv^2,$$
$$\mathrm{II} = e\, du^2 + 2f\, du\, dv + g\, dv^2$$

have the property that $F = f = 0$.

(c) Use the structure equations for the dual forms in part (b) of Exercise 4.24 to show that

$$\bar{\omega}_2^1 = \frac{1}{2\sqrt{EG}}\,(E_v\, du - G_u\, dv).$$

(d) Show that

$$\bar{\omega}_1^3 = \frac{e}{E}\,\bar{\omega}^1 = \frac{e}{\sqrt{E}}\, du,$$
$$\bar{\omega}_2^3 = \frac{g}{G}\,\bar{\omega}^2 = \frac{g}{\sqrt{G}}\, dv.$$

(e) Show that the Gauss equation is equivalent to

$$\frac{eg}{EG} = -\frac{1}{2\sqrt{EG}}\left[\left(\frac{E_v}{\sqrt{EG}}\right)_v + \left(\frac{G_u}{\sqrt{EG}}\right)_u\right]. \tag{4.11}$$

(f) Show that the Codazzi equations are equivalent to

$$
\begin{aligned}
e_v &= \frac{1}{2}E_v\left(\frac{e}{E}+\frac{g}{G}\right),\\
g_u &= \frac{1}{2}G_u\left(\frac{e}{E}+\frac{g}{G}\right).
\end{aligned}
\tag{4.12}
$$

While isolated umbilic points on a surface can be problematic, it is natural to ask whether we can categorize those surfaces that are *totally umbilic*, i.e., surfaces with the property that every point is an umbilic point.

Exercise 4.42. Suppose that the surface $\mathbf{x} : U \to \mathbb{E}^3$ is totally umbilic.

(a) Show that any adapted frame field $(\mathbf{e}_1(\mathbf{u}), \mathbf{e}_2(\mathbf{u}), \mathbf{e}_3(\mathbf{u}))$ is a principal adapted frame field.

(b) Let $(\bar{\omega}^i, \bar{\omega}^i_j)$ be the Maurer-Cartan forms for an adapted frame field on $\Sigma = \mathbf{x}(U)$. Show that there exists a smooth function $\lambda : U \to \mathbb{R}$ such that

$$\bar{\omega}^3_1 = \lambda\bar{\omega}^1, \qquad \bar{\omega}^3_2 = \lambda\bar{\omega}^2. \tag{4.13}$$

Conclude that the second fundamental form of Σ is a scalar multiple of the first fundamental form, i.e., that

$$\mathrm{II} = \lambda\mathrm{I}.$$

(c) Prove that λ is constant. (Hint: Use the structure equations to differentiate equations (4.13), taking into account the fact that we must have

$$d\lambda = \lambda_1\bar{\omega}^1 + \lambda_2\bar{\omega}^2$$

for some functions λ_1, λ_2 on U. Then use Cartan's lemma.)

(d) Show that if $\lambda = 0$, then $d\mathbf{e}_3 = 0$. Conclude that the normal vector field of Σ is constant and that Σ is therefore contained in a plane.

(e) Show that if $\lambda \neq 0$, then $d(\mathbf{x}+\frac{1}{\lambda}\mathbf{e}_3) = 0$. Conclude that the vector-valued function $\mathbf{x} + \frac{1}{\lambda}\mathbf{e}_3 : U \to \mathbb{E}^3$ is equal to some constant point $\mathbf{q} \in \mathbb{E}^3$ and that Σ is therefore contained in the sphere of radius $\frac{1}{|\lambda|}$ centered at $\mathbf{q}$.

Thus, the only totally umbilic surfaces in $\mathbb{E}^3$ are (open subsets of) planes and spheres.

One of the conclusions of Exercise 4.42 is that if the principal curvatures κ_1, κ_2 of a regular surface Σ are equal at every point of Σ, then they must in fact be constant. This suggests a related question: Can we categorize those surfaces for which κ_1, κ_2 are constants, but not necessarily equal?

Exercise 4.43. Suppose that the surface $\mathbf{x} : U \to \mathbb{E}^3$ has the property that both principal curvatures κ_1, κ_2 are constants. We know from Exercise 4.42 that if $\kappa_1 = \kappa_2$, then $\Sigma = \mathbf{x}(U)$ is contained in either a plane or a sphere, so assume that $\kappa_1 \neq \kappa_2$. Let $(\mathbf{e}_1(\mathbf{u}), \mathbf{e}_2(\mathbf{u}), \mathbf{e}_3(\mathbf{u}))$ be a principal adapted frame field on Σ, and let $\tilde{\mathbf{x}} : U \to E(3)$ denote the corresponding lifting of $\mathbf{x} : U \to \mathbb{E}^3$, with associated Maurer-Cartan forms $(\bar{\omega}^i, \bar{\omega}^i_j)$. Then we have

$$\bar{\omega}^3_1 = \kappa_1 \bar{\omega}^1, \qquad \bar{\omega}^3_2 = \kappa_2 \bar{\omega}^2. \tag{4.14}$$

(a) Differentiate equations (4.14) to obtain

$$(\kappa_1 - \kappa_2)\bar{\omega}^1_2 \wedge \bar{\omega}^1 = (\kappa_1 - \kappa_2)\bar{\omega}^1_2 \wedge \bar{\omega}^2 = 0.$$

Use Cartan's lemma to conclude that $\bar{\omega}^1_2 = 0$.

(b) Differentiate the equation $\bar{\omega}^1_2 = 0$ and show that $\kappa_1\kappa_2 = 0$. Without loss of generality, we may assume that $\kappa_1 = 0$, $\kappa_2 \neq 0$.

In the remainder of this exercise, we will see how the structure equations can be *integrated* in order to determine the surface Σ.

(c) Show that $d\bar{\omega}^1 = d\bar{\omega}^2 = 0$. Apply the Poincaré lemma (cf. Theorem 2.31) to conclude that there exist functions u, v on U such that

$$\bar{\omega}^1 = du, \qquad \bar{\omega}^2 = dv.$$

Thus the pullbacks of the structure equations (3.1) to U via $\tilde{\mathbf{x}}$ can be written as

$$\begin{aligned} d\mathbf{x} &= \mathbf{e}_1\, du + \mathbf{e}_2\, dv, \\ d\mathbf{e}_1 &= 0, \\ d\mathbf{e}_2 &= \mathbf{e}_3(\kappa_2\, dv), \\ d\mathbf{e}_3 &= -\mathbf{e}_2(\kappa_2\, dv). \end{aligned} \tag{4.15}$$

(d) Integrate equations (4.15) (beginning with the equations for $d\mathbf{e}_1, d\mathbf{e}_2, d\mathbf{e}_3$ and working backwards) to show that there exist *constant* vectors $\bar{\mathbf{e}}_1$, $\bar{\mathbf{e}}_2$, $\bar{\mathbf{e}}_3$, $\bar{\mathbf{x}} \in \mathbb{E}^3$ such that

$$\begin{aligned} \mathbf{e}_1(u,v) &= \bar{\mathbf{e}}_1, \\ \mathbf{e}_2(u,v) &= \cos(\kappa_2 v)\,\bar{\mathbf{e}}_2 + \sin(\kappa_2 v)\,\bar{\mathbf{e}}_3, \\ \mathbf{e}_3(u,v) &= -\sin(\kappa_2 v)\,\bar{\mathbf{e}}_2 + \cos(\kappa_2 v)\,\bar{\mathbf{e}}_3, \\ \mathbf{x}(u,v) &= \bar{\mathbf{x}} + u\,\bar{\mathbf{e}}_1 + \frac{1}{\kappa_2}\sin(\kappa_2 v)\,\bar{\mathbf{e}}_2 - \frac{1}{\kappa_2}\cos(\kappa_2 v)\,\bar{\mathbf{e}}_3. \end{aligned} \tag{4.16}$$

(e) Use equations (4.16) and the fact that $(\mathbf{e}_1(\mathbf{u}), \mathbf{e}_2(\mathbf{u}), \mathbf{e}_3(\mathbf{u}))$ is an orthonormal frame field to show that $(\bar{\mathbf{e}}_1, \bar{\mathbf{e}}_2, \bar{\mathbf{e}}_3)$ is an orthonormal frame.

Conclude that via a Euclidean transformation, we can arrange that $\bar{\mathbf{x}} = \mathbf{0}$ and

$$\bar{\mathbf{e}}_1 = \begin{bmatrix} 1 \\ 0 \\ 0 \end{bmatrix}, \qquad \bar{\mathbf{e}}_2 = \begin{bmatrix} 0 \\ 1 \\ 0 \end{bmatrix}, \qquad \bar{\mathbf{e}}_3 = \begin{bmatrix} 0 \\ 0 \\ 1 \end{bmatrix},$$

and hence that

$$\mathbf{x}(u, v) = \begin{bmatrix} u \\ \frac{1}{\kappa_2}\sin(\kappa_2 v) \\ -\frac{1}{\kappa_2}\cos(\kappa_2 v) \end{bmatrix}. \tag{4.17}$$

Equation (4.17) describes a parametrization for the cylinder of radius $\frac{1}{|\kappa_2|}$ centered along the x^1-axis; therefore, $\Sigma = \mathbf{x}(U)$ is contained in a cylinder of radius $\frac{1}{|\kappa_2|}$.

Together, Exercises 4.42 and 4.43 prove the following classification theorem:

Theorem 4.44. *Let Σ be a connected, regular surface in $\mathbb{E}^3$ whose principal curvatures are constant. Then Σ is contained in either a plane, sphere, or cylinder.*

Any invariant of an immersed surface $\mathbf{x} : U \to \mathbb{E}^3$ that can be expressed purely in terms of the first fundamental form

$$\mathrm{I} = (\bar{\omega}^1)^2 + (\bar{\omega}^2)^2$$

is called an *intrinsic* invariant of the surface. For instance, arc length and area are intrinsic quantities on $\Sigma = \mathbf{x}(U)$. The principal curvatures κ_1, κ_2, however, are not intrinsic; they depend not only on the metric, but also on how the surface is immersed.

Two important notions of curvature for surfaces are given in the following definition:

Definition 4.45. The function $K = \kappa_1\kappa_2$ on Σ is called the *Gauss curvature* of Σ. The function $H = \frac{1}{2}(\kappa_1 + \kappa_2)$ on Σ is called the *mean curvature* of Σ.

Remark 4.46. It is not necessary that an adapted frame field be principal in order to compute the Gauss and mean curvatures. For any adapted frame field on Σ with associated matrix $[h_{ij}]$, we have

$$K = \det[h_{ij}], \qquad H = \tfrac{1}{2}\mathrm{tr}[h_{ij}].$$

Even though κ_1, κ_2 are not intrinsic quantities, Gauss's "Theorema Egregium" states that their product K is, in fact, intrinsic. (The mean curvature H, however, is *not* intrinsic.) In the following exercise, we will prove this in several steps. (For simplicity, we will assume that the surface is oriented, meaning that a choice of $\mathbf{e}_3$ has been specified.)

Exercise 4.47. Let $\mathbf{x} : U \to \mathbb{E}^3$ be an immersed surface. The 1-forms $(\bar{\omega}^1, \bar{\omega}^2)$ are determined by the first fundamental form of $\mathbf{x}$ up to a transformation of the form

$$\begin{bmatrix} \tilde{\bar{\omega}}^1 \\ \tilde{\bar{\omega}}^2 \end{bmatrix} = \begin{bmatrix} \cos(\theta) & \sin(\theta) \\ -\sin(\theta) & \cos(\theta) \end{bmatrix} \begin{bmatrix} \bar{\omega}^1 \\ \bar{\omega}^2 \end{bmatrix}$$

for some function θ on U.

(a) Show that the area form

$$dA = \bar{\omega}^1 \wedge \bar{\omega}^2$$

is an intrinsic quantity, i.e., that

$$d\tilde{A} = \tilde{\bar{\omega}}^1 \wedge \tilde{\bar{\omega}}^2 = dA.$$

(Note: The notation dA for the area form is traditional, but the d does *not* signify that dA is the exterior derivative of some 1-form.)

(b) Show that if $\bar{\omega}^1_2$ is the Levi-Civita connection form corresponding to $(\bar{\omega}^1, \bar{\omega}^2)$, then

$$\tilde{\bar{\omega}}^1_2 = \bar{\omega}^1_2 - d\theta.$$

Conclude that $d\bar{\omega}^1_2$ is an intrinsic quantity.

(c) Show that

$$d\bar{\omega}^1_2 = \bar{\omega}^3_1 \wedge \bar{\omega}^3_2 = K\,\bar{\omega}^1 \wedge \bar{\omega}^2 = K\,dA.$$

Conclude that K must be an intrinsic quantity. (Note that this is simply another version of equation (4.11), which expresses the Gauss curvature $K = \frac{eg}{EG}$ as a function of E, G, and their derivatives.)

Surfaces for which the mean curvature H is identically zero are called *minimal surfaces*; these surfaces are of considerable interest and will be treated in detail in Chapter 8. Surfaces for which the Gauss curvature K is identically zero are called *flat*, and we will conclude this chapter with a brief exploration of their local theory.

Because the Gauss curvature of a regular surface Σ is an intrinsic quantity, it is not changed by any deformation of Σ that preserves the first fundamental form of Σ. Intuitively, this means that the surface may be smoothly bent and/or twisted, but not stretched or contracted. So for instance, any surface that can be obtained by smoothly bending a sheet of paper must be flat.

Since any flat surface Σ must have $K = \kappa_1\kappa_2 = 0$, one of the principal curvatures κ_1, κ_2 must be identically zero on Σ. As we saw in Exercise 4.42 that any surface with $\kappa_1 = \kappa_2 = 0$ must be contained in a plane, we will disregard this case and, to keep things simple, we will assume that Σ has no umbilic points. (In practice, this simply means that we restrict our attention to the open subset of Σ consisting of the non-umbilic points.)

Exercise 4.48. Suppose that the surface $\mathbf{x} : U \to \mathbb{E}^3$ is flat and has no umbilic points. Without loss of generality, we may assume that the principal curvatures of $\Sigma = \mathbf{x}(U)$ satisfy $\kappa_1 = 0$, $\kappa_2 \neq 0$. Let $(\mathbf{e}_1(\mathbf{u}), \mathbf{e}_2(\mathbf{u}), \mathbf{e}_3(\mathbf{u}))$ be a principal adapted frame field on Σ, and let $\tilde{\mathbf{x}} : U \to E(3)$ denote the corresponding lifting of $\mathbf{x} : U \to \mathbb{E}^3$, with associated Maurer-Cartan forms $(\bar{\omega}^i, \bar{\omega}^i_j)$. Then since $\kappa_1 = 0$, we have

$$\bar{\omega}^3_1 = 0, \qquad \bar{\omega}^3_2 = \kappa_2\bar{\omega}^2. \tag{4.18}$$

(a) Differentiate the equation $\bar{\omega}^3_1 = 0$ and use Cartan's lemma to conclude that

$$\bar{\omega}^1_2 = \mu\,\bar{\omega}^2 \tag{4.19}$$

for some function $\mu : U \to \mathbb{R}$.

(b) Show that $d\bar{\omega}^1 = 0$, and apply the Poincaré lemma (cf. Theorem 2.31) to conclude that there exists a function u on U such that $\bar{\omega}^1 = du$.

(c) Use the structure equation for $d\bar{\omega}^2$ and the Frobenius theorem (cf. Theorem 2.33) to conclude that for any point $\mathbf{u} \in U$, there exist a neighborhood $V \subset U$ of $\mathbf{u}$ and differentiable functions $\lambda, v : V \to \mathbb{R}$ (with $\lambda \neq 0$) such that the restriction of $\bar{\omega}^2$ to V is given by $\bar{\omega}^2 = \lambda\, dv$. (For simplicity, we will shrink U if necessary and assume that these functions are defined on the entire open set U.)

(d) Since $\bar{\omega}^1 \wedge \bar{\omega}^2 \neq 0$, the functions (u, v) form a local coordinate system on U, and we may regard λ, μ as functions of u and v. Show that the structure equation for $d\bar{\omega}^2$ implies that $\mu = -\frac{\lambda_u}{\lambda}$, and therefore

$$\bar{\omega}^1_2 = -\lambda_u\, dv. \tag{4.20}$$

(e) Use the structure equation for $d\bar{\omega}^1_2$ to conclude that $\lambda_{uu} = 0$, and therefore

$$\lambda(u, v) = u f_1(v) + f_0(v) \tag{4.21}$$

for some smooth functions $f_0(v), f_1(v)$.

(f) Use the structure equation for $d\bar{\omega}^3_2$ to conclude that

$$(\kappa_2\lambda)_u = 0.$$

Integrate and use equation (4.21) to conclude that

$$\kappa_2(u,v) = \frac{f_2(v)}{uf_1(v) + f_0(v)} \tag{4.22}$$

for some smooth functions $f_0(v), f_1(v), f_2(v)$.

(g) The pullbacks of the structure equations (3.1) to U via $\tilde{\mathbf{x}}$ can now be written as

$$\begin{aligned} d\mathbf{x} &= \mathbf{e}_1\, du + \mathbf{e}_2\, (uf_1(v) + f_0(v))dv, \\ d\mathbf{e}_1 &= \mathbf{e}_2\, f_1(v)\, dv, \\ d\mathbf{e}_2 &= -\mathbf{e}_1\, f_1(v)\, dv + \mathbf{e}_3\, f_2(v)\, dv, \\ d\mathbf{e}_3 &= -\mathbf{e}_2\, f_2(v)\, dv. \end{aligned} \tag{4.23}$$

Conclude that the u-parameter curves are straight line segments (and that u is an arc-length parameter along these curves) and hence that Σ is a ruled surface.

We have now proved the following theorem, keeping in mind that with the notation of Exercise 4.48, the mean curvature H of Σ is given by $H = \frac{1}{2}\kappa_2$ with κ_2 as in equation (4.22):

Theorem 4.49. *Let Σ be a flat surface whose mean curvature H is nonzero everywhere. Then for each point $\mathbf{x} \in \Sigma$, there exists a unique straight line $\ell_{\mathbf{x}}$ in $\mathbb{E}^3$ such that $\mathbf{x} \in \ell_{\mathbf{x}}$ and $\ell_{\mathbf{x}} \cap \Sigma$ is an open neighborhood of $\mathbf{x}$ in $\ell_{\mathbf{x}}$. Moreover, the restriction of the function $\frac{1}{H}$ to the open interval $\ell_{\mathbf{x}} \cap \Sigma$ is an affine linear function of the arc length parameter along this interval.*

A more traditional proof of this result is given in [**MR05**].

4.6. Maple computations

In order to get set up to use Maple for some of the exercises in this chapter, begin by loading the `Cartan` and `LinearAlgebra` packages into Maple:

```
> with(Cartan);
> with(LinearAlgebra);
```

Next, introduce the Maurer-Cartan forms on the frame bundle $\mathcal{F}(\mathbb{E}^3)$; these need to be declared so that Maple will recognize them as 1-forms. It suffices to declare a linearly independent subset; we'll define the others in terms of these shortly.

```
> Form(omega[1], omega[2], omega[3]);
  Form(omega[1,2], omega[3,1], omega[3,2]);
```

Next, tell MAPLE about the symmetries in the connection forms:

```
> omega[1,1]:= 0;
  omega[2,2]:= 0;
  omega[3,3]:= 0;
  omega[2,1]:= -omega[1,2];
  omega[1,3]:= -omega[3,1];
  omega[2,3]:= -omega[3,2];
```

Tell MAPLE how to differentiate these forms according to the Cartan structure equations (3.8):

```
> for i from 1 to 3 do
    d(omega[i]):= -add('omega[i,j] &^ omega[j]', j=1..3);
  end do;

  d(omega[1,2]):= -add('omega[1,k] &^ omega[k,2]', k=1..3);
  d(omega[3,1]):= -add('omega[3,k] &^ omega[k,1]', k=1..3);
  d(omega[3,2]):= -add('omega[3,k] &^ omega[k,2]', k=1..3);
```

Now consider the pullbacks of the Maurer-Cartan forms to the surface via an adapted frame field. The first condition that these forms must satisfy is $\bar{\omega}^3 = 0$.

In MAPLE, it's often useful to impose such conditions via a substitution rather than by simply setting $\bar{\omega}^3$ equal to zero. The reason for this is that if we make the assignment $\bar{\omega}^3 = 0$, we lose the ability to use the structure equation for $d\bar{\omega}^3$ because MAPLE will just evaluate $d\bar{\omega}^3$ as $d(0) = 0$. Using a substitution allows us to choose when we want MAPLE to be aware that $\bar{\omega}^3 = 0$ and when we don't.

So, introduce the following substitution for the Maurer-Cartan forms associated to an adapted frame field. (We'll add more information to this substitution as we learn more about the Maurer-Cartan forms.)

```
> adaptedsub1:= [omega[3]=0];
```

Now, since we have $\bar{\omega}^3 = 0$, we must have $d\bar{\omega}^3 = 0$ as well. So the following quantity must be zero:

```
> zero1:= Simf(subs(adaptedsub1, Simf(d(omega[3]))));
```

$$zero1 := (\omega_1) \mathbin{\&\hat{}} (\omega_{31}) + (\omega_2) \mathbin{\&\hat{}} (\omega_{32})$$

Note that we first computed $d\bar{\omega}^3$ and *then* applied the substitution to tell MAPLE that $\bar{\omega}^3 = 0$. In this case the knowledge that $\bar{\omega}^3 = 0$ didn't affect the computation of $d\bar{\omega}^3$, but it's a good idea to get in the habit of applying such substitutions when you intend for them to be in effect.

Applying Cartan's lemma tells us that $(\bar{\omega}^3_1, \bar{\omega}^3_2)$ must be symmetric linear combinations of $(\bar{\omega}^1, \bar{\omega}^2)$, so we add this information to our substitution:

```
> adaptedsub1:= [op(adaptedsub1),
  omega[3,1] = h[1,1]*omega[1] + h[1,2]*omega[2],
  omega[3,2] = h[1,2]*omega[1] + h[2,2]*omega[2]];
```

EXERCISE 4.28: In order to keep up with both the original Maurer-Cartan forms $(\bar{\omega}^i, \bar{\omega}^i_j)$ and the transformed forms $(\tilde{\bar{\omega}}^i, \tilde{\bar{\omega}}^i_j)$, introduce new 1-forms to represent the transformed forms, with the same symmetry conditions as the original forms:

```
> Form(Omega[1], Omega[2], Omega[3]);
  Form(Omega[1,2], Omega[3,1], Omega[3,2]);
  Omega[1,1]:= 0;
  Omega[2,2]:= 0;
  Omega[3,3]:= 0;
  Omega[2,1]:= -Omega[1,2];
  Omega[1,3]:= -Omega[3,1];
  Omega[2,3]:= -Omega[3,2];
```

(It won't be necessary to assign their exterior derivatives because these will be computed in terms of the exterior derivatives of the original forms when needed.)

We can introduce the relations (4.5), (4.6) via the following substitution:

```
> framechangesub:= [
  Omega[1]  = cos(theta)*omega[1] + sin(theta)*omega[2],
  Omega[2]  = -sin(theta)*omega[1] + cos(theta)*omega[2],
  Omega[3,1]  = cos(theta)*omega[3,1] + sin(theta)*omega[3,2],
  Omega[3,2]  = -sin(theta)*omega[3,1] + cos(theta)*omega[3,2]];
```

We'll also need the reverse substitution so that we can go back and forth between the two sets of Maurer-Cartan forms:

```
> framechangebacksub:= makebacksub(framechangesub);
```

In order to compare the functions $(\tilde{h}_{ij})$ associated to the transformed forms to the functions (h_{ij}) associated to the original forms, introduce another substitution describing the adaptations of the transformed frame:

```
> adaptedsub2:= [Omega[3]=0,
  Omega[3,1]  = H[1,1]*Omega[1]  + H[1,2]*Omega[2],
  Omega[3,2]  = H[1,2]*Omega[1]  + H[2,2]*Omega[2]];
```

Now combine all these substitutions to see how the $(\tilde{h}_{ij})$ are expressed in terms of the (h_{ij}): First, write $\tilde{\bar{\omega}}_1^3$ in terms of $(\bar{\omega}_1^3, \bar{\omega}_2^3)$:

```
> Simf(subs(framechangesub, Omega[3,1]));
```

$$\cos(\theta)\,\omega_{3,1} + \sin(\theta)\,\omega_{3,2}$$

Next, convert this to an expression in terms of the (h_{ij}) and $(\bar{\omega}^1, \bar{\omega}^2)$:

```
> Simf(subs(adaptedsub1, %));
```

$$(\cos(\theta)\,h_{1,1} + \sin(\theta)\,h_{1,2})\,\omega_1 + (\cos(\theta)\,h_{1,2} + \sin(\theta)\,h_{2,2})\,\omega_2$$

Finally, convert this to an expression in terms of $(\tilde{\bar{\omega}}^1, \tilde{\bar{\omega}}^2)$:

```
> Simf(subs(framechangebacksub, %));
```

$$(\cos(\theta)^2\,h_{1,1} + 2\cos(\theta)\sin(\theta)\,h_{1,2} + h_{2,2} - h_{2,2}\cos(\theta)^2)\,\Omega_1 \\ + (-\cos(\theta)\sin(\theta)\,h_{1,1} + \cos(\theta)\sin(\theta)\,h_{2,2} + 2\cos(\theta)^2\,h_{1,2} - h_{1,2})\Omega_2$$

Of course, this sequence of operations can be combined into a single command:

```
> Simf(subs(framechangebacksub, Simf(subs(adaptedsub1,
  Simf(subs(framechangesub, Omega[3,1]))))));
```

Now, the coefficients of $(\tilde{\bar{\omega}}^1, \tilde{\bar{\omega}}^2)$ in the output are, of course, equal to $\tilde{h}_{11}, \tilde{h}_{12}$, respectively. But in order to illustrate how we might handle a slightly more complicated situation, we will let MAPLE do the work of comparing this expression to our original expression for $\tilde{\bar{\omega}}_1^3$:

```
> zero2:= Simf(subs(adaptedsub2, Omega[3,1]) - %);
```

The coefficients of this expression must both be zero, which gives us two equations that can be solved for $\tilde{h}_{11}$ and $\tilde{h}_{12}$. These equations can be extracted as follows:

```
> eqns:= {op(ScalarForm(zero2))};
```

Before solving these equations, we might as well compute the analogous equations that result from consideration of $\tilde{\bar{\omega}}_2^3$. We can add these to our system of equations as follows:

```
> zero3:= Simf(subs(adaptedsub2, Omega[3,2])
  - Simf(subs(framechangebacksub, Simf(subs(adaptedsub1,
  Simf(subs(framechangesub, Omega[3,2])))))));
> eqns:= eqns union {op(ScalarForm(zero3))};
```

Now solve these equations for the functions $(\tilde{h}_{ij})$:

```
> solve(eqns, {H[1,1], H[1,2], H[2,2]});
```

Then, we might as well actually assign these values to the $(\tilde{h}_{ij})$:

```
> assign(%);
```

Now, it's not entirely obvious how to recognize these expressions as those of equation (4.7), but we can at least check that our computations are consistent with these expressions. First, define matrices $[h_{ij}], [\tilde{h}_{ij}], B$ as follows:

```
> hmatrix:= Matrix([[h[1,1], h[1,2]], [h[1,2], h[2,2]]]);
  Hmatrix:= Matrix([[H[1,1], H[1,2]], [H[1,2], H[2,2]]]);
  B:= Matrix([[cos(theta), -sin(theta)],
  [sin(theta), cos(theta)]]);
```

If everything has gone according to plan, the following matrix should be zero:

```
> Hmatrix - simplify(Transpose(B).hmatrix.B);
```

EXERCISE 4.40: Set up a substitution for the forms that we know from part (a), together with an expression for $\bar{\omega}_2^1$ with coefficients to be determined later:

```
> examplesub:= [omega[1] = cosh(v)*d(u),
  omega[2] = cosh(v)*d(v), omega[3]=0,
  omega[3,1] = d(u)/cosh(v), omega[3,2] = -d(v)/cosh(v),
  omega[1,2] = a*d(u) + b*d(v)];
```

Now, compute $d\bar{\omega}^1$ in two ways: by first making the substitution into $\bar{\omega}^1$ and then differentiating, and by applying the structure equations and then making the substitution. Then the difference of the resulting expressions must be equal to zero:

```
> Simf(d(Simf(subs(examplesub, omega[1])))
  - subs(examplesub, Simf(d(omega[1]))));
```

$$(\sinh(v) - \cosh(v)\,a)\,d(v)\ \&\hat{}\ d(u)$$

```
> a:= solve(%, a);
```

$$a := \frac{\sinh(v)}{\cosh(v)}$$

An analogous computation for $d\bar{\omega}^2$ yields $b = 0$; once we have made this assignment, we will have $\bar{\omega}_2^1 = \tanh(v)\, du$, as expected.

Finally, verifying the Gauss and Codazzi equations simply involves checking that both ways of computing the structure equations for each of the $(d\bar{\omega}_j^i)$ yield the same result:

```
> Simf(d(Simf(subs(examplesub, omega[1,2])))
  - subs(examplesub, Simf(d(omega[1,2]))));
```

$$0$$

```
> Simf(d(Simf(subs(examplesub, omega[3,1])))
  - subs(examplesub, Simf(d(omega[3,1]))));
```

$$0$$

```
> Simf(d(Simf(subs(examplesub, omega[3,2])))
  - subs(examplesub, Simf(d(omega[3,2]))));
```

$$0$$

EXERCISE 4.41: For a principal parametrization as in Exercise 4.24, we have

$$\bar{\omega}^1 = \sqrt{E}\, du, \qquad \bar{\omega}^2 = \sqrt{G}\, dv.$$

Then, in order for the second fundamental form to have the desired form, we must have

$$\bar{\omega}_1^3 = \frac{e}{\sqrt{E}}\, du, \qquad \bar{\omega}_2^3 = \frac{g}{\sqrt{G}}\, dv.$$

Moreover, $\bar{\omega}_2^1$ must be equal to some linear combination of du and dv.

Start by unassigning the variables a, b so that we can use them again and declaring that E, G, e, g are functions of u and v. (This declaration isn't strictly necessary, but it will make the output of some computations look nicer.)

```
> unassign('a', 'b');
> PDETools[declare](E(u,v), G(u,v), e(u,v), g(u,v));
```

Introduce a substitution for the Maurer-Cartan forms in terms of the coordinate 1-forms:

```
> coordsub:= [omega[3]=0, omega[1] = sqrt(E(u,v))*d(u),
  omega[2] = sqrt(G(u,v))*d(v),
```

```
  omega[3,1] = (e(u,v)/sqrt(E(u,v)))*d(u),
  omega[3,2] = (g(u,v)/sqrt(G(u,v)))*d(v),
  omega[1,2] = a*d(u) + b*d(v)];
```

Compute the coefficients in $\bar{\omega}_2^1$ as we did in the previous exercise:

```
> Simf(d(Simf(subs(coordsub, omega[1])))
  - subs(coordsub, Simf(d(omega[1]))));
> a:= solve(%, a);
> Simf(d(Simf(subs(coordsub, omega[2])))
  - subs(coordsub, Simf(d(omega[2]))));
> b:= solve(%, b);
```

The Gauss equation comes from comparing the two expressions for $d\bar{\omega}_2^1$:

```
> Simf(d(Simf(subs(coordsub, omega[1,2])))
  - subs(coordsub, Simf(d(omega[1,2]))));
> Gausseq1:= pick(%, d(u), d(v));
```

Now, you'll probably notice that this expression doesn't look quite like the one in part (e) of the exercise. But we can ask MAPLE to compare the two expressions to confirm that they are, in fact, equivalent. First, give a name to the expression that results from moving all the terms in equation (4.11) to the left-hand side:

```
> Gausseq2:= (e(u,v)*g(u,v))/(E(u,v)*G(u,v))
  + (1/(2*sqrt(E(u,v)*G(u,v))))*
  (diff(diff(E(u,v), v)/sqrt(E(u,v)*G(u,v)), v)
  + diff(diff(G(u,v), u)/sqrt(E(u,v)*G(u,v)), u));
```

Solve this equation for one of the variables (say, g), and then substitute this expression into the first version of the Gauss equation. If the two equations are equivalent, then the result should be zero.

```
> solve(Gausseq2, {g(u,v)});
> Simf(subs(%, Gausseq1));
```

$$0$$

Similar manipulations involving $d\bar{\omega}_1^3$ and $d\bar{\omega}_2^3$ will confirm that their structure equations are equivalent to the Codazzi equations (4.12).

Now, in fact, there's nothing special about assuming that $F = f = 0$, except that it makes the computations simpler. For a challenge, you might try redoing this exercise without this assumption. You'll need to start by applying the Gram-Schmidt algorithm to the basis $(\mathbf{x}_u, \mathbf{x}_v)$ in order to obtain an orthonormal frame field and then compute the dual forms $(\bar{\omega}^1, \bar{\omega}^2)$ for this frame field. (This part isn't too bad to do by hand.) Details are given in the MAPLE worksheet for this chapter on the AMS webpage.

Chapter 5

Curves and surfaces in Minkowski space

5.1. Introduction

In physics, the study of relativity generally begins with *special relativity*, which is primarily the study of timelike curves in the Minkowski space $\mathbb{M}^{1,3}$. As discussed in §3.5, such curves represent the world lines of particles in spacetime, and special relativity describes how particles behave in the absence of a gravitational field. The effects of gravity are considered in *general relativity*, where the Minkowski space $\mathbb{M}^{1,3}$ is replaced by a more general 4-dimensional manifold with a pseudo-Riemannian metric of signature $(1,3)$ and the strength of the gravitational field is reflected in the curvature of the metric. An excellent introduction to these topics from a geometric point of view is given in [**Cal00**].

In keeping with our general treatment of 3-dimensional homogeneous spaces, we will confine our attention to the study of curves and surfaces in the Minkowski space $\mathbb{M}^{1,2}$. In §5.2, we will apply the method of moving frames to study the geometry of timelike curves in $\mathbb{M}^{1,2}$, and many of the features of special relativity will already be apparent here. In §5.3, we will study the geometry of *timelike surfaces* in $\mathbb{M}^{1,2}$, i.e., surfaces for which the restriction of the Minkowski metric on $\mathbb{M}^{1,2}$ to each tangent plane has signature $(1,1)$. (By contrast, *spacelike* surfaces are those for which the restriction of the Minkowski metric to each tangent plane has signature $(0,2)$.) Such a surface may be regarded as a 2-dimensional model for the 4-dimensional pseudo-Riemannian manifolds that are studied in general relativity.

5.2. Moving frames for timelike curves in $\mathbb{M}^{1,2}$

Consider a smooth, parametrized timelike curve $\alpha : I \to \mathbb{M}^{1,2}$, where $I \subset \mathbb{R}$ is some open interval. $\mathbb{M}^{1,2}$ has the structure of the homogeneous space $M(1,2)/SO^+(1,2)$, so an adapted frame field along α should be a lifting $\tilde{\alpha} : I \to M(1,2)$. Any such lifting can be written as

$$\tilde{\alpha}(t) = (\alpha(t); \mathbf{e}_0(t), \mathbf{e}_1(t), \mathbf{e}_2(t)),$$

where for each $t \in I$, $(\mathbf{e}_0(t), \mathbf{e}_1(t), \mathbf{e}_2(t))$ is an oriented, orthonormal basis for the tangent space $T_{\alpha(t)}\mathbb{M}^{1,2}$. (Recall that, by convention, we require that $\mathbf{e}_0(t)$ be timelike and that $\mathbf{e}_1(t)$ and $\mathbf{e}_2(t)$ be spacelike; cf. Exercise 3.39.) As in the Euclidean case, such an adapted frame field is usually called an *orthonormal frame field* along α.

As in the Euclidean case, we say that α is *regular* if $\alpha'(t) \neq \mathbf{0}$ for every $t \in I$; henceforth, we will only consider regular curves. The construction of an orthonormal frame field along α is very similar to that for a curve in $\mathbb{E}^3$; we begin by setting

$$\mathbf{e}_0(t) = \frac{\alpha'(t)}{|\alpha'(t)|},$$

where $|\alpha'(t)| = \sqrt{\langle \alpha'(t), \alpha'(t) \rangle}$ is computed using the Minkowski inner product; i.e., we take $\mathbf{e}_0(t)$ to be the unit tangent vector to the curve at $\alpha(t)$. (Note that, since α is a timelike curve, $\mathbf{e}_0(t)$ is a timelike vector, as desired.) The Minkowski analog of arc length is the following:

Definition 5.1. Given a fixed point $t_0 \in I$, the *proper time* function along α is

$$\tau(t) = \int_{t_0}^{t} |\alpha'(u)|\, du.$$

This terminology arises from the key fact that in relativity, time is not an absolute quantity; different observers may measure the passage of time differently, and the proper time along a world line $\alpha(t)$ represents how time passes for an observer traveling along α.

Remark 5.2. The use of the letter τ to denote proper time is traditional in relativity; however, it should not be confused with the torsion of a nondegenerate curve in $\mathbb{E}^3$, which is also denoted by τ! Hopefully it will be clear from the context which quantity is intended. This convention will also necessitate a departure from the traditional geometric notation for curves in $\mathbb{E}^3$; thus, we will denote the analogs of curvature and torsion for timelike curves in $\mathbb{M}^{1,2}$ by $\varkappa_1, \varkappa_2$.

Exercise 5.3. Show that $\tau(t)$ is invariant under the action of $M(1,2)$; that is, for any $g \in M(1,2)$, the curves α and $g \cdot \alpha$ have the same proper time function.

***Exercise 5.4.** Consider two particles with world lines $\alpha, \beta : I \to \mathbb{M}^{1,2}$ given by

$$\alpha(t) = {}^t[t,\, 0,\, 0], \qquad \beta(t) = {}^t[t,\, vt,\, 0],$$

so that α represents a particle that remains stationary at the origin in $\mathbb{R}^2$ and β represents a particle traveling in the positive direction along the x^1-axis with speed $v < 1$. (Recall that the inner product on $\mathbb{M}^{1,2}$ is normalized so that the speed of light is $c = 1$; so this just means that v is less than the speed of light.)

(a) Show that the proper time function for α (with $t_0 = 0$) is $\tau_\alpha(t) = t$ and the proper time function for β is $\tau_\beta(t) = t\sqrt{1 - v^2}$.

(b) Reparametrize β according to its proper time function:

$$\beta(\tau) = \beta(t(\tau)) = {}^t\!\left[\frac{\tau}{\sqrt{1 - v^2}},\, \frac{v\tau}{\sqrt{1 - v^2}},\, 0\right].$$

Let $u = \tanh^{-1}(v)$, so that $v = \tanh(u)$, and consider the Lorentz transformation $T : \mathbb{M}^{1,2} \to \mathbb{M}^{1,2}$ defined by

$$\begin{bmatrix} \tilde{x}^0 \\ \tilde{x}^1 \\ \tilde{x}^2 \end{bmatrix} = T\left(\begin{bmatrix} x^0 \\ x^1 \\ x^2 \end{bmatrix} \right) = \begin{bmatrix} \cosh(u) & -\sinh(u) & 0 \\ -\sinh(u) & \cosh(u) & 0 \\ 0 & 0 & 1 \end{bmatrix} \begin{bmatrix} x^0 \\ x^1 \\ x^2 \end{bmatrix}.$$

Show that with respect to the $(\tilde{x}^0, \tilde{x}^1, \tilde{x}^2)$-coordinates,

$$\alpha(t) = {}^t\!\left[\frac{t}{\sqrt{1 - v^2}},\, \frac{-vt}{\sqrt{1 - v^2}},\, 0\right], \qquad \beta(\tau) = {}^t[\tau,\, 0,\, 0].$$

So in this new coordinate system, β represents a stationary particle and α represents a particle traveling in the negative direction along the x^1-axis with speed v. (Hint: You will need to make use of the hyperbolic trig identity $\operatorname{sech}^2(u) = 1 - \tanh^2(u)$.)

(c) Let $t_1 = \tau_1 = T$. Show that in either coordinate system, the points $\alpha(t_1)$ and $\beta(\tau_1)$ both lie on the hyperboloid

$$\{{}^t[x^0, x^1, x^2] \in \mathbb{M}^{1,2} \mid (x^0)^2 - (x^1)^2 - (x^2)^2 = T^2\}.$$

This hyperboloid represents the points that can be reached from the origin in proper time $\tau = T$ by particles traveling with any constant spatial velocity $\mathbf{v} = v^1 \frac{\partial}{\partial x^1} + v^2 \frac{\partial}{\partial x^2}$ with $(v^1)^2 + (v^2)^2 = v^2 < 1$.

Henceforth, we will assume that α is parametrized by proper time τ, and so $\mathbf{e}_0(\tau) = \alpha'(\tau)$.

As in the Euclidean case, we will say that a timelike curve α is *nondegenerate* if α is regular and, in addition, $\mathbf{e}_0'(\tau) \neq \mathbf{0}$ for all $\tau \in I$. In this case, differentiating the equation

$$\langle \mathbf{e}_0(\tau), \mathbf{e}_0(\tau) \rangle = 1$$

yields

$$\langle \mathbf{e}_0'(\tau), \mathbf{e}_0(\tau) \rangle = 0.$$

Thus, $\mathbf{e}_0'(\tau)$ is orthogonal to $\mathbf{e}_0(\tau)$, and we define

$$\mathbf{e}_1(\tau) = \frac{\mathbf{e}_0'(\tau)}{|\mathbf{e}_0'(\tau)|}.$$

This vector will be called the *unit normal vector* to the curve at $\alpha(\tau)$; note that $\mathbf{e}_1(\tau)$ is a spacelike vector (cf. Exercise 3.39).

By analogy with the Euclidean case, at this point we would like to define

$$\mathbf{e}_2(\tau) = \mathbf{e}_0(\tau) \times \mathbf{e}_1(\tau).$$

But first, we have to define an appropriate notion of the cross product in $\mathbb{M}^{1,2}$ and check that it has the desired properties. Recall that the cross product for two vectors $\mathbf{v} = {}^t[v^1, v^2, v^3]$, $\mathbf{w} = {}^t[w^1, w^2, w^3]$ in $\mathbb{E}^3$ is

$$\mathbf{v} \times \mathbf{w} = {}^t[v^2w^3 - v^3w^2,\ v^3w^1 - v^1w^3,\ v^1w^2 - v^2w^1],$$

often written schematically as

$$\mathbf{v} \times \mathbf{w} = \begin{vmatrix} \underline{\mathbf{e}}_1 & \underline{\mathbf{e}}_2 & \underline{\mathbf{e}}_3 \\ v^1 & v^2 & v^3 \\ w^1 & w^2 & w^3 \end{vmatrix}.$$

The analogous notion for Minkowski space is

Definition 5.5. Let $\mathbf{v} = {}^t[v^0, v^1, v^2]$, $\mathbf{w} = {}^t[w^0, w^1, w^2] \in \mathbb{M}^{1,2}$. The *Minkowski cross product* $\mathbf{v} \times \mathbf{w}$ is

$$\mathbf{v} \times \mathbf{w} = {}^t[v^1w^2 - v^2w^1,\ v^0w^2 - v^2w^0,\ v^1w^0 - v^0w^1],$$

written schematically as

$$\mathbf{v} \times \mathbf{w} = \begin{vmatrix} \underline{\mathbf{e}}_0 & \underline{\mathbf{e}}_1 & \underline{\mathbf{e}}_2 \\ v^0 & -v^1 & -v^2 \\ w^0 & -w^1 & -w^2 \end{vmatrix}.$$

The following exercise shows that the Minkowski cross product has properties analogous to those of the Euclidean cross product.

Exercise 5.6. (a) Show that with respect to the Minkowski inner product, $\mathbf{v} \times \mathbf{w}$ is orthogonal to both $\mathbf{v}$ and $\mathbf{w}$.

(b) Show that

$$\langle \mathbf{v} \times \mathbf{w}, \mathbf{v} \times \mathbf{w} \rangle = \langle \mathbf{v}, \mathbf{v} \rangle \langle \mathbf{w}, \mathbf{w} \rangle - \langle \mathbf{v}, \mathbf{w} \rangle^2.$$

Conclude that if $\mathbf{v}, \mathbf{w}$ are orthogonal vectors of Minkowski norm 1, then $\mathbf{v} \times \mathbf{w}$ also has Minkowski norm equal to 1.

(c) Show that for any three vectors $\mathbf{u}, \mathbf{v}, \mathbf{w} \in \mathbb{M}^{1,2}$,

$$\langle \mathbf{u}, \mathbf{v} \times \mathbf{w} \rangle = \det \begin{bmatrix} \mathbf{u} & \mathbf{v} & \mathbf{w} \end{bmatrix}.$$

(This is an orientation condition that dictates how the sign of $\mathbf{v} \times \mathbf{w}$ should be chosen.)

As a consequence of Exercise 5.6, we can complete our adapted frame field along α by defining

$$\mathbf{e}_2(\tau) = \mathbf{e}_0(\tau) \times \mathbf{e}_1(\tau).$$

Exercise 5.7. Prove that this choice of frame field $(\mathbf{e}_0(\tau), \mathbf{e}_1(\tau), \mathbf{e}_2(\tau))$ is equivariant under the action of $M(1,2)$: If we replace α by $g \cdot \alpha$ for some $g \in M(1,2)$, then $\mathbf{e}_\alpha(\tau) \in T_{\alpha(\tau)}\mathbb{M}^{1,2}$ will be replaced by $(L_g)_*(\mathbf{e}_\alpha(\tau)) \in T_{g\cdot\alpha(\tau)}\mathbb{M}^{1,2}$.

We now have a canonical adapted frame field $(\mathbf{e}_0(\tau), \mathbf{e}_1(\tau), \mathbf{e}_2(\tau))$ defined at each point of $\alpha(\tau)$, which in turn defines a canonical, left-invariant lifting $\tilde{\alpha} : I \to M(1,2)$ for any nondegenerate timelike curve α, given by

$$\tilde{\alpha}(\tau) = (\alpha(\tau); \mathbf{e}_0(\tau), \mathbf{e}_1(\tau), \mathbf{e}_2(\tau)).$$

Now consider the pullbacks of equations (3.1) to I via $\tilde{\alpha}$; these can be written as

$$\begin{aligned} \alpha'(\tau)d\tau &= \mathbf{e}_\alpha(\tau)\,\bar{\omega}^\alpha, \\ \mathbf{e}'_\alpha(\tau)d\tau &= \mathbf{e}_\beta(\tau)\,\bar{\omega}^\beta_\alpha. \end{aligned} \tag{5.1}$$

Recall that we constructed our adapted frame field so that $\alpha'(\tau) = \mathbf{e}_0(\tau)$; therefore, the first equation in (5.1) implies that

$$\bar{\omega}^0 = d\tau, \qquad \bar{\omega}^1 = \bar{\omega}^2 = 0.$$

Then we chose $\mathbf{e}_1(\tau)$ so that $\mathbf{e}'_0(\tau)$ is a multiple of $\mathbf{e}_1(\tau)$, say $\mathbf{e}'_0(\tau) = \kappa_1(\tau)\mathbf{e}_1(\tau)$. The function $\kappa_1(\tau)$ is the analog of the curvature $\kappa(s)$ for curves in $\mathbb{E}^3$; note that α is nondegenerate if and only if $\kappa_1(\tau) > 0$ for all $\tau \in I$. We will call $\kappa_1(\tau)$ the *Minkowski curvature* of α. So the equation for $\mathbf{e}'_0(\tau)$ in (5.1) implies that

$$\bar{\omega}^1_0 = \kappa_1(\tau)d\tau, \qquad \bar{\omega}^2_0 = 0.$$

(Recall from Exercise 3.50 that $\bar{\omega}_0^0 = 0$.) The only remaining Maurer-Cartan form is $\bar{\omega}_2^1$; it must be equal to some multiple of $d\tau$, so define a function $\kappa_2(\tau)$ by the condition that

$$\bar{\omega}_2^1 = \kappa_2(\tau)d\tau.$$

The function $\kappa_2(\tau)$ is the analog of the torsion $\tau(s)$ for curves in $\mathbb{E}^3$. We will call $\kappa_2(\tau)$ the *Minkowski torsion* of α.

Using the symmetry relations between the $(\bar{\omega}_\beta^\alpha)$ from Exercise 3.50, we have the Minkowski analog of the Frenet equations for timelike curves:

$$\left[\alpha'(\tau)\,\mathbf{e}_0'(\tau)\,\mathbf{e}_1'(\tau)\,\mathbf{e}_2'(\tau)\right] = \left[\alpha(\tau)\,\mathbf{e}_0(\tau)\,\mathbf{e}_1(\tau)\,\mathbf{e}_2(\tau)\right] \begin{bmatrix} 0 & 0 & 0 & 0 \\ 1 & 0 & \kappa_1(\tau) & 0 \\ 0 & \kappa_1(\tau) & 0 & \kappa_2(\tau) \\ 0 & 0 & -\kappa_2(\tau) & 0 \end{bmatrix}.$$

Applying Lemma 4.2 yields the following theorem:

Theorem 5.8. *Two nondegenerate timelike curves $\alpha_1, \alpha_2 : I \to \mathbb{M}^{1,2}$ parametrized by proper time differ by a Lorentz transformation if and only if they have the same Minkowski curvature $\kappa_1(\tau)$ and Minkowski torsion $\kappa_2(\tau)$.*

***Exercise 5.9.** Consider the world line α of a particle moving along the x^1-axis, with acceleration proportional to its distance from the origin and directed away from the origin and with initial conditions

$$x^1(0) = \frac{1}{a}, \qquad (x^1)'(0) = 0,$$

where $a > 0$ is the constant of proportionality for which

$$(x^1)''(\tau) = a^2 x^1(\tau).$$

(Note that α and all its derivatives will be contained in the plane spanned by $(\underline{\mathbf{e}}_0, \underline{\mathbf{e}}_1)$.)

(a) Show that

$$\alpha(\tau) = {}^t\!\left[\frac{1}{a}\sinh(a\tau),\ \frac{1}{a}\cosh(a\tau),\ 0\right].$$

(Hint: Let $\alpha(\tau) = {}^t[x^0(\tau), x^1(\tau), 0]$. Solve the given initial value problem for $x^1(\tau)$, and then use the condition that α is timelike and $\|\alpha'(\tau)\| = 1$ to solve for $x^0(\tau)$.)

(b) Show that the world line of α lies on a hyperbola in the (x^0, x^1)-plane.

(c) Show that the Minkowski curvature and torsion of α are

$$\kappa_1(\tau) = a, \qquad \kappa_2(\tau) = 0.$$

Thus, α is the Minkowski analog of a circle in $\mathbb{E}^3$, i.e., a nondegenerate curve with constant nonzero curvature and zero torsion.

***Exercise 5.10.** Consider the world line α of a particle moving along a circle of radius r in the (x^1, x^2)-plane with what a stationary observer believes to be constant angular velocity $k > 0$. This means that α has a parametrization of the form

$$\alpha(t) = {}^t[t,\, r\cos(kt),\, r\sin(kt)].$$

(a) Show that the proper time function for α is

$$\tau(t) = t\sqrt{1-k^2r^2};$$

therefore, the proper time parametrization for α is

$$\alpha(\tau) = {}^t\left[\frac{\tau}{\sqrt{1-k^2r^2}},\, r\cos\left(\frac{k\tau}{\sqrt{1-k^2r^2}}\right),\, r\sin\left(\frac{k\tau}{\sqrt{1-k^2r^2}}\right)\right].$$

(Note that this means that an observer traveling along α believes its angular velocity to be $\tilde{k} = \frac{k}{\sqrt{1-k^2r^2}}$.)

(b) Compute the Frenet frame $(\mathbf{e}_0(\tau), \mathbf{e}_1(\tau), \mathbf{e}_2(\tau))$ for α and show that the Minkowski curvature and torsion of α are

$$\kappa_1(\tau) = \frac{k^2 r}{1-k^2r^2}, \qquad \kappa_2(\tau) = \frac{k}{1-k^2r^2}.$$

Thus, α is the Minkowski analog of a helix in $\mathbb{E}^3$, i.e., a nondegenerate curve with constant nonzero curvature and torsion. (This shouldn't be too surprising since the world line of α *is* a helix in $\mathbb{M}^{1,2}$!)

5.3. Moving frames for timelike surfaces in $\mathbb{M}^{1,2}$

Now, let U be an open set in $\mathbb{R}^2$, and let $\mathbf{x} : U \to \mathbb{M}^{1,2}$ be an immersion whose image is a timelike surface $\Sigma = \mathbf{x}(U)$. Just as for curves, an adapted frame field along Σ is a lifting $\tilde{\mathbf{x}} : U \to M(1,2)$ of the form

$$\tilde{\mathbf{x}}(\mathbf{u}) = (\mathbf{x}(\mathbf{u}); \mathbf{e}_0(\mathbf{u}), \mathbf{e}_1(\mathbf{u}), \mathbf{e}_2(\mathbf{u})),$$

where for each $\mathbf{u} \in U$, $(\mathbf{e}_0(\mathbf{u}), \mathbf{e}_1(\mathbf{u}), \mathbf{e}_2(\mathbf{u}))$ is an oriented, orthonormal basis for the tangent space $T_{\mathbf{x}(\mathbf{u})}\mathbb{M}^{1,2}$.

As in the Euclidean case, we begin by choosing $\mathbf{e}_2(\mathbf{u})$ to be orthogonal to the tangent plane $T_{\mathbf{x}(\mathbf{u})}\Sigma$. (Note that, since Σ is assumed to be timelike, this implies that $\mathbf{e}_2(\mathbf{u})$ is a spacelike vector of Minkowski norm 1.)

Exercise 5.11. Show that this choice of $\mathbf{e}_2(\mathbf{u})$ is equivariant (up to sign) under the action of $M(1,2)$.

The vectors $(\mathbf{e}_0(\mathbf{u}), \mathbf{e}_1(\mathbf{u}))$ must now form a basis for the tangent space $T_{\mathbf{x}(\mathbf{u})}\Sigma$. We will assume that $\mathbf{e}_0(\mathbf{u})$ is timelike and $\mathbf{e}_1(\mathbf{u})$ is spacelike, but otherwise $(\mathbf{e}_0(\mathbf{u}), \mathbf{e}_1(\mathbf{u}))$ is allowed to be an arbitrary orthonormal basis of $T_{\mathbf{x}(\mathbf{u})}\Sigma$. An orthonormal frame field satisfying this condition will be called *adapted.*

Let $(\bar{\omega}^\alpha, \bar{\omega}^\alpha_\beta)$ represent the pulled-back forms $(\tilde{\mathbf{x}}^*\omega^\alpha, \tilde{\mathbf{x}}^*\omega^\alpha_\beta)$ on U. Precisely the same reasoning as in the Euclidean case can be used to prove the following:

Proposition 5.12. *Let $U \subset \mathbb{R}^2$ be an open set, and let $\mathbf{x} : U \to \mathbb{M}^{1,2}$ be a timelike immersion. For any adapted orthonormal frame field $(\mathbf{e}_0(\mathbf{u}),$ $\mathbf{e}_1(\mathbf{u}),$ $\mathbf{e}_2(\mathbf{u}))$ along $\Sigma = \mathbf{x}(U)$, the associated dual and connection forms $(\bar{\omega}^\alpha, \bar{\omega}^\alpha_\beta)$ have the property that $\bar{\omega}^2 = 0$. Moreover, $(\bar{\omega}^0, \bar{\omega}^1)$ form a basis for the 1-forms on U.*

The metric properties of the surface Σ are once again contained in the first fundamental form of the surface:

Definition 5.13. Let $U \subset \mathbb{R}^2$ be an open set, and let $\mathbf{x} : U \to \mathbb{M}^{1,2}$ be a timelike immersion. The *first fundamental form* of $\Sigma = \mathbf{x}(U)$ is the quadratic form I on TU defined by

$$\mathrm{I}(\mathbf{v}) = \langle d\mathbf{x}(\mathbf{v}), d\mathbf{x}(\mathbf{v})\rangle$$

for $\mathbf{v} \in T_\mathbf{u}U$, where $\langle \cdot, \cdot \rangle$ is the Minkowski inner product.

***Exercise 5.14.** Show that for any $\mathbf{v} \in T_\mathbf{u}U$,

$$\mathrm{I}(\mathbf{v}) = \left(\bar{\omega}^0(\mathbf{v})\right)^2 - \left(\bar{\omega}^1(\mathbf{v})\right)^2.$$

This is often written more concisely as

$$\mathrm{I} = (\bar{\omega}^0)^2 - (\bar{\omega}^1)^2.$$

Note that I is a quadratic form of signature $(1,1)$ rather than a positive definite quadratic form. This reflects the fact that Σ is a timelike surface; if Σ were spacelike, then its first fundamental form would have signature $(0,2)$.

As in the Euclidean case, differentiating the equation $\bar{\omega}^2 = 0$ yields

$$d\bar{\omega}^2 = -\bar{\omega}^2_0 \wedge \bar{\omega}^0 - \bar{\omega}^2_1 \wedge \bar{\omega}^1 = 0,$$

and Cartan's lemma (cf. Lemma 2.49) implies that there exist functions h_{00}, h_{01}, h_{11} on U such that

$$\begin{bmatrix} \bar{\omega}^2_0 \\ \bar{\omega}^2_1 \end{bmatrix} = -\begin{bmatrix} h_{00} & h_{01} \\ h_{01} & h_{11} \end{bmatrix} \begin{bmatrix} \bar{\omega}^0 \\ \bar{\omega}^1 \end{bmatrix}.$$

(The minus sign is included here in order to minimize minus signs in the second fundamental form below.)

Once again, the functions $(h_{\alpha\beta})$ are related to the differential of the Gauss map. The Gauss map and the second fundamental form are defined essentially as before:

Definition 5.15. Let $U \subset \mathbb{R}^2$ be an open set, and let $\mathbf{x} : U \to \mathbb{M}^{1,2}$ be a timelike immersion with image $\Sigma = \mathbf{x}(U)$. The *Gauss map* of $\Sigma = \mathbf{x}(U)$ is the map $N : \Sigma \to \mathbb{M}^{1,2}$ defined by

$$N(\mathbf{x}(\mathbf{u})) = \mathbf{e}_2(\mathbf{u}),$$

where $(\mathbf{e}_0(\mathbf{u}), \mathbf{e}_1(\mathbf{u}), \mathbf{e}_2(\mathbf{u}))$ is any adapted frame field on $\Sigma = \mathbf{x}(U)$. (Note that N takes values in the "sphere" $\mathbb{S}_{-1}$.)

Definition 5.16. Let $U \subset \mathbb{R}^2$ be an open set, and let $\mathbf{x} : U \to \mathbb{M}^{1,2}$ be a timelike immersion. The *second fundamental form* of $\Sigma = \mathbf{x}(U)$ is the quadratic form II on TU defined by

$$\mathrm{II}(\mathbf{v}) = -\langle d\mathbf{e}_2(\mathbf{v}), d\mathbf{x}(\mathbf{v})\rangle$$

for $\mathbf{v} \in T_{\mathbf{u}}U$, where $(\mathbf{e}_0(\mathbf{u}), \mathbf{e}_1(\mathbf{u}), \mathbf{e}_2(\mathbf{u}))$ is any adapted frame field on $\Sigma = \mathbf{x}(U)$.

***Exercise 5.17.** Show that for any $\mathbf{v} \in T_{\mathbf{u}}U$,

$$\begin{aligned}\mathrm{II}(\mathbf{v}) &= -\left(\bar{\omega}_0^2(\mathbf{v})\,\bar{\omega}^0(\mathbf{v}) + \bar{\omega}_1^2(\mathbf{v})\,\bar{\omega}^1(\mathbf{v})\right)\\ &= h_{00}\,(\bar{\omega}^0(\mathbf{v}))^2 + 2h_{01}\,\bar{\omega}^0(\mathbf{v})\,\bar{\omega}^1(\mathbf{v}) + h_{11}\,(\bar{\omega}^1(\mathbf{v}))^2.\end{aligned}$$

This is often written more concisely as

$$\mathrm{II} = -(\bar{\omega}_0^2\,\bar{\omega}^0 + \bar{\omega}_1^2\,\bar{\omega}^1) = h_{00}\,(\bar{\omega}^0)^2 + 2h_{01}\,\bar{\omega}^0\,\bar{\omega}^1 + h_{11}\,(\bar{\omega}^1)^2.$$

(Hint: Although the result looks the same as in the Euclidean case, there are some sign differences in the details of the computation. Note that, because of the slightly different symmetries in the $(\bar{\omega}_\beta^\alpha)$, we have

$$d\mathbf{e}_2 = \mathbf{e}_0\omega_2^0 + \mathbf{e}_1\omega_2^1 = \mathbf{e}_0\omega_0^2 - \mathbf{e}_1\omega_1^2.$$

And don't forget to use the Minkowski inner product in the definition of II!)

Now we will examine how the matrix $[h_{\alpha\beta}]$ changes if we vary the frame. Let $(\mathbf{e}_0(\mathbf{u}), \mathbf{e}_1(\mathbf{u}), \mathbf{e}_2(\mathbf{u}))$ be any adapted frame field along Σ, with associated Maurer-Cartan forms $(\bar{\omega}^\alpha, \bar{\omega}_\beta^\alpha)$. Any other adapted frame field $(\tilde{\mathbf{e}}_0(\mathbf{u}), \tilde{\mathbf{e}}_1(\mathbf{u}), \tilde{\mathbf{e}}_2(\mathbf{u}))$ has the form (up to sign)

(5.2)

$$\begin{bmatrix}\tilde{\mathbf{e}}_0(\mathbf{u}) & \tilde{\mathbf{e}}_1(\mathbf{u}) & \tilde{\mathbf{e}}_2(\mathbf{u})\end{bmatrix} = \begin{bmatrix}\mathbf{e}_0(\mathbf{u}) & \mathbf{e}_1(\mathbf{u}) & \mathbf{e}_2(\mathbf{u})\end{bmatrix}\begin{bmatrix}\cosh(\theta) & \sinh(\theta) & 0\\ \sinh(\theta) & \cosh(\theta) & 0\\ 0 & 0 & 1\end{bmatrix}$$

for some function θ on U. Let $(\tilde{\bar{\omega}}^\alpha, \tilde{\bar{\omega}}^\alpha_\beta)$ be the Maurer-Cartan forms associated to the new frame field, and set $B = \begin{bmatrix} \cosh(\theta) & \sinh(\theta) \\ \sinh(\theta) & \cosh(\theta) \end{bmatrix}$.

***Exercise 5.18** (Cf. Exercise 4.28). (a) Show that

$$\begin{bmatrix} \tilde{\bar{\omega}}^0 \\ \tilde{\bar{\omega}}^1 \end{bmatrix} = B^{-1} \begin{bmatrix} \bar{\omega}^0 \\ \bar{\omega}^1 \end{bmatrix}.$$

(b) Show that

$$\begin{bmatrix} \tilde{\bar{\omega}}^0_2 \\ \tilde{\bar{\omega}}^1_2 \end{bmatrix} = B^{-1} \begin{bmatrix} \bar{\omega}^0_2 \\ \bar{\omega}^1_2 \end{bmatrix}.$$

(Hint: Use the equation for $d\mathbf{e}_2$ in (3.1).)

(c) Cartan's lemma implies that there exist functions $\tilde{h}_{00}, \tilde{h}_{01}, \tilde{h}_{11}$ on U such that

$$\begin{bmatrix} \tilde{\bar{\omega}}^2_0 \\ \tilde{\bar{\omega}}^2_1 \end{bmatrix} = - \begin{bmatrix} \tilde{h}_{00} & \tilde{h}_{01} \\ \tilde{h}_{01} & \tilde{h}_{11} \end{bmatrix} \begin{bmatrix} \tilde{\bar{\omega}}^0 \\ \tilde{\bar{\omega}}^1 \end{bmatrix}.$$

Show that

$$\begin{bmatrix} -\tilde{h}_{00} & -\tilde{h}_{01} \\ \tilde{h}_{01} & \tilde{h}_{11} \end{bmatrix} = B^{-1} \begin{bmatrix} -h_{00} & -h_{01} \\ h_{01} & h_{11} \end{bmatrix} B.$$

(Hint: Use the fact that $\bar{\omega}^0_2 = \bar{\omega}^2_0$, $\bar{\omega}^1_2 = -\bar{\omega}^2_1$.)

(d) Use part (c) to show that

$$\begin{bmatrix} \tilde{h}_{00} & \tilde{h}_{01} \\ \tilde{h}_{01} & \tilde{h}_{11} \end{bmatrix} = {}^tB \begin{bmatrix} h_{00} & h_{01} \\ h_{01} & h_{11} \end{bmatrix} B. \tag{5.3}$$

(Hint: Note that B^{-1} is *not* equal to ${}^tB$.)

Recall that at this point in the Euclidean case, we used the fact that any symmetric matrix has an orthogonal basis of eigenvectors to conclude that we could choose B so as to diagonalize the matrix $[h_{ij}]$. The key fact is not so much that the matrix $[h_{ij}]$ is symmetric (which depends on the fact that it is expressed relative to an orthonormal basis $(\mathbf{e}_1, \mathbf{e}_2)$), but rather that the linear operator $d\mathbf{e}_3 : T_{\mathbf{x}(\mathbf{u})}\Sigma \to T_{\mathbf{x}(\mathbf{u})}\Sigma$ is *self-adjoint*; i.e., for any two vectors $\mathbf{v}, \mathbf{w} \in T_{\mathbf{x}(\mathbf{u})}\Sigma$, we have

$$\langle d\mathbf{e}_3(\mathbf{v}), \mathbf{w}\rangle = \langle \mathbf{v}, d\mathbf{e}_3(\mathbf{w})\rangle.$$

The formal statement of the linear algebra theorem is that a self-adjoint operator on $\mathbb{E}^n$ has all real eigenvalues and an orthogonal basis of eigenvectors.

Unfortunately, this result is not true in Minkowski space! The following exercise shows how this property can fail in the Minkowski setting.

***Exercise 5.19.** (a) Show that

$$d\mathbf{e}_2 = \mathbf{e}_0\bar{\omega}_2^0 + \mathbf{e}_1\bar{\omega}_2^1 = \mathbf{e}_0(-h_{00}\bar{\omega}^0 - h_{01}\bar{\omega}^1) + \mathbf{e}_1(h_{01}\bar{\omega}^0 + h_{11}\bar{\omega}^1).$$

(You probably did this as part of Exercise 5.18.) Conclude that the matrix of the linear transformation $d\mathbf{e}_2 : T_{\mathbf{x}(\mathbf{u})}\Sigma \to T_{\mathbf{x}(\mathbf{u})}\Sigma$ with respect to the basis $(\mathbf{e}_0(\mathbf{u}), \mathbf{e}_1(\mathbf{u}))$ for $T_{\mathbf{x}(\mathbf{u})}\Sigma$ is

$$\begin{bmatrix} -h_{00} & -h_{01} \\ h_{01} & h_{11} \end{bmatrix}.$$

(b) Show that the linear transformation $d\mathbf{e}_2 : T_{\mathbf{x}(\mathbf{u})}\Sigma \to T_{\mathbf{x}(\mathbf{u})}\Sigma$ is self-adjoint with respect to the Minkowski metric; that is, for any two vectors $\mathbf{v}, \mathbf{w} \in T_{\mathbf{x}(\mathbf{u})}\Sigma$, we have

$$\langle d\mathbf{e}_2(\mathbf{v}), \mathbf{w}\rangle = \langle \mathbf{v}, d\mathbf{e}_2(\mathbf{w})\rangle.$$

Conclude that the matrix S that expresses a self-adjoint operator on a (1+1)-dimensional Minkowski vector space V relative to an orthonormal basis $(\mathbf{e}_0, \mathbf{e}_1)$ for V has the property that the matrix $S + {}^tS$ is diagonal. (Contrast this with the Euclidean case, where any self-adjoint operator is expressed by a *symmetric* matrix S relative to an orthonormal basis.)

(c) Show that:

- If $|h_{00} + h_{11}| > 2|h_{01}|$, then $d\mathbf{e}_2$ has distinct, real eigenvalues and an orthogonal basis of eigenvectors, one timelike and one spacelike.
- If $|h_{00} + h_{11}| < 2|h_{01}|$, then $d\mathbf{e}_2$ has complex eigenvalues and no real eigenvectors.
- If $|h_{00} + h_{11}| = 2|h_{01}| = 0$, then $d\mathbf{e}_2$ is a multiple of the identity transformation.
- If $|h_{00} + h_{11}| = 2|h_{01}| \neq 0$, then $d\mathbf{e}_2$ has a repeated real eigenvalue and a 1-dimensional, lightlike eigenspace.

In order to make sense of the result of Exercise 5.19, we introduce the following analogs of the Euclidean notions of Gauss and mean curvature.

Definition 5.20. The function $K = \det(d\mathbf{e}_2)$ on Σ is called the *Gauss curvature* of Σ. The function $H = -\frac{1}{2}\mathrm{tr}(d\mathbf{e}_2)$ on Σ is called the *mean curvature* of Σ.

***Exercise 5.21.** Show that $K = h_{01}^2 - h_{00}h_{11}$ and $H = \frac{1}{2}(h_{00} - h_{11})$.

Definition 5.22. The function $H' = \sqrt{H^2 - K}$ on Σ is called the *skew curvature* of Σ. (By convention, if $H^2 - K < 0$, then H' is chosen so that $iH' < 0$.)

***Exercise 5.23.** (a) Show that for a regular surface Σ in Euclidean space $\mathbb{E}^3$, the skew curvature H' is always real and $H' = 0$ precisely at umbilic points.

(b) Show that for a timelike, regular surface in Minkowski space $\mathbb{M}^{1,2}$,

$$H' = \tfrac{1}{2}\sqrt{(h_{00} + h_{11})^2 - 4h_{01}^2}.$$

(c) Conclude from the result of Exercise 5.19 that at any point $\mathbf{x}(\mathbf{u}) \in \Sigma$:

- If $H'(\mathbf{u})$ is real and $H'(\mathbf{u}) > 0$, then $d(\mathbf{e}_2)_{\mathbf{x}(\mathbf{u})}$ has distinct, real eigenvalues and an orthogonal basis of eigenvectors, one timelike and one spacelike.
- If $H'(\mathbf{u})$ is imaginary, then $d(\mathbf{e}_2)_{\mathbf{x}(\mathbf{u})}$ has complex eigenvalues and no real eigenvectors.
- If $H'(\mathbf{u}) = 0$ and $h_{01} = 0$, then $d(\mathbf{e}_2)_{\mathbf{x}(\mathbf{u})}$ is a multiple of the identity transformation.
- If $H'(\mathbf{u}) = 0$ and $h_{01} \neq 0$, then $d(\mathbf{e}_2)_{\mathbf{x}(\mathbf{u})}$ has a repeated real eigenvalue and a 1-dimensional, lightlike eigenspace.

In light of this result, we will need to divide into cases based on the skew curvature in order to make further refinements to our adapted frame field.

5.3.1. Case 1: $H'(\mathbf{u})$ is real and $d(\mathbf{e}_2)_{\mathbf{x}(\mathbf{u})}$ has an orthogonal basis of eigenvectors for all $\mathbf{u} \in U$. This assumption covers the first and third cases in Exercise 5.23, and the frame adaptation proceeds much as it did for surfaces in $\mathbb{E}^3$. In this case, there exists an adapted frame $(\mathbf{e}_0(\mathbf{u}), \mathbf{e}_1(\mathbf{u}), \mathbf{e}_2(\mathbf{u}))$ at each point $\mathbf{x}(\mathbf{u}) \in \Sigma$ with the property that

$$\begin{bmatrix} -h_{00}(\mathbf{u}) & -h_{01}(\mathbf{u}) \\ h_{01}(\mathbf{u}) & h_{11}(\mathbf{u}) \end{bmatrix} = \begin{bmatrix} -\kappa_0(\mathbf{u}) & 0 \\ 0 & -\kappa_1(\mathbf{u}) \end{bmatrix}$$

for some real numbers $\kappa_0(\mathbf{u}), \kappa_1(\mathbf{u})$. As in the Euclidean case, $\mathbf{e}_0(\mathbf{u})$ and $\mathbf{e}_1(\mathbf{u})$ are eigenvectors for $d(\mathbf{e}_2)_{\mathbf{u}}$, with

$$d(\mathbf{e}_2)_{\mathbf{u}}(\mathbf{e}_0(\mathbf{u})) = -\kappa_0\mathbf{e}_0(\mathbf{u}), \qquad d(\mathbf{e}_2)_{\mathbf{u}}(\mathbf{e}_1(\mathbf{u})) = -\kappa_1\mathbf{e}_1(\mathbf{u}). \tag{5.4}$$

We have the following analog of Definition 4.30:

Definition 5.24. The vectors $\mathbf{e}_0(\mathbf{u})$ and $\mathbf{e}_1(\mathbf{u})$ in equations (5.4) are called *principal vectors* or *principal directions* at $\mathbf{x}(\mathbf{u}) \in \Sigma$, and $\kappa_0(\mathbf{u})$, $\kappa_1(\mathbf{u})$ are called the *principal curvatures* of Σ at $\mathbf{x}(\mathbf{u})$. Such a frame $(\mathbf{e}_0(\mathbf{u}),\ \mathbf{e}_1(\mathbf{u}),\ \mathbf{e}_2(\mathbf{u}))$ is called a *principal adapted frame* at the point $\mathbf{x}(\mathbf{u}) \in \Sigma$, and an adapted frame field on Σ which has this property at every point $\mathbf{x}(\mathbf{u}) \in \Sigma$ is called a *principal adapted frame field* on Σ.

Exercise 5.25. Show that $K = \kappa_0\kappa_1$ and $H = \frac{1}{2}(\kappa_0 + \kappa_1)$.

Definition 5.26. If $\kappa_0(\mathbf{u}) = \kappa_1(\mathbf{u})$ for some point $\mathbf{u} \in U$, then the corresponding point $\mathbf{x}(\mathbf{u})$ of Σ is called an *umbilic point* of Σ.

If Σ has no umbilic points, then a principal adapted frame field along Σ is determined uniquely (up to sign) since $\mathbf{e}_0$ must be timelike and $\mathbf{e}_1$ must be spacelike.

***Exercise 5.27.** Show that the umbilic points of Σ are precisely those points where $H' = 0$.

***Exercise 5.28.** (a) Show that if Σ has no umbilic points, then a principal adapted frame field $(\mathbf{e}_0(\mathbf{u}), \mathbf{e}_1(\mathbf{u}), \mathbf{e}_2(\mathbf{u}))$ is equivariant (up to sign) under the action of $M(1,2)$.

(b) Show that for a principal adapted frame field, the second fundamental form is

$$\mathrm{II} = \kappa_0(\bar{\omega}^0)^2 - \kappa_1(\bar{\omega}^1)^2.$$

The analog of Bonnet's theorem (cf. Theorem 4.35) is the following:

Theorem 5.29. *Let $U \subset \mathbb{R}^2$ be an open set. Two timelike immersions $\mathbf{x}_1, \mathbf{x}_2 : U \to \mathbb{M}^{1,2}$ for which the differential of the Gauss map has two linearly independent real eigenvectors differ by a Lorentz transformation if and only if they have the same first and second fundamental forms.*

The proof of this theorem (at least for surfaces with no umbilic points) is very similar to that in the Euclidean case.

***Exercise 5.30.** Prove Theorem 5.29 for surfaces without umbilic points as follows: Let $\mathbf{x}_1, \mathbf{x}_2 : U \to \mathbb{M}^{1,2}$ be immersions as in Theorem 5.29, with the same first and second fundamental forms and $\kappa_0(\mathbf{u}) \neq \kappa_1(\mathbf{u})$ for every $\mathbf{u} \in U$.

(a) Let $\tilde{\mathbf{x}}_1, \tilde{\mathbf{x}}_2 : U \to M(1,2)$ be principal adapted frame fields for $\mathbf{x}_1, \mathbf{x}_2$. Let $(\bar{\omega}^\alpha, \bar{\omega}^\alpha_\beta)$ denote the pulled-back dual and connection forms for $\mathbf{x}_1$ and let $(\bar{\Omega}^\alpha, \bar{\Omega}^\alpha_\beta)$ denote those for $\mathbf{x}_2$. Show that up to sign,

$$\bar{\Omega}^\alpha = \bar{\omega}^\alpha, \qquad \bar{\Omega}^\alpha_\beta = \bar{\omega}^\alpha_\beta.$$

Moreover, by making appropriate sign changes in the principal adapted frame fields, we can arrange that these equations hold exactly.

(b) Use Lemma 4.2 to conclude that $\mathbf{x}_1$ and $\mathbf{x}_2$ differ by a Lorentz transformation.

***Exercise 5.31.** Let $\mathbf{x} : U \to \mathbb{M}^{1,2}$ be a timelike immersion as in Theorem 5.29, and let (u, v) be local coordinates on U. We can write the first and second fundamental forms as

$$\begin{aligned} \mathrm{I} &= E\,du^2 + 2F\,du\,dv + G\,dv^2, \\ \mathrm{II} &= e\,du^2 + 2f\,du\,dv + g\,dv^2 \end{aligned}$$

for some functions E, F, G, e, f, g on U. The fact that $\Sigma = \mathbf{x}(U)$ is timelike implies that $EG - F^2 < 0$.

Given a surface Σ of this type with no umbilic points, there exists a parametrization $\mathbf{x} : U \to \mathbb{M}^{1,2}$ of Σ (at least locally) with the property that the coordinate curves of $\mathbf{x}$ are all principal curves, and such a parametrization has the property that $F = f = 0$, as in the Euclidean case. Without loss of generality, we may assume that $E > 0$ and $G < 0$. (Show that this condition implies that $\mathbf{x}_u$ is timelike and $\mathbf{x}_v$ is spacelike.) Follow the outline of Exercises 4.24 and 4.27 to derive the analogs of the Gauss and Codazzi equations for such surfaces.

***Exercise 5.32.** In this exercise, we will classify the totally umbilic timelike surfaces in $\mathbb{M}^{1,2}$. So, suppose that $\mathbf{x} : U \to \mathbb{M}^{1,2}$ is a timelike immersion as in Theorem 5.29, with the property that every point of $\Sigma = \mathbf{x}(U)$ is umbilic.

(a) Show that any adapted frame field $(\mathbf{e}_0(\mathbf{u}), \mathbf{e}_1(\mathbf{u}), \mathbf{e}_2(\mathbf{u}))$ is a principal adapted frame field.

(b) Let $(\bar{\omega}^\alpha, \bar{\omega}^\alpha_\beta)$ be the Maurer-Cartan forms for an adapted frame field on $\Sigma = \mathbf{x}(U)$. Show that there exists a smooth function $\lambda : U \to \mathbb{R}$ such that

$$\bar{\omega}^2_0 = -\lambda\bar{\omega}^0, \qquad \bar{\omega}^2_1 = \lambda\bar{\omega}^1.$$

Conclude that the second fundamental form of Σ is a scalar multiple of the first fundamental form, i.e., that

$$\mathrm{II} = \lambda\mathrm{I}.$$

(c) Prove that λ is constant. (Hint: Use the structure equations to differentiate the equations above, taking into account the fact that we must have

$$d\lambda = \lambda_0 \bar{\omega}^0 + \lambda_1 \bar{\omega}^1$$

for some functions λ_0, λ_1 on U. Then use Cartan's lemma.)

(d) Show that if $\lambda = 0$, then $d\mathbf{e}_2 = 0$. Conclude that the normal vector field of Σ is constant and that Σ is therefore contained in a plane.

(e) Show that if $\lambda \neq 0$, then $d(\mathbf{x}+\frac{1}{\lambda}\mathbf{e}_2) = 0$. Conclude that the vector-valued function $\mathbf{x} + \frac{1}{\lambda}\mathbf{e}_2 : U \to \mathbb{M}^{1,2}$ is equal to some constant point $\mathbf{q} \in \mathbb{M}^{1,2}$ and that Σ is therefore contained in the hyperboloid of one sheet defined by the equation

$$\langle \mathbf{x} - \mathbf{q}, \mathbf{x} - \mathbf{q} \rangle = -\frac{1}{|\lambda|^2}.$$

(Note that the minus sign on the right-hand side is due to the fact that $\mathbf{e}_2$ is a spacelike vector and that this hyperboloid is actually the "sphere" $\mathbb{S}_{-\frac{1}{\lambda^2}}$.)

Thus, the only totally umbilic timelike surfaces are (open subsets of) planes and hyperboloids of one sheet.

Example 5.33 (De Sitter spacetime)**.** In relativity, the de Sitter spacetime dS_4 is perhaps the simplest non-flat model for general relativity. It can be defined as the set of spacelike unit vectors in the Minkowski space $\mathbb{M}^{1,4}$:

$$dS_4 = \{\mathbf{x} \in \mathbb{M}^{1,4} \mid \langle \mathbf{x}, \mathbf{x} \rangle = -1\}.$$

De Sitter space is the maximally symmetric vacuum solution of Einstein's field equations with a positive cosmological constant. It represents a universe that is spatially homogeneous, diffeomorphic to a 3-dimensional sphere (and hence compact), and expanding in size for $t > 0$.

A 2-dimensional analog is the de Sitter spacetime dS_2, which is the "sphere" $\mathbb{S}_{-1}$ of spacelike unit vectors in $\mathbb{M}^{1,2}$ and is a totally umbilic surface as in Exercise 5.32. It represents a 1-dimensional universe that is diffeomorphic to a circle.

***Exercise 5.34.** We can parametrize dS_2 by the map $\mathbf{x} : \mathbb{R}^2 \to \mathbb{M}^{1,2}$ given by

$$\mathbf{x}(u, v) = {}^t[\sinh(u),\ \cosh(u)\cos(v),\ \cosh(u)\sin(v)], \tag{5.5}$$

where u should be thought of as a time parameter and v as a spatial parameter.

(a) Show that the metric on dS_2 corresponding to the parametrization (5.5) has

$$E = 1, \qquad F = 0, \qquad G = -\cosh^2(u).$$

(Cf. Exercise 5.31.)

(b) Show that the circumference of the space at time $u = \tau$ is

$$\int_{-\pi}^{\pi} \|\mathbf{x}_v(\tau, v)\| \, dv = 2\pi \cosh(\tau).$$

Thus, the "radius" of the space at time τ is $\cosh(\tau)$, which grows exponentially as τ increases for $\tau > 0$.

Now consider the possible world lines for particles traveling in the de Sitter space dS_2. Since dS_2 is spatially finite, we might think that a moving particle should be able to "circumnavigate" the space and return to its original position. But in fact this is not possible, as the following exercise shows.

***Exercise 5.35.** Consider the path β of a photon emitted at the point $\mathbf{x}(0,0)$ and traveling in the positive v direction. We can parametrize the photon's world line as

$$\beta(t) = \mathbf{x}(t, v(t))$$

for some function $v(t)$ which is determined by the conditions that $v(0) = 0$, $v'(t) > 0$, and $\beta'(t)$ is a lightlike vector.

(a) Show that the lightlike tangent directions to dS_2 at the point $\mathbf{x}(u_0, v_0)$ are spanned by the vectors $\mathbf{v}_\pm = \cosh(u_0)\, \mathbf{x}_u \pm \mathbf{x}_v$.

(b) Show that the function $v(t)$ above must satisfy the differential equation

$$v'(t) = \operatorname{sech}(t)$$

and that the solution to this equation with $v(0) = 0$ and $v'(t) > 0$ is

$$v(t) = 2 \tan^{-1}\left(e^t\right) - \frac{\pi}{2}.$$

(c) Show that

$$\lim_{t \to +\infty} v(t) = \frac{\pi}{2}, \qquad \lim_{t \to -\infty} v(t) = -\frac{\pi}{2}.$$

Therefore, no photon ever travels more than halfway around the circle. (And, of course, no other particle can travel farther than a photon!)

5.3.2. Case 2: $H'(\mathbf{u})$ is imaginary for all $\mathbf{u} \in U$. This assumption covers the second case in Exercise 5.23.

***Exercise 5.36.** (a) Show that under a transformation of the form (5.3),

$$\tilde{h}_{00} + \tilde{h}_{11} = \cosh(2\theta)(h_{00} + h_{11}) + 2\sinh(2\theta) h_{01}.$$

Conclude that there exists an adapted frame field $(\mathbf{e}_0(\mathbf{u}), \mathbf{e}_1(\mathbf{u}), \mathbf{e}_2(\mathbf{u}))$ with the property that

$$h_{00} + h_{11} = 0,$$

and therefore,

$$\begin{bmatrix} h_{00} & h_{01} \\ h_{01} & h_{11} \end{bmatrix} = \begin{bmatrix} \lambda_0 & \lambda_1 \\ \lambda_1 & -\lambda_0 \end{bmatrix}$$

for some functions λ_0, λ_1 on U with $|\lambda_1| > 0$ (cf. Exercise 5.18).

(b) Show that the Maurer-Cartan forms associated to such a frame field satisfy

$$\bar{\omega}_0^2 = -\lambda_0\bar{\omega}^0 - \lambda_1\bar{\omega}^1,$$
$$\bar{\omega}_1^2 = -\lambda_1\bar{\omega}^0 + \lambda_0\bar{\omega}^1.$$

(c) Show that for such a frame field, the second fundamental form is

$$\mathrm{II} = \lambda_0\left((\bar{\omega}^0)^2 - (\bar{\omega}^1)^2\right) + 2\lambda_1\bar{\omega}^0\bar{\omega}^1 = \lambda_0\mathrm{I} + 2\lambda_1\bar{\omega}^0\bar{\omega}^1.$$

(d) What goes wrong with part (a) when $H' = 0$ and $h_{01} \neq 0$?

In order to get a feel for what such a surface Σ might look like, consider the case where Σ is a graph of the form

$$x^2 = f(x^0, x^1).$$

(Any timelike surface can locally be described in this way, possibly after a rotation in the (x^1, x^2)-plane.) Consider a parametrization $\mathbf{x} : U \to \mathbb{M}^{1,2}$ of the form

$$\mathbf{x}(u, v) = {}^t[u, v, f(u, v)].$$

Exercise 5.37. Show that Σ is timelike if and only if $|f_u| < 1$.

Now choose a point $(u_0, v_0) \in U$, and "rotate" the surface (via a Lorentzian transformation) if necessary so that

$$f_u(u_0, v_0) = f_v(u_0, v_0) = 0,$$

so that the tangent plane to Σ at the point $\mathbf{x}_0 = \mathbf{x}(u_0, v_0)$ is parallel to the (x^0, x^1)-plane.

Exercise 5.38. (a) Show that the normal vector field along Σ is

$$\mathbf{e}_2(u, v) = \frac{1}{\sqrt{1 + f_v^2 - f_u^2}}\,{}^t[-f_u,\, f_v,\, -1].$$

In particular,

$$\mathbf{e}_2(u_0, v_0) = {}^t[0,\, 0,\, -1].$$

(b) Choose an adapted frame field along Σ so that

$$\mathbf{e}_0(u_0, v_0) = {}^t[1,\, 0,\, 0], \qquad \mathbf{e}_1(u_0, v_0) = {}^t[0,\, 1,\, 0],$$

and let $(\bar{\omega}^\alpha, \bar{\omega}^\alpha_\beta)$ be the associated Maurer-Cartan forms. Show that at the point $\mathbf{x}_0$,

$$\bar{\omega}^0 = du, \qquad \bar{\omega}^1 = dv,$$
$$\bar{\omega}^2_0 = -f_{uu}\, du - f_{uv}\, dv, \qquad \bar{\omega}^2_1 = -f_{uv}\, du - f_{vv}\, dv.$$

Therefore,

$$\begin{bmatrix} h_{00} & h_{01} \\ h_{01} & h_{11} \end{bmatrix}\Bigg|_{(u_0,v_0)} = \begin{bmatrix} f_{uu}(u_0,v_0) & f_{uv}(u_0,v_0) \\ f_{uv}(u_0,v_0) & f_{vv}(u_0,v_0) \end{bmatrix}.$$

(c) Conclude that the graph of the function

$$f(u, v) = a(u^2 - v^2) + 2buv,$$

where a, b are constants with $|b| > 0$, is one example of a surface of this type. (See Figure 5.1; note that the $\underline{\mathbf{e}}_0$-axis is drawn as the vertical axis.)

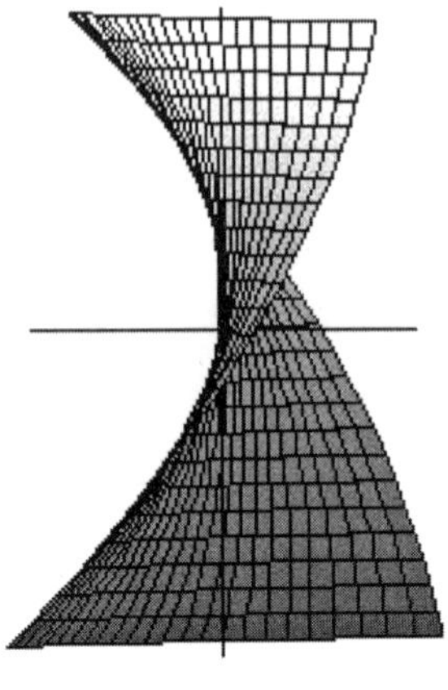

Figure 5.1. A timelike surface with H' imaginary

5.3.3. Case 3: $H'(\mathbf{u}) = 0$ and $h_{01}(\mathbf{u}) \neq 0$ for all $\mathbf{u} \in U$. Note that this is a degenerate condition, and generically it will only hold along a closed subset of Σ, similar to the set of umbilic points for a surface in Case 1.

Definition 5.39. A point $\mathbf{x}(\mathbf{u})$ in a timelike surface $\Sigma = \mathbf{x}(U)$ in $\mathbb{M}^{1,2}$ will be called a *quasi-umbilic* point if $H'(\mathbf{u}) = 0$ and $h_{01}(\mathbf{u}) \neq 0$. If every point of Σ is quasi-umbilic, then Σ is called *totally quasi-umbilic.*

Exercise 5.40. Suppose that $\mathbf{x} : U \to \mathbb{M}^{1,2}$ is a totally quasi-umbilic timelike surface.

(a) Show that

$$|h_{01}| = \tfrac{1}{2}|h_{00} + h_{11}| \neq 0.$$

(b) Show that under a transformation of the form (5.3),

$$\tilde{h}_{01} = e^{2\theta} h_{01}.$$

Conclude that there exists an adapted frame field $(\mathbf{e}_0(\mathbf{u}), \mathbf{e}_1(\mathbf{u}), \mathbf{e}_2(\mathbf{u}))$ with the property that

$$|h_{01}| = \tfrac{1}{2}|h_{00} + h_{11}| = 1,$$

and therefore,

$$\begin{bmatrix} h_{00} & h_{01} \\ h_{01} & h_{11} \end{bmatrix} = \begin{bmatrix} \varepsilon_1 + \lambda & \varepsilon_2 \\ \varepsilon_2 & \varepsilon_1 - \lambda \end{bmatrix},$$

where λ is a function on U and $\varepsilon_i = \pm 1$, $i = 1, 2$. (In fact, λ is equal to the mean curvature H of Σ.)

(c) Show that the Maurer-Cartan forms associated to such a frame field satisfy

$$\begin{aligned} \bar{\omega}_0^2 &= -(\varepsilon_1 + \lambda)\bar{\omega}^0 - \varepsilon_2 \bar{\omega}^1, \\ \bar{\omega}_1^2 &= -\varepsilon_2 \bar{\omega}^0 - (\varepsilon_1 - \lambda)\bar{\omega}^1. \end{aligned}$$

(d) Show that for such a frame field, the second fundamental form is

$$\begin{aligned} \mathrm{II} &= \lambda\left((\bar{\omega}^0)^2 - (\bar{\omega}^1)^2\right) + \varepsilon_1\left((\bar{\omega}^0)^2 + (\bar{\omega}^1)^2\right) + 2\varepsilon_2 \bar{\omega}^0 \bar{\omega}^1 \\ &= \lambda \mathrm{I} + \varepsilon_1\left((\bar{\omega}^0)^2 + (\bar{\omega}^1)^2\right) + 2\varepsilon_2 \bar{\omega}^0 \bar{\omega}^1. \end{aligned}$$

By analogy with the totally umbilic surfaces, one might expect that the totally quasi-umbilic surfaces would form a fairly limited set. But in fact, the analysis of totally quasi-umbilic surfaces is considerably more involved than that for totally umbilic surfaces, and it turns out that there is an infinite-dimensional family of such surfaces! We will explore some examples in Exercise 5.49.

5.4. An alternate construction for timelike surfaces

In §5.3, we used orthonormal frame fields to study timelike surfaces in $\mathbb{M}^{1,2}$. This is in keeping with our general perspective that a choice of frame field along a surface $\mathbf{x} : U \to G/H$ in a homogeneous space G/H should define a lifting $\tilde{\mathbf{x}} : U \to G$ to a surface in the Lie group G. However, this is not always the most geometrically natural way to choose frames, and in this section, we will explore an alternate method for constructing moving frames for timelike surfaces in $\mathbb{M}^{1,2}$.

Let $\Sigma = \mathbf{x}(U)$ be a timelike surface in $\mathbb{M}^{1,2}$. Since the first fundamental form of a timelike surface has signature $(1, 1)$, there are two linearly independent null (i.e., lightlike) directions in each tangent space $T_{\mathbf{x}(\mathbf{u})}\Sigma$. In this section, we will construct frame fields $(\mathbf{f}_1(\mathbf{u}), \mathbf{f}_2(\mathbf{u}), \mathbf{f}_3(\mathbf{u}))$ along $\Sigma = \mathbf{x}(U)$ with the property that $\mathbf{f}_1(\mathbf{u})$ and $\mathbf{f}_2(\mathbf{u})$ are null vectors.

Exercise 5.41. Suppose that $(\mathbf{e}_0(\mathbf{u}), \mathbf{e}_1(\mathbf{u}), \mathbf{e}_2(\mathbf{u}))$ is an orthonormal frame field along $\Sigma = \mathbf{x}(U)$. Show that the vectors

$$\mathbf{f}_1(\mathbf{u}) = \tfrac{1}{\sqrt{2}}(\mathbf{e}_0(\mathbf{u}) + \mathbf{e}_1(\mathbf{u})), \qquad \mathbf{f}_2(\mathbf{u}) = \tfrac{1}{\sqrt{2}}(\mathbf{e}_0(\mathbf{u}) - \mathbf{e}_1(\mathbf{u}))$$

are linearly independent null vectors and that

$$\langle \mathbf{f}_1(\mathbf{u}), \mathbf{f}_2(\mathbf{u})\rangle = 1.$$

Definition 5.42. A *null adapted frame field* $(\mathbf{f}_1(\mathbf{u}), \mathbf{f}_2(\mathbf{u}), \mathbf{f}_3(\mathbf{u}))$ along a timelike surface $\Sigma = \mathbf{x}(U)$ is a basis for the tangent space $T_{\mathbf{x}(\mathbf{u})}\mathbb{M}^{1,2}$ with the properties that

(1) $(\mathbf{f}_1(\mathbf{u}), \mathbf{f}_2(\mathbf{u}))$ are null vectors that span the tangent space $T_{\mathbf{x}(\mathbf{u})}\Sigma$ at each point $\mathbf{x}(\mathbf{u}) \in \Sigma$;

(2) $\langle \mathbf{f}_1(\mathbf{u}), \mathbf{f}_2(\mathbf{u})\rangle = 1$;

(3) $\mathbf{f}_3(\mathbf{u})$ is the normal vector $\mathbf{f}_3(\mathbf{u}) = \mathbf{f}_1(\mathbf{u}) \times \mathbf{f}_2(\mathbf{u})$ to $T_{\mathbf{x}(\mathbf{u})}\Sigma$ at each point $\mathbf{x}(\mathbf{u}) \in \Sigma$.

Now, let $(\bar{\eta}^i, \bar{\eta}^i_j)$ be the Maurer-Cartan forms associated to a null adapted frame field along Σ. These forms are defined just as in (3.1):

$$d\mathbf{x} = \mathbf{f}_i \bar{\eta}^i,$$
$$d\mathbf{f}_i = \mathbf{f}_j \bar{\eta}^j_i,$$

where $1 \le i, j \le 3$, and they still satisfy the structure equations

$$d\bar{\eta}^i = -\bar{\eta}^i_j \wedge \bar{\eta}^j,$$
$$d\bar{\eta}^i_j = -\bar{\eta}^i_k \wedge \bar{\eta}^k_j.$$

The main difference is that the vectors $(\mathbf{f}_1, \mathbf{f}_2, \mathbf{f}_3)$ satisfy the somewhat unusual inner product relations

$$\begin{aligned} &\langle \mathbf{f}_1, \mathbf{f}_1\rangle = 0, \qquad \langle \mathbf{f}_2, \mathbf{f}_2\rangle = 0, \qquad \langle \mathbf{f}_3, \mathbf{f}_3\rangle = -1, \\ &\langle \mathbf{f}_1, \mathbf{f}_2\rangle = 1, \qquad \langle \mathbf{f}_1, \mathbf{f}_3\rangle = 0, \qquad \langle \mathbf{f}_2, \mathbf{f}_3\rangle = 0. \end{aligned} \tag{5.6}$$

***Exercise 5.43.** Differentiate the inner product relations (5.6) to obtain the following relations among the $(\bar{\eta}^i_j)$:

$$\begin{aligned} &\bar{\eta}^1_2 = \bar{\eta}^2_1 = \bar{\eta}^3_3 = 0, \\ &\bar{\eta}^2_2 = -\bar{\eta}^1_1, \qquad \bar{\eta}^1_3 = \bar{\eta}^3_2, \qquad \bar{\eta}^2_3 = \bar{\eta}^3_1. \end{aligned} \tag{5.7}$$

The same reasoning as before can be used to prove the following:

Proposition 5.44. *Let $U \subset \mathbb{R}^2$ be an open set, and let $\mathbf{x} : U \to \mathbb{M}^{1,2}$ be a timelike immersion. For any null adapted frame field $(\mathbf{f}_1(\mathbf{u}), \mathbf{f}_2(\mathbf{u}), \mathbf{f}_3(\mathbf{u}))$ along $\Sigma = \mathbf{x}(U)$, the associated dual and connection forms $(\bar{\eta}^i, \bar{\eta}^i_j)$ have the property that $\bar{\eta}^3 = 0$. Moreover, $(\bar{\eta}^1, \bar{\eta}^2)$ form a basis for the 1-forms on U.*

***Exercise 5.45.** Show that the first fundamental form of $\Sigma = \mathbf{x}(U)$ (cf. Definition 5.13) is

$$\mathrm{I} = 2\bar{\eta}^1\bar{\eta}^2.$$

***Exercise 5.46.** (a) Differentiate the equation $\bar{\eta}^3 = 0$ and use Cartan's lemma to conclude that there exist functions k_{11}, k_{12}, k_{22} on U such that

$$\begin{bmatrix} \bar{\eta}^3_1 \\ \bar{\eta}^3_2 \end{bmatrix} = - \begin{bmatrix} k_{11} & k_{12} \\ k_{12} & k_{22} \end{bmatrix} \begin{bmatrix} \bar{\eta}^1 \\ \bar{\eta}^2 \end{bmatrix}. \tag{5.8}$$

(b) The Gauss map (cf. Definition 5.15) of $\Sigma = \mathbf{x}(U)$ is the map $N : \Sigma \to \mathbb{M}^{1,2}$ defined by

$$N(\mathbf{x}(\mathbf{u})) = \mathbf{f}_3(\mathbf{u}).$$

Show that

$$d\mathbf{f}_3 = \mathbf{f}_1\bar{\eta}^1_3 + \mathbf{f}_2\bar{\eta}^2_3 = -\mathbf{f}_1(k_{12}\bar{\eta}^1 + k_{22}\bar{\eta}^2) - \mathbf{f}_2(k_{11}\bar{\eta}^1 + k_{12}\bar{\eta}^2).$$

Conclude that with respect to the basis $(\mathbf{f}_1(\mathbf{u}), \mathbf{f}_2(\mathbf{u}))$ for $T_{\mathbf{x}(\mathbf{u})}\Sigma$, the matrix of the linear transformation $d\mathbf{f}_3 : T_{\mathbf{x}(\mathbf{u})}\Sigma \to T_{\mathbf{x}(\mathbf{u})}\Sigma$ is

$$- \begin{bmatrix} k_{12} & k_{22} \\ k_{11} & k_{12} \end{bmatrix}.$$

(c) The second fundamental form (cf. Definition 5.16) of $\Sigma = \mathbf{x}(U)$ is defined by

$$\mathrm{II}(\mathbf{v}) = -\langle d\mathbf{f}_3(\mathbf{v}), d\mathbf{x}(\mathbf{v})\rangle$$

for $\mathbf{v} \in T_{\mathbf{u}}U$. Show that

$$\mathrm{II} = k_{11}(\bar{\eta}^1)^2 + 2k_{12}\bar{\eta}^1\bar{\eta}^2 + k_{22}(\bar{\eta}^2)^2.$$

(d) Show that the Gauss and mean curvatures (cf. Definition 5.20) of Σ are

$$K = k_{12}^2 - k_{11}k_{22}, \qquad H = k_{12}$$

and that the skew curvature (cf. Definition 5.22) is

$$H' = \sqrt{k_{11}k_{22}}.$$

Conclude that the analogous conditions to those in Exercise 5.19(c) are

- $k_{11}k_{22} > 0$ (H' real and $H' > 0$);
- $k_{11}k_{22} < 0$ (H' imaginary);
- $k_{11} = k_{22} = 0$ (umbilic points);
- $k_{11}k_{22} = 0$, and exactly one of k_{11}, k_{22} vanishes (quasi-umbilic points).

Now we will examine how the matrix $[k_{ij}]$ changes if we vary the frame.

***Exercise 5.47.** Let $(\mathbf{f}_1(\mathbf{u}), \mathbf{f}_2(\mathbf{u}), \mathbf{f}_3(\mathbf{u}))$ be any null adapted frame field along Σ, with associated Maurer-Cartan forms $(\bar{\eta}^i, \bar{\eta}^i_j)$.

(a) Show that any other null adapted frame field $(\tilde{\mathbf{f}}_1(\mathbf{u}), \tilde{\mathbf{f}}_2(\mathbf{u}), \tilde{\mathbf{f}}_3(\mathbf{u}))$ has the form (up to sign)

$$\begin{bmatrix}\tilde{\mathbf{f}}_1(\mathbf{u}) & \tilde{\mathbf{f}}_2(\mathbf{u}) & \tilde{\mathbf{f}}_3(\mathbf{u})\end{bmatrix} = \begin{bmatrix}\mathbf{f}_1(\mathbf{u}) & \mathbf{f}_2(\mathbf{u}) & \mathbf{f}_3(\mathbf{u})\end{bmatrix} \begin{bmatrix} e^{\theta} & 0 & 0 \\ 0 & e^{-\theta} & 0 \\ 0 & 0 & 1 \end{bmatrix} \tag{5.9}$$

for some function θ on U. (Hint: Consider the transformation (5.2) for the adapted orthonormal frame field

$$\mathbf{e}_0(\mathbf{u}) = \tfrac{1}{\sqrt{2}}(\mathbf{f}_1(\mathbf{u}) + \mathbf{f}_2(\mathbf{u})), \quad \mathbf{e}_1(\mathbf{u}) = \tfrac{1}{\sqrt{2}}(\mathbf{f}_1(\mathbf{u}) - \mathbf{f}_2(\mathbf{u})), \quad \mathbf{e}_2(\mathbf{u}) = \mathbf{f}_3(\mathbf{u}).)$$

(b) Let $(\tilde{\bar{\eta}}^i, \tilde{\bar{\eta}}^i_j)$ be the Maurer-Cartan forms associated to the new frame field, and set $B = \begin{bmatrix} e^{\theta} & 0 \\ 0 & e^{-\theta} \end{bmatrix}$. Show that

$$\begin{bmatrix} \tilde{\bar{\eta}}^1 \\ \tilde{\bar{\eta}}^2 \end{bmatrix} = B^{-1} \begin{bmatrix} \bar{\eta}^1 \\ \bar{\eta}^2 \end{bmatrix} = \begin{bmatrix} e^{-\theta}\bar{\eta}^1 \\ e^{\theta}\bar{\eta}^2 \end{bmatrix}, \qquad \begin{bmatrix} \tilde{\bar{\eta}}^1_3 \\ \tilde{\bar{\eta}}^2_3 \end{bmatrix} = B^{-1} \begin{bmatrix} \bar{\eta}^1_3 \\ \bar{\eta}^2_3 \end{bmatrix} = \begin{bmatrix} e^{-\theta}\bar{\eta}^1_3 \\ e^{\theta}\bar{\eta}^2_3 \end{bmatrix}. \tag{5.10}$$

(c) Cartan's lemma implies that there exist functions $\tilde{k}_{11}, \tilde{k}_{12}, \tilde{k}_{22}$ on U such that

$$\begin{bmatrix} \tilde{\bar{\eta}}^3_1 \\ \tilde{\bar{\eta}}^3_2 \end{bmatrix} = - \begin{bmatrix} \tilde{k}_{11} & \tilde{k}_{12} \\ \tilde{k}_{12} & \tilde{k}_{22} \end{bmatrix} \begin{bmatrix} \tilde{\bar{\eta}}^1 \\ \tilde{\bar{\eta}}^2 \end{bmatrix}.$$

Show that

$$\begin{bmatrix} \tilde{k}_{12} & \tilde{k}_{22} \\ \tilde{k}_{11} & \tilde{k}_{12} \end{bmatrix} = B^{-1} \begin{bmatrix} k_{12} & k_{22} \\ k_{11} & k_{12} \end{bmatrix} B$$

and that

$$\begin{bmatrix} \tilde{k}_{11} & \tilde{k}_{12} \\ \tilde{k}_{12} & \tilde{k}_{22} \end{bmatrix} = {}^tB \begin{bmatrix} k_{11} & k_{12} \\ k_{12} & k_{22} \end{bmatrix} B = \begin{bmatrix} e^{2\theta} k_{11} & k_{12} \\ k_{12} & e^{-2\theta} k_{22} \end{bmatrix}. \tag{5.11}$$

The result of Exercise 5.47 provides a guide as to how to choose a canonical null adapted frame field in most cases:

(1) If $H' \neq 0$, then there exists a null adapted frame field $(\mathbf{f}_1(\mathbf{u}), \mathbf{f}_2(\mathbf{u}), \mathbf{f}_3(\mathbf{u}))$ (unique up to sign) along Σ with the property that $k_{11} = \pm k_{22}$. The sign is positive if H' is real and negative if H' is imaginary.

(2) It may not be possible to choose such a frame field continuously near umbilic or quasi-umbilic points. But in a totally quasi-umbilic *neighborhood* where, without loss of generality, $k_{22} = 0$ and $k_{11} \neq 0$, there exists a null adapted frame field $(\mathbf{f}_1(\mathbf{u}), \mathbf{f}_2(\mathbf{u}), \mathbf{f}_3(\mathbf{u}))$ (unique up to sign) along Σ with the property that $k_{11} = \pm 1$.

***Exercise 5.48** (Cf. Exercise 5.31)**.** A timelike surface Σ always has a local parametrization $\mathbf{x} : U \to \mathbb{M}^{1,2}$ in terms of *null coordinates.* Such a parametrization has the property that

$$\mathrm{I} = 2F\,du\,dv$$

for some function $F > 0$ on U. We can write the second fundamental form as

$$\mathrm{II} = e\,du^2 + 2f\,du\,dv + g\,dv^2$$

for some functions e, f, g on U. Follow the outline of Exercises 4.24 and 4.27 to derive the analogs of the Gauss and Codazzi equations for timelike surfaces in null coordinates.

Exercise 5.49. Let $\alpha : I \to \mathbb{M}^{1,2}$ be a nondegenerate *null curve*, i.e., a curve with the property that $\alpha'(u)$ is a null vector for all $u \in I$. (The condition that α is nondegenerate means that $\alpha'(u)$ and $\alpha''(u)$ are linearly independent for all $u \in I$.) Let $\mathbf{f}^0$ be any null vector that is linearly independent from $\alpha'(u)$ for all $u \in I$, and consider the cylindrical surface Σ with parametrization $\mathbf{x} : I \times \mathbb{R} \to \mathbb{M}^{1,2}$ given by

$$\mathbf{x}(u, v) = \alpha(u) + v\mathbf{f}^0.$$

(a) Show that α can be reparametrized so as to satisfy the condition

$$\langle \alpha'(u), \mathbf{f}^0 \rangle = 1.$$

(b) With α as in part (a), show that the vector fields

$$\mathbf{f}_1(u,v) = \alpha'(u), \qquad \mathbf{f}_2(u,v) = \mathbf{f}^0, \qquad \mathbf{f}_3(u,v) = \mathbf{f}_1(u,v) \times \mathbf{f}_2(u,v)$$

form a null adapted frame field along Σ.

(c) Show by direct computation of $d\mathbf{x}$, $d\mathbf{f}_2$, and $d\mathbf{f}_1$ (preferably in that order!) that the Maurer-Cartan forms associated to this frame field satisfy

$$\begin{aligned} \bar{\eta}^1 &= du, & \bar{\eta}^2 &= dv, \\ \bar{\eta}^1_3 = \bar{\eta}^3_2 &= 0, & \bar{\eta}^2_3 = \bar{\eta}^3_1 &= \lambda\, du \end{aligned}$$

for some nonvanishing function $\lambda : I \to \mathbb{R}$.

(d) Conclude that $k_{11} = -\lambda$, $k_{12} = 0$, $k_{22} = 0$ and that Σ is therefore totally quasi-umbilic.

(e) Show that Σ has Gauss and mean curvatures

$$K \equiv 0, \qquad H \equiv 0.$$

This exercise shows that, unlike in the Euclidean case, there exist non-planar timelike surfaces in $\mathbb{M}^{1,2}$ whose Gauss and mean curvatures are identically zero! Conversely, it can be shown that if $\Sigma \subset \mathbb{M}^{1,2}$ is a timelike surface with $K \equiv H \equiv 0$, then every point of Σ is either umbilic or quasi-umbilic; see [**Cle12**] for details.

5.5. Maple computations

The Maple setup for the orthonormal frame approach in this chapter is much the same as in Chapter 4, except that indices now range from 0 to 2 and the symmetries of the connection forms are slightly different. Likewise, Exercise 5.18 is very similar to Exercise 4.28; aside from the adjustments in the index ranges, the only change is in the matrix B. We leave these modifications as an exercise for the reader. (More details are given in the Maple worksheet for this chapter on the AMS webpage.

Here, we will explore how to do analogous computations for a null adapted frame field along a timelike surface in $\mathbb{M}^{1,2}$. After loading the `Cartan` and `LinearAlgebra` packages into Maple, begin by declaring the necessary 1-forms:

```
> Form(eta[1], eta[2], eta[3]);
  Form(eta[1,1], eta[3,1], eta[3,2]);
```

Tell MAPLE about the symmetries in the connection forms:

```
> eta[1,2]:= 0;
  eta[2,1]:= 0;
  eta[3,3]:= 0;
  eta[2,2]:= -eta[1,1];
  eta[1,3]:= eta[3,2];
  eta[2,3]:= eta[3,1];
```

Tell MAPLE how to differentiate these forms according to the Cartan structure equations (3.8):

```
> for i from 1 to 3 do
    d(eta[i]):= -add('eta[i,j] &^ eta[j]', j=1..3);
  end do;

  d(eta[1,1]):= -add('eta[1,k] &^ eta[k,1]', k=1..3);
  d(eta[3,1]):= -add('eta[3,k] &^ eta[k,1]', k=1..3);
  d(eta[3,2]):= -add('eta[3,k] &^ eta[k,2]', k=1..3);
```

Set up a substitution for the Maurer-Cartan forms of a null adapted frame field, taking into account the relations (5.8) that result from computing $d\bar{\eta}^3 = 0$:

```
> adaptedsub1:= [eta[3]=0,
  eta[3,1]  = -(k[1,1]*eta[1]  + k[1,2]*eta[2]),
  eta[3,2]  = -(k[1,2]*eta[1]  + k[2,2]*eta[2])];
```

EXERCISE 5.47: Introduce new 1-forms to represent the transformed forms, with the same symmetry conditions as the original forms:

```
> Form(Eta[1], Eta[2], Eta[3]);
  Form(Eta[1,1], Eta[3,1], Eta[3,2]);
  Eta[1,2]:= 0;
  Eta[2,1]:= 0;
  Eta[3,3]:= 0;
  Eta[2,2]:= -Eta[1,1];
  Eta[1,3]:= Eta[3,2];
  Eta[2,3]:= Eta[3,1];
```

Introduce the relations (5.10) via the following substitution, along with its reverse substitution:

```
> framechangesub:= [Eta[1]  = exp(-theta)*eta[1],
  Eta[2]  = exp(theta)*eta[2], Eta[3,1]  = exp(theta)*eta[3,1],
  Eta[3,2]  = exp(-theta)*eta[3,2]];
> framechangebacksub:= makebacksub(framechangesub);
```

In order to compare the functions $(\tilde{k}_{ij})$ to the functions (k_{ij}), introduce another substitution describing the adaptations of the transformed frame:

```
> adaptedsub2:= [Eta[3]=0,
  Eta[3,1] = -(K[1,1]*Eta[1] + K[1,2]*Eta[2]),
  Eta[3,2] = -(K[1,2]*Eta[1] + K[2,2]*Eta[2])];
```

Now combine all these substitutions to see how the $(\tilde{k}_{ij})$ are expressed in terms of the (k_{ij}):

```
> zero2:= Simf(subs(adaptedsub2, Eta[3,1])
  - Simf(subs(framechangebacksub, Simf(subs(adaptedsub1,
  Simf(subs(framechangesub, Eta[3,1])))))));
> eqns:= {op(ScalarForm(zero2))};
> zero3:= Simf(subs(adaptedsub2, Eta[3,2])
  - Simf(subs(framechangebacksub, Simf(subs(adaptedsub1,
  Simf(subs(framechangesub, Eta[3,2])))))));
> eqns:= eqns union {op(ScalarForm(zero3))};
> solve(eqns, {K[1,1], K[1,2], K[2,2]});
```

EXERCISE 5.48: Start by declaring that F, e, f, g are functions of u and v:

```
> PDETools[declare](F(u,v), e(u,v), f(u,v), g(u,v));
```

For a null parametrization $\mathbf{x} : U \to \mathbb{M}^{1,2}$ of a timelike surface with $F > 0$, we can define a null adapted frame field by

$$\mathbf{f}_1(\mathbf{u}) = \frac{1}{\sqrt{F}}\mathbf{x}_u, \qquad \mathbf{f}_2(\mathbf{u}) = \frac{1}{\sqrt{F}}\mathbf{x}_v, \qquad \mathbf{f}_3(\mathbf{u}) = \mathbf{f}_1(\mathbf{u}) \times \mathbf{f}_2(\mathbf{u}).$$

The associated dual forms are

$$\bar{\eta}^1 = \sqrt{F}\,du, \qquad \bar{\eta}^2 = \sqrt{F}\,dv, \qquad \bar{\eta}^3 = 0,$$

and in order for the second fundamental form to have the desired form, we must have

$$\bar{\eta}^3_1 = -\left(\frac{e}{\sqrt{F}}\,du + \frac{f}{\sqrt{F}}dv\right), \qquad \bar{\eta}^3_2 = -\left(\frac{f}{\sqrt{F}}\,du + \frac{g}{\sqrt{F}}dv\right).$$

Introduce a substitution for the Maurer-Cartan forms in terms of the coordinate 1-forms:

```
> coordsub:= [eta[3]=0, eta[1] = sqrt(F(u,v))*d(u),
  eta[2] = sqrt(F(u,v))*d(v),
  eta[3,1] = -((e(u,v)/sqrt(F(u,v)))*d(u)
  + (f(u,v)/sqrt(F(u,v)))*d(v)),
  eta[3,2] = -((f(u,v)/sqrt(F(u,v)))*d(u)
```

```
  + (g(u,v)/sqrt(F(u,v)))*d(v)),
  eta[1,1] = a*d(u) + b*d(v)];
```

Use the structure equations for $d\bar{\eta}^1$ and $d\bar{\eta}^2$ to compute the coefficients in η_1^1:

```
> Simf(d(Simf(subs(coordsub, eta[1])))
  - subs(coordsub, Simf(d(eta[1]))));
> b:= solve(%, b);
> Simf(d(Simf(subs(coordsub, eta[2])))
  - subs(coordsub, Simf(d(eta[2]))));
> a:= solve(%, a);
```

The Gauss equation comes from comparing the two expressions for $d\bar{\eta}_1^1$:

```
> Simf(d(Simf(subs(coordsub, eta[1,1])))
  - subs(coordsub, Simf(d(eta[1,1]))));
> Gausseq:= pick(%, d(u), d(v));
```

The Gauss curvature is defined to be $K = \frac{f^2-eg}{F^2}$, and from the output of the last computation we see that the Gauss equation takes the form

$$K = \frac{1}{F^3}\left(FF_{uv} - F_uF_v\right) = \frac{(\ln F)_{uv}}{F}.$$

Similar manipulations involving $d\bar{\eta}_1^3$ and $d\bar{\eta}_2^3$ show that the Codazzi equations take the form

$$e_v = \frac{Ff_u - fF_u}{F} = F\left(\frac{f}{F}\right)_u, \qquad g_u = \frac{Ff_v - fF_v}{F} = F\left(\frac{f}{F}\right)_v.$$

Chapter 6

Curves and surfaces in equi-affine space

6.1. Introduction

Now we will apply the method of moving frames to study the geometry of curves and surfaces in equi-affine space. The lack of a metric structure will lead to results that may seem less intuitive than those of the last two chapters, but the general procedure for constructing invariants remains the same. And rather than rediscovering the familiar invariants for curves and surfaces in Euclidean space, we will see how this method naturally leads to an alternative set of invariants that are preserved under the action of the entire equi-affine group.

Exercise 6.1. Show by example that an equi-affine transformation $T : \mathbb{R}^3 \to \mathbb{R}^3$ does not necessarily preserve the curvature $\kappa(s)$ and torsion $\tau(s)$ of a curve $\alpha : I \to \mathbb{E}^3$. (Hint: Let α be something simple, like a helix, and consider a diagonal transformation such as

$$T(x^1, x^2, x^3) = {}^t\big[2x^1,\ \tfrac{1}{2}x^2,\ x^3\big]\,.)$$

A remark about the notation: A curve in $\mathbb{R}^3$ may be regarded as a curve in *either* $\mathbb{E}^3$ or $\mathbb{A}^3$, depending on what symmetry group we allow to act on $\mathbb{R}^3$. In general, an equi-affine transformation acting on a curve in $\mathbb{E}^3$ will transform the curve to another curve that is *not* equivalent to the original curve under the action of the Euclidean group.

6.2. Moving frames for curves in $\mathbb{A}^3$

Consider a smooth, parametrized curve $\alpha : I \to \mathbb{A}^3$ that maps some open interval $I \subset \mathbb{R}$ into equi-affine space. $\mathbb{A}^3$ has the structure of the homogeneous space $A(3)/SL(3)$, so an adapted frame field along α should be a lifting $\tilde{\alpha} : I \to A(3)$. Any such lifting can be written as

$$\tilde{\alpha}(t) = (\alpha(t); \mathbf{e}_1(t), \mathbf{e}_2(t), \mathbf{e}_3(t)),$$

where for each $t \in I$, $(\mathbf{e}_1(t), \mathbf{e}_2(t), \mathbf{e}_3(t))$ is a unimodular basis for the tangent space $T_{\alpha(t)}\mathbb{A}^3$. Such an adapted frame field is called a *unimodular frame field* along α. The situation is quite different from that of either Euclidean or Minkowski space; for instance, there is no obvious notion of arc length for a curve that is invariant under the action of $A(3)$. Moreover, we have much greater freedom in choosing our frame; the only requirement is that $\det[\mathbf{e}_1(t)\ \mathbf{e}_2(t)\ \mathbf{e}_3(t)] = 1$. So, how should we proceed?

In the Euclidean case, we used the first derivative of α to choose $\mathbf{e}_1(t)$ and the second derivative of α to choose $\mathbf{e}_2(t)$, pausing along the way to normalize according to arc length so that the frame would be orthonormal. These choices determined $\mathbf{e}_3(t)$ uniquely, but it is clear from the structure equations that $\mathbf{e}_3(t)$ is related to the third derivative of α. In order for this procedure to work, we had to assume that α was "nondegenerate", i.e., that the vectors $(\alpha'(t), \alpha''(t))$ were linearly independent for each $t \in I$.

So, how might we use similar reasoning to construct a canonical adapted frame field in the equi-affine case, and what would be the right notion of "nondegenerate" for equi-affine curves? Since orthonormality is no longer required, our first guess towards constructing an adapted frame field might be to take

$$\mathbf{e}_1(t) = \alpha'(t),$$

$$\mathbf{e}_2(t) = \alpha''(t),$$

$$\mathbf{e}_3(t) = \alpha'''(t).$$

In order for this to work, we must assume that the vectors $(\alpha'(t), \alpha''(t), \alpha'''(t))$ are linearly independent for each $t \in I$; such a curve will be called *nondegenerate*. (Note that this word means something different for curves in equi-affine space than for curves in Euclidean space!) For nondegenerate curves, the only problem with this choice of frame field is that it is not necessarily unimodular. But we could fix this by defining the adapted frame field to be

$$\begin{aligned}\mathbf{e}_1(t) &= \frac{\alpha'(t)}{\sqrt[3]{\det[\alpha'(t)\ \alpha''(t)\ \alpha'''(t)]}},\\ \mathbf{e}_2(t) &= \frac{\alpha''(t)}{\sqrt[3]{\det[\alpha'(t)\ \alpha''(t)\ \alpha'''(t)]}},\\ \mathbf{e}_3(t) &= \frac{\alpha'''(t)}{\sqrt[3]{\det[\alpha'(t)\ \alpha''(t)\ \alpha'''(t)]}}.\end{aligned} \tag{6.1}$$

Exercise 6.2. Prove that this choice of frame field $(\mathbf{e}_1(t), \mathbf{e}_2(t), \mathbf{e}_3(t))$ is equivariant under the action of $A(3)$: If we replace α by $g \cdot \alpha$ for some $g \in A(3)$, then $\mathbf{e}_i(t) \in T_{\alpha(t)}\mathbb{A}^3$ will be replaced by $(L_g)_*(\mathbf{e}_i(t)) \in T_{g\cdot\alpha(t)}$.

Now, wouldn't it be nice to get rid of that ugly denominator? In the Euclidean case, we were able to get rid of the denominator for $\mathbf{e}_1(t)$ by reparametrizing according to arc length. So let's see if we can find a suitable reparametrization to do the trick here; specifically, we would like to find a reparametrization $\alpha(s)$ of α for which

$$\det[\alpha'(s)\ \alpha''(s)\ \alpha'''(s)] = 1.$$

This will be a bit more complicated than in the Euclidean case since the denominators in (6.1) involve the first three derivatives of α instead of just the first derivative.

Suppose that we reparametrize the curve by setting $\alpha(s) = \alpha(t(s))$ for some invertible function $t(s)$. Then

$$\begin{aligned}\frac{d\alpha}{ds} &= t'(s)\frac{d\alpha}{dt},\\ \frac{d^2\alpha}{ds^2} &\equiv t'(s)^2\,\frac{d^2\alpha}{dt^2} \quad \mod \frac{d\alpha}{ds},\\ \frac{d^3\alpha}{ds^3} &\equiv t'(s)^3\,\frac{d^3\alpha}{dt^3} \quad \mod \frac{d\alpha}{ds}, \frac{d^2\alpha}{ds^2}.\end{aligned} \tag{6.2}$$

Note that it doesn't matter what multiple of $\frac{d\alpha}{ds}$ we are ignoring in $\frac{d^2\alpha}{ds^2}$—and similarly for the $\frac{d\alpha}{ds}, \frac{d^2\alpha}{ds}$ terms in $\frac{d^3\alpha}{ds^2}$—because they won't affect the determinant $\det[\alpha'(s)\ \alpha''(s)\ \alpha'''(s)]$. Therefore,

$$\det[\alpha'(s)\ \alpha''(s)\ \alpha'''(s)] = t'(s)^6 \det[\alpha'(t(s))\ \alpha''(t(s))\ \alpha'''(t(s))].$$

Note that the sign of $\det[\alpha'(s)\ \alpha''(s)\ \alpha'''(s)]$ is fixed, so the best that we can hope for is to arrange that $\det[\alpha'(s)\ \alpha''(s)\ \alpha'''(s)] = \pm 1$. This suggests that we make the following definition:

Definition 6.3. Let $\alpha : I \to \mathbb{A}^3$ be a nondegenerate curve, and fix $t_0 \in I$. The function

$$s(t) = \int_{t_0}^{t} \sqrt[6]{\left| \det[\alpha'(u)\ \alpha''(u)\ \alpha'''(u)] \right|}\, du$$

is called the *equi-affine arc length* function along α.

***Exercise 6.4.** (a) Show that any nondegenerate curve $\alpha : I \to \mathbb{A}^3$ can be smoothly reparametrized by its equi-affine arc length.

(b) Show that if α is parametrized by equi-affine arc length s, then

$$\det[\alpha'(s)\ \alpha''(s)\ \alpha'''(s)] = \pm 1.$$

Remark 6.5. As for curves in Euclidean space, it may or may not be possible to find an *explicit* parametrization by equi-affine arc length. The existence of such a parametrization is of more importance for developing the geometric theory than for working with explicit examples.

Exercise 6.6. Show that for a nondegenerate curve $\alpha : I \to \mathbb{A}^3$, the equi-affine arc length $s(t)$ is invariant under the action of $A(3)$; that is, for any $g \in A(3)$, the curves α and $g \cdot \alpha$ have the same equi-affine arc length function.

Equi-affine arc length is a very different notion from Euclidean arc length. Some of the most notable differences are:

(1) Unlike Euclidean arc length, which depends only on the first derivative of α, the equi-affine arc length depends on the first *three* derivatives of α. In fact, this number is dependent on the dimension of the ambient equi-affine space: The equi-affine arc length of a curve $\alpha : I \to \mathbb{A}^n$ depends on the first n derivatives of α.

(2) The equi-affine arc length is only nonzero for nondegenerate curves; so, for instance, any curve contained in a plane in $\mathbb{A}^3$ has equi-affine arc length zero according to this definition. It could, however, have nonzero equi-affine arc length when regarded as a curve in $\mathbb{A}^2$.

(3) There is no ambient metric—Euclidean or otherwise—on $\mathbb{A}^3$ whose restriction to a curve gives its equi-affine arc length, as there is for curves in Euclidean space. This makes it difficult to develop any real intuition for what the equi-affine arc length represents, but we will at least make an attempt in the following exercise.

***Exercise 6.7.** Let $\alpha : I \to \mathbb{A}^3$ be a nondegenerate curve, parametrized by its *Euclidean* arc length $\bar{s}$. Show that the equi-affine arc length of α is given by

$$s(\bar{s}) = \int_{\bar{s}_0}^{\bar{s}} \sqrt[6]{\left|\kappa(u)^2\,\tau(u)\right|}\,du, \tag{6.3}$$

where κ and τ represent the Euclidean curvature and torsion, respectively, of α. So, the greater a curve's Euclidean curvature and torsion, the greater its equi-affine arc length. (Hint: Use the Frenet equations for α to compute $\alpha'(\bar{s}), \alpha''(\bar{s}), \alpha'''(\bar{s})$.)

Remark 6.8. This exercise implies that, while the curvature κ and torsion τ of a unit-speed curve in $\mathbb{E}^3$ are not preserved by the action of the equi-affine group, the 1-form $\sqrt[6]{|\kappa^2\tau|}\,d\bar{s}$ (where $\bar{s}$ is the Euclidean arc length of the curve) *is* invariant under the action of this group!

By Exercise 6.4, we can assume that α is parametrized by equi-affine arc length, so that

$$\det[\alpha'(s)\ \alpha''(s)\ \alpha'''(s)] = \pm 1.$$

Then the frame field

$$\mathbf{e}_1(s) = \pm\alpha'(s), \qquad \mathbf{e}_2(s) = \pm\alpha''(s), \qquad \mathbf{e}_3(s) = \pm\alpha'''(s) \tag{6.4}$$

(with signs chosen to agree with the sign of $\det[\alpha'(s)\ \alpha''(s)\ \alpha'''(s)]$) is unimodular and equivariant under the action of $A(3)$. Moreover, this frame field is canonical, in that it is uniquely determined by the geometry of α.

Definition 6.9. Let $\alpha : I \to \mathbb{A}^3$ be a nondegenerate curve, parametrized by equi-affine arc length s. The unimodular frame field (6.4) is called the *equi-affine Frenet frame* of α.

Remark 6.10. The vectors $(\mathbf{e}_1(s), \mathbf{e}_2(s), \mathbf{e}_3(s))$ of the equi-affine Frenet frame (6.4) may not be exactly equal to those of the frame field (6.1); they will differ by precisely the terms that we ignored in the chain rule computations (6.2).

In order to avoid a proliferation of sign ambiguities in what follows, for the rest of this section we will assume that $\det[\alpha'(s)\ \alpha''(s)\ \alpha'''(s)] = 1$, so that the signs in equation (6.4) are all positive. Inserting the appropriate minus signs when $\det[\alpha'(s)\ \alpha''(s)\ \alpha'''(s)] = -1$ is left as an exercise for the reader.

Now that we have a canonical adapted frame field along α, we can compute invariants for nondegenerate equi-affine curves in $\mathbb{A}^3$. As in the Euclidean case, the pullbacks of equations (3.1) to I via α can be written as

$$\begin{aligned}\alpha'(s)ds &= \mathbf{e}_i(s)\,\bar{\omega}^i,\\ \mathbf{e}_i'(s)ds &= \mathbf{e}_j(s)\,\bar{\omega}_i^j.\end{aligned} \tag{6.5}$$

But we constructed our adapted frame field so that

$$\alpha'(s) = \mathbf{e}_1(s), \qquad \mathbf{e}_1'(s) = \mathbf{e}_2(s), \qquad \mathbf{e}_2'(s) = \mathbf{e}_3(s).$$

So, the first equation in (6.5) implies that

$$\bar{\omega}^1 = ds, \qquad \bar{\omega}^2 = \bar{\omega}^3 = 0,$$

and the equations for $\mathbf{e}_1'(s), \mathbf{e}_2'(s)$ in (6.5) imply that

$$\bar{\omega}_1^2 = \bar{\omega}_2^3 = ds, \qquad \bar{\omega}_1^1 = \bar{\omega}_1^3 = \bar{\omega}_2^1 = \bar{\omega}_2^2 = 0.$$

Finally, $\mathbf{e}_3'(s)$ must satisfy

$$\mathbf{e}_3'(s) = \kappa_1(s)\,\mathbf{e}_1(s) + \kappa_2(s)\,\mathbf{e}_2(s)$$

for some functions $\kappa_1(s), \kappa_2(s)$. These functions are called the *equi-affine curvatures* of α at s.

***Exercise 6.11.** Why is there no $\mathbf{e}_3(s)$ term in the equation for $\mathbf{e}_3'(s)$? (Hint: What does the equation for $\mathbf{e}_3'(s)$ in (6.5) tell you about $\bar{\omega}_3^1$, $\bar{\omega}_3^2$, and $\bar{\omega}_3^3$? What relations must the $(\bar{\omega}_j^i)$ satisfy?)

Thus, the equi-affine analog of the Frenet equations is

$$\left[\alpha'(s)\ \mathbf{e}_1'(s)\ \mathbf{e}_2'(s)\ \mathbf{e}_3'(s)\right] = \left[\alpha(s)\ \mathbf{e}_1(s)\ \mathbf{e}_2(s)\ \mathbf{e}_3(s)\right]\begin{bmatrix}0 & 0 & 0 & 0\\ 1 & 0 & 0 & \kappa_1(s)\\ 0 & 1 & 0 & \kappa_2(s)\\ 0 & 0 & 1 & 0\end{bmatrix}.$$

Applying Lemma 4.2 yields the following theorem:

Theorem 6.12. *Two nondegenerate equi-affine curves $\alpha_1, \alpha_2 : I \to \mathbb{A}^3$ parametrized by equi-affine arc length differ by an equi-affine transformation if and only if they have the same equi-affine curvatures $\kappa_1(s), \kappa_2(s)$.*

***Exercise 6.13.** Let $\alpha : I \to \mathbb{A}^2$ be a curve in the equi-affine plane $\mathbb{A}^2$.

(a) How would you define an adapted frame field $(\mathbf{e}_1(t), \mathbf{e}_2(t))$ at each point of the curve?

(b) When should a curve $\alpha : I \to \mathbb{A}^2$ be called "nondegenerate"?

(c) How would you define equi-affine arc length for a nondegenerate curve?

(d) Use the pullbacks of equations (3.1) to find a complete set of invariants for nondegenerate equi-affine curves $\alpha : I \to \mathbb{A}^2$ parametrized by equi-affine arc length.

***Exercise 6.14.** Hopefully you discovered a single invariant $\kappa(s)$, called the *equi-affine curvature* of α, in Exercise 6.13. Suppose that $\alpha : I \to \mathbb{A}^2$ is nondegenerate and that its equi-affine curvature $\kappa(s)$ is equal to a constant κ. Show that:

(a) If $\kappa = 0$, then α is a parabola.

(b) If $\kappa > 0$, then α is a hyperbola.

(c) If $\kappa < 0$, then α is a circle or an ellipse.

(Hint: In each case, you should be able to show that $\alpha(s)$ satisfies a differential equation whose general solution is not difficult to find. Since each of these conditions is preserved under equi-affine transformations, you can perform an equi-affine transformation to eliminate most of the arbitrary constants in the general solution. Finally, eliminate the parameter s to show that α lies on the appropriate conic section.)

***Exercise 6.15.** Let $\alpha : I \to \mathbb{A}^2$ be a curve, parametrized by its Euclidean arc length $\bar{s}$, and let $\bar{\kappa}(\bar{s})$ denote the Euclidean curvature of α.

(a) Show that the equi-affine arc length of α is given by

$$s(\bar{s}) = \int_{\bar{s}_0} \sqrt[3]{\bar{\kappa}(u)^2}\, du. \tag{6.6}$$

(b) Let $(\bar{\mathbf{e}}_1(\bar{s}), \bar{\mathbf{e}}_2(\bar{s}))$ denote the Euclidean Frenet frame field along α. Show that the equi-affine Frenet frame field along α is given by

$$\begin{aligned} \mathbf{e}_1(s) &= \bar{\kappa}(\bar{s}(s))^{-1/3}\bar{\mathbf{e}}_1(\bar{s}(s)), \\ \mathbf{e}_2(s) &= -\frac{\bar{\kappa}'(\bar{s}(s))}{3\bar{\kappa}(\bar{s}(s))^{5/3}}\bar{\mathbf{e}}_1(\bar{s}(s)) + \bar{\kappa}(\bar{s}(s))^{1/3}\bar{\mathbf{e}}_2(\bar{s}(s)), \end{aligned} \tag{6.7}$$

where $\bar{s}(s)$ is the inverse function of the equi-affine arc length function (6.6) and primes denote derivatives of $\bar{\kappa}(\bar{s})$ with respect to $\bar{s}$. (Hint: Use the chain rule very carefully!)

(c) Differentiate the expression (6.7) for $\mathbf{e}_2(s)$ (again, being very careful with the chain rule) and conclude that the equi-affine curvature of α is given by

$$\kappa(s) = -\frac{1}{9\bar{\kappa}(\bar{s}(s))^{8/3}}\left(3\bar{\kappa}(\bar{s}(s))\bar{\kappa}''(\bar{s}(s)) - 5(\bar{\kappa}'(\bar{s}(s)))^2 + 9\bar{\kappa}(\bar{s}(s))^4\right). \tag{6.8}$$

This shows that, while Euclidean curvature is a second-order invariant of α (i.e., it may be expressed in terms of α and its derivatives up to order 2), the equi-affine curvature is a fourth-order invariant of α. (And, like the equi-affine arc length, the order of this invariant is dependent on the dimension of the ambient space.) This expression for the equi-affine curvature in terms of the Euclidean curvature is originally due to Blaschke [**Bla23**]; for an alternative approach, see [**Kog03**] or [**Olv11b**].

Exercise 6.16. Now that you know how things work for curves in $\mathbb{A}^2$ and $\mathbb{A}^3$, what sorts of invariants would you expect to appear for curves in $\mathbb{A}^n$?

***Exercise 6.17.** Let $\alpha : I \to \mathbb{A}^3$ be a nondegenerate curve parametrized by equi-affine arc length s, and suppose that its equi-affine curvatures $\kappa_1(s)$, $\kappa_2(s)$ are both identically equal to zero.

(a) Show that there exist vectors $\mathbf{v}_0, \mathbf{v}_1, \mathbf{v}_2, \mathbf{v}_3 \in \mathbb{A}^3$, with $\det[\mathbf{v}_1\ \mathbf{v}_2\ \mathbf{v}_3] = 1$, such that

$$\alpha(s) = \mathbf{v}_0 + s\mathbf{v}_1 + \tfrac{1}{2}s^2\mathbf{v}_2 + \tfrac{1}{6}s^3\mathbf{v}_3.$$

(b) Show that there exists an equi-affine transformation $g \in A(3)$ such that

$$g \cdot \alpha(s) = {}^t\!\left[s,\ \tfrac{1}{2}s^2,\ \tfrac{1}{6}s^3\right].$$

This curve is called the *rational normal curve* of degree 3; we will encounter it again in a slightly different form in Chapter 7 (cf. Exercise 7.31).

6.3. Moving frames for surfaces in $\mathbb{A}^3$

Now, let U be an open set in $\mathbb{R}^2$, and let $\mathbf{x} : U \to \mathbb{A}^3$ be an immersion whose image is a surface $\Sigma = \mathbf{x}(U)$. Just as for curves, an adapted frame field along Σ is a lifting $\tilde{\mathbf{x}} : U \to A(3)$ of the form

$$\tilde{\mathbf{x}}(\mathbf{u}) = (\mathbf{x}(\mathbf{u}); \mathbf{e}_1(\mathbf{u}), \mathbf{e}_2(\mathbf{u}), \mathbf{e}_3(\mathbf{u})),$$

where for each $\mathbf{u} \in U$, $(\mathbf{e}_1(\mathbf{u}), \mathbf{e}_2(\mathbf{u}), \mathbf{e}_3(\mathbf{u}))$ is a unimodular basis for the tangent space $T_{\mathbf{x}(\mathbf{u})}\mathbb{A}^3$.

In the Euclidean case, we began by choosing $\mathbf{e}_3(\mathbf{u})$ to be orthogonal to the tangent plane $T_{\mathbf{x}(\mathbf{u})}\Sigma$; having done so, it followed that $(\mathbf{e}_1(\mathbf{u}), \mathbf{e}_2(\mathbf{u}))$ must be a basis for $T_{\mathbf{x}(\mathbf{u})}\Sigma$. But in equi-affine space, there is no notion of orthogonality, so we don't have any obvious way to normalize $\mathbf{e}_3(\mathbf{u})$ immediately. However, we can still make our first adaptation by requiring that $(\mathbf{e}_1(\mathbf{u}), \mathbf{e}_2(\mathbf{u}))$ span $T_{\mathbf{x}(\mathbf{u})}\Sigma$. A frame field satisfying this condition will be called 0-*adapted.* (The 0 reflects the fact that we will be making further refinements to the frame later on; these frame fields will be called 1-adapted, 2-adapted, etc.)

Exercise 6.18. Show that the condition that $(\mathbf{e}_1(\mathbf{u}), \mathbf{e}_2(\mathbf{u}))$ span $T_{\mathbf{x}(\mathbf{u})}\Sigma$ is equivariant under the action of $A(3)$.

Let $(\bar{\omega}^i, \bar{\omega}^i_j)$ represent the pulled-back forms $(\tilde{\mathbf{x}}^*\omega^i, \tilde{\mathbf{x}}^*\omega^i_j)$ on U. Precisely the same reasoning as in the Euclidean case can be used to prove the following:

Proposition 6.19. *Let $U \subset \mathbb{R}^2$ be an open set, and let $\mathbf{x} : U \to \mathbb{A}^3$ be an immersion. For any 0-adapted frame field $(\mathbf{e}_1(\mathbf{u}), \mathbf{e}_2(\mathbf{u}), \mathbf{e}_3(\mathbf{u}))$ along $\Sigma = \mathbf{x}(U)$, the associated dual and connection forms $(\bar{\omega}^i, \bar{\omega}^i_j)$ have the property that $\bar{\omega}^3 = 0$. Moreover, $(\bar{\omega}^1, \bar{\omega}^2)$ form a basis for the 1-forms on U.*

As in the Euclidean case, differentiating the equation $\bar{\omega}^3 = 0$ yields

$$d\bar{\omega}^3 = -\bar{\omega}^3_1 \wedge \bar{\omega}^1 - \bar{\omega}^3_2 \wedge \bar{\omega}^2 = 0,$$

and Cartan's lemma (cf. Lemma 2.49) implies that there exist functions h_{11}, h_{12}, h_{22} on U such that

$$\begin{bmatrix} \bar{\omega}^3_1 \\ \bar{\omega}^3_2 \end{bmatrix} = \begin{bmatrix} h_{11} & h_{12} \\ h_{12} & h_{22} \end{bmatrix} \begin{bmatrix} \bar{\omega}^1 \\ \bar{\omega}^2 \end{bmatrix}. \tag{6.9}$$

Once again, with an eye towards making further adaptations to our frame field, we will investigate how the matrix $[h_{ij}]$ changes if we vary the frame. At this point things begin to look different from the Euclidean case because we have a larger symmetry group to make use of.

Let $(\mathbf{e}_1(\mathbf{u}), \mathbf{e}_2(\mathbf{u}), \mathbf{e}_3(\mathbf{u}))$ be any 0-adapted frame field, with associated Maurer-Cartan forms $(\bar{\omega}^i, \bar{\omega}^i_j)$. Any other 0-adapted frame field $(\tilde{\mathbf{e}}_1(\mathbf{u}),\ \tilde{\mathbf{e}}_2(\mathbf{u}),\ \tilde{\mathbf{e}}_3(\mathbf{u}))$ must have the properties that

$$\operatorname{span}(\tilde{\mathbf{e}}_1(\mathbf{u}), \tilde{\mathbf{e}}_2(\mathbf{u})) = \operatorname{span}(\mathbf{e}_1(\mathbf{u}), \mathbf{e}_2(\mathbf{u}))$$

and

$$\det\left[\tilde{\mathbf{e}}_1(\mathbf{u})\ \tilde{\mathbf{e}}_2(\mathbf{u})\ \tilde{\mathbf{e}}_3(\mathbf{u})\right] = \det\left[\mathbf{e}_1(\mathbf{u})\ \mathbf{e}_2(\mathbf{u})\ \mathbf{e}_3(\mathbf{u})\right].$$

This will be true if and only if

$$\left[\tilde{\mathbf{e}}_1(\mathbf{u})\ \tilde{\mathbf{e}}_2(\mathbf{u})\ \tilde{\mathbf{e}}_3(\mathbf{u})\right] = \left[\mathbf{e}_1(\mathbf{u})\ \mathbf{e}_2(\mathbf{u})\ \mathbf{e}_3(\mathbf{u})\right] \begin{bmatrix} \multicolumn{2}{c}{\multirow{2}{*}{B}} & r_1 \\ & & r_2 \\ 0 & 0 & (\det B)^{-1} \end{bmatrix} \tag{6.10}$$

for some $GL(2)$-valued function B and real-valued functions r_1, r_2 on U. Let $(\tilde{\bar{\omega}}^i, \tilde{\bar{\omega}}^i_j)$ be the Maurer-Cartan forms associated to the new frame field.

***Exercise 6.20.** (a) Show that

$$\begin{bmatrix} \tilde{\bar{\omega}}^1 \\ \tilde{\bar{\omega}}^2 \end{bmatrix} = B^{-1} \begin{bmatrix} \bar{\omega}^1 \\ \bar{\omega}^2 \end{bmatrix}. \tag{6.11}$$

(b) Show that

$$\begin{bmatrix} \tilde{\bar{\omega}}_1^3 \\ \tilde{\bar{\omega}}_2^3 \end{bmatrix} = (\det B)\, {}^t\!B \begin{bmatrix} \bar{\omega}_1^3 \\ \bar{\omega}_2^3 \end{bmatrix}. \tag{6.12}$$

(Hint: It is no longer true that $\bar{\omega}_i^3 = -\bar{\omega}_3^i$, so you must use the equations for $d\mathbf{e}_1$ and $d\mathbf{e}_2$ in (3.1) to compute $\bar{\omega}_1^3$ and $\bar{\omega}_2^3$. It will be simplest to keep track of everything if you write them in the form

$$d\begin{bmatrix} \mathbf{e}_1 & \mathbf{e}_2 \end{bmatrix} = \begin{bmatrix} \mathbf{e}_1 & \mathbf{e}_2 & \mathbf{e}_3 \end{bmatrix} \begin{bmatrix} \bar{\omega}_1^1 & \bar{\omega}_2^1 \\ \bar{\omega}_1^2 & \bar{\omega}_2^2 \\ \bar{\omega}_1^3 & \bar{\omega}_2^3 \end{bmatrix}.)$$

(c) Cartan's lemma implies that there exist functions $\tilde{h}_{11}, \tilde{h}_{12}, \tilde{h}_{22}$ on U such that

$$\begin{bmatrix} \tilde{\bar{\omega}}_1^3 \\ \tilde{\bar{\omega}}_2^3 \end{bmatrix} = \begin{bmatrix} \tilde{h}_{11} & \tilde{h}_{12} \\ \tilde{h}_{12} & \tilde{h}_{22} \end{bmatrix} \begin{bmatrix} \tilde{\bar{\omega}}^1 \\ \tilde{\bar{\omega}}^2 \end{bmatrix}.$$

Show that

$$\begin{bmatrix} \tilde{h}_{11} & \tilde{h}_{12} \\ \tilde{h}_{12} & \tilde{h}_{22} \end{bmatrix} = (\det B)\, {}^t\!B \begin{bmatrix} h_{11} & h_{12} \\ h_{12} & h_{22} \end{bmatrix} B. \tag{6.13}$$

The transformation (6.13) has the property that

$$\det[\tilde{h}_{ij}] = (\det B)^4 \det[h_{ij}],$$

so the sign of $\det[h_{ij}]$ is fixed.

Definition 6.21. Assume that the matrix $[h_{ij}]$ is nonsingular at every point of U. The surface $\Sigma = \mathbf{x}(U)$ is called

(1) *elliptic* if $\det[h_{ij}] > 0$ at every point of U;

(2) *hyperbolic* if $\det[h_{ij}] < 0$ at every point of U.

For the remainder of this section, we will assume that Σ is elliptic; the hyperbolic case will be treated in Exercise 6.42.

The transformation (6.13) acts transitively on the set of 2×2 matrices with positive determinant; therefore, there exists a choice of 0-adapted frame field $(\mathbf{e}_1(\mathbf{u}), \mathbf{e}_2(\mathbf{u}), \mathbf{e}_3(\mathbf{u}))$ for which

$$\begin{bmatrix} h_{11} & h_{12} \\ h_{12} & h_{22} \end{bmatrix} = \begin{bmatrix} 1 & 0 \\ 0 & 1 \end{bmatrix}.$$

Such a frame field will be called 1-*adapted*.

***Exercise 6.22.** Let $(\mathbf{e}_1(\mathbf{u}), \mathbf{e}_2(\mathbf{u}), \mathbf{e}_3(\mathbf{u}))$ be any 1-adapted frame field for an elliptic equi-affine surface $\Sigma = \mathbf{x}(U) \subset \mathbb{A}^3$.

(a) Show that any other 1-adapted frame field $(\tilde{\mathbf{e}}_1(\mathbf{u}), \tilde{\mathbf{e}}_2(\mathbf{u}), \tilde{\mathbf{e}}_3(\mathbf{u}))$ for Σ must have the form

$$\left[\tilde{\mathbf{e}}_1(\mathbf{u})\ \tilde{\mathbf{e}}_2(\mathbf{u})\ \tilde{\mathbf{e}}_3(\mathbf{u})\right] = \left[\mathbf{e}_1(\mathbf{u})\ \mathbf{e}_2(\mathbf{u})\ \mathbf{e}_3(\mathbf{u})\right] \begin{bmatrix} \multicolumn{2}{c}{\multirow{2}{*}{B}} & r_1 \\ & & r_2 \\ 0 & 0 & (\det B)^{-1} \end{bmatrix} \tag{6.14}$$

for some $SO(2)$-valued function B and real-valued functions r_1, r_2 on U.

(b) Let $(\bar{\omega}^i, \bar{\omega}^i_j)$ be the Maurer-Cartan forms associated to a 1-adapted frame field $(\mathbf{e}_1(\mathbf{u}), \mathbf{e}_2(\mathbf{u}), \mathbf{e}_3(\mathbf{u}))$, and let $(\tilde{\bar{\omega}}^i, \tilde{\bar{\omega}}^i_j)$ be the Maurer-Cartan forms associated to the 1-adapted frame field (6.14). Show that

$$\tilde{\bar{\omega}}^3_1\,\tilde{\bar{\omega}}^1 + \tilde{\bar{\omega}}^3_2\,\tilde{\bar{\omega}}^2 = (\tilde{\bar{\omega}}^1)^2 + (\tilde{\bar{\omega}}^2)^2 = (\bar{\omega}^1)^2 + (\bar{\omega}^2)^2 = \bar{\omega}^3_1\,\bar{\omega}^1 + \bar{\omega}^3_2\,\bar{\omega}^2$$

(cf. Exercise 6.20).

Definition 6.23. Let $U \subset \mathbb{R}^2$ be an open set, and let $\mathbf{x} : U \to \mathbb{A}^3$ be an immersion. Let $(\mathbf{e}_1(\mathbf{u}), \mathbf{e}_2(\mathbf{u}), \mathbf{e}_3(\mathbf{u}))$ be any 1-adapted frame field along $\Sigma = \mathbf{x}(U)$, with associated Maurer-Cartan forms $(\bar{\omega}^i, \bar{\omega}^i_j)$. The quadratic form

$$\mathrm{I} = \bar{\omega}^3_1\,\bar{\omega}^1 + \bar{\omega}^3_2\,\bar{\omega}^2 = (\bar{\omega}^1)^2 + (\bar{\omega}^2)^2$$

is called the *equi-affine first fundamental form* of Σ.

The result of Exercise 6.22 implies that the equi-affine first fundamental form is well-defined, independent of the choice of a particular 1-adapted frame field on Σ.

Exercise 6.24. Show that the equi-affine first fundamental form is invariant under the action of $A(3)$.

Because the equi-affine first fundamental form is a positive definite quadratic form at each point $\mathbf{u} \in U$, it defines a metric on Σ. But unlike in the Euclidean case, this metric is *not* the restriction of any ambient metric on $\mathbb{A}^3$ to Σ. Moreover, this metric depends on first *and* second derivatives of $\mathbf{x}$, whereas in the Euclidean case, the first fundamental form depends only on first derivatives of $\mathbf{x}$.

Remark 6.25. Suppose that Σ is an elliptic surface in $\mathbb{A}^3$ and that $\alpha : I \to \mathbb{A}^3$ is a nondegenerate curve such that $\alpha(I) \subset \Sigma$. We now have two ways of defining an arc length function on α: We can use the equi-affine arc length function from §6.2 or we can use the restriction of the metric given by the equi-affine first fundamental form on Σ to α. For curves in either Euclidean or Minkowski space, these two notions of arc length would agree since both come from the restriction of the same ambient metric on the entire space. But there is no reason why they should be the same for curves in equi-affine space, especially considering the fact that the equi-affine arc length of a curve $\alpha : I \to \mathbb{A}^3$ depends on the first three derivatives of α, whereas the equi-affine first fundamental form of a surface $\mathbf{x} : U \to \mathbb{A}^3$ depends only on the first two derivatives of $\mathbf{x}$. In fact, as the following exercise will show, these two notions of arc length do *not* necessarily agree!

Exercise 6.26. Let $U = (0, 2\pi) \times (-\frac{\pi}{2}, \frac{\pi}{2}) \subset \mathbb{R}^2$, and let $\mathbf{x} : U \to \mathbb{A}^3$ be the standard parametrization of the unit sphere $\mathbb{S}^2$:

$$\mathbf{x}(u, v) = {}^t[\cos(u)\cos(v), \sin(u)\cos(v), \sin(v)] .$$

Let $\alpha : (-\frac{\pi}{2}, \frac{\pi}{2}) \to \mathbb{A}^3$ be the curve

$$\alpha(t) = \mathbf{x}(t, t) = {}^t\left[\cos^2(t), \sin(t)\cos(t), \sin(t)\right] .$$

(a) Show that the equi-affine arc length function for α is

$$s(t) = \int_0^t \sqrt[6]{6\cos(u)}\, du.$$

(b) Verify that, because $\mathbb{S}^2$ (regarded as a surface in $\mathbb{E}^3$) is totally umbilic with all principal curvatures equal to 1, the Maurer-Cartan forms associated to any adapted orthonormal frame field $(\mathbf{e}_1(\mathbf{u}), \mathbf{e}_2(\mathbf{u}), \mathbf{e}_3(\mathbf{u}))$ on $\mathbb{S}^2$ satisfy

$$\bar{\omega}_1^3 = \bar{\omega}^1, \qquad \bar{\omega}_2^3 = \bar{\omega}^2.$$

This means that any such frame field is *also* a 1-adapted frame field for $\mathbb{S}^2$ regarded as a surface in $\mathbb{A}^3$. Therefore, the equi-affine first fundamental form of $\mathbb{S}^2$ is the same as its Euclidean first fundamental form. (Moreover, the unit sphere is the *only* surface in $\mathbb{A}^3$ with this property!)

(c) By part (b), the restriction of the equi-affine first fundamental form on $\mathbb{S}^2$ to α is the same as the restriction of the Euclidean first fundamental form on $\mathbb{S}^2$ to α, which in turn is the same as the Euclidean arc length of α. Show that this arc length function for α is given by

$$s(t) = \int_0^t \sqrt{\cos^2(u) + 1}\, du.$$

(d) (MAPLE recommended) Plot both arc length functions for α, and compute the arc length of α with respect to both arc length functions. What happens if you change the pitch of α, i.e., set $\alpha(t) = \mathbf{x}(ct, t)$ for various choices of c?

This raises the question: Given an elliptic surface $\Sigma \subset \mathbb{A}^3$, are there any curves in Σ for which these two notions of arc length *do* agree? This question is explored in [**CEM**$^+$**14**].

Remark 6.33. The quantity $L = \frac{1}{2}(\ell_{11} + \ell_{22})$ is called the *equi-affine mean curvature* of Σ. It has the property (similar to that of the mean curvature H in the Euclidean case) that it vanishes identically if and only if Σ is a critical point of the equi-affine area functional.We will explore this further in Chapter 8.

We still have a bit more differentiating to do:

***Exercise 6.34.** (a) Differentiate the equations

$$\bar\omega_1^3 = \bar\omega^1, \qquad \bar\omega_2^3 = \bar\omega^2$$

and use Cartan's lemma to conclude that there exist functions h_{111}, h_{112}, h_{122}, h_{222} on U such that

$$\begin{bmatrix} 2\bar\omega_1^1 \\ \bar\omega_2^1 + \bar\omega_1^2 \\ 2\bar\omega_2^2 \end{bmatrix} = \begin{bmatrix} h_{111} & h_{112} \\ h_{112} & h_{122} \\ h_{122} & h_{222} \end{bmatrix} \begin{bmatrix} \bar\omega^1 \\ \bar\omega^2 \end{bmatrix}. \tag{6.19}$$

(b) Use the fact that $\bar\omega_1^1 + \bar\omega_2^2 = 0$ (Why is this true for a 2-adapted frame field? Hint: It is *not* true for a 1-adapted frame field.) to show that

$$h_{122} = -h_{111}, \qquad h_{112} = -h_{222}. \tag{6.20}$$

Remark 6.35. If we were to define functions (h_{ijk}) with $i, j, k = 1, 2$ by the equations

$$\begin{aligned} 2\bar\omega_1^1 &= h_{111}\bar\omega^1 + h_{112}\bar\omega^2, \\ \bar\omega_2^1 + \bar\omega_1^2 &= h_{121}\bar\omega^1 + h_{122}\bar\omega^2 = h_{211}\bar\omega^1 + h_{212}\bar\omega^2, \\ 2\bar\omega_2^2 &= h_{221}\bar\omega^1 + h_{222}\bar\omega^2 \end{aligned}$$

(where the second line makes sense because the expression on the left is symmetric in the first two indices), then the result of Exercise 6.34 says that the (h_{ijk}) are symmetric in all their indices.

Definition 6.36. Let $U \subset \mathbb{R}^2$ be an open set, and let $\mathbf{x} : U \to \mathbb{A}^3$ be an immersion. Let $(\mathbf{e}_1(\mathbf{u}), \mathbf{e}_2(\mathbf{u}), \mathbf{e}_3(\mathbf{u}))$ be any 2-adapted frame field along $\Sigma = \mathbf{x}(U)$, with associated Maurer-Cartan forms $(\bar\omega^i, \bar\omega_j^i)$. The cubic form

$$\begin{aligned} P = h_{ijk}\bar\omega^i\bar\omega^j\bar\omega^k &= h_{111}(\bar\omega^1)^3 + 3h_{112}(\bar\omega^1)^2\bar\omega^2 + 3h_{122}\bar\omega^1(\bar\omega^2)^2 + h_{222}(\bar\omega^2)^3 \\ &= h_{111}\left((\bar\omega^1)^3 - 3\bar\omega^1(\bar\omega^2)^2\right) + h_{222}\left((\bar\omega^2)^3 - 3(\bar\omega^1)^2\bar\omega^2\right) \end{aligned}$$

is called the *Fubini-Pick form* of Σ. P is also known as the *cubic form* of Σ.

***Exercise 6.37.** Let $(\mathbf{e}_1(\mathbf{u}), \mathbf{e}_2(\mathbf{u}), \mathbf{e}_3(\mathbf{u}))$ be any 2-adapted frame field along Σ with associated Maurer-Cartan forms $(\bar\omega^i, \bar\omega_j^i)$, and let $(\tilde{\mathbf{e}}_1(\mathbf{u}), \tilde{\mathbf{e}}_2(\mathbf{u}),$

$\tilde{\mathbf{e}}_3(\mathbf{u})$) be any other 2-adapted frame field of the form (6.15), with associated Maurer-Cartan forms $(\tilde{\bar{\omega}}^i, \tilde{\bar{\omega}}^i_j)$. We can write

$$B = \begin{bmatrix} \cos(\theta) & -\sin(\theta) \\ \sin(\theta) & \cos(\theta) \end{bmatrix}$$

for some function θ on U.

(a) Show that

$$\begin{bmatrix} \tilde{\bar{\omega}}^1_1 & \tilde{\bar{\omega}}^1_2 \\ \tilde{\bar{\omega}}^2_1 & \tilde{\bar{\omega}}^2_2 \end{bmatrix} = B^{-1} \begin{bmatrix} \bar{\omega}^1_1 & \bar{\omega}^1_2 \\ \bar{\omega}^2_1 & \bar{\omega}^2_2 \end{bmatrix} B + \begin{bmatrix} 0 & -d\theta \\ d\theta & 0 \end{bmatrix}. \tag{6.21}$$

More explicitly:

$$\begin{aligned} \tilde{\bar{\omega}}^1_1 &= \cos^2(\theta)\,\bar{\omega}^1_1 + \sin(\theta)\cos(\theta)\,(\bar{\omega}^1_2 + \bar{\omega}^2_1) + \sin^2(\theta)\,\bar{\omega}^2_2, \\ \tilde{\bar{\omega}}^1_2 &= \cos^2(\theta)\,\bar{\omega}^1_2 + \sin(\theta)\cos(\theta)\,(\bar{\omega}^2_2 - \bar{\omega}^1_1) - \sin^2(\theta)\,\bar{\omega}^2_1 - d\theta, \\ \tilde{\bar{\omega}}^2_1 &= \cos^2(\theta)\,\bar{\omega}^2_1 + \sin(\theta)\cos(\theta)\,(\bar{\omega}^2_2 - \bar{\omega}^1_1) - \sin^2(\theta)\,\bar{\omega}^1_2 + d\theta, \\ \tilde{\bar{\omega}}^2_2 &= \cos^2(\theta)\,\bar{\omega}^2_2 - \sin(\theta)\cos(\theta)\,(\bar{\omega}^1_2 + \bar{\omega}^2_1) + (\sin^2\theta)\bar{\omega}^1_1. \end{aligned} \tag{6.22}$$

(b) Show that

$$\begin{bmatrix} \tilde{h}_{111} \\ \tilde{h}_{222} \end{bmatrix} = B^3 \begin{bmatrix} h_{111} \\ h_{222} \end{bmatrix}.$$

Conclude that the quantity $h_{111}^2 + h_{222}^2$ is well-defined, independent of the choice of 2-adapted frame field and associated Maurer-Cartan forms. The quantity $J = \frac{1}{6}(h_{111}^2 + h_{222}^2)$ is called the *Pick invariant* of Σ.

(c) Show that the Fubini-Pick form of $\mathbf{x}$ is well-defined, independent of the choice of 2-adapted frame field and associated Maurer-Cartan forms.

***Exercise 6.38.** Suppose that the equi-affine second fundamental form is not a multiple of the equi-affine first fundamental form at any point of U. (This is equivalent to assuming that the matrix $[\ell_{ij}]$ is not a multiple of the identity at any point of U, similar to the assumption that a surface in $\mathbb{E}^3$ has no umbilic points.)

(a) Show that there exists a 2-adapted frame field on Σ for which $[\ell_{ij}]$ is a diagonal matrix and that this choice of frame field is unique up to signs (and possibly exchanging $\mathbf{e}_1(\mathbf{u})$ and $\mathbf{e}_2(\mathbf{u})$). We will call such a frame field an *equi-affine principal adapted frame field* on Σ. (Cf. Exercise 4.32.)

(b) Show that prescribing the first and second equi-affine fundamental forms and the Fubini-Pick form for an equi-affine principal adapted frame field uniquely determines all the connection forms $(\bar{\omega}^i_j)$ up to sign. (Cf. Exercise 4.38.)

(c) Conclude that any two elliptic surfaces $\Sigma_1, \Sigma_2 \subset \mathbb{A}^3$ with the same first and second equi-affine fundamental forms and Fubini-Pick form differ by an equi-affine transformation. In other words, the first and second equi-affine fundamental forms, together with the Fubini-Pick form, form a complete set of local invariants for elliptic equi-affine surfaces with no umbilic points.

Exercise 6.39 (MAPLE recommended). Let Σ be an elliptic equi-affine surface as in Exercise 6.38, and suppose that $\mathbf{x} : U \to \mathbb{A}^3$ is a parametrization of Σ with the property that the coordinate curves are all equi-affine principal curves. (This means that $\mathbf{x}_u, \mathbf{x}_v$ are multiples of the vectors $\mathbf{e}_1(\mathbf{u})$, $\mathbf{e}_2(\mathbf{u})$ in an equi-affine principal adapted frame field for Σ.) The equi-affine first and second fundamental forms and the Fubini-Pick form of Σ can be written in terms of such a parametrization as

$$\begin{gathered} \mathrm{I} = M\,du^2 + N\,dv^2, \\ \mathrm{II} = m\,du^2 + n\,dv^2, \\ P = p_{111}\,du^3 + 3p_{112}\,du^2\,dv + 3p_{122}\,du\,dv^2 + p_{222}\,dv^3 \end{gathered} \tag{6.23}$$

for some functions M, N, m, n, p_{ijk} on U.

(a) Show that the equi-affine principal vectors tangent to Σ are

$$\mathbf{e}_1(\mathbf{u}) = \frac{1}{\sqrt{M}}\mathbf{x}_u, \qquad \mathbf{e}_2(\mathbf{u}) = \frac{1}{\sqrt{N}}\mathbf{x}_v.$$

(b) Show that the Maurer-Cartan forms associated to this equi-affine principal adapted frame field are given by

$$\begin{aligned} \bar{\omega}^1 = \bar{\omega}^3_1 &= \sqrt{M}\,du, & \bar{\omega}^2 = \bar{\omega}^3_2 &= \sqrt{N}\,dv, \\ \bar{\omega}^1_3 &= \frac{m}{\sqrt{M}}\,du, & \bar{\omega}^2_3 &= \frac{n}{\sqrt{N}}\,dv, \\ \bar{\omega}^1_1 &= \frac{p_{111}}{2M}du + \frac{p_{112}}{2M}dv, & \bar{\omega}^2_2 &= \frac{p_{122}}{2N}du + \frac{p_{222}}{2N}dv, \end{aligned}$$

$$\bar{\omega}^1_2 = \left(\frac{p_{112}}{2\sqrt{MN}} + s_1\right) du + \left(\frac{p_{122}}{2\sqrt{MN}} + s_2\right) dv,$$

$$\bar{\omega}^2_1 = \left(\frac{p_{112}}{2\sqrt{MN}} - s_1\right) du + \left(\frac{p_{122}}{2\sqrt{MN}} - s_2\right) dv$$

for some functions s_1, s_2 on U and that

$$Mp_{122} + Np_{111} = Mp_{222} + Np_{112} = 0. \tag{6.24}$$

(Hint: Use the result of Exercise 6.34.)

(c) Use the Cartan structure equations for $d\bar{\omega}^1$ and $d\bar{\omega}^2$ to find the functions s_1, s_2.

(d) Suppose that the functions M, N, m, n, p_{ijk} are prescribed arbitrarily, subject to the conditions that $M, N > 0$ and that equation (6.24) is satisfied. Use the Cartan structure equations (3.8) to determine the PDE system that must be satisfied by these functions in order to guarantee the local existence of an elliptic equi-affine surface with equi-affine first and second fundamental forms and Fubini-Pick form given by (6.23). This PDE system is the analog of the Gauss-Codazzi system for elliptic surfaces in $\mathbb{A}^3$.

Exercise 6.40 (MAPLE recommended). Let $\mathbf{x} : U \to \mathbb{A}^3$ be an elliptic equi-affine surface, and suppose that $\mathbf{x}$ has the property that, when regarded as a surface in Euclidean space, the coordinate curves are principal curves (cf. Exercise 4.41). Let

$$\kappa_1 = \frac{e}{E}, \qquad \kappa_2 = \frac{g}{G}$$

denote the principal curvatures of the Euclidean surface, where E, G, e, g are as in Exercise 4.41.

(a) Let $(\mathbf{e}_1(\mathbf{u}), \mathbf{e}_2(\mathbf{u}), \mathbf{e}_3(\mathbf{u}))$ be the Euclidean adapted frame field of Exercise 4.41. Show that the frame field

$$\tilde{\mathbf{e}}_1(\mathbf{u}) = \left(\frac{\kappa_1}{\kappa_2^3}\right)^{1/8} \mathbf{e}_1(\mathbf{u}), \qquad \tilde{\mathbf{e}}_2(\mathbf{u}) = \left(\frac{\kappa_2}{\kappa_1^3}\right)^{1/8} \mathbf{e}_2(\mathbf{u}),$$

$$\tilde{\mathbf{e}}_3(\mathbf{u}) = (\kappa_1\kappa_2)^{1/4}\mathbf{e}_3(\mathbf{u}) - \left(\frac{(\kappa_1\kappa_2)_u}{4\sqrt{E}\kappa_1^{7/4}\kappa_2^{3/4}}\right)\mathbf{e}_1(\mathbf{u}) - \left(\frac{(\kappa_1\kappa_2)_v}{4\sqrt{G}\kappa_1^{3/4}\kappa_2^{7/4}}\right)\mathbf{e}_2(\mathbf{u})$$

is a 2-adapted equi-affine frame field along Σ.

(b) Use the Maurer-Cartan equation (3.1) to compute the Maurer-Cartan forms $(\tilde{\bar{\omega}}^i, \tilde{\bar{\omega}}^i_j)$ associated to this frame field. (Hint: The first equation in (3.1) implies that

$$d\mathbf{x} = \mathbf{e}_1\bar{\omega}^1 + \mathbf{e}_2\bar{\omega}^2 = \tilde{\mathbf{e}}_1\tilde{\bar{\omega}}^1 + \tilde{\mathbf{e}}_2\tilde{\bar{\omega}}^2,$$

so the relationship between $(\tilde{\bar{\omega}}^1, \tilde{\bar{\omega}}^2)$ and $(\bar{\omega}^1, \bar{\omega}^2)$ is closely related to the relationship between $(\tilde{\mathbf{e}}_1, \tilde{\mathbf{e}}_2)$ and $(\mathbf{e}_1, \mathbf{e}_2)$.)

(c) Use the result of part (b) to compute the first and second equi-affine fundamental forms and Fubini-Pick form of Σ in terms of the Euclidean invariants of Σ. In particular, note that

$$\mathrm{I}_{\mathit{aff}} = K^{-1/4}\mathrm{II}_{Euc},$$

where K is the Euclidean Gauss curvature of Σ.

Exercise 6.41. Let $\mathbf{x} : U \to \mathbb{A}^3$ be an elliptic equi-affine surface, and suppose that the equi-affine second fundamental form is a multiple of the equi-affine first fundamental form, so that the Maurer-Cartan forms $(\bar{\omega}^i, \bar{\omega}^i_j)$ associated to a 2-adapted frame field satisfy the conditions

$$\bar{\omega}^1_3 = \lambda\,\bar{\omega}^1, \qquad \bar{\omega}^2_3 = \lambda\,\bar{\omega}^2$$

for some function λ on U. (This is the analog of assuming that a surface in $\mathbb{E}^3$ is totally umbilic.)

(a) Prove that λ is constant. (Hint: Use the structure equations to differentiate the equations above, taking into account the fact that we must have

$$d\lambda = \lambda_1\bar{\omega}^1 + \lambda_2\bar{\omega}^2$$

for some functions λ_1, λ_2 on U. Then use Cartan's lemma.)

(b) Show that if $\lambda = 0$, then $d\mathbf{e}_3 = 0$, and therefore the equi-affine normals of Σ are all parallel. Such surfaces are called *improper equi-affine spheres.*

(c) Show that if $\lambda \neq 0$, then $d(\mathbf{x}(\mathbf{u}) - \frac{1}{\lambda}\mathbf{e}_3(\mathbf{u})) = 0$. Therefore, all the equi-affine normals of Σ intersect at the point $\mathbf{q} = \mathbf{x}(\mathbf{u}) - \frac{1}{\lambda}\mathbf{e}_3(\mathbf{u})$. Such surfaces are called *proper equi-affine spheres.*

Equi-affine spheres, both proper and improper, are much more plentiful than spheres in Euclidean space; in fact, there is an infinite-dimensional family of such surfaces.

Exercise 6.42. In this exercise, we will explore the frame adaptation process for *hyperbolic* equi-affine surfaces. Suppose that $\mathbf{x} : U \to \mathbb{A}^3$ is a parametrization for a hyperbolic equi-affine surface $\Sigma = \mathbf{x}(U) \subset \mathbb{A}^3$. Then the matrix $[h_{ij}]$ in (6.9) has $\det[h_{ij}] < 0$.

(a) Show that there exists a 0-adapted frame field $(\mathbf{f}_1(\mathbf{u}), \mathbf{f}_2(\mathbf{u}), \mathbf{f}_3(\mathbf{u}))$ for which

$$\begin{bmatrix} h_{11} & h_{12} \\ h_{12} & h_{22} \end{bmatrix} = \begin{bmatrix} 0 & 1 \\ 1 & 0 \end{bmatrix}.$$

Such a frame field will be called a 1-*adapted null frame field* on $\Sigma = \mathbf{x}(U)$. The associated Maurer-Cartan forms $(\bar{\eta}^i, \bar{\eta}^i_j)$ satisfy

$$\bar{\eta}^3_1 = \bar{\eta}^2, \qquad \bar{\eta}^3_2 = \bar{\eta}^1,$$

and the equi-affine first fundamental form becomes

$$\mathrm{I} = 2\,\bar{\eta}^1\bar{\eta}^2.$$

Thus, it defines an *indefinite* metric on Σ. (Based on this observation, we might expect that the geometry of hyperbolic equi-affine surfaces will be somewhat reminiscent of the geometry of timelike surfaces in Minkowski space!)

(b) Let $(\mathbf{f}_1(\mathbf{u}), \mathbf{f}_2(\mathbf{u}), \mathbf{f}_3(\mathbf{u}))$ be any 1-adapted null frame field for a hyperbolic equi-affine surface $\Sigma = \mathbf{x}(U) \subset \mathbb{A}^3$. Show that any other 1-adapted null frame field $(\tilde{\mathbf{f}}_1(\mathbf{u}), \tilde{\mathbf{f}}_2(\mathbf{u}), \tilde{\mathbf{f}}_3(\mathbf{u}))$ for Σ must have the form

$$\begin{bmatrix}\tilde{\mathbf{f}}_1(\mathbf{u}) & \tilde{\mathbf{f}}_2(\mathbf{u}) & \tilde{\mathbf{f}}_3(\mathbf{u})\end{bmatrix} = \begin{bmatrix}\mathbf{f}_1(\mathbf{u}) & \mathbf{f}_2(\mathbf{u}) & \mathbf{f}_3(\mathbf{u})\end{bmatrix}\begin{bmatrix} e^{\theta} & 0 & r_1 \\ 0 & e^{-\theta} & r_2 \\ 0 & 0 & 1\end{bmatrix} \tag{6.25}$$

for some functions θ, r_1, r_2 on U.

(c) Show that under a transformation of the form (6.25), we have

$$\tilde{\bar{\eta}}^3_3 = \bar{\eta}^3_3 + r_2\bar{\eta}^1 + r_1\bar{\eta}^2.$$

(No, that's not a typo in the indices!) Conclude that there is a unique choice of r_1, r_2 for which $\bar{\eta}^3_3 = 0$. A 1-adapted null frame field satisfying the condition $\bar{\eta}^3_3 = 0$ will be called a 2-*adapted null frame field.*

For the remainder of this exercise, assume that $(\mathbf{f}_1(\mathbf{u}), \mathbf{f}_2(\mathbf{u}), \mathbf{f}_3(\mathbf{u}))$ is a 2-adapted null frame field along Σ.

(d) Differentiate the equation $\bar{\eta}^3_3 = 0$ and use Cartan's lemma to conclude that there exist functions $\ell, \ell_{12}, \ell_{21}$ on U such that

$$\begin{bmatrix}\bar{\eta}^1_3 \\ \bar{\eta}^2_3\end{bmatrix} = \begin{bmatrix}\ell & \ell_{12} \\ \ell_{21} & \ell\end{bmatrix}\begin{bmatrix}\bar{\eta}^1 \\ \bar{\eta}^2\end{bmatrix}.$$

The equi-affine second fundamental form of Σ is given by

$$\begin{aligned}\mathrm{II} &= \bar{\eta}^1_3\bar{\eta}^3_1 + \bar{\eta}^2_3\bar{\eta}^3_2 \\ &= \bar{\eta}^1_3\bar{\eta}^2 + \bar{\eta}^2_3\bar{\eta}^2 \\ &= \ell_{21}(\bar{\eta}^1)^2 + 2\ell\bar{\eta}^1\bar{\eta}^2 + \ell_{12}(\bar{\eta}^2)^2.\end{aligned}$$

(e) Differentiate the equations

$$\bar{\eta}_1^3 = \bar{\eta}^2, \qquad \bar{\eta}_2^3 = \bar{\eta}^1$$

and use Cartan's lemma to conclude that there exist functions h_{111}, h_{112}, h_{122}, h_{222} on U such that

$$\begin{bmatrix} 2\bar{\eta}_1^2 \\ \bar{\eta}_1^1 + \bar{\eta}_2^2 \\ 2\bar{\eta}_2^1 \end{bmatrix} = \begin{bmatrix} h_{111} & h_{112} \\ h_{112} & h_{122} \\ h_{122} & h_{222} \end{bmatrix} \begin{bmatrix} \bar{\eta}^1 \\ \bar{\eta}^2 \end{bmatrix}.$$

(f) Show that

$$\bar{\eta}_1^1 + \bar{\eta}_2^2 = 0$$

(Hint: The reasoning is the same as for a 2-adapted frame field for an elliptic surface.), and conclude that

$$h_{112} = h_{122} = 0.$$

The Fubini-Pick form of Σ is

$$P = h_{111}(\bar{\eta}^1)^3 + h_{222}(\bar{\eta}^2)^3. \tag{6.26}$$

Further adaptations could be made to the frame in order to further normalize the equi-affine second fundamental form; this would require considering several cases similar to those that arose when we considered the second fundamental form for timelike surfaces in Minkowski space.

6.4. MAPLE computations

Once again, the MAPLE setup is similar to that of Chapter 4, but since there are fewer relations among the connection forms, we need to declare more of them. (In fact, we'll go ahead and declare them all; it turns out to be convenient to wait until after we define the structure equations to tell MAPLE about the single relation among the connection forms because it allows us to use a for loop to define the structure equations without creating redundant assignments.) So, after loading the `Cartan` and `LinearAlgebra` packages into MAPLE, begin by declaring the necessary 1-forms:

```
> Form(omega[1], omega[2], omega[3]);
  Form(omega[1,1], omega[1,2], omega[1,3],
  omega[2,1], omega[2,2], omega[2,3],
  omega[3,1], omega[3,2], omega[3,3]);
```

Tell MAPLE how to differentiate these forms according to the Cartan structure equations (3.8):

```
> for i from 1 to 3 do
    d(omega[i]):= -add('omega[i,j] &^ omega[j]', j=1..3);
  end do;

  for i from 1 to 3 do
    for j from 1 to 3 do
      d(omega[i,j]):= -add('omega[i,k] &^ omega[k,j]',
      k=1..3);
    end do;
  end do;
```

Now tell MAPLE about the relation between the connection forms:

```
> omega[3,3]:= -(omega[1,1] + omega[2,2]);
```

Set up a substitution for the Maurer-Cartan forms associated to a 0-adapted frame field, taking into account the relations (6.9) that result from computing $d\bar{\omega}^3 = 0$:

```
> adaptedsub1:= [omega[3]=0,
  omega[3,1] = h[1,1]*omega[1] + h[1,2]*omega[2],
  omega[3,2] = h[1,2]*omega[1] + h[2,2]*omega[2]];
```

EXERCISE 6.20: Introduce new 1-forms to represent the transformed forms, with the same relation as the original forms:

```
> Form(Omega[1], Omega[2], Omega[3]);
  Form(Omega[1,1], Omega[1,2], Omega[1,3],
  Omega[2,1], Omega[2,2], Omega[2,3],
  Omega[3,1], Omega[3,2], Omega[3,3]);
  Omega[3,3]:= -(Omega[1,1] + Omega[2,2]);
```

Under a transformation of the form (6.10), we have the relations (6.11), (6.12). Since B is now an arbitrary matrix in $GL(2)$, the corresponding substitution requires a bit more typing than in previous chapters:

```
> framechangesub:= [
  Omega[1] = (1/(b[1,1]*b[2,2] - b[1,2]*b[2,1]))*
  (b[2,2]*omega[1] - b[1,2]*omega[2]),
  Omega[2] = (1/(b[1,1]*b[2,2] - b[1,2]*b[2,1]))*
  (-b[2,1]*omega[1] + b[1,1]*omega[2]),
  Omega[3,1] = (b[1,1]*b[2,2] - b[1,2]*b[2,1])*
  (b[1,1]*omega[3,1] + b[2,1]*omega[3,2]),
  Omega[3,2] = (b[1,1]*b[2,2] - b[1,2]*b[2,1])*
```

```
  (b[1,2]*omega[3,1] + b[2,2]*omega[3,2])];
> framechangebacksub:= makebacksub(framechangesub);
```

In order to compare the functions $(\tilde{h}_{ij})$ to the functions (h_{ij}), introduce another substitution describing the adaptations for the transformed frame:

```
> adaptedsub2:= [Omega[3]=0,
  Omega[3,1] = H[1,1]*Omega[1] + H[1,2]*Omega[2],
  Omega[3,2] = H[1,2]*Omega[1] + H[2,2]*Omega[2]];
```

Now combine all these substitutions to see how the $(\tilde{h}_{ij})$ are expressed in terms of the (h_{ij}):

```
> zero2:= Simf(subs(adaptedsub2, Omega[3,1])
  - Simf(subs(framechangebacksub, Simf(subs(adaptedsub1,
  Simf(subs(framechangesub, Omega[3,1])))))));
> eqns:= {op(ScalarForm(zero2))};
> zero3:= Simf(subs(adaptedsub2, Omega[3,2])
  - Simf(subs(framechangebacksub, Simf(subs(adaptedsub1,
  Simf(subs(framechangesub, Omega[3,2])))))));
> eqns:= eqns union {op(ScalarForm(zero3))};
> solve(eqns, {H[1,1], H[1,2], H[2,2]});
> assign(%);
```

Check that these expressions for the $(\tilde{h}_{ij})$ agree with those in equation (6.13):

```
> hmatrix:= Matrix([[h[1,1], h[1,2]], [h[1,2], h[2,2]]]);
  Hmatrix:= Matrix([[H[1,1], H[1,2]], [H[1,2], H[2,2]]]);
  B:= Matrix([[b[1,1], b[1,2]], [b[2,1], b[2,2]]]);
> Hmatrix - simplify(Determinant(B)*Transpose(B).hmatrix.B);
```

$$0$$

EXERCISE 6.32: Now suppose that Σ is an elliptic surface and that we have chosen a 2-adapted frame field, so that $[h_{ij}]$ is the identity matrix, $B \in SO(2)$, and the equi-affine normal vector field $\mathbf{e}_3(\mathbf{u})$ is well-defined. Since we now wish to explore transformations among 2-adapted frame fields, assign these conditions for both (h_{ij}) and $(\tilde{h}_{ij})$:

```
> h[1,1]:= 1;
  h[1,2]:= 0;
  h[2,2]:= 1;

  H[1,1]:= 1;
  H[1,2]:= 0;
  H[2,2]:= 1;
```

```
b[1,1]:= cos(theta);
b[1,2]:= -sin(theta);
b[2,1]:= sin(theta);
b[2,2]:= cos(theta);
```

Note that our adapted frame substitution is now as it should be:

```
> Simf(adaptedsub1);
```

$$[\omega_3 = 0,\ \omega_{3,1} = \omega_1,\ \omega_{3,2} = \omega_2]$$

Similarly, our frame change substitution has simplified considerably:

```
> Simf(framechangesub);
```

$$[\Omega_1 = \cos(\theta)\,\omega_1 + \sin(\theta)\,\omega_2,\ \Omega_2 = -\sin(\theta)\,\omega_1 + \cos(\theta)\,\omega_2,$$
$$\Omega_{3,1} = \cos(\theta)\,\omega_{3,1} + \sin(\theta)\,\omega_{3,2},\ \Omega_{3,2} = -\sin(\theta)\,\omega_{3,1} + \cos(\theta)\,\omega_{3,2}]$$

At this point, we know that $\bar{\omega}_3^3 = -(\bar{\omega}_1^1 + \bar{\omega}_2^2) = 0$, and from equation (6.16) we have expressions for $\bar{\omega}_3^1$ and $\bar{\omega}_3^2$. Add these relations to our substitution (we will use `ell` for the letter ℓ in order to avoid confusing it with the number 1):

```
> adaptedsub1:= [op(adaptedsub1), omega[2,2] = -omega[1,1],
  omega[1,3] = ell[1,1]*omega[1] + ell[1,2]*omega[2],
  omega[2,3] = ell[1,2]*omega[1] + ell[2,2]*omega[2]];
```

Add similar relations to the substitution for the transformed forms (we can't use `L` because it is a command in the `Cartan` package, so use `LL` instead):

```
> adaptedsub2:= [op(adaptedsub2), Omega[2,2] = -Omega[1,1],
  Omega[1,3] = LL[1,1]*Omega[1] + LL[1,2]*Omega[2],
  Omega[2,3] = LL[1,2]*Omega[1] + LL[2,2]*Omega[2]];
```

Next, add the relations (6.17) to our frame change substitution (keeping in mind that we now know that $B \in SO(2)$):

```
> framechangesub:= Simf([op(framechangesub),
  Omega[1,3] = cos(theta)*omega[1,3] + sin(theta)*omega[2,3],
  Omega[2,3] = -sin(theta)*omega[1,3]
  + cos(theta)*omega[2,3]]);
> framechangebacksub:= makebacksub(framechangesub);
```

Now combine all these substitutions to see how the $(\tilde{\ell}_{ij})$ are expressed in terms of the (ℓ_{ij}):

```
> zero4:= Simf(subs(adaptedsub2, Omega[1,3])
  - Simf(subs(framechangebacksub, Simf(subs(adaptedsub1,
```

```
  Simf(subs(framechangesub, Omega[1,3])))))));
> eqns:= {op(ScalarForm(zero4))};
> zero5:= Simf(subs(adaptedsub2, Omega[2,3])
  - Simf(subs(framechangebacksub, Simf(subs(adaptedsub1,
  Simf(subs(framechangesub, Omega[2,3])))))));
> eqns:= eqns union {op(ScalarForm(zero5))};
> solve(eqns, {LL[1,1], LL[1,2], LL[2,2]});
> assign(%);
```

Check that these expressions for the $(\tilde{\ell}_{ij})$ agree with those in equation (6.18):

```
> ellmatrix:= Matrix([[ell[1,1], ell[1,2]],
  [ell[1,2], ell[2,2]]]);
  LLmatrix:= Matrix([[LL[1,1], LL[1,2]], [LL[1,2], LL[2,2]]]);
> LLmatrix - simplify(Transpose(B).ellmatrix.B);
```

$$0$$

EXERCISE 6.34: Continuing on, we now need to compute the quantities

$$d(\bar{\omega}_1^3 - \bar{\omega}^1), \qquad d(\bar{\omega}_2^3 - \bar{\omega}^2),$$

both of which should be equal to zero, and then substitute in what we already know:

```
> zero6:= Simf(subs(adaptedsub1,
  Simf(d(omega[3,1] - omega[1]))));
```

$$-2\,(\omega_1)\ \&\hat{}\ (\omega_{1,1}) - (\omega_2)\ \&\hat{}\ (\omega_{1,2}) - (\omega_2)\ \&\hat{}\ (\omega_{2,1})$$

(See the MAPLE worksheet for this chapter on the AMS webpage for an illustration of why the inner `Simf` command is a good idea—or experiment for yourself and see what happens without it!)

We can recognize this as an expression of the form

$$\phi_1 \wedge \bar{\omega}^1 + \phi_2 \wedge \bar{\omega}^2,$$

to which we should apply Cartan's lemma. If you want MAPLE to help you identify ϕ_1 and ϕ_2, you can use the `pick` command:

```
> pick(zero6, omega[1]);
```

$$2\,\omega_{1,1}$$

```
> pick(zero6, omega[2]);
```

$$\omega_{1,2} + \omega_{2,1}$$

Similarly for $d(\bar{\omega}_2^3 - \bar{\omega}^2)$:

```
> zero7:= Simf(subs(adaptedsub1,
  Simf(d(omega[3,2] - omega[2]))));
```

$$-(\omega_1)\ \&\hat{}\ (\omega_{2,1}) + 2\,(\omega_2)\ \&\hat{}\ (\omega_{1,1}) - (\omega_1)\ \&\hat{}\ (\omega_{1,2})$$

```
> pick(zero7, omega[1]);
```

$$\omega_{1,2} + \omega_{2,1}$$

```
> pick(zero7, omega[2]);
```

$$-2\,\omega_{1,1}$$

Applying Cartan's lemma to both of these expressions yields equations (6.19), (6.20). We need to add these expressions to adaptedsub1 (say, by solving for $\bar{\omega}_1^1$ and $\bar{\omega}_1^2$), but we must do so carefully. Because these forms already occur in some of the right-hand sides of equations in adaptedsub1, we need to add them in two steps: First, substitute the new expressions into adaptedsub1 as it currently is, and *then* add them as new list items in adaptedsub1:

```
> adaptedsub1:= Simf(subs([
  omega[1,1] = (1/2)*(h[1,1,1]*omega[1] - h[2,2,2]*omega[2]),
  omega[2,1] = -omega[1,2] - h[2,2,2]*omega[1]
  - h[1,1,1]*omega[2]], adaptedsub1));
> adaptedsub1:= [op(adaptedsub1),
  omega[1,1] = (1/2)*(h[1,1,1]*omega[1] - h[2,2,2]*omega[2]),
  omega[2,1] = -omega[1,2] - h[2,2,2]*omega[1]
  - h[1,1,1]*omega[2]];
```

EXERCISE 6.37: First, we need to make the additions to adaptedsub2 corresponding to those that we just made to adaptedsub1:

```
> adaptedsub2:= Simf(subs([
  Omega[1,1] = (1/2)*(H[1,1,1]*Omega[1] - H[2,2,2]*Omega[2]),
  Omega[2,1] = -Omega[1,2] - H[2,2,2]*Omega[1]
  - H[1,1,1]*Omega[2]], adaptedsub2));
> adaptedsub2:= [op(adaptedsub2),
  Omega[1,1] = (1/2)*(H[1,1,1]*Omega[1] - H[2,2,2]*Omega[2]),
  Omega[2,1] = -Omega[1,2] - H[2,2,2]*Omega[1]
  - H[1,1,1]*Omega[2]];
```

We can get MAPLE to help us expand equation (6.21) into the expressions (6.22):

```
> omegamatrix:= Matrix([[omega[1,1], omega[1,2]],
  [omega[2,1], omega[2,2]]]);
```

```
> Omegamatrix:= simplify(MatrixInverse(B).omegamatrix.B
  + Matrix([[0, -d(theta)], [d(theta), 0]]));
```

Then we need to add these expressions to our frame change substitution:

```
> framechangesub:= Simf([op(framechangesub),
  Omega[1,1] = Omegamatrix[1,1],
  Omega[1,2] = Omegamatrix[1,2],
  Omega[2,1] = Omegamatrix[2,1],
  Omega[2,2] = Omegamatrix[2,2]]);
> framechangebacksub:= makebacksub(framechangesub);
```

Now combine all these substitutions to see how the $(\tilde{h}_{ijk})$ are expressed in terms of the (h_{ijk}):

```
> zero8:= Simf(subs(adaptedsub2, Omega[1,1])
  - Simf(subs(framechangebacksub, Simf(subs(adaptedsub1,
  Simf(subs(framechangesub, Omega[1,1])))))));
> eqns:= {op(ScalarForm(zero8))};
> solve(eqns, {H[1,1,1], H[2,2,2]});
> assign(%);
```

If you like, you can perform the analogous computations for $\tilde{\omega}^1_2$ and $\tilde{\omega}^2_1$ and confirm that they are now identically zero. But be warned that sometimes MAPLE can be clumsy about substitutions and require that they be applied repeatedly. For instance, performing the same computation as above for $\tilde{\omega}^1_2$ now yields an apparently nonzero expression, but applying `adaptedsub2` to this expression yet *again* does, in fact, yield zero.

Finally, check that these expressions for the $(\tilde{h}_{ijk})$ agree with those in equation (6.22):

```
> hvector:= Vector([h[1,1,1], h[2,2,2]]);
> Hvector:= Vector([H[1,1,1], H[2,2,2]]);
> simplify(Hvector - B.B.B.hvector);
```

$$\begin{bmatrix} 0 \\ 0 \end{bmatrix}$$

EXERCISE 6.40: First declare the functions associated to the Euclidean first and second fundamental forms of Σ. Since we want to express everything in terms of κ_1, κ_2, declare these functions and then express e, g in terms of them:

```
> PDETools[declare](E(u,v), G(u,v), kappa1(u,v), kappa2(u,v));
> e(u,v):= E(u,v)*kappa1(u,v);
  g(u,v):= G(u,v)*kappa2(u,v);
```

For MAPLE purposes, use $(\bar{\eta}^i, \bar{\eta}^i_j)$ to denote the Maurer-Cartan forms associated to the Euclidean adapted frame field $(\mathbf{e}_1(\mathbf{u}), \mathbf{e}_2(\mathbf{u}), \mathbf{e}_3(\mathbf{u}))$ and use $(\bar{\omega}^i, \bar{\omega}^i_j)$ to denote those associated to the 2-adapted equi-affine frame field $(\tilde{\mathbf{e}}_1(\mathbf{u}),\ \tilde{\mathbf{e}}_2(\mathbf{u}),\ \tilde{\mathbf{e}}_3(\mathbf{u}))$. We have computed $(\bar{\eta}^i, \bar{\eta}^i_j)$ previously, and we can simply assign them:

```
> eta[1]:= E(u,v)^(1/2)*d(u);
  eta[2]:= G(u,v)^(1/2)*d(v);
  eta[1,2]:= (diff(E(u,v), v)*d(u) - diff(G(u,v), u)*d(v))/
  (2*E(u,v)^(1/2)*G(u,v)^(1/2));
  eta[3,1]:= (e(u,v)/E(u,v)^(1/2))*d(u);
  eta[3,2]:= (g(u,v)/G(u,v)^(1/2))*d(v);
  eta[2,1]:= -eta[1,2];
  eta[1,3]:= -eta[3,1];
  eta[2,3]:= -eta[3,2];
```

Next, we need to introduce variables for the vector fields $(\mathbf{e}_i(\mathbf{u}))$ and $(\tilde{\mathbf{e}}_i(\mathbf{u}))$ so that we can use their structure equations to compute the associated Maurer-Cartan forms. There's no good way to tell MAPLE that these functions are vector-valued, but it won't really matter.

First we tell MAPLE the structure equations for the exterior derivatives of the Euclidean frame field, the components of which we call (e01, e02, e03):

```
> d(e01):= e02*eta[2,1] + e03*eta[3,1];
  d(e02):= e01*eta[1,2] + e03*eta[3,2];
  d(e03):= e01*eta[1,3] + e02*eta[2,3];
```

We'll need the Gauss and Codazzi equations later, and these can be computed directly from the equations $d(d\mathbf{e}_i) = 0$. Start with $d(d\mathbf{e}_3)$:

```
zero9:= Simf(d(d(e03)));
```

Pick off the scalar coefficient of $du \wedge dv$, and collect terms in $(\mathbf{e}_1(\mathbf{u}), \mathbf{e}_2(\mathbf{u}))$:

```
> zero9a:= collect(pick(zero9, d(u), d(v)), {e01, e02});
```

Since $(\mathbf{e}_1(\mathbf{u}), \mathbf{e}_2(\mathbf{u}))$ are linearly independent, both coefficients must be zero; indeed, these are the Codazzi equations. We can collect these into a convenient substitution:

```
> Codazzisub:= [
  op(solve({coeff(zero9a, e01), coeff(zero9a, e02)},
  {diff(kappa1(u,v), v), diff(kappa2(u,v), u)}))];
```

Now compute $d(d\mathbf{e}_1)$, taking the Codazzi equations into account:

```
> zero10:= Simf(subs(Codazzisub, Simf(d(d(e01)))));
```

The coefficient of $du \wedge dv$ in the resulting expression is a multiple of $\mathbf{e}_2(\mathbf{u})$, and the scalar multiple is precisely the Gauss equation. We can add this equation to our substitution by solving for any term we like, but since we want to have things expressed in terms of κ_1, κ_2 later, it's best to solve for something else. The other choices are the functions E, G, and their assorted derivatives. As a general rule, the safest thing to do is to solve for one of the highest-order derivatives in the expression; in this case, G_{uu} will work.

```
> GaussCodazzisub:= [op(Codazzisub),
  op(solve(coeff(pick(zero10, d(u), d(v)), e02),
  {diff(G(u,v), u,u)}))];
```

Check that there are no additional conditions lurking in the equations $d(d\mathbf{e}_i) = 0$:

```
> Simf(subs(GaussCodazzisub, Simf(d(d(e01)))));
  Simf(subs(GaussCodazzisub, Simf(d(d(e02)))));
  Simf(subs(GaussCodazzisub, Simf(d(d(e03)))));
```

$$0$$

$$0$$

$$0$$

Rather than just computing the Maurer-Cartan forms for the frame field given in Exercise 6.40(a), let's explore where these expressions came from. First, the fact that the coordinate curves of Σ are principal curves means that the Maurer-Cartan forms associated to this Euclidean frame field have the property that $\bar{\eta}_1^3$ is a multiple of $\bar{\eta}^1$ and $\bar{\eta}_2^3$ is a multiple of $\bar{\eta}^2$. This means that this Euclidean frame field is actually a 0-adapted equi-affine frame field for which the matrix $[h_{ij}]$ is diagonal. Based on the transformation rule (6.13), this suggests that we should be able to create a 2-adapted equi-affine frame field for which $(\tilde{\mathbf{e}}_1(\mathbf{u}), \tilde{\mathbf{e}}_2(\mathbf{u}))$ are scalar multiples of $(\mathbf{e}_1(\mathbf{u}), \mathbf{e}_2(\mathbf{u}))$, respectively. So, set

$$\tilde{\mathbf{e}}_1(\mathbf{u}) = \lambda_1 \mathbf{e}_1(\mathbf{u}), \qquad \tilde{\mathbf{e}}_2(\mathbf{u}) = \lambda_2 \mathbf{e}_2(\mathbf{u})$$

for some (unknown) functions λ_1, λ_2. Then in order to keep the frame unimodular, we must have

$$\tilde{\mathbf{e}}_3(\mathbf{u}) = r_1 \mathbf{e}_1(\mathbf{u}) + r_2 \mathbf{e}_2(\mathbf{u}) + \frac{1}{\lambda_1 \lambda_2} \mathbf{e}_3$$

for some functions r_1, r_2. For MAPLE purposes, we will call this frame field (e1, e2, e3):

```
> e1:= lambda1(u,v)*e01;
  e2:= lambda2(u,v)*e02;
  e3:= r1(u,v)*e01 + r2(u,v)*e02
  + e03/(lambda1(u,v)*lambda2(u,v));
```

In order for this frame field to be 2-adapted, the corresponding Maurer-Cartan forms must satisfy the conditions

$$\bar{\omega}^3_1 = \bar{\omega}^1, \qquad \bar{\omega}^3_2 = \bar{\omega}^2, \qquad \bar{\omega}^1_1 + \bar{\omega}^2_2 = 0.$$

Moreover, we must have

$$\bar{\omega}^1 = \frac{1}{\lambda_1}\bar{\eta}^1 = \frac{\sqrt{E}}{\lambda_1}du, \qquad \bar{\omega}^2 = \frac{1}{\lambda_2}\bar{\eta}^2 = \frac{\sqrt{G}}{\lambda_2}dv.$$

So, set up a substitution with these conditions:

```
> affineframesub:= [omega[3] = 0,
  omega[1] = eta[1]/lambda1(u,v),
  omega[2] = eta[2]/lambda2(u,v),
  omega[3,1] = eta[1]/lambda1(u,v),
  omega[3,2] = eta[2]/lambda2(u,v),
  omega[2,2] = -omega[1,1]];
```

We can now determine the functions $\lambda_1, \lambda_2, r_1, r_2$ by using the structure equations for $d\tilde{\mathbf{e}}_1, d\tilde{\mathbf{e}}_2, d\tilde{\mathbf{e}}_3$:

```
> zero11:= collect(Simf(subs(affineframesub, Simf(d(e1)
  - (e1*omega[1,1] + e2*omega[2,1] + e3*omega[3,1])))),
  {e01, e02, e03});
  zero12:= collect(Simf(subs(affineframesub, Simf(d(e2)
  - (e1*omega[1,2] + e2*omega[2,2] + e3*omega[3,2])))),
  {e01, e02, e03});
  zero13:= collect(Simf(subs(affineframesub, Simf(d(e3)
  - (e1*omega[1,3] + e2*omega[2,3] + e3*omega[3,3])))),
  {e01, e02, e03});
```

If you look at the output from these computations, you'll see that the $\mathbf{e}_1(\mathbf{u})$ and $\mathbf{e}_2(\mathbf{u})$ terms are rather a mess because they involve Maurer-Cartan forms that haven't been computed yet. But the $\mathbf{e}_3(\mathbf{u})$ terms are manageable, and we can use them to solve for the unknown functions. It turns out to be best to do this in two steps, mainly because MAPLE is very clumsy at handling algebraic expressions. So, start by using the $\mathbf{e}_3(\mathbf{u})$ terms in $d\tilde{\mathbf{e}}_1$ and $d\tilde{\mathbf{e}}_2$ to solve for λ_1, λ_2:

```
> solve({pick(coeff(zero11, e03), d(u)),
  pick(coeff(zero12, e03), d(v))},
  {lambda1(u,v), lambda2(u,v)});
```

At this point, MAPLE returns the rather obtuse expression

$$\left\{\lambda_1(u,v) = \frac{1}{RootOf(-\kappa 1 + \_Z^8 \kappa 2^3)^3 \kappa 2},\right.$$
$$\left.\lambda_2(u,v) = RootOf(-\kappa 1 + \_Z^8 \kappa 2^3)\right\}$$

and we will need to intervene by hand. The expression

$$RootOf(-\kappa 1 + \_Z^8 \kappa 2^3)$$

means any one of the values obtained by setting the expression in parentheses equal to zero and solving for $\_Z$. Since we know that we want everything to be positive-valued, we can easily see by inspection that the desired solution is

$$\_Z = \frac{\kappa_1^{(1/8)}}{\kappa_2^{(3/8)}},$$

which leads to the solution

$$\lambda_1 = \frac{\kappa_2^{(1/8)}}{\kappa_1^{(3/8)}}, \qquad \lambda_2 = \frac{\kappa_1^{(1/8)}}{\kappa_2^{(3/8)}}.$$

So make these assignments:

```
> lambda1(u,v):= kappa2(u,v)^(1/8)/kappa1(u,v)^(3/8);
  lambda2(u,v):= kappa1(u,v)^(1/8)/kappa2(u,v)^(3/8);
```

Now we can use the $\mathbf{e}_3(\mathbf{u})$ term in $d\tilde{\mathbf{e}}_3$ to solve for r_1, r_2:

```
> solve({op(ScalarForm(Simf(coeff(zero13, e03))))},
  {r1(u,v), r2(u,v)});
> assign(%);
```

You can now inspect the following expressions to see that they agree with those in Exercise 6.40(a):

```
> Simf(e1);
  Simf(e2);
  Simf(e3);
```

Moreover, the remaining Maurer-Cartan forms can now be computed from the $\mathbf{e}_1(\mathbf{u})$ and $\mathbf{e}_2(\mathbf{u})$ terms in `zero11, zero12, zero13`. This is left as an exercise for the reader; details may be found in the MAPLE worksheet for this chapter on the AMS webpage.

Chapter 7

Curves and surfaces in projective space

7.1. Introduction

Applying the method of moving frames to curves and surfaces in projective space leads to invariants that are a bit more complicated than those that we have encountered thus far. For nondegenerate curves in the projective space $\mathbb{P}^n$, there is no projectively invariant notion of arc length—not even an unconventional one such as the equi-affine arc length that we defined for curves in $\mathbb{A}^n$. (It may, however, be possible to define an arc length function with some additional restrictions on the curve; cf. Remark 7.13.) Instead, a nondegenerate curve in $\mathbb{P}^n$ carries a canonical *projective structure*. Moreover, additional invariants associated to the curve (i.e., the projective analogs of curvature, torsion, etc.) are no longer real-valued functions on the curve; rather, they are more general geometric objects.

In order to keep the complexity to a minimum while introducing these concepts, we will start one dimension lower than in previous chapters and begin by investigating curves in $\mathbb{P}^2$. This is already a topic with important applications, notably in areas such as computer graphics. From there, we will move on to curves and surfaces in $\mathbb{P}^3$.

7.2. Moving frames for curves in $\mathbb{P}^2$

Consider a smooth, parametrized curve $[\alpha] : I \to \mathbb{P}^2$ that maps some open interval $I \subset \mathbb{R}$ into the projective plane. $\mathbb{P}^2$ has the structure of the homogeneous space $SL(3)/H_{[\mathbf{x}_0]}$, where $H_{[\mathbf{x}_0]}$ is the isotropy group from equation (3.17), so an adapted frame field along $[\alpha]$ should be a lifting $\tilde{\alpha} : I \to SL(3)$. Any such lifting can be written as

$$\tilde{\alpha}(t) = (\mathbf{e}_0(t), \mathbf{e}_1(t), \mathbf{e}_2(t)),$$

where for each $t \in I$,

$$[\mathbf{e}_0(t)] = [\alpha(t)] \in \mathbb{P}^2$$

and

$$\det \left[\mathbf{e}_0(t)\ \mathbf{e}_1(t)\ \mathbf{e}_2(t)\right] = 1.$$

Such an adapted frame field is called a *projective frame field* along $[\alpha]$. How should we choose such a lifting in some geometrically natural way?

As in the equi-affine case, it seems natural to choose the vectors $(\mathbf{e}_0(t), \mathbf{e}_1(t), \mathbf{e}_2(t))$ so that

$$\mathbf{e}_1(t) = \mathbf{e}_0'(t), \qquad \mathbf{e}_2(t) = \mathbf{e}_1'(t).$$

This suggests that we look for a lifting $\alpha : I \to \mathbb{R}^3$ of $[\alpha]$ with the property that

$$\det \left[\alpha(t)\ \alpha'(t)\ \alpha''(t)\right] = 1. \tag{7.1}$$

In particular, the vectors $(\alpha(t), \alpha'(t), \alpha''(t))$ should be linearly independent for all $t \in I$.

Exercise 7.1. Given a curve $[\alpha] : I \to \mathbb{P}^2$, let $\alpha_1, \alpha_2 : I \to \mathbb{R}^3$ be two liftings of $[\alpha]$, so that

$$[\alpha_1(t)] = [\alpha_2(t)] = [\alpha(t)] \in \mathbb{P}^2$$

for all $t \in I$. Show that the vectors $(\alpha_1(t), \alpha_1'(t), \alpha_1''(t))$ are linearly independent if and only if the vectors $(\alpha_2(t), \alpha_2'(t), \alpha_2''(t))$ are linearly independent. (Hint: $\alpha_2(t) = \lambda(t)\alpha_1(t)$ for some nonvanishing, real-valued function $\lambda(t)$.)

Exercise 7.1 implies that the linear dependence or independence of the vectors $(\alpha(t), \alpha'(t), \alpha''(t))$ does not depend on the choice of lifting $\alpha : I \to \mathbb{R}^3$.

Therefore, the following definition makes sense:

Definition 7.2. A regular curve $[\alpha] : I \to \mathbb{P}^2$ will be called *nondegenerate* if for any lifting $\alpha : I \to \mathbb{R}^3$ of $[\alpha]$, the vectors $(\alpha(t), \alpha'(t), \alpha''(t))$ are linearly independent for all $t \in I$.

***Exercise 7.3.** Let $[\alpha] : I \to \mathbb{P}^2$ be a nondegenerate curve. Show that $[\alpha]$ has a unique lifting $\alpha : I \to \mathbb{R}^3$ satisfying the determinant condition (7.1). (Hint: Let $\alpha_0 : I \to \mathbb{R}^3$ be any lifting of $[\alpha]$, and consider an arbitrary lifting $\alpha(t) = \lambda(t)\alpha_0(t)$. Show that (7.1) uniquely determines the function $\lambda(t)$.)

Definition 7.4. Let $[\alpha] : I \to \mathbb{P}^2$ be a nondegenerate curve. The lifting $\alpha : I \to \mathbb{R}^3$ of Exercise 7.3 will be called the *canonical lifting* of $[\alpha]$. The projective frame field

$$\mathbf{e}_0(t) = \alpha(t), \qquad \mathbf{e}_1(t) = \alpha'(t), \qquad \mathbf{e}_2(t) = \alpha''(t)$$

will be called the *canonical projective frame field* associated to $[\alpha]$.

Exercise 7.5. Prove that the canonical projective frame field $(\mathbf{e}_0(t), \mathbf{e}_1(t), \mathbf{e}_2(t))$ associated to $[\alpha]$ is equivariant under the action of $SL(3)$ on $\mathbb{P}^2$: If we replace $[\alpha]$ by $g \cdot [\alpha]$ for some $g \in SL(3)$, then $\mathbf{e}_\gamma(t) \in T_{\alpha(t)}\mathbb{R}^3$ will be replaced by $g \cdot \mathbf{e}_\gamma(t) \in T_{g\cdot\alpha(t)}\mathbb{R}^3$ for $\gamma = 0, 1, 2$.

Now things start to get interesting. The canonical projective frame field depends on the parametrization of $[\alpha]$. How will it change if we reparametrize $[\alpha]$? Is there some particular parametrization of $[\alpha]$—something akin to the arc length parametrization for curves in Euclidean space—that is somehow geometrically natural? In order to answer this question, we will first compute invariants for parametrized nondegenerate curves. Then we will investigate how these invariants transform under a change of parametrization for $[\alpha]$ and look for special parametrizations that normalize the invariants in some natural way.

The pullbacks of equations (3.1) to I via α can be written as

$$\mathbf{e}'_\gamma(t)dt = \mathbf{e}_\delta(t)\,\bar{\omega}^\delta_\gamma, \tag{7.2}$$

where the indices γ, δ range from 0 to 2. But we constructed the canonical projective frame field so that

$$\mathbf{e}'_0(t) = \mathbf{e}_1(t), \qquad \mathbf{e}'_1(t) = \mathbf{e}_2(t).$$

The first of these equations implies that

$$\bar{\omega}_0^1 = dt, \qquad \bar{\omega}_0^0 = \bar{\omega}_0^2 = 0,$$

and the second equation implies that

$$\bar{\omega}_1^2 = dt, \qquad \bar{\omega}_1^0 = \bar{\omega}_1^1 = 0.$$

Finally, $\mathbf{e}_2(t)$ must satisfy

$$\mathbf{e}_2'(t) = \kappa_0(t)\,\mathbf{e}_0(t) + \kappa_1(t)\,\mathbf{e}_1(t)$$

for some functions $\kappa_0(t), \kappa_1(t)$. These functions are called the *Wilczynski invariants* of the parametrized curve $[\alpha] : I \to \mathbb{P}^2$. They were first introduced by Wilczynski in the context of studying invariants for linear differential operators; see [**Wil62**] for details.

***Exercise 7.6.** Why is there no $\mathbf{e}_2(t)$ term in the equation for $\mathbf{e}_2'(t)$? (Cf. Exercise 6.11.)

Next, we investigate how the Wilczynski invariants transform under a reparametrization of α.

***Exercise 7.7** (MAPLE recommended)**.** Let $J \subset \mathbb{R}$, and let $[\beta] : J \to \mathbb{P}^2$ be a reparametrization of $[\alpha]$ given by $[\beta(s)] = [\alpha(t(s))]$, with $t'(s) \neq 0$.

(a) Show that the canonical lifting of $[\beta]$ is given by

$$\beta(s) = \frac{1}{t'(s)}\alpha(t(s)),$$

where $\alpha(t)$ is the canonical lifting of $[\alpha]$.

(b) Show that the canonical projective frame field $(\tilde{\mathbf{e}}_0(s), \tilde{\mathbf{e}}_1(s), \tilde{\mathbf{e}}_2(s))$ associated to $[\beta]$ is given by

$$\begin{aligned}
\tilde{\mathbf{e}}_0(s) &= \frac{1}{t'(s)}\mathbf{e}_0(t(s)),\\
\tilde{\mathbf{e}}_1(s) &= -\frac{t''(s)}{(t'(s))^2}\mathbf{e}_0(t(s)) + \mathbf{e}_1(t(s)),\\
\tilde{\mathbf{e}}_2(s) &= -\left(\frac{t'''(s)}{(t'(s))^2} - 2\frac{(t''(s))^2}{(t'(s))^3}\right)\mathbf{e}_0(t(s)) - \frac{t''(s)}{t'(s)}\mathbf{e}_1(t(s)) + t'(s)\mathbf{e}_2(t(s)),
\end{aligned}$$

where $(\mathbf{e}_0(t), \mathbf{e}_1(t), \mathbf{e}_2(t))$ is the canonical projective frame field associated to $[\alpha]$.

(c) Show that the Wilczynski invariants $\tilde{\kappa}_0(s), \tilde{\kappa}_1(s)$ associated to $[\beta]$ are given by

$$\tag{7.3} \begin{aligned} \tilde{\kappa}_0(s) = &-\frac{t^{(4)}(s)}{t'(s)} + 4\frac{t'''(s)t''(s)}{(t'(s))^2} - 3\left(\frac{t''(s)}{t'(s)}\right)^3 \\ &+ (t'(s))^3\kappa_0(t(s)) + t'(s)t''(s)\kappa_1(t(s)), \end{aligned}$$

$$\tag{7.4} \tilde{\kappa}_1(s) = -2\frac{t'''(s)}{t'(s)} + 3\left(\frac{t''(s)}{t'(s)}\right)^2 + (t'(s))^2\kappa_1(t(s)),$$

where $\kappa_0(t), \kappa_1(t)$ are the Wilczynski invariants associated to $[\alpha]$.

The transformation rule (7.4) for $\tilde{\kappa}_1$ is simpler than the transformation rule (7.3) for $\tilde{\kappa}_0$, in that (7.4) involves only κ_1, whereas (7.3) involves both κ_0 and κ_1. So, first consider equation (7.4). Local existence theory for ordinary differential equations guarantees that, given any smooth function $\kappa_1(t)$ on an interval $I \subset \mathbb{R}$, any point $t_0 \in I$, and any point $s_0 \in \mathbb{R}$, there exists an interval $J \subset \mathbb{R}$ containing the point s_0 and a function $t : J \to I$ with the properties that

(1) $t(s_0) = t_0$;

(2) $t'(s) \neq 0$ for all $s \in J$ (indeed, $t'(s_0) \neq 0$ can be prescribed arbitrarily, and then J can be chosen to be small enough so that $t'(s) \neq 0$ on the entire interval);

(3) $t(s)$ satisfies the differential equation

$$\tag{7.5} -2\frac{t'''(s)}{t'(s)} + 3\left(\frac{t''(s)}{t'(s)}\right)^2 + (t'(s))^2\kappa_1(t(s)) = 0.$$

According to (7.4), the corresponding reparametrization $[\beta(s)]$ will have $\tilde{\kappa}_1(s) \equiv 0$.

Definition 7.8. Let $I \subset \mathbb{R}$, and let $[\alpha] : I \to \mathbb{P}^2$ be a nondegenerate curve. $[\alpha]$ is called a *projective parametrization* if the Wilczynski invariants $\kappa_0(t), \kappa_1(t)$ associated to $[\alpha]$ have the property that $\kappa_1(t) \equiv 0$. In this case, the parameter $t \in I$ is called a *projective parameter* for $[\alpha]$.

A projective parametrization for a nondegenerate curve in $\mathbb{P}^2$ is the natural analog of an arc length parametrization for a nondegenerate curve in $\mathbb{E}^n, \mathbb{M}^{1,n}$, or $\mathbb{A}^n$. This is certainly a less intuitive notion than that of an arc length parametrization, and as such it merits further investigation.

For instance, does a nondegenerate curve in $\mathbb{P}^2$ have a unique projective parametrization? If not, how much flexibility does this notion admit?

In order to address this question, suppose that $[\alpha] : I \to \mathbb{P}^2$ is a nondegenerate, projectively parametrized curve, and suppose that $[\beta(s)] = [\alpha(t(s))]$ is a reparametrization of $[\alpha]$ that is also a projective parametrization. This means that the Wilczynski invariants $\kappa_1(t), \tilde{\kappa}_1(s)$ associated to $[\alpha], [\beta]$, respectively, are both identically zero. According to (7.4), it follows that the function $t(s)$ must satisfy the differential equation

$$-2\frac{t'''(s)}{t'(s)} + 3\left(\frac{t''(s)}{t'(s)}\right)^2 = 0. \tag{7.6}$$

The expression on the left-hand side of (7.6) is equal to -2 times the *Schwarzian derivative* $S(t)$ of the function $t(s)$, defined by

$$S(t) = \frac{t'''}{t'} - \frac{3}{2}\left(\frac{t''}{t'}\right)^2. \tag{7.7}$$

The Schwarzian derivative is the fundamental invariant differential operator of projective differential geometry; it has the property that for any differentiable function $f : \mathbb{P}^1 \to \mathbb{P}^1$ and any projective transformation $g : \mathbb{P}^1 \to \mathbb{P}^1$,

$$S(g \circ f) = S(f).$$

Moreover, $S(f) = 0$ if and only if f is a projective transformation.

***Exercise 7.9.** Recall that a projective transformation $g : \mathbb{P}^1 \to \mathbb{P}^1$ has the form

$$g([x^0 : x^1]) = [dx^0 + cx^1 : bx^0 + ax^1],$$

where

$$\det \begin{bmatrix} a & b \\ c & d \end{bmatrix} = 1.$$

In terms of the affine coordinate $\bar{x} = \frac{x^1}{x^0}$ on $\mathbb{P}^1$, g is given by a linear fractional transformation

$$g(\bar{x}) = \frac{a\bar{x} + b}{c\bar{x} + d}.$$

(a) Show by direct computation that if g is a linear fractional transformation, then $S(g) = 0$.

(b) Suppose that a differentiable function $f : \mathbb{P}^1 \to \mathbb{P}^1$ satisfies $S(f) = 0$. Show that f is a linear fractional transformation. (Hint: First show that

$S(f)$ can be rewritten as

$$S(f) = \left(\frac{f''}{f'}\right)' - \frac{1}{2}\left(\frac{f''}{f'}\right)^2.$$

Then solve the differential equation

$$z' - \tfrac{1}{2}z^2 = 0$$

for the function $z = \frac{f''}{f'}$.)

As a consequence of Exercise 7.9, any projective reparametrization of a projectively parametrized nondegenerate curve $[\alpha(t)]$ must have the form $[\beta(s)] = [\alpha(t(s))]$, where

$$t(s) = \frac{as + b}{cs + d}$$

with $ad - bc = 1$. Thus, a given nondegenerate curve in $\mathbb{P}^2$ does not have a unique projective parametrization, but rather a 3-parameter family of projective parametrizations related to each other by linear fractional transformations.

Remark 7.10. Lest this seem too bizarre, note that a curve in Euclidean space does not actually have a unique arc length parameter s. Given an arc length parameter s along a curve $\alpha : I \to \mathbb{E}^n$, any reparametrization of the form

$$\tilde{s} = s + c, \tag{7.8}$$

where $c \in \mathbb{R}$ is constant, yields an equally valid arc length parameter for α. It is even possible—for instance, when α is a closed curve—that there is no single, continuous arc length parameter defined on the entire curve. Rather, the curve may be covered with open intervals, each with its own arc length parameter, and transition functions of the form (7.8) defined on the regions where intervals overlap. (Recall the discussion of transition functions between systems of local coordinates on manifolds in §1.1.) What we *really* have is not so much an arc length parameter as a *metric structure* on the curve, defined by the condition that the transition functions between systems of local coordinates are isometries.

Similarly, the projective parameter along a nondegenerate curve in $\mathbb{P}^2$ given by the solution of (7.5) may not be well-defined along the entire curve, as only local existence of a solution is guaranteed. However, a local projective parameter may be found in a neighborhood of any point on the curve, and the transition functions between projective parameters on overlapping intervals must be linear fractional transformations. Thus, what we really have is a well-defined *projective structure* on the curve, analogous to the metric structure on a curve in $\mathbb{E}^n$.

Now, assume that $[\alpha] : I \to \mathbb{P}^2$ is projectively parametrized, and consider the remaining invariant function $\kappa_0(t)$. How does it transform under a projective reparametrization?

***Exercise 7.11.** Let $[\alpha] : I \to \mathbb{P}^2$ be a nondegenerate, projectively parametrized curve, and let $[\beta(s)] = [\alpha(t(s))]$ be a projective reparametrization of $[\alpha]$ given by a linear fractional transformation

$$t(s) = \frac{as+b}{cs+d} \tag{7.9}$$

with $ad - bc = 1$. Let $\kappa_0(t), \tilde{\kappa}_0(s)$ be the invariants associated to $[\alpha], [\beta]$, respectively. Use equation (7.3) to show that

$$\tilde{\kappa}_0(s) = \frac{\kappa_0(t(s))}{(cs+d)^6} = \kappa_0(t(s)) \left(\frac{dt}{ds}\right)^3.$$

Conclude that under the transformation (7.9), we have

$$\tilde{\kappa}_0(s)(ds)^3 = \kappa_0(t)(dt)^3.$$

Thus, the curvature "function" κ_0 is not really a well-defined function on the curve at all, but rather a "cubic form" that can be represented in terms of any projective parameter t as

$$\hat{\kappa}_0 = \kappa_0(t)(dt)^3. \tag{7.10}$$

The cubic form (7.10) is called the *projective curvature form* of $[\alpha]$. In the following exercise, we show that the projective curvature form is a well-defined rank 3 symmetric tensor along $[\alpha]$.

***Exercise 7.12.** Let $[\alpha] : I \to \mathbb{P}^2$ be a nondegenerate curve (not necessarily projectively parametrized), and let $[\beta(s)] = [\alpha(t(s))]$ be an arbitrary reparametrization of $[\alpha]$. Let $\kappa_0(t), \kappa_1(t)$ be the invariants associated to $[\alpha]$ and let $\tilde{\kappa}_0(s), \tilde{\kappa}_1(s)$ be the invariants associated to $[\beta]$. Use the transformation rules (7.3), (7.4) to show that

$$\left(\tilde{\kappa}_0(s) - \tfrac{1}{2}\tilde{\kappa}_1'(s)\right)(ds)^3 = \left(\kappa_0(t) - \tfrac{1}{2}\kappa_1'(t)\right)(dt)^3.$$

Conclude that the cubic form

$$\hat{\kappa}_0 = \left(\kappa_0(t) - \tfrac{1}{2}\kappa_1'(t)\right)(dt)^3 \tag{7.11}$$

is a well-defined rank 3 tensor along $[\alpha]$ that has the form (7.10) for any projective parametrization of $[\alpha]$.

Remark 7.13. Our terminology here is somewhat nonstandard. In the case where $\hat{\kappa}_0$ is nonvanishing, the 1-form

$$\sqrt[3]{\hat{\kappa}_0} = \left(\kappa_0(t) - \tfrac{1}{2}\kappa_1'(t)\right)^{\frac{1}{3}}\, dt$$

is sometimes referred to as the *projective arc length element* and the function

$$s(t) = \int_{t_0}^{t} \sqrt[3]{\hat{\kappa}_0}$$

is sometimes referred to as the *projective arc length* of $[\alpha]$. Parametrizing $[\alpha]$ according to this projective arc length function then leads to a notion of projective curvature that is somewhat different from ours. This approach has the advantage that the projective curvature so obtained is a well-defined function on $[\alpha]$ rather than a tensor; the disadvantage is that, while any nondegenerate curve can be projectively parametrized, not all nondegenerate curves in $\mathbb{P}^2$ can be parametrized by projective arc length, so this approach is more restrictive. (For instance, conic curves have projective arc length identically equal to zero.)

Likewise, the canonical projective frame field associated to a projectively parametrized nondegenerate curve is not quite invariant under a projective reparametrization.

***Exercise 7.14.** Let $[\alpha] : I \to \mathbb{P}^2$ be a nondegenerate, projectively parametrized curve, and let $[\beta(s)] = [\alpha(t(s))]$ be a projective reparametrization of $[\alpha]$ given by a linear fractional transformation as in equation (7.9). Let $(\mathbf{e}_0(t), \mathbf{e}_1(t), \mathbf{e}_2(t))$ be the canonical projective frame field associated to $[\alpha]$. Use the result of Exercise 7.7(b) to show that the canonical projective frame field $(\tilde{\mathbf{e}}_0(s), \tilde{\mathbf{e}}_1(s), \tilde{\mathbf{e}}_2(s))$ associated to $[\beta]$ is given by

$$\begin{aligned}
\tilde{\mathbf{e}}_0(s) &= (cs+d)^2\mathbf{e}_0(t(s)),\\
\tilde{\mathbf{e}}_1(s) &= 2c(cs+d)\mathbf{e}_0(t(s)) + \mathbf{e}_1(t(s)),\\
\tilde{\mathbf{e}}_2(s) &= 2c^2\mathbf{e}_0(t(s)) + \frac{2c}{cs+d}\mathbf{e}_1(t(s)) + \frac{1}{(cs+d)^2}\mathbf{e}_2(t(s)).
\end{aligned}$$

Thus, the projective structure gives rise to a 2-parameter family of canonical projective frame fields along a nondegenerate curve $[\alpha] : I \to \mathbb{P}^2$. Note that this is *not* the same thing as a rank 2 principal bundle of projective frames along the curve: In a rank 2 principal bundle, the choice of a frame at each point amounts to choosing two arbitrary functions along the curve. Here we have only two *constant* parameters' worth of choices for a projective frame field along the curve, and choosing a specific projective frame at any point of the curve uniquely determines the canonical projective frame field along the entire curve.

For any canonical projective frame field $(\mathbf{e}_0(t), \mathbf{e}_1(t), \mathbf{e}_2(t))$ associated to a nondegenerate, projectively parametrized curve $[\alpha] : I \to \mathbb{P}^2$, we have the following analog of the Frenet equations:

$$\begin{bmatrix}\mathbf{e}_0'(t) & \mathbf{e}_1'(t) & \mathbf{e}_2'(t)\end{bmatrix} = \begin{bmatrix}\mathbf{e}_0(t) & \mathbf{e}_1(t) & \mathbf{e}_2(t)\end{bmatrix} \begin{bmatrix} 0 & 0 & \kappa_0(t) \\ 1 & 0 & 0 \\ 0 & 1 & 0 \end{bmatrix}.$$

Applying Lemma 4.2 yields the following theorem:

Theorem 7.15. *Two nondegenerate, projectively parametrized curves $[\alpha_1]$, $[\alpha_2] : I \to \mathbb{P}^2$ differ by a projective transformation if and only if they have the same projective curvature form $\hat{\kappa}_0 = \kappa_0(t)(dt)^3$.*

***Exercise 7.16.** Suppose that a nondegenerate, projectively parametrized curve $[\alpha] : I \to \mathbb{P}^2$ has projective curvature $\kappa_0(t) = 0$. (Note that this condition is invariant under projective reparametrization.)

(a) Show that the canonical lifting $\alpha : I \to \mathbb{R}^3$ of $[\alpha]$ is part of a parabola contained in a plane in $\mathbb{R}^3$.

(b) Conclude that all conic sections in $\mathbb{P}^2$ have projective curvature identically equal to zero.

Exercise 7.17. This exercise demonstrates the result of Exercise 7.16 explicitly in the case where $[\alpha]$ is a circle. Let $[\alpha] : \mathbb{R} \to \mathbb{P}^2$ be the circle parametrized in homogeneous coordinates as

$$[\alpha(t)] = [1 : \cos(t) : \sin(t)].$$

(a) Show that the canonical lifting $\alpha : \mathbb{R} \to \mathbb{R}^3$ of $[\alpha]$ is

$$\alpha(t) = {}^t[1, \cos(t), \sin(t)].$$

(b) Show that the invariants $\kappa_0(t), \kappa_1(t)$ associated to $[\alpha]$ are

$$\kappa_0(t) = 0, \qquad \kappa_1(t) = -1.$$

Conclude that $[\alpha]$ is *not* a projective parametrization.

(c) Show that the function

$$t(s) = 2\tan^{-1}(s)$$

satisfies equation (7.5); thus, the curve

$$[\beta(s)] = [\alpha(2\tan^{-1}(s))]$$

is a projective reparametrization of $[\alpha]$.

(d) Show that the canonical lifting of $[\beta]$ is

$$\beta(s) = \tfrac{1}{2}\begin{bmatrix} 1+s^2 \\ 1-s^2 \\ 2s \end{bmatrix} = \tfrac{1}{2}s^2\begin{bmatrix} 1 \\ -1 \\ 0 \end{bmatrix} + s\begin{bmatrix} 0 \\ 0 \\ 1 \end{bmatrix} + \tfrac{1}{2}\begin{bmatrix} 1 \\ 1 \\ 0 \end{bmatrix}.$$

Therefore, β is a parabola in the plane spanned by the vectors ${}^t[1,\,-1,\,0]$, ${}^t[0,\,0,\,1]$ and passing through the point ${}^t[\frac{1}{2},\,\frac{1}{2},\,0]$.

Exercise 7.18. We like to think of $\mathbb{P}^2$ as the plane $\mathbb{R}^2$, represented as the open set

$$V_0 = \left\{\left[1 : \bar{x}^1 : \bar{x}^2\right]\right\} \subset \mathbb{P}^2,$$

plus points "at infinity". Thus, we like to think of a curve $[\alpha] : I \to \mathbb{P}^2$ as having a parametrization of the form

$$[\alpha(t)] = [1 : \bar{x}^1(t) : \bar{x}^2(t)] \tag{7.12}$$

that is valid at all but finitely many points on the curve.

(a) Show that any nondegenerate curve $[\alpha] : I \to \mathbb{P}^2$ of the form (7.12) has a reparametrization $[\beta(\bar{s})] = [\alpha(t(\bar{s}))]$ whose canonical lifting is

$$\beta(\bar{s}) = {}^t[1,\, \bar{x}^1(t(\bar{s})),\, \bar{x}^2(t(\bar{s}))].$$

(b) Show that the invariants $\kappa_0(\bar{s}), \kappa_1(\bar{s})$ associated to $[\beta]$ have the property that $\kappa_0(\bar{s}) \equiv 0$.

(c) Conclude that $[\beta(\bar{s})]$ is *not* a projective parametrization unless $[\alpha]$ is a conic section. It follows that for non-conic curves, the canonical lifting of a projective parametrization is never contained in a plane in $\mathbb{R}^3$.

(d) Show that, while $[\beta(\bar{s})]$ is not a projective parametrization, it has the property that the curve $\bar{\beta} : I \to \mathbb{A}^2$ defined by

$$\bar{\beta}(\bar{s}) = {}^t[\bar{x}^1(t(\bar{s})),\, \bar{x}^2(t(\bar{s}))]$$

is parametrized by its *equi-affine* arc length, and the Wilczynski invariant $\kappa_1(\bar{s})$ of $[\beta]$ is equal to the equi-affine curvature $\kappa(\bar{s})$ of $\bar{\beta}$. This is related to the fact that the equi-affine transformation group $A(2)$ can be realized as the subgroup of the projective transformation group $SL(3)$ that preserves the plane $\{{}^t[x^0, x^1, x^2] \in \mathbb{R}^3 \mid x^0 = 1\}$ in $\mathbb{R}^3$.

(e) Let $[\gamma(s)] = [\beta(\bar{s}(s))]$ be a projective reparametrization of $[\beta(\bar{s})]$, so that the Wilczynski invariants $\tilde{\kappa}_0(s), \tilde{\kappa}_1(s)$ associated to $[\gamma]$ satisfy $\tilde{\kappa}_1(s) = 0$. Use the formulas (7.3), (7.4) and the fact that $\kappa_0(\bar{s}) = 0$ to show that

$$\tilde{\kappa}_0(s) = -\tfrac{1}{2}(\bar{s}'(s))^3\kappa'(\bar{s}(s)),$$

where $\kappa(\bar{s})$ is the equi-affine curvature of $\bar{\beta}$ and $\kappa'(\bar{s}(s))$ denotes the derivative of $\kappa(\bar{s})$ with respect to $\bar{s}$, evaluated at $\bar{s} = \bar{s}(s)$.

(f) Conclude that the projective curvature form of the original curve $[\alpha]$ may be expressed as

$$\hat{\kappa}_0 = \tilde{\kappa}_0(s)(ds)^3 = -\tfrac{1}{2}\kappa'(\bar{s})(d\bar{s})^3,$$

where $\bar{s}$ is the equi-affine arc length of the curve $\bar{\beta} : I \to \mathbb{A}^2$ and $\kappa(\bar{s})$ is the equi-affine curvature of $\bar{\beta}$. Comparing with the result of Exercise 6.15, we see that the projective curvature form is a fifth-order invariant of $[\alpha]$.

7.3. Moving frames for curves in $\mathbb{P}^3$

Now we will extend the ideas developed in Section 7.2 to curves in $\mathbb{P}^3$. Consider a smooth parametrized curve $[\alpha] : I \to \mathbb{P}^3$ that maps some open interval $I \subset \mathbb{R}$ into the projective space $\mathbb{P}^3$. The construction of the canonical projective frame field associated to $[\alpha]$ is analogous to that for curves in $\mathbb{P}^2$: A projective frame field along $[\alpha]$ is a lifting $\tilde{\alpha} : I \to SL(4)$, written as

$$\tilde{\alpha}(t) = (\mathbf{e}_0(t), \mathbf{e}_1(t), \mathbf{e}_2(t), \mathbf{e}_3(t)),$$

with the properties that

$$[\mathbf{e}_0(t)] = [\alpha(t)] \in \mathbb{P}^3$$

and

$$\det\left[\mathbf{e}_0(t)\ \mathbf{e}_1(t)\ \mathbf{e}_2(t)\ \mathbf{e}_3(t)\right] = 1. \tag{7.13}$$

As for curves in $\mathbb{P}^2$, this suggests that we should look for a lifting $\alpha : I \to \mathbb{R}^4$ with the property that

$$\det\left[\alpha(t)\ \alpha'(t)\ \alpha''(t)\ \alpha'''(t)\right] = 1. \tag{7.14}$$

In particular, the vectors $(\alpha(t), \alpha'(t), \alpha''(t), \alpha'''(t))$ should be linearly independent for all $t \in I$.

Definition 7.19. A regular curve $[\alpha] : I \to \mathbb{P}^3$ will be called *nondegenerate* if for any lifting $\alpha : I \to \mathbb{R}^4$ of $[\alpha]$, the vectors $(\alpha(t), \alpha'(t), \alpha''(t), \alpha'''(t))$ are linearly independent for all $t \in I$.

Exercise 7.20 (Cf. Exercise 7.1). Show that Definition 7.19 makes sense, as follows. Given a curve $[\alpha] : I \to \mathbb{P}^3$, let $\alpha_1, \alpha_2 : I \to \mathbb{R}^4$ be two liftings of $[\alpha]$, so that

$$[\alpha_1(t)] = [\alpha_2(t)] = [\alpha(t)] \in \mathbb{P}^3$$

for all $t \in I$. Show that the vectors $(\alpha_1(t), \alpha_1'(t), \alpha_1''(t), \alpha_1'''(t))$ are linearly independent if and only if the vectors $(\alpha_2(t), \alpha_2'(t), \alpha_2''(t), \alpha_2'''(t))$ are linearly independent.

***Exercise 7.21** (Cf. Exercise 7.3). (a) Let $[\alpha] : I \to \mathbb{P}^3$ be a nondegenerate curve. Show that $[\alpha]$ has a lifting $\alpha : I \to \mathbb{R}^4$ satisfying the determinant condition

$$\det \left[\alpha(t)\ \alpha'(t)\ \alpha''(t)\ \alpha'''(t)\right] = \pm 1$$

and that this lifting is uniquely determined up to sign (i.e., α may be replaced by $-\alpha$).

(b) Explain the ambiguities of sign in this result, which did not appear in Exercise 7.3. (Hint: Cf. Exercise 3.66.)

In order to simplify the exposition in the remainder of this section, we will assume that

$$\det \left[\alpha(t)\ \alpha'(t)\ \alpha''(t)\ \alpha'''(t)\right] = 1.$$

(This can be achieved by replacing $[\alpha]$ by its reflection through a plane in $\mathbb{P}^3$ if necessary.)

Definition 7.22. Let $[\alpha] : I \to \mathbb{P}^3$ be a nondegenerate curve. The lifting $\alpha : I \to \mathbb{R}^4$ of Exercise 7.21 will be called the *canonical lifting* of $[\alpha]$. The projective frame field

$$\mathbf{e}_0(t) = \alpha(t), \qquad \mathbf{e}_1(t) = \alpha'(t), \qquad \mathbf{e}_2(t) = \alpha''(t), \qquad \mathbf{e}_3(t) = \alpha'''(t)$$

will be called the *canonical projective frame field* associated to $[\alpha]$.

Exercise 7.23. Prove that the canonical projective frame field $(\mathbf{e}_0(t), \mathbf{e}_1(t), \mathbf{e}_2(t), \mathbf{e}_3(t))$ associated to $[\alpha]$ is equivariant under the action of $SL(4)$ on $\mathbb{P}^3$: If we replace $[\alpha]$ by $g \cdot [\alpha]$ for some $g \in SL(4)$, then $\mathbf{e}_\gamma(t) \in T_{\alpha(t)}\mathbb{R}^4$ will be replaced by $g \cdot \mathbf{e}_\gamma(t) \in T_{g\cdot\alpha(t)}\mathbb{R}^4$ for $\gamma = 0, 1, 2, 3$.

As in Section 7.2, we will proceed by computing invariants for parametrized nondegenerate curves and then investigating how these invariants transform under reparametrizations.

***Exercise 7.24.** Show that the canonical projective frame field associated to a parametrized nondegenerate curve $[\alpha] : I \to \mathbb{P}^3$ satisfies the structure equations

$$\begin{gathered}\mathbf{e}_0'(t) = \mathbf{e}_1(t), \qquad \mathbf{e}_1'(t) = \mathbf{e}_2(t), \qquad \mathbf{e}_2'(t) = \mathbf{e}_3(t),\\ \mathbf{e}_3'(t) = \kappa_0(t)\,\mathbf{e}_0(t) + \kappa_1(t)\,\mathbf{e}_1(t) + \kappa_2(t)\,\mathbf{e}_2(t)\end{gathered}$$

for some functions $\kappa_0(t), \kappa_1(t), \kappa_2(t)$.

As for curves in $\mathbb{P}^2$, the functions $\kappa_0(t), \kappa_1(t), \kappa_2(t)$ are called the *Wilczynski invariants* of the parametrized curve $[\alpha] : I \to \mathbb{P}^3$. Next, we investigate how the Wilczynski invariants transform under a reparametrization of $[\alpha]$.

***Exercise 7.25** (MAPLE recommended; cf. Exercise 7.7)**.** Let $J \subset \mathbb{R}$, and let $[\beta] : J \to \mathbb{P}^3$ be a reparametrization of $[\alpha]$ given by $[\beta(s)] = [\alpha(t(s))]$, with $t'(s) > 0$. (The sign of $t'(s)$ is not particularly important, but choosing it to be positive will simplify the following computations.)

(a) Show that the canonical lifting of $[\beta]$ is given by

$$\beta(s) = \frac{1}{t'(s)}\alpha(t(s)),$$

where $\alpha(t)$ is the canonical lifting of $[\alpha]$.

(b) Show that the canonical projective frame field $(\tilde{\mathbf{e}}_0(s), \tilde{\mathbf{e}}_1(s), \tilde{\mathbf{e}}_2(s), \tilde{\mathbf{e}}_3(s))$ associated to $[\beta]$ is given by

$$\begin{aligned}
\tilde{\mathbf{e}}_0(s) &= \frac{1}{(t'(s))^{3/2}}\mathbf{e}_0(t(s)),\\
\tilde{\mathbf{e}}_1(s) &= -\frac{3t''(s)}{(2t'(s))^{5/2}}\mathbf{e}_0(t(s)) + \frac{1}{(t'(s))^{1/2}}\mathbf{e}_1(t(s)),\\
\tilde{\mathbf{e}}_2(s) &= -\left(\frac{3t'''(s)}{2(t'(s))^{5/2}} - \frac{15(t''(s))^2}{4(t'(s))^{7/2}}\right)\mathbf{e}_0(t(s)) - 2\frac{t''(s)}{(t'(s))^{3/2}}\mathbf{e}_1(t(s))\\
&\quad + (t'(s))^{1/2}\mathbf{e}_2(t(s)),\\
\tilde{\mathbf{e}}_3(s) &= -\left(\frac{3t^{(4)}(s)}{2(t'(s))^{5/2}} - \frac{45t'''(s)t''(s)}{4(t'(s))^{7/2}} + \frac{105(t''(s))^3}{8(t'(s))^{9/2}}\right)\mathbf{e}_0(t(s))\\
&\quad - \left(\frac{7t'''(s)}{2(t'(s))^{3/2}} - \frac{27(t''(s))^2}{4(t'(s))^{5/2}}\right)\mathbf{e}_1(t(s)) - \frac{3t''(s)}{2(t'(s))^{1/2}}\mathbf{e}_2(t(s))\\
&\quad + (t'(s))^{3/2}\mathbf{e}_3(t(s)),
\end{aligned}$$

where $(\mathbf{e}_0(t), \mathbf{e}_1(t), \mathbf{e}_2(t), \mathbf{e}_3(t))$ is the canonical projective frame field associated to $[\alpha]$.

(c) Show that the Wilczynski invariants $\tilde{\kappa}_0(s), \tilde{\kappa}_1(s), \tilde{\kappa}_2(s)$ associated to $[\beta]$ are given by

$$
\begin{aligned}
\tilde{\kappa}_0(s) = \frac{1}{16}\Bigg(&-24\frac{t^{(5)}(s)}{t'(s)} + 120\frac{t^{(4)}(s)t''(s)}{(t'(s))^2} + 60\left(\frac{t'''(s)}{t'(s)}\right)^2 \\
&-300\frac{t'''(s)(t''(s))^2}{(t'(s))^3} + 135\left(\frac{t''(s)}{t'(s)}\right)^4\Bigg) \\
&+ (t'(s))^4\kappa_0(t(s)) + \frac{3}{2}t''(s)(t'(s))^2\kappa_1(t(s)) \\
&+ \left(\frac{3}{2}t'''(s)t'(s) - \frac{3}{4}(t''(s))^2\right)\kappa_2(t(s)),
\end{aligned} \tag{7.15}
$$

$$
\begin{aligned}
\tilde{\kappa}_1(s) = 5&\left(-\frac{t^{(4)}(s)}{t'(s)} + 4\frac{t'''(s)t''(s)}{(t'(s))^2} - 3\left(\frac{t''(s)}{t'(s)}\right)^3\right) \\
&+ (t'(s))^3\kappa_1(t(s)) + 2t'(s)t''(s)\kappa_2(t(s)),
\end{aligned} \tag{7.16}
$$

$$
\tilde{\kappa}_2(s) = \frac{5}{2}\left(-2\frac{t'''(s)}{t'(s)} + 3\left(\frac{t''(s)}{t'(s)}\right)^2\right) + (t'(s))^2\kappa_2(t(s)), \tag{7.17}
$$

where $\kappa_0(t), \kappa_1(t), \kappa_2(t)$ are the Wilczynski invariants associated to $[\alpha]$.

Observe that equation (7.17) is very similar to equation (7.4). Precisely the same argument that we used in Section 7.2 can be applied to equation (7.17) to show that there exists a reparametrization $[\beta(s)]$ for which $\tilde{\kappa}_2(s) \equiv 0$. Thus, we have the analog of Definition 7.8:

Definition 7.26. Let $I \subset \mathbb{R}$, and let $[\alpha] : I \to \mathbb{P}^3$ be a nondegenerate curve. $[\alpha]$ is called a *projective parametrization* if the Wilczynski invariants $\kappa_0(t), \kappa_1(t), \kappa_2(t)$ associated to $[\alpha]$ have the property that $\kappa_2(t) \equiv 0$. In this case, the parameter $t \in I$ is called a *projective parameter* for $[\alpha]$.

As for curves in $\mathbb{P}^2$, a projective parametrization is unique up to reparametrizations of the form $[\beta(s)] = [\alpha(t(s))]$, where $t(s)$ is a linear fractional transformation. Therefore, any nondegenerate curve in $\mathbb{P}^3$ has a well-defined projective structure.

Now, assume that $[\alpha] : I \to \mathbb{P}^3$ is projectively parametrized, and consider the remaining invariant functions $\kappa_0(t), \kappa_1(t)$.

***Exercise 7.27** (Cf. Exercise 7.11). Let $[\alpha] : I \to \mathbb{P}^3$ be a nondegenerate, projectively parametrized curve, and let $[\beta(s)] = [\alpha(t(s))]$ be a projective reparametrization of $[\alpha]$ given by a linear fractional transformation as in equation (7.9). Let $\kappa_0(t), \kappa_1(t)$ and $\tilde{\kappa}_0(s), \tilde{\kappa}_1(s)$ be the invariants associated to $[\alpha], [\beta]$, respectively.

(a) Use equations (7.15) and (7.16) to show that

$$\tilde{\kappa}_1(s) = \frac{\kappa_1(t(s))}{(cs+d)^6},$$

$$\tilde{\kappa}_0(s) = \frac{\kappa_0(t(s))}{(cs+d)^8} - \frac{3c\,\kappa_1(t(s))}{(cs+d)^7}.$$

(b) Show that under the transformation (7.9), we have

$$\tilde{\kappa}_1(s)(ds)^3 = \kappa_1(t)(dt)^3,$$

$$\left(\tilde{\kappa}_0(s) - \tfrac{1}{2}\tilde{\kappa}_1'(s)\right)(ds)^4 = \left(\kappa_0(t) - \tfrac{1}{2}\kappa_1'(t)\right)(dt)^4.$$

Therefore, the cubic form

$$\hat{\kappa}_1 = \kappa_1(t)(dt)^3 \tag{7.18}$$

and the quartic form

$$\hat{\kappa}_0 = \left(\kappa_0(t) - \tfrac{1}{2}\kappa_1'(t)\right)(dt)^4 \tag{7.19}$$

are invariant under projective transformations. We will call these forms the *projective curvature forms* of $[\alpha]$.

***Exercise 7.28.** Let $[\alpha] : I \to \mathbb{P}^3$ be a nondegenerate curve (not necessarily projectively parametrized), and let $[\beta(s)] = [\alpha(t(s))]$ be an arbitrary reparametrization of $[\alpha]$. Let $\kappa_0(t), \kappa_1(t), \kappa_2(t)$ be the invariants associated to $[\alpha]$, and let $\tilde{\kappa}_0(s), \tilde{\kappa}_1(s), \tilde{\kappa}_2(s)$ be the invariants associated to $[\beta]$.

(a) Use the transformation rules (7.15), (7.16), (7.17) to show that

$$\left(\tilde{\kappa}_1(s) - \tilde{\kappa}_2'(s)\right)(ds)^3 = \left(\kappa_1(t) - \kappa_2'(t)\right)(dt)^3.$$

Conclude that the cubic form

$$\hat{\kappa}_1 = \left(\kappa_1(t) - \kappa_2'(t)\right)(dt)^3 \tag{7.20}$$

is a well-defined rank 3 tensor along $[\alpha]$ that has the form (7.18) for any projective parametrization of $[\alpha]$.

(b) Use the transformation rules (7.15), (7.16), (7.17) to show that

$$\begin{aligned}&\left(\tilde{\kappa}_0(s) - \tfrac{1}{2}\tilde{\kappa}_1'(s) + \tfrac{1}{5}\tilde{\kappa}_2''(s) + \tfrac{9}{100}(\tilde{\kappa}_2(s))^2\right)(ds)^4 \\ &\qquad = \left(\kappa_0(t) - \tfrac{1}{2}\kappa_1'(t) + \tfrac{1}{5}\kappa_2''(t) + \tfrac{9}{100}(\kappa_2(t))^2\right)(dt)^4.\end{aligned}$$

Conclude that the quartic form

$$\hat{\kappa}_0 = \left(\kappa_0(t) - \tfrac{1}{2}\kappa_1'(t) + \tfrac{1}{5}\kappa_2''(t) + \tfrac{9}{100}(\kappa_2(t))^2\right)(dt)^4$$

is a well-defined rank 4 tensor along $[\alpha]$ that has the form (7.19) for any projective parametrization of $[\alpha]$.

As for curves in $\mathbb{P}^2$, the canonical projective frame field associated to a projectively parametrized nondegenerate curve in $\mathbb{P}^3$ is not quite invariant under a projective reparametrization:

***Exercise 7.29** (Cf. Exercise 7.14). Let $[\alpha] : I \to \mathbb{P}^3$ be a nondegenerate, projectively parametrized curve, and let $[\beta(s)] = [\alpha(t(s))]$ be a projective reparametrization of $[\alpha]$ given by a linear fractional transformation as in equation (7.9). Let $(\mathbf{e}_0(t), \mathbf{e}_1(t), \mathbf{e}_2(t), \mathbf{e}_3(t))$ be the canonical projective frame field associated to $[\alpha]$. Use the result of Exercise 7.25(b) to show that the canonical projective frame field $(\tilde{\mathbf{e}}_0(s), \tilde{\mathbf{e}}_1(s), \tilde{\mathbf{e}}_2(s), \tilde{\mathbf{e}}_3(s))$ associated to $[\beta]$ is given by

$$\begin{aligned}
\tilde{\mathbf{e}}_0(s) &= (cs+d)^3\mathbf{e}_0(t(s)),\\
\tilde{\mathbf{e}}_1(s) &= 3c(cs+d)^2\mathbf{e}_0(t(s)) + (cs+d)\mathbf{e}_1(t(s)),\\
\tilde{\mathbf{e}}_2(s) &= 6c^2(cs+d)\mathbf{e}_0(t(s)) + 4c\mathbf{e}_1(t(s)) + \frac{1}{(cs+d)}\mathbf{e}_2(t(s)),\\
\tilde{\mathbf{e}}_3(s) &= 6c^3\mathbf{e}_0(t(s)) + \frac{6c^2}{(cs+d)}\mathbf{e}_1(t(s)) + \frac{3c}{(cs+d)^2}\mathbf{e}_2(t(s))\\
&\quad + \frac{1}{(cs+d)^3}\mathbf{e}_3(t(s)).
\end{aligned}$$

Thus, the projective structure gives rise to a 2-parameter family of canonical projective frame fields along a nondegenerate curve $[\alpha] : I \to \mathbb{P}^3$, just as it did for curves in $\mathbb{P}^2$. For any canonical projective frame field $(\mathbf{e}_0(t)$, $\mathbf{e}_1(t)$, $\mathbf{e}_2(t)$, $\mathbf{e}_3(t))$ associated to such a curve, we have the following analog of the Frenet equations:

$$\begin{bmatrix}\mathbf{e}_0'(t) & \mathbf{e}_1'(t) & \mathbf{e}_2'(t) & \mathbf{e}_3'(t)\end{bmatrix} = \begin{bmatrix}\mathbf{e}_0(t) & \mathbf{e}_1(t) & \mathbf{e}_2(t) & \mathbf{e}_3(t)\end{bmatrix}\begin{bmatrix}0 & 0 & 0 & \kappa_0(t)\\ 1 & 0 & 0 & \kappa_1(t)\\ 0 & 1 & 0 & 0\\ 0 & 0 & 1 & 0\end{bmatrix}.$$

Applying Lemma 4.2 yields the following theorem:

Theorem 7.30. *Two nondegenerate, projectively parametrized curves $[\alpha_1]$, $[\alpha_2] : I \to \mathbb{P}^3$ differ by a projective transformation if and only if they have* *the same projective curvature forms*

$$\hat{\kappa}_1 = \kappa_1(t)(dt)^3, \qquad \hat{\kappa}_0 = \left(\kappa_0(t) - \tfrac{1}{2}\kappa_1'(t)\right)(dt)^4.$$

***Exercise 7.31.** Suppose that a nondegenerate, projectively parametrized curve $[\alpha] : I \to \mathbb{P}^3$ has projective curvatures $\kappa_0(t) = \kappa_1(t) = 0$. (Note that this condition is invariant under projective reparametrization.)

(a) Show that the canonical lifting $\alpha : I \to \mathbb{R}^4$ of $[\alpha]$ has the form

$$\alpha(t) = \mathbf{v}_0 + \mathbf{v}_1 t + \tfrac{1}{2}\mathbf{v}_2 t^2 + \tfrac{1}{6}\mathbf{v}_3 t^3,$$

where $(\mathbf{v}_0, \mathbf{v}_1, \mathbf{v}_2, \mathbf{v}_3)$ are linearly independent vectors in $\mathbb{R}^4$ with

$$\det\left[\mathbf{v}_0\ \mathbf{v}_1\ \mathbf{v}_2\ \mathbf{v}_3\right] = 1.$$

In particular, note that α is contained in a 3-dimensional affine hyperplane in $\mathbb{R}^4$—namely, the plane passing through $\mathbf{v}_0$ and spanned by the vectors $(\mathbf{v}_1, \mathbf{v}_2, \mathbf{v}_3)$.

(b) Show that $[\alpha]$ is projectively equivalent to the curve $[\beta]$ with canonical lifting

$$\beta(t) = {}^t[1,\, t,\, \tfrac{1}{2}t^2,\, \tfrac{1}{6}t^3].$$

$[\beta]$ is called the *rational normal curve* of degree 3 (cf. Exercise 6.17). In general, the rational normal curve $[\beta]$ of degree n in $\mathbb{P}^n$ is the curve parametrized in homogeneous coordinates as

$$[\beta(t)] = [1 : t : \tfrac{1}{2}t^2 : \cdots : \tfrac{1}{n!}t^n].$$

7.4. Moving frames for surfaces in $\mathbb{P}^3$

Now, let U be an open set in $\mathbb{R}^2$, and let $[\mathbf{x}] : U \to \mathbb{P}^3$ be an immersion whose image is a surface $[\Sigma] = [\mathbf{x}(U)]$. Just as for curves in $\mathbb{P}^3$, an adapted frame field along $[\Sigma]$ is a lifting $\tilde{\mathbf{x}} : U \to SL(4)$ of the form

$$\tilde{\mathbf{x}}(\mathbf{u}) = \left(\mathbf{e}_0(\mathbf{u}), \mathbf{e}_1(\mathbf{u}), \mathbf{e}_2(\mathbf{u}), \mathbf{e}_3(\mathbf{u})\right),$$

where for each $\mathbf{u} \in U$, we have

$$[\mathbf{e}_0(\mathbf{u})] = [\mathbf{x}(\mathbf{u})] \in \mathbb{P}^3$$

and

$$\det\left[\mathbf{e}_0(\mathbf{u})\ \mathbf{e}_1(\mathbf{u})\ \mathbf{e}_2(\mathbf{u})\ \mathbf{e}_3(\mathbf{u})\right] = 1.$$

Any choice for the function $\mathbf{e}_0(\mathbf{u})$ defines a lifting $\mathbf{x} : U \to \mathbb{R}^4$ of $[\mathbf{x}]$, given by

$$\mathbf{x}(\mathbf{u}) = \mathbf{e}_0(\mathbf{u}),$$

whose image is a surface $\Sigma \subset \mathbb{R}^4$.

We might like to proceed as in the equi-affine case, by choosing $(\mathbf{e}_1(\mathbf{u}), \mathbf{e}_2(\mathbf{u}))$ so that they span the tangent space $T_{\mathbf{x}(\mathbf{u})}\Sigma$. However, this choice is complicated by the fact that this space is not well-defined: If we replace the lifting $\mathbf{x}$ by an alternate lifting

$$\hat{\mathbf{x}}(\mathbf{u}) = \lambda(\mathbf{u})\mathbf{x}(\mathbf{u})$$

for some nonvanishing function $\lambda : U \to \mathbb{R}$, then the tangent vectors $\mathbf{x}_u, \mathbf{x}_v$, which span $T_{\mathbf{x}(\mathbf{u})}\Sigma$, are replaced by the tangent vectors

$$\hat{\mathbf{x}}_u = \lambda\mathbf{x}_u + \lambda_u\mathbf{x}, \qquad \hat{\mathbf{x}}_v = \lambda\mathbf{x}_v + \lambda_v\mathbf{x}.$$

Since the function λ and its partial derivatives at any point are arbitrary (aside from the requirement that $\lambda \neq 0$), the tangent space $T_{\mathbf{x}(\mathbf{u})}\Sigma$ is only well-defined modulo the vector $\mathbf{x}(\mathbf{u}) = \mathbf{e}_0(\mathbf{u})$. Consequently, we can only require that $(\mathbf{e}_1(\mathbf{u}), \mathbf{e}_2(\mathbf{u}))$ span $T_{\mathbf{x}(\mathbf{u})}\Sigma$ modulo $\mathbf{e}_0(\mathbf{u})$. More concretely, this means that we will require that the three vectors $(\mathbf{e}_0(\mathbf{u}), \mathbf{e}_1(\mathbf{u}), \mathbf{e}_2(\mathbf{u}))$ span the 3-dimensional subspace of $\mathbb{R}^4$ determined by the position vector $\mathbf{x}(\mathbf{u})$ and the tangent plane $T_{\mathbf{x}(\mathbf{u})}\Sigma$. In particular, we will require that this space be 3-dimensional, which means that the position vector of Σ cannot be contained in its tangent plane at any point. (In fact, this condition is precisely what it means for the smooth map $[\mathbf{x}] : U \to \mathbb{P}^3$ to be an immersion.) A projective frame field $(\mathbf{e}_0(\mathbf{u}), \mathbf{e}_1(\mathbf{u}), \mathbf{e}_2(\mathbf{u}), \mathbf{e}_3(\mathbf{u}))$ along $[\Sigma]$ satisfying these conditions will be called 0-*adapted*.

Exercise 7.32. Show that, given a smooth map $[\mathbf{x}] : U \to \mathbb{P}^3$, the condition that $\mathrm{span}(\mathbf{x}, \mathbf{x}_u, \mathbf{x}_v)$ be 3-dimensional is independent of the choice of lifting $\mathbf{x} : U \to \mathbb{R}^4$ of $[\mathbf{x}]$.

Now, suppose that $\tilde{\mathbf{x}} : U \to SL(4)$ is a 0-adapted frame field along $[\Sigma] = [\mathbf{x}(U)]$, and let $(\bar{\omega}^\alpha_\beta)$ represent the pulled-back forms $(\tilde{\mathbf{x}}^*\omega^\alpha_\beta)$ on U. Recall from §3.7 that the 1-forms $(\omega^1_0, \omega^2_0, \omega^3_0)$ are the semi-basic forms for the projection $\pi : SL(4) \to \mathbb{P}^3$. Thus, these should be regarded as the dual forms, and the remaining (ω^α_β) as the connection forms.

Precisely the same reasoning as in the Euclidean and equi-affine cases can be used to prove the following:

Proposition 7.33. *Let $U \subset \mathbb{R}^2$ be an open set, and let $[\mathbf{x}] : U \to \mathbb{P}^3$ be an immersion. For any 0-adapted frame field $(\mathbf{e}_0(\mathbf{u}), \mathbf{e}_1(\mathbf{u}), \mathbf{e}_2(\mathbf{u}), \mathbf{e}_3(\mathbf{u}))$ along $[\Sigma] = [\mathbf{x}(U)]$, the associated dual and connection forms $(\bar{\omega}^\alpha_\beta)$ have the property that $\bar{\omega}^3_0 = 0$. Moreover, $(\bar{\omega}^1_0, \bar{\omega}^2_0)$ form a basis for the 1-forms on U.*

Differentiating the equation $\bar{\omega}^3_0 = 0$ yields

$$d\bar{\omega}^3_0 = -\bar{\omega}^3_1 \wedge \bar{\omega}^1_0 - \bar{\omega}^3_2 \wedge \bar{\omega}^2_0 = 0,$$

and Cartan's lemma (cf. Lemma 2.49) implies that there exist functions h_{11}, h_{12}, h_{22} on U such that

$$\begin{bmatrix} \bar{\omega}_1^3 \\ \bar{\omega}_2^3 \end{bmatrix} = \begin{bmatrix} h_{11} & h_{12} \\ h_{12} & h_{22} \end{bmatrix} \begin{bmatrix} \bar{\omega}_0^1 \\ \bar{\omega}_0^2 \end{bmatrix}. \tag{7.21}$$

The next step is to investigate how the matrix $[h_{ij}]$ changes if we vary the frame. First, we must identify the group of transformations between 0-adapted frames.

Let $(\mathbf{e}_0(\mathbf{u}), \mathbf{e}_1(\mathbf{u}), \mathbf{e}_2(\mathbf{u}), \mathbf{e}_3(\mathbf{u}))$ be any 0-adapted frame field, with associated Maurer-Cartan forms $(\bar{\omega}^\alpha_\beta)$. Any other 0-adapted frame field $(\tilde{\mathbf{e}}_0(\mathbf{u}),\ \tilde{\mathbf{e}}_1(\mathbf{u}),\ \tilde{\mathbf{e}}_2(\mathbf{u}),\ \tilde{\mathbf{e}}_3(\mathbf{u}))$ must have the properties that

$$\operatorname{span}(\tilde{\mathbf{e}}_0(\mathbf{u})) = \operatorname{span}(\mathbf{e}_0(\mathbf{u})),$$
$$\operatorname{span}(\tilde{\mathbf{e}}_0(\mathbf{u}), \tilde{\mathbf{e}}_1(\mathbf{u}), \tilde{\mathbf{e}}_2(\mathbf{u})) = \operatorname{span}(\mathbf{e}_0(\mathbf{u}), \mathbf{e}_1(\mathbf{u}), \mathbf{e}_2(\mathbf{u})),$$

and

$$\det\left[\tilde{\mathbf{e}}_0(\mathbf{u})\ \tilde{\mathbf{e}}_1(\mathbf{u})\ \tilde{\mathbf{e}}_2(\mathbf{u})\ \tilde{\mathbf{e}}_3(\mathbf{u})\right] = \det\left[\mathbf{e}_0(\mathbf{u})\ \mathbf{e}_1(\mathbf{u})\ \mathbf{e}_2(\mathbf{u})\ \mathbf{e}_3(\mathbf{u})\right] = 1.$$

This will be true if and only if

$$\begin{aligned} &\left[\tilde{\mathbf{e}}_0(\mathbf{u})\ \tilde{\mathbf{e}}_1(\mathbf{u})\ \tilde{\mathbf{e}}_2(\mathbf{u})\ \tilde{\mathbf{e}}_3(\mathbf{u})\right] \\ &\quad = \left[\mathbf{e}_0(\mathbf{u})\ \mathbf{e}_1(\mathbf{u})\ \mathbf{e}_2(\mathbf{u})\ \mathbf{e}_3(\mathbf{u})\right] \begin{bmatrix} \lambda & r_1 & r_2 & s_0 \\ 0 & & & s_1 \\ & & B & \\ 0 & & & s_2 \\ 0 & 0 & 0 & (\lambda \det B)^{-1} \end{bmatrix} \end{aligned} \tag{7.22}$$

for some $GL(2)$-valued function B and real-valued functions $\lambda, r_1, r_2, s_0, s_1, s_2$ on U, with $\lambda \neq 0$. Let $(\tilde{\bar{\omega}}^\alpha_\beta)$ be the Maurer-Cartan forms associated to the new frame field.

The computations in the following exercise should feel familiar from the equi-affine case:

***Exercise 7.34** (Cf. Exercise 6.20). (a) Show that

$$\begin{bmatrix} \tilde{\bar{\omega}}_0^1 \\ \tilde{\bar{\omega}}_0^2 \end{bmatrix} = \lambda B^{-1} \begin{bmatrix} \bar{\omega}_0^1 \\ \bar{\omega}_0^2 \end{bmatrix}.$$

(b) Show that

$$\begin{bmatrix} \tilde{\bar{\omega}}_1^3 \\ \tilde{\bar{\omega}}_2^3 \end{bmatrix} = \lambda(\det B)\, {}^t\!B \begin{bmatrix} \bar{\omega}_1^3 \\ \bar{\omega}_2^3 \end{bmatrix}.$$

(c) Cartan's lemma implies that there exist functions $\tilde{h}_{11}, \tilde{h}_{12}, \tilde{h}_{22}$ on U such that

$$\begin{bmatrix} \tilde{\omega}_1^3 \\ \tilde{\omega}_2^3 \end{bmatrix} = \begin{bmatrix} \tilde{h}_{11} & \tilde{h}_{12} \\ \tilde{h}_{12} & \tilde{h}_{22} \end{bmatrix} \begin{bmatrix} \tilde{\omega}_0^1 \\ \tilde{\omega}_0^2 \end{bmatrix}.$$

Show that

$$\begin{bmatrix} \tilde{h}_{11} & \tilde{h}_{12} \\ \tilde{h}_{12} & \tilde{h}_{22} \end{bmatrix} = (\det B)\, {}^t B \begin{bmatrix} h_{11} & h_{12} \\ h_{12} & h_{22} \end{bmatrix} B. \tag{7.23}$$

The transformation (7.23) is precisely the same as in the equi-affine case, so we use the same terminology here:

Definition 7.35. Assume that the matrix $[h_{ij}]$ is nonsingular at every point of U. The surface $[\Sigma] = [\mathbf{x}(U)]$ is called

(1) *elliptic* if $\det[h_{ij}] > 0$ at every point of U;

(2) *hyperbolic* if $\det[h_{ij}] < 0$ at every point of U.

For the remainder of this section, we will assume that $[\Sigma]$ is elliptic. The hyperbolic case was treated in detail by Cartan in [**Car20**]; we will explore this case in Exercise 7.50.

The transformation (7.23) acts transitively on the set of 2×2 matrices with positive determinant; therefore, there exists a choice of 0-adapted frame field $(\mathbf{e}_0(\mathbf{u}), \mathbf{e}_1(\mathbf{u}), \mathbf{e}_2(\mathbf{u}), \mathbf{e}_3(\mathbf{u}))$ for which

$$\begin{bmatrix} h_{11} & h_{12} \\ h_{12} & h_{22} \end{bmatrix} = \begin{bmatrix} 1 & 0 \\ 0 & 1 \end{bmatrix}.$$

Such a frame field will be called 1-*adapted.*

***Exercise 7.36.** Let $(\mathbf{e}_0(\mathbf{u}), \mathbf{e}_1(\mathbf{u}), \mathbf{e}_2(\mathbf{u}), \mathbf{e}_3(\mathbf{u}))$ be any 1-adapted frame field for an elliptic projective surface $[\Sigma] = [\mathbf{x}(U)] \subset \mathbb{P}^3$.

(a) Show that any other 1-adapted frame field $(\tilde{\mathbf{e}}_0(\mathbf{u}), \tilde{\mathbf{e}}_1(\mathbf{u}), \tilde{\mathbf{e}}_2(\mathbf{u}), \tilde{\mathbf{e}}_3(\mathbf{u}))$ for $[\Sigma]$ must have the form

$$\begin{aligned} &\begin{bmatrix} \tilde{\mathbf{e}}_0(\mathbf{u}) & \tilde{\mathbf{e}}_1(\mathbf{u}) & \tilde{\mathbf{e}}_2(\mathbf{u}) & \tilde{\mathbf{e}}_3(\mathbf{u}) \end{bmatrix} \\ &= \begin{bmatrix} \mathbf{e}_0(\mathbf{u}) & \mathbf{e}_1(\mathbf{u}) & \mathbf{e}_2(\mathbf{u}) & \mathbf{e}_3(\mathbf{u}) \end{bmatrix} \begin{bmatrix} \lambda & r_1 & r_2 & s_0 \\ 0 & \multicolumn{2}{c}{\multirow{2}{*}{B}} & s_1 \\ 0 & & & s_2 \\ 0 & 0 & 0 & (\lambda \det B)^{-1} \end{bmatrix} \end{aligned} \tag{7.24}$$

for some $SO(2)$-valued function B and real-valued functions $\lambda, r_1, r_2, s_0, s_1, s_2$ on U, with $\lambda \neq 0$.

(b) Let $(\bar{\omega}^\alpha_\beta)$ be the Maurer-Cartan forms associated to a 1-adapted frame field $(\mathbf{e}_0(\mathbf{u}), \mathbf{e}_1(\mathbf{u}), \mathbf{e}_2(\mathbf{u}), \mathbf{e}_3(\mathbf{u}))$, and let $(\tilde{\bar{\omega}}^\alpha_\beta)$ be the Maurer-Cartan forms associated to the 1-adapted frame field (7.24). Show that

$$\tilde{\bar{\omega}}^3_1 \tilde{\bar{\omega}}^1_0 + \tilde{\bar{\omega}}^3_2 \tilde{\bar{\omega}}^2_0 = (\tilde{\bar{\omega}}^1_0)^2 + (\tilde{\bar{\omega}}^2_0)^2 = \lambda^2 \left((\bar{\omega}^1_0)^2 + (\bar{\omega}^2_0)^2\right) = \lambda^2 \left(\bar{\omega}^3_1 \bar{\omega}^1_0 + \bar{\omega}^3_2 \bar{\omega}^2_0\right).$$

(Cf. Exercise 7.34.)

Note that the result of Exercise 7.36(b) is slightly different from the equi-affine case. Now the quadratic form

$$\mathrm{I} = \bar{\omega}^3_1 \bar{\omega}^1_0 + \bar{\omega}^3_2 \bar{\omega}^2_0 = (\bar{\omega}^1_0)^2 + (\bar{\omega}^2_0)^2 \tag{7.25}$$

is well-defined only up to a scalar multiple. Rather than defining a metric, it defines a *conformal structure* on $[\Sigma]$, in which angles between tangent vectors are well-defined, but lengths of tangent vectors are not.

Exercise 7.37. Show that the conformal structure I is invariant (up to a scalar multiple) under the action of $SL(4)$.

In order to continue making further adaptations, our next step is to differentiate the equations

$$\bar{\omega}^3_1 = \bar{\omega}^1_0, \qquad \bar{\omega}^3_2 = \bar{\omega}^2_0, \tag{7.26}$$

which hold for the Maurer-Cartan forms associated to any 1-adapted frame field.

***Exercise 7.38.** Differentiate the equations (7.26) and use Cartan's lemma to conclude that there exist functions h_{111}, h_{112}, h_{122}, h_{222} on U such that

$$\begin{bmatrix} 2\bar{\omega}^1_1 - (\bar{\omega}^0_0 + \bar{\omega}^3_3) \\ \bar{\omega}^1_2 + \bar{\omega}^2_1 \\ 2\bar{\omega}^2_2 - (\bar{\omega}^0_0 + \bar{\omega}^3_3) \end{bmatrix} = \begin{bmatrix} h_{111} & h_{112} \\ h_{112} & h_{122} \\ h_{122} & h_{222} \end{bmatrix} \begin{bmatrix} \bar{\omega}^1_0 \\ \bar{\omega}^2_0 \end{bmatrix}. \tag{7.27}$$

Next, we need to compute how the functions (h_{ijk}) vary under a transformation of the form (7.24). This computation gets rather complicated, but we can make it simpler by breaking it down into two steps.

***Exercise 7.39.** Consider transformations of the form (7.24) with B equal to the identity matrix and $\lambda = 1$, so that

$$\left[\tilde{\mathbf{e}}_0(\mathbf{u})\ \tilde{\mathbf{e}}_1(\mathbf{u})\ \tilde{\mathbf{e}}_2(\mathbf{u})\ \tilde{\mathbf{e}}_3(\mathbf{u})\right] = \left[\mathbf{e}_0(\mathbf{u})\ \mathbf{e}_1(\mathbf{u})\ \mathbf{e}_2(\mathbf{u})\ \mathbf{e}_3(\mathbf{u})\right] \begin{bmatrix} 1 & r_1 & r_2 & s_0 \\ 0 & 1 & 0 & s_1 \\ 0 & 0 & 1 & s_2 \\ 0 & 0 & 0 & 1 \end{bmatrix}.$$

Show that under such a transformation, we have

$$\begin{aligned}\tilde{h}_{111} &= h_{111} + 3(r_1 - s_1),\\ \tilde{h}_{112} &= h_{112} + (r_2 - s_2),\\ \tilde{h}_{122} &= h_{122} + (r_1 - s_1),\\ \tilde{h}_{222} &= h_{222} + 3(r_2 - s_2).\end{aligned}$$

Therefore,

$$\begin{aligned}\tilde{h}_{111} + \tilde{h}_{122} &= h_{111} + h_{122} + 4(r_1 - s_1),\\ \tilde{h}_{112} + \tilde{h}_{222} &= h_{112} + h_{222} + 4(r_2 - s_2),\end{aligned}$$

so there exists a choice of 1-adapted frame field $(\mathbf{e}_0(\mathbf{u}), \mathbf{e}_1(\mathbf{u}), \mathbf{e}_2(\mathbf{u}), \mathbf{e}_3(\mathbf{u}))$ for which

$$h_{111} + h_{122} = h_{112} + h_{222} = 0. \tag{7.28}$$

Any 1-adapted frame field satisfying the condition (7.28) will be called *2-adapted.*

Remark 7.40. This is certainly not the only way that one might choose to normalize these functions! But our experience with the equi-affine case—where a similar relation held for any 2-adapted frame field—suggests that this might turn out to be a nice choice.

***Exercise 7.41.** Let $(\mathbf{e}_0(\mathbf{u}), \mathbf{e}_1(\mathbf{u}), \mathbf{e}_2(\mathbf{u}), \mathbf{e}_3(\mathbf{u}))$ be any 2-adapted frame field for an elliptic projective surface $[\Sigma] = [\mathbf{x}(U)] \subset \mathbb{P}^3$.

(a) Show that any other 2-adapted frame field $(\tilde{\mathbf{e}}_0(\mathbf{u}), \tilde{\mathbf{e}}_1(\mathbf{u}), \tilde{\mathbf{e}}_2(\mathbf{u}), \tilde{\mathbf{e}}_3(\mathbf{u}))$ for $[\Sigma]$ must have the form (7.24), where

$$\begin{bmatrix} r_1 \\ r_2 \end{bmatrix} = \lambda\, {}^tB \begin{bmatrix} s_1 \\ s_2 \end{bmatrix}. \tag{7.29}$$

(b) Show that under a transformation of the form (7.24) satisfying equation (7.29), we have

$$\begin{bmatrix} \tilde{h}_{111} \\ \tilde{h}_{222} \end{bmatrix} = \lambda^{-1} B^3 \begin{bmatrix} h_{111} \\ h_{222} \end{bmatrix}. \tag{7.30}$$

(c) Let $(\bar{\omega}^\alpha_\beta)$ be the Maurer-Cartan forms associated to the 2-adapted frame field $(\mathbf{e}_0(\mathbf{u}), \mathbf{e}_1(\mathbf{u}), \mathbf{e}_2(\mathbf{u}), \mathbf{e}_3(\mathbf{u}))$, and let $(\tilde{\bar{\omega}}^\alpha_\beta)$ be the Maurer-Cartan forms associated to the 2-adapted frame field $(\tilde{\mathbf{e}}_0(\mathbf{u}), \tilde{\mathbf{e}}_1(\mathbf{u}), \tilde{\mathbf{e}}_2(\mathbf{u}), \tilde{\mathbf{e}}_3(\mathbf{u}))$. Show that

$$\tilde{h}_{111}\left((\tilde{\bar{\omega}}_0^1)^3 - 3\tilde{\bar{\omega}}_0^1(\tilde{\bar{\omega}}_0^2)^2\right) + \tilde{h}_{222}\left((\tilde{\bar{\omega}}_0^2)^3 - 3(\tilde{\bar{\omega}}_0^1)^2\tilde{\bar{\omega}}_0^2\right)$$
$$= \lambda^2\left[h_{111}\left((\bar{\omega}_0^1)^3 - 3\bar{\omega}_0^1(\bar{\omega}_0^2)^2\right) + h_{222}\left((\bar{\omega}_0^2)^3 - 3(\bar{\omega}_0^1)^2\bar{\omega}_0^2\right)\right].$$

Conclude that the cubic form

$$\begin{aligned} P &= h_{ijk}\,\bar{\omega}_0^i\,\bar{\omega}_0^j\,\bar{\omega}_0^k \\ &= h_{111}(\bar{\omega}_0^1)^3 + 3h_{112}(\bar{\omega}_0^1)^2\bar{\omega}_0^2 + 3h_{122}\bar{\omega}_0^1(\bar{\omega}_0^2)^2 + h_{222}(\bar{\omega}_0^2)^3 \qquad (7.31) \\ &= h_{111}\left((\bar{\omega}_0^1)^3 - 3\,\bar{\omega}_0^1(\bar{\omega}_0^2)^2\right) + h_{222}\left((\bar{\omega}_0^2)^3 - 3(\bar{\omega}_0^1)^2\bar{\omega}_0^2\right) \end{aligned}$$

is well-defined up to a scalar multiple, independent of the choice of a particular 2-adapted frame field on $[\Sigma]$. (P is the projective analog of the Fubini-Pick form for surfaces in equi-affine space.)

According to (7.30), under a transformation from one 2-adapted frame to another, the vector ${}^t\!\left[h_{111}\ h_{222}\right]$ transforms by a rotation and a scaling. This vector is an example of a *relative invariant*: If it vanishes for any 2-adapted frame based at a point $\mathbf{u} \in U$, then it vanishes for every 2-adapted frame based at $\mathbf{u}$. On the other hand, if this vector is nonzero for some 2-adapted frame based at $\mathbf{u}$, then we can find a 2-adapted frame based at $\mathbf{u}$ for which it is equal to any nonzero vector we choose. Any point $[\mathbf{x}(\mathbf{u})] \in [\Sigma]$ at which $h_{111} = h_{222} = 0$ will be called an *umbilic point* of Σ.

For the remainder of this section, we will assume that the vector ${}^t\!\left[h_{111}\ h_{222}\right]$ is nonzero for every 2-adapted frame based at each point $\mathbf{u} \in U$, i.e., that Σ contains no umbilic points. (We will treat the *totally umbilic* case, where $h_{111} \equiv h_{222} \equiv 0$, in Exercises 7.48 and 7.49.) With this assumption, equation (7.30) implies that there exists a choice of 2-adapted frame field for which

$$h_{111} = 2, \qquad h_{222} = 0.$$

Such a frame field will be called *3-adapted.*

***Exercise 7.42.** (a) Let $(\mathbf{e}_0(\mathbf{u}), \mathbf{e}_1(\mathbf{u}), \mathbf{e}_2(\mathbf{u}), \mathbf{e}_3(\mathbf{u}))$ be any 3-adapted frame field for an elliptic projective surface $[\Sigma] = [\mathbf{x}(U)] \subset \mathbb{P}^3$ with no umbilic points. Show that any other 3-adapted frame field $(\tilde{\mathbf{e}}_0(\mathbf{u}),\ \tilde{\mathbf{e}}_1(\mathbf{u}),\ \tilde{\mathbf{e}}_2(\mathbf{u}),\ \tilde{\mathbf{e}}_3(\mathbf{u}))$ for $[\Sigma]$ must have the form described in Exercise 7.41, where either $\lambda = 1$ and B is one of the three matrices

$$\begin{bmatrix} 1 & 0 \\ 0 & 1 \end{bmatrix}, \qquad \frac{1}{2}\begin{bmatrix} -1 & -\sqrt{3} \\ \sqrt{3} & -1 \end{bmatrix}, \qquad \frac{1}{2}\begin{bmatrix} -1 & \sqrt{3} \\ -\sqrt{3} & -1 \end{bmatrix}$$

or $\lambda = -1$ and B is one of the three matrices

$$\begin{bmatrix} -1 & 0 \\ 0 & -1 \end{bmatrix}, \qquad \frac{1}{2}\begin{bmatrix} 1 & -\sqrt{3} \\ \sqrt{3} & 1 \end{bmatrix}, \qquad \frac{1}{2}\begin{bmatrix} 1 & \sqrt{3} \\ -\sqrt{3} & 1 \end{bmatrix}.$$

(b) Let $(\bar{\omega}^\alpha_\beta)$ be the Maurer-Cartan forms associated to a 3-adapted frame field $(\mathbf{e}_0(\mathbf{u}), \mathbf{e}_1(\mathbf{u}), \mathbf{e}_2(\mathbf{u}), \mathbf{e}_3(\mathbf{u}))$, and let $(\tilde{\bar{\omega}}^\alpha_\beta)$ be the Maurer-Cartan forms associated to any other 3-adapted frame field $(\tilde{\mathbf{e}}_0(\mathbf{u}), \tilde{\mathbf{e}}_1(\mathbf{u}), \tilde{\mathbf{e}}_2(\mathbf{u}), \tilde{\mathbf{e}}_3(\mathbf{u}))$. Show that

$$(\tilde{\bar{\omega}}^1_0)^2 + (\tilde{\bar{\omega}}^2_0)^2 = (\bar{\omega}^1_0)^2 + (\bar{\omega}^2_0)^2.$$

Therefore, the conformal structure (7.25) associated to a 3-adapted frame field provides a well-defined metric for an elliptic surface in $\mathbb{P}^3$ with no umbilic points.

Definition 7.43. Let $U \subset \mathbb{R}^2$ be an open set, and let $[\mathbf{x}] : U \to \mathbb{P}^3$ be an elliptic immersion with no umbilic points. Let $(\mathbf{e}_0(\mathbf{u}), \mathbf{e}_1(\mathbf{u}), \mathbf{e}_2(\mathbf{u}), \mathbf{e}_3(\mathbf{u}))$ be any 3-adapted frame field along $[\Sigma] = [\mathbf{x}(U)]$, with associated Maurer-Cartan forms $(\bar{\omega}^\alpha_\beta)$. The quadratic form

$$\mathrm{I} = (\bar{\omega}^1_0)^2 + (\bar{\omega}^2_0)^2$$

is called the *projective first fundamental form* of $[\Sigma]$.

Remark 7.44. For a 3-adapted frame field, the cubic form (7.31) becomes

$$P = 2\left((\bar{\omega}^1_0)^3 - 3\,\bar{\omega}^1_0(\bar{\omega}^2_0)^2\right).$$

The 3-adapted frames have the property that the tangent vector $\mathbf{e}_2(\mathbf{u})$ is one of the three null directions for P at the point $[\mathbf{x}(\mathbf{u})] \in [\Sigma]$. These directions are called the *Darboux tangents* at the point $[\mathbf{x}(\mathbf{u})] \in [\Sigma]$; for more details, see [**Su83**].

For simplicity, we will restrict our attention to transformations between 3-adapted frames with B equal to the identity matrix; thus, we will assume that any two 3-adapted frame fields vary by a transformation of the form (7.32)

$$\begin{bmatrix}\tilde{\mathbf{e}}_0(\mathbf{u}) & \tilde{\mathbf{e}}_1(\mathbf{u}) & \tilde{\mathbf{e}}_2(\mathbf{u}) & \tilde{\mathbf{e}}_3(\mathbf{u})\end{bmatrix} = \begin{bmatrix}\mathbf{e}_0(\mathbf{u}) & \mathbf{e}_1(\mathbf{u}) & \mathbf{e}_2(\mathbf{u}) & \mathbf{e}_3(\mathbf{u})\end{bmatrix} \begin{bmatrix} 1 & s_1 & s_2 & s_0 \\ 0 & 1 & 0 & s_1 \\ 0 & 0 & 1 & s_2 \\ 0 & 0 & 0 & 1 \end{bmatrix}.$$

***Exercise 7.45.** (a) Show that the Maurer-Cartan forms associated to any 3-adapted frame field satisfy the equations

$$\bar{\omega}^0_0 + \bar{\omega}^3_3 = \bar{\omega}^1_1 + \bar{\omega}^2_2 = 0.$$

(Hint: Apply the result of Exercise 7.38, keeping in mind condition (7.28), and recall the defining condition for the Lie algebra $\mathfrak{sl}(4)$.)

(b) Differentiate the equation

$$\bar{\omega}_0^0 + \bar{\omega}_3^3 = 0$$

and use Cartan's lemma to conclude that there exist functions $\ell_{11}, \ell_{12}, \ell_{22}$ on U such that

$$\begin{bmatrix} \bar{\omega}_1^0 - \bar{\omega}_3^1 \\ \bar{\omega}_2^0 - \bar{\omega}_3^2 \end{bmatrix} = \begin{bmatrix} \ell_{11} & \ell_{12} \\ \ell_{12} & \ell_{22} \end{bmatrix} \begin{bmatrix} \bar{\omega}_0^1 \\ \bar{\omega}_0^2 \end{bmatrix}. \tag{7.33}$$

***Exercise 7.46.** (a) Show that under a transformation of the form (7.32), we have

$$\begin{aligned}
\tilde{\ell}_{11} &= \ell_{11} + s_1^2 + s_2^2 - 2s_0 - 2s_1, \\
\tilde{\ell}_{12} &= \ell_{12} + 2s_2, \\
\tilde{\ell}_{22} &= \ell_{22} + s_1^2 + s_2^2 - 2s_0 + 2s_1.
\end{aligned}$$

(b) Conclude that there exists a choice of 3-adapted frame field for which

$$\ell_{11} = \ell_{12} = \ell_{22} = 0.$$

Any frame field satisfying this condition will be called 4-*adapted.*

(c) Show that any two 4-adapted frame fields vary by a transformation of the form

$$\begin{bmatrix} \tilde{\mathbf{e}}_0(\mathbf{u}) & \tilde{\mathbf{e}}_1(\mathbf{u}) & \tilde{\mathbf{e}}_2(\mathbf{u}) & \tilde{\mathbf{e}}_3(\mathbf{u}) \end{bmatrix} = \begin{bmatrix} \mathbf{e}_0(\mathbf{u}) & \mathbf{e}_1(\mathbf{u}) & \mathbf{e}_2(\mathbf{u}) & \mathbf{e}_3(\mathbf{u}) \end{bmatrix} \begin{bmatrix} \lambda & 0 \quad 0 & 0 \\ 0 & & 0 \\ & B & \\ 0 & & 0 \\ 0 & 0 \quad 0 & \lambda^{-1} \end{bmatrix},$$

where (λ, B) are one of the pairs from Exercise 7.42. Therefore, the 4-adapted frames at each point form a discrete set, and a continuous 4-adapted frame field along $[\Sigma]$ is uniquely determined by its value at any point $\mathbf{u} \in U$.

***Exercise 7.47.** (a) Show that the Maurer-Cartan forms associated to a 4-adapted frame field satisfy the equations

$$\begin{gathered}
\bar{\omega}_1^1 = \bar{\omega}_0^1, \qquad \bar{\omega}_2^2 = -\bar{\omega}_0^1, \qquad \bar{\omega}_2^1 + \bar{\omega}_1^2 = -2\bar{\omega}_0^2, \\
\bar{\omega}_1^0 = \bar{\omega}_3^1, \qquad \bar{\omega}_2^0 = \bar{\omega}_3^2.
\end{gathered} \tag{7.34}$$

(b) Differentiate the equations in part (a) and use Cartan's lemma to conclude that there exist functions $p_1, p_2, q_{11}, q_{12}, q_{21}, q_{22}$ such that

$$\begin{aligned}
\bar{\omega}_2^1 &= p_1\bar{\omega}_0^1 + (p_2 - 1)\bar{\omega}_0^2, \\
\bar{\omega}_1^2 &= -p_1\bar{\omega}_0^1 - (p_2 + 1)\bar{\omega}_0^2, \\
\bar{\omega}_3^1 &= q_{11}\bar{\omega}_0^1 + q_{12}\bar{\omega}_0^2, \\
\bar{\omega}_3^2 &= q_{21}\bar{\omega}_0^1 + q_{22}\bar{\omega}_0^2, \\
\bar{\omega}_0^0 = -\bar{\omega}_3^3 &= 3p_2\bar{\omega}_0^1 - 3p_1\bar{\omega}_0^2, \\
\bar{\omega}_3^0 &= (q_{11} - q_{22})\bar{\omega}_0^1 - (q_{12} + q_{21})\bar{\omega}_0^2.
\end{aligned} \tag{7.35}$$

The functions $p_1, p_2, q_{11}, q_{12}, q_{21}, q_{22}$ are the fundamental invariants for elliptic surfaces in $\mathbb{P}^3$ with no umbilic points.

In the next two exercises, we consider the totally umbilic case.

Exercise 7.48 (Projective spheres, Part 1). Let $(\mathbf{e}_0(\mathbf{u}), \mathbf{e}_1(\mathbf{u}), \mathbf{e}_2(\mathbf{u}), \mathbf{e}_3(\mathbf{u}))$ be any 2-adapted frame field for an elliptic projective surface $[\Sigma] = [\mathbf{x}(U)] \subset \mathbb{P}^3$, and suppose that $[\Sigma]$ is totally umbilic, i.e., that $h_{111} = h_{222} = 0$.

(a) Use the result of Exercise 7.38 to show that the Maurer-Cartan forms associated to this frame field satisfy

$$\bar{\omega}_1^1 = \bar{\omega}_2^2 = \bar{\omega}_2^1 + \bar{\omega}_1^2 = \bar{\omega}_0^0 + \bar{\omega}_3^3 = 0. \tag{7.36}$$

(b) Differentiate equations (7.36) to obtain

$$\begin{aligned}
(\bar{\omega}_1^0 - \bar{\omega}_3^1) \wedge \bar{\omega}_0^1 &= 0, \\
(\bar{\omega}_2^0 - \bar{\omega}_3^2) \wedge \bar{\omega}_0^2 &= 0, \\
(\bar{\omega}_2^0 - \bar{\omega}_3^2) \wedge \bar{\omega}_0^1 + (\bar{\omega}_1^0 - \bar{\omega}_3^1) \wedge \bar{\omega}_0^2 &= 0.
\end{aligned}$$

Use Cartan's lemma to conclude that there exists a function σ such that

$$\begin{aligned}
\bar{\omega}_1^0 - \bar{\omega}_3^1 &= \sigma\bar{\omega}_0^1, \\
\bar{\omega}_2^0 - \bar{\omega}_3^2 &= \sigma\bar{\omega}_0^2.
\end{aligned}$$

(c) Show that under a transformation of the form

$$\begin{bmatrix}\tilde{\mathbf{e}}_0(\mathbf{u}) & \tilde{\mathbf{e}}_1(\mathbf{u}) & \tilde{\mathbf{e}}_2(\mathbf{u}) & \tilde{\mathbf{e}}_3(\mathbf{u})\end{bmatrix} = \begin{bmatrix}\mathbf{e}_0(\mathbf{u}) & \mathbf{e}_1(\mathbf{u}) & \mathbf{e}_2(\mathbf{u}) & \mathbf{e}_3(\mathbf{u})\end{bmatrix} \begin{bmatrix} 1 & 0 & 0 & s_0 \\ 0 & 1 & 0 & 0 \\ 0 & 0 & 1 & 0 \\ 0 & 0 & 0 & 1 \end{bmatrix},$$

we have

$$\tilde{\sigma} = \sigma - 2s_0.$$

Conclude that there exists a 2-adapted frame field with $\sigma = 0$ and that for such a frame field we have

$$\bar{\omega}^1_3 = \bar{\omega}^0_1, \qquad \bar{\omega}^2_3 = \bar{\omega}^0_2. \tag{7.37}$$

We will call such a frame field *fully adapted.*

(d) Show that any two fully adapted frame fields differ by a transformation of the form

$$\begin{aligned} &[\tilde{\mathbf{e}}_0(\mathbf{u})\ \tilde{\mathbf{e}}_1(\mathbf{u})\ \tilde{\mathbf{e}}_2(\mathbf{u})\ \tilde{\mathbf{e}}_3(\mathbf{u})] \\ &\quad = [\mathbf{e}_0(\mathbf{u})\ \mathbf{e}_1(\mathbf{u})\ \mathbf{e}_2(\mathbf{u})\ \mathbf{e}_3(\mathbf{u})] \begin{bmatrix} \lambda & r_1 & r_2 & s_0 \\ 0 & \multicolumn{2}{c}{\multirow{2}{*}{B}} & s_1 \\ 0 & & & s_2 \\ 0 & 0 & 0 & \lambda^{-1} \end{bmatrix}, \end{aligned} \tag{7.38}$$

where

$$B \in SO(2), \qquad [r_1 \ \ r_2] = \lambda[s_1 \ \ s_2]B, \qquad s_0 = \frac{1}{2}\lambda(s_1^2 + s_2^2).$$

(e) Differentiate equations (7.37) to obtain

$$\begin{aligned} 2\bar{\omega}^0_3 \wedge \bar{\omega}^1_0 &= 0, \\ 2\bar{\omega}^0_3 \wedge \bar{\omega}^2_0 &= 0. \end{aligned}$$

Use Cartan's lemma to conclude that $\bar{\omega}^0_3 = 0$.

(f) Show that differentiating the equation $\bar{\omega}^0_3 = 0$ yields an identity.

At this point, the Maurer-Cartan form for the bundle of fully adapted frames over $[\Sigma]$ is

$$\bar{\omega} = \begin{bmatrix} \bar{\omega}^0_0 & \bar{\omega}^0_1 & \bar{\omega}^0_2 & 0 \\ \bar{\omega}^1_0 & 0 & \bar{\omega}^1_2 & \bar{\omega}^0_1 \\ \bar{\omega}^2_0 & -\bar{\omega}^1_2 & 0 & \bar{\omega}^0_2 \\ 0 & \bar{\omega}^1_0 & \bar{\omega}^2_0 & -\bar{\omega}^0_0 \end{bmatrix}.$$

We have not yet found a unique frame over each point of $[\Sigma]$, but since differentiating the structure equations yields no further relations, this is as far as the frame bundle can be reduced. This means that the bundle of fully adapted frames over $[\Sigma]$ is *itself* a Lie group G whose Lie algebra $\mathfrak{g}$ is the space of matrices with the symmetries of the Maurer-Cartan form above and that $[\Sigma]$ is a homogeneous space of the form G/H, where H is the transformation group in equation (7.38). All that remains is to identify the group G and the quotient space G/H.

Exercise 7.49 (Projective spheres, Part 2). Let Q be the matrix

$$Q = \begin{bmatrix} 0 & 0 & 0 & -1 \\ 0 & 1 & 0 & 0 \\ 0 & 0 & 1 & 0 \\ -1 & 0 & 0 & 0 \end{bmatrix}.$$

Q represents the quadratic form

$$Q(\mathbf{x}) = (x^1)^2 + (x^2)^2 - 2x^0x^3,$$

which is an indefinite quadratic form of signature $(3, 1)$ on $\mathbb{R}^4$. The Lie group $SO(Q)$ is defined to be the group of matrices of determinant 1 that preserve this quadratic form; i.e.,

$$SO(Q) = \{A \in SL(4) \mid {}^tAQA = Q\}.$$

The group $SO(Q)$ is isomorphic to the Lie group $SO(3, 1)$.

(a) Show that the Lie algebra $\mathfrak{so}(Q)$ is defined by

$$\mathfrak{so}(Q) = \{B \in \mathfrak{sl}(4) \mid {}^tBQ + QB = 0\}.$$

(Hint: Let $A(t)$ be a curve in $SO(Q)$ with $A(0) = I$. Differentiate the equation

$${}^tA(t)QA(t) = Q$$

and evaluate at $t = 0$.)

(b) Show that $\mathfrak{so}(Q)$ consists of all matrices of the form

$$B = \begin{bmatrix} a_0^0 & a_1^0 & a_2^0 & 0 \\ a_0^1 & 0 & a_2^1 & a_1^0 \\ a_0^2 & -a_2^1 & 0 & a_2^0 \\ 0 & a_0^1 & a_0^2 & -a_0^0 \end{bmatrix}.$$

Conclude that the reduced Maurer-Cartan form on the bundle of fully adapted frames in Exercise 7.48 takes values in $\mathfrak{so}(Q)$.

(c) Recall that for any fully adapted frame field

$$\tilde{\mathbf{x}}(\mathbf{u}) = (\mathbf{e}_0(\mathbf{u}), \mathbf{e}_1(\mathbf{u}), \mathbf{e}_2(\mathbf{u}), \mathbf{e}_3(\mathbf{u}))$$

as in Exercise 7.48, the pullback of the Maurer-Cartan form $\bar{\omega}$ to U via $\tilde{\mathbf{x}}$ is given by $\tilde{\mathbf{x}}^*(\bar{\omega}) = \tilde{\mathbf{x}}^{-1}d\tilde{\mathbf{x}}$. Use this fact to show that any fully adapted frame field has the form

$$\tilde{\mathbf{x}}(\mathbf{u}) = A_0A(\mathbf{u}),$$

where $A_0 \in SL(4)$ is a constant matrix and $A(\mathbf{u}) \in SO(Q)$ for all $\mathbf{u} \in U$.

Since $A_0 \in SL(4)$, the frame field $\tilde{\mathbf{x}}(\mathbf{u})$ is projectively equivalent to the frame field $A_0^{-1}\tilde{\mathbf{x}}(\mathbf{u})$. (This equivalence replaces the surface

$$[\Sigma] = [\mathbf{x}(U)] = [\mathbf{e}_0(U)] \subset \mathbb{P}^3$$

with the projectively equivalent surface

$$[A_0^{-1}\Sigma] = [A_0^{-1}\mathbf{x}(U)] = [A_0^{-1}\mathbf{e}_0(U)] \subset \mathbb{P}^3.)$$

Therefore, up to projective equivalence, we can assume that $\tilde{\mathbf{x}}(\mathbf{u}) \in SO(Q)$ for all $\mathbf{u} \in U$.

(d) Show that for any matrix $\begin{bmatrix}\mathbf{e}_0 & \mathbf{e}_1 & \mathbf{e}_2 & \mathbf{e}_3\end{bmatrix} \in SO(Q)$, the vector $\mathbf{e}_0$ is a null vector for Q; i.e., $Q(\mathbf{e}_0) = 0$. Therefore, the surface $\mathbf{x}(U) = \mathbf{e}_0(U) \subset \mathbb{R}^4 \setminus \{\mathbf{0}\}$ must be contained in the 3-dimensional hypersurface $S \subset \mathbb{R}^4 \setminus \{\mathbf{0}\}$ defined by the equation $Q(\mathbf{x}) = 0$.

(e) Show that S consists of all lines through the origin passing through all points of the form

$$\mathbf{x} = {}^t[1,\, x^1,\, x^2,\, x^3]$$

with $(x^1)^2 + (x^2)^2 + (x^3)^2 = 1$. Thus, the image of S under the quotient map

$$\mathbb{R}^4 \setminus \{\mathbf{0}\} \to \mathbb{P}^3$$

may be regarded as a sphere in $\mathbb{P}^3$, and $[\Sigma] = [\mathbf{x}(U)]$ must be an open subset of this sphere.

Exercise 7.50. In this exercise, we explore the frame adaptation process for hyperbolic projective surfaces. Let $[\mathbf{x}] : U \to \mathbb{P}^3$ be a smooth immersion whose image is a hyperbolic projective surface $[\Sigma]$. Then the matrix $[h_{ij}]$ in (7.21) has $\det[h_{ij}] < 0$.

(a) Show that there exists a 0-adapted frame field $(\mathbf{f}_0(\mathbf{u}), \mathbf{f}_1(\mathbf{u}), \mathbf{f}_2(\mathbf{u}), \mathbf{f}_3(\mathbf{u}))$ along Σ for which

$$\begin{bmatrix} h_{11} & h_{12} \\ h_{12} & h_{22} \end{bmatrix} = \begin{bmatrix} 0 & 1 \\ 1 & 0 \end{bmatrix}.$$

Such a frame field will be called a 1-*adapted null frame field* along $[\Sigma]$. The associated Maurer-Cartan forms $(\bar{\eta}^\alpha_\beta)$ satisfy

$$\bar{\eta}^3_1 = \bar{\eta}^2_0, \qquad \bar{\eta}^3_2 = \bar{\eta}^1_0, \tag{7.39}$$

and the conformal structure on $[\Sigma]$ is the indefinite quadratic form

$$\mathrm{I} = \bar{\eta}^3_1\,\bar{\eta}^1_0 + \bar{\eta}^3_2\,\bar{\eta}^2_0 = 2\bar{\eta}^1_0\,\bar{\eta}^2_0.$$

(b) Let $(\mathbf{f}_0(\mathbf{u}), \mathbf{f}_1(\mathbf{u}), \mathbf{f}_2(\mathbf{u}), \mathbf{f}_3(\mathbf{u}))$ be any 1-adapted null frame field for a hyperbolic projective surface $[\Sigma] = [\mathbf{x}(U)] \subset \mathbb{P}^3$. Show that any other 1-adapted null frame field $(\tilde{\mathbf{f}}_0(\mathbf{u}), \tilde{\mathbf{f}}_1(\mathbf{u}), \tilde{\mathbf{f}}_2(\mathbf{u}), \tilde{\mathbf{f}}_3(\mathbf{u}))$ for Σ must have the form (7.40)

$$\begin{bmatrix} \tilde{\mathbf{f}}_0(\mathbf{u}) & \tilde{\mathbf{f}}_1(\mathbf{u}) & \tilde{\mathbf{f}}_2(\mathbf{u}) & \tilde{\mathbf{f}}_3(\mathbf{u}) \end{bmatrix} = \begin{bmatrix} \mathbf{f}_0(\mathbf{u}) & \mathbf{f}_1(\mathbf{u}) & \mathbf{f}_2(\mathbf{u}) & \mathbf{f}_3(\mathbf{u}) \end{bmatrix} \begin{bmatrix} \lambda & r_1 & r_2 & s_0 \\ 0 & e^{\theta} & 0 & s_1 \\ 0 & 0 & e^{-\theta} & s_2 \\ 0 & 0 & 0 & \lambda^{-1} \end{bmatrix}$$

for some real-valued functions $\lambda, \theta, r_1, r_2, s_0, s_1, s_2$ on U, with $\lambda \neq 0$.

(c) Differentiate equations (7.39) and use Cartan's lemma to conclude that there exist functions h_{111}, h_{112}, h_{122}, h_{222} on U such that

$$\begin{bmatrix} 2\bar{\eta}_1^2 \\ (\bar{\eta}_1^1 + \bar{\eta}_2^2) - (\bar{\eta}_0^0 + \bar{\eta}_3^3) \\ 2\bar{\eta}_2^1 \end{bmatrix} = \begin{bmatrix} h_{111} & h_{112} \\ h_{112} & h_{122} \\ h_{122} & h_{222} \end{bmatrix} \begin{bmatrix} \bar{\eta}_0^1 \\ \bar{\eta}_0^2 \end{bmatrix}.$$

(d) Show that under a transformation of the form (7.40), we have

$$\begin{aligned} \tilde{h}_{111} &= \lambda^{-1} e^{3\theta} h_{111}, \\ \tilde{h}_{112} &= \lambda^{-1} e^{\theta} h_{112} + 2(\lambda^{-1} r_1 - e^{\theta} s_2), \\ \tilde{h}_{122} &= \lambda^{-1} e^{-\theta} h_{122} + 2(\lambda^{-1} r_2 - e^{-\theta} s_1), \\ \tilde{h}_{222} &= \lambda^{-1} e^{-3\theta} h_{222}. \end{aligned} \tag{7.41}$$

Therefore, there exists a choice of 1-adapted null frame field $(\mathbf{f}_0(\mathbf{u}), \mathbf{f}_1(\mathbf{u}), \mathbf{f}_2(\mathbf{u}), \mathbf{f}_3(\mathbf{u}))$ for which

$$h_{112} = h_{122} = 0.$$

Any 1-adapted null frame field satisfying this condition will be called a *2-adapted null frame field.*

(e) Show that any other 2-adapted null frame field $(\tilde{\mathbf{f}}_0(\mathbf{u}), \tilde{\mathbf{f}}_1(\mathbf{u}), \tilde{\mathbf{f}}_2(\mathbf{u}), \tilde{\mathbf{f}}_3(\mathbf{u}))$ for $[\Sigma]$ must have the form (7.40), where

$$r_1 = \lambda e^{\theta} s_2, \qquad r_2 = \lambda e^{-\theta} s_1. \tag{7.42}$$

Equations (7.41) imply that the functions h_{111}, h_{222} are relative invariants; for the remainder of this exercise, we will assume that both h_{111}, h_{222} are nonzero. (Cartan showed in [**Car20**] that if either of these functions is identically zero, then $[\Sigma]$ is a ruled surface in $\mathbb{P}^3$.) Equations (7.41) then imply that there exists a choice of 2-adapted null frame field for which

$$h_{111} = h_{222} = 2.$$

(This may require a slight extension of the transformation (7.40) to include transformations of the form

$$\mathbf{f}_\alpha(\mathbf{u}) \to -\mathbf{f}_\alpha(\mathbf{u})$$

for an even number of indices α.) Any 2-adapted null frame field satisfying this condition will be called a 3-*adapted null frame field.*

(f) Show that, up to signs, any two 3-adapted null frame fields vary by a transformation of the form

$$(7.43)\quad \left[\tilde{\mathbf{f}}_0(\mathbf{u})\ \tilde{\mathbf{f}}_1(\mathbf{u})\ \tilde{\mathbf{f}}_2(\mathbf{u})\ \tilde{\mathbf{f}}_3(\mathbf{u})\right] = \left[\mathbf{f}_0(\mathbf{u})\ \mathbf{f}_1(\mathbf{u})\ \mathbf{f}_2(\mathbf{u})\ \mathbf{f}_3(\mathbf{u})\right] \begin{bmatrix} 1 & s_2 & s_1 & s_0 \\ 0 & 1 & 0 & s_1 \\ 0 & 0 & 1 & s_2 \\ 0 & 0 & 0 & 1 \end{bmatrix}.$$

(g) Show that the Maurer-Cartan forms associated to any 3-adapted null frame field satisfy the equations

$$\bar{\eta}^0_0 + \bar{\eta}^3_3 = \bar{\eta}^1_1 + \bar{\eta}^2_2 = 0.$$

Differentiate the equation

$$\bar{\eta}^0_0 + \bar{\eta}^3_3 = 0$$

and use Cartan's lemma to conclude that there exist functions $\ell_{11}, \ell_{12}, \ell_{22}$ on U such that

$$\begin{bmatrix} \bar{\eta}^0_1 - \bar{\eta}^2_3 \\ \bar{\eta}^0_2 - \bar{\eta}^1_3 \end{bmatrix} = \begin{bmatrix} \ell_{11} & \ell_{12} \\ \ell_{12} & \ell_{22} \end{bmatrix} \begin{bmatrix} \bar{\eta}^1_0 \\ \bar{\eta}^2_0 \end{bmatrix}.$$

(h) Show that under a transformation of the form (7.43), we have

$$\begin{aligned} \tilde{\ell}_{11} &= \ell_{11} - 2s_1, \\ \tilde{\ell}_{12} &= \ell_{12} + 2s_1 s_2 - 2s_0, \\ \tilde{\ell}_{22} &= \ell_{22} - 2s_2. \end{aligned}$$

Conclude that there exists a choice of 3-adapted null frame field for which

$$\ell_{11} = \ell_{12} = \ell_{22} = 0.$$

Any frame field satisfying this condition will be called a 4-*adapted null frame field.* The 4-adapted null frames at each point form a discrete set (they are unique up to signs), and a continuous 4-adapted null frame field along $[\Sigma]$ is uniquely determined by its value at any point $\mathbf{u} \in U$.

(i) Show that the Maurer-Cartan forms associated to a 4-adapted null frame field satisfy the equations

$$\bar{\eta}^2_1 = \bar{\eta}^1_0, \qquad \bar{\eta}^1_2 = \bar{\eta}^2_0, \qquad \bar{\eta}^0_1 = \bar{\eta}^2_3, \qquad \bar{\eta}^0_2 = \bar{\eta}^1_3.$$

Differentiate these equations and use Cartan's lemma to conclude that there exist functions $p_1, p_2, q_{11}, q_{12}, q_{21}, q_{22}$ such that

$$\begin{aligned}
\bar\eta^1_1 &= -\bar\eta^2_2 = p_1\bar\eta^1_0 + p_2\bar\eta^2_0,\\
\bar\eta^0_0 &= -\bar\eta^3_3 = -3p_1\bar\eta^1_0 + 3p_2\bar\eta^2_0,\\
\bar\eta^1_3 &= q_{11}\bar\eta^1_0 + q_{12}\bar\eta^2_0,\\
\bar\eta^2_3 &= q_{21}\bar\eta^1_0 + q_{22}\bar\eta^2_0,\\
\bar\eta^0_3 &= q_{12}\bar\eta^1_0 + q_{21}\bar\eta^2_0.
\end{aligned}$$

The functions $p_1, p_2, q_{11}, q_{12}, q_{21}, q_{22}$ are the fundamental invariants for non-ruled, hyperbolic surfaces in $\mathbb{P}^3$.

7.5. MAPLE computations

In this chapter, the computations for curves in $\mathbb{P}^2$ and $\mathbb{P}^3$ are already sufficiently complicated that some assistance from MAPLE may be appreciated. Here we will explore some of the exercises for curves in $\mathbb{P}^3$; the case of curves in $\mathbb{P}^2$ is left as an exercise for the reader. As usual, begin by loading the `Cartan` and `LinearAlgebra` packages into MAPLE.

EXERCISE 7.25: The canonical lifting $\tilde{\mathbf{e}}_0(s)$ of $[\beta(s)]$ must be equal to a scalar multiple of $\mathbf{e}_0(t(s))$; say

$$\tilde{\mathbf{e}}_0(s) = \lambda(s)\,\mathbf{e}_0(t(s)).$$

For MAPLE purposes, we will denote the canonical projective frame field for $[\beta(s)]$ by `(f0(s), f1(s), f2(s), f3(s))`.

```
> f0(s):= lambda(s)*e0(t(s));
```

Compute the first three derivatives of $\tilde{\mathbf{e}}_0(s)$:

```
> f1(s):= diff(f0(s), s);
  f2(s):= diff(f1(s), s);
  f3(s):= diff(f2(s), s);
```

MAPLE expresses the output from these computations in terms of the derivatives of $\mathbf{e}_0(t(s))$; for instance, it returns

$$f1(s) := \lambda_s\, e0(t(s)) + \lambda(s)\, D(e0)(t(s))\, t_s$$

(The expression $D(e0)(t(s))$ refers to the derivative $\frac{d}{dt}\left(\mathbf{e}_0(t)\right)$ evaluated at $t = t(s)$.)

We don't have any good way of telling MAPLE that these derivatives are all linearly independent, but we can still collect terms and extract the coefficients needed to compute the necessary determinant.

```
> collect(f1(s), {e0(t(s)), D(e0)(t(s))});
  collect(f2(s), {e0(t(s)), D(e0)(t(s)), D(D(e0))(t(s))});
  collect(f3(s), {e0(t(s)), D(e0)(t(s)), D(D(e0))(t(s)),
  D(D(D(e0)))(t(s))});
```

Upon inspection, it becomes clear that $\det \left[\tilde{\mathbf{e}}_0(s)\ \tilde{\mathbf{e}}_1(s)\ \tilde{\mathbf{e}}_2(s)\ \tilde{\mathbf{e}}_3(s)\right]$ can be computed as follows:

```
> detf:= coeff(f0(s), e0(t(s)))*coeff(f1(s), D(e0)(t(s)))*
  coeff(f2(s), D(D(e0))(t(s)))*coeff(f3(s), D(D(D(e0)))(t(s)));
```

$$detf := \lambda(s)^4 t_s^6$$

Since this determinant must be equal to 1, we must have

$$\lambda(s) = \frac{1}{(t'(s))^{(3/2)}}.$$

Now computing the rest of the canonical projective frame field for $[\beta]$ is simply a matter of differentiating the expression

$$\tilde{\mathbf{e}}_0(s) = \frac{1}{(t'(s))^{(3/2)}}\mathbf{e}_0(t(s)).$$

But keep in mind that we still need to compute the Wilczynski invariants, which requires expressing $\tilde{\mathbf{e}}_3'(s)$ as a linear combination of $(\tilde{\mathbf{e}}_0(s), \tilde{\mathbf{e}}_1(s), \tilde{\mathbf{e}}_2(s))$. This can get a bit messy, but we can keep things better organized if, instead of explicitly expressing everything as functions of t and s, we instead work with appropriately chosen differential forms. In this context, we assign the derivatives (and the Wilczynski invariants) of the original frame field as follows:

```
> d(e0):= e1*d(t);
  d(e1):= e2*d(t);
  d(e2):= e3*d(t);
  d(e3):= e0*kappa0*d(t) + e1*kappa1*d(t) + e2*kappa2*d(t);
```

Next, observe that the function $t(s)$ never appears explicitly; all that matters is its derivative $t'(s)$. So, let $r(s)$ denote $t'(s)$, and set up substitutions that relate dt to ds and $\tilde{\mathbf{e}}_0$ to $\mathbf{e}_0$:

```
> PDETools[declare](r(s));
  tsub:= [d(t) = r(s)*d(s)];
  fsub:= [f0 = e0/r(s)^(3/2)];
```

Compute the derivative of $\tilde{\mathbf{e}}_0$ with respect to s by computing $d\tilde{\mathbf{e}}_0$ and then substituting for dt in terms of ds:

```
> Simf(subs(tsub, Simf(d(Simf(subs(fsub, f0))))));
```

The coefficient of ds in this expression should be $\tilde{\mathbf{e}}_1(s)$, so add it to `fsub`:

```
> fsub:= [op(fsub), f1 = pick(%, d(s))];
```

Similarly for $\tilde{\mathbf{e}}_2(s)$ and $\tilde{\mathbf{e}}_3(s)$:

```
> Simf(subs(tsub, Simf(d(Simf(subs(fsub, f1))))));
  fsub:= [op(fsub), f2 = pick(%, d(s))];
  Simf(subs(tsub, Simf(d(Simf(subs(fsub, f2))))));
  fsub:= [op(fsub), f3 = pick(%, d(s))];
```

In order to compute the Wilczinksi invariants for the reparametrized curve, we need to express $d\tilde{\mathbf{e}}_3$ in terms of $(\tilde{\mathbf{e}}_0, \tilde{\mathbf{e}}_1, \tilde{\mathbf{e}}_2)$. The `makebacksub` command won't work here because $(\mathbf{e}_0, \mathbf{e}_1, \mathbf{e}_2, \mathbf{e}_3)$ are not 1-forms, but we can create the reverse substitution as follows:

```
> esub:= [op(solve({op(fsub)}, {e0, e1, e2, e3}))];
```

Now we can compute $d\tilde{\mathbf{e}}_3$ and express the result in terms of the canonical projective frame field for $[\beta]$:

```
> Simf(subs(esub, Simf(subs(tsub,
  Simf(d(Simf(subs(fsub, f3))))))));
  df3:= collect(pick(%, d(s)), {f0, f1, f2, f3});
```

As expected, we see that the output contains no $\tilde{\mathbf{e}}_3$ term, and the invariants $\tilde{\kappa}_0(s), \tilde{\kappa}_1(s), \tilde{\kappa}_2(s)$ associated to $[\beta]$ are given by

```
> Kappa0:= coeff(df3, f0);
  Kappa1:= coeff(df3, f1);
  Kappa2:= coeff(df3, f2);
```

EXERCISE 7.28: First, it will be helpful to introduce variables to represent the derivatives of $\kappa_0(t), \kappa_1(t), \kappa_2(t)$:

```
> d(kappa0):= kappa0_t*d(t);
  d(kappa1):= kappa1_t*d(t);
  d(kappa2):= kappa2_t*d(t);

  d(kappa0_t):= kappa0_tt*d(t);
  d(kappa1_t):= kappa1_tt*d(t);
  d(kappa2_t):= kappa2_tt*d(t);
```

Next, we will need to express some of the derivatives of $\tilde{\kappa}_0(s), \tilde{\kappa}_1(s), \tilde{\kappa}_2(s)$ (with respect to s) in terms of $\kappa_0(t), \kappa_1(t), \kappa_2(t)$ and their derivatives (with respect to t). As it turns out, these are the ones that we will need:

```
> Kappa1_s:= Simf(pick(Simf(subs(tsub, d(Kappa1))), d(s)));
  Kappa2_s:= Simf(pick(Simf(subs(tsub, d(Kappa2))), d(s)));
  Kappa2_ss:= Simf(pick(Simf(subs(tsub, d(Kappa2_s))), d(s)));
```

In order to identify invariant forms, the trick is to look for combinations of $\tilde{\kappa}_0(s), \tilde{\kappa}_1(s), \tilde{\kappa}_2(s)$ and their derivatives that don't contain any derivatives of $r(s)$, so that they transform tensorially under the change of variables $t \to t(s)$. The first one is relatively easy to spot: The computations above show that

$$\tilde{\kappa}_1(s) = r(s)^3\kappa_1(t) + 2\,r(s)r'(s)\kappa_2(t) + \frac{1}{r(s)^3}\left(20\,r(s)r'(s)r''(s) - 5\,r(s)^2r'''(s) - 15\,r'(s)^3\right)$$

(compare with equation (7.16)), while

$$\tilde{\kappa}_2'(s) = r(s)^3\kappa_2'(t) + 2\,r(s)r'(s)\kappa_2(t) + \frac{1}{r(s)^3}\left(20\,r(s)r'(s)r''(s) - 5\,r(s)^2r'''(s) - 15\,r'(s)^3\right).$$

Therefore, we have

$$\tilde{\kappa}_1(s) - \tilde{\kappa}_2'(s) = r(s)^3\left(\kappa_1(t) - \kappa_2'(t)\right),$$

and hence

$$\left(\tilde{\kappa}_1(s) - \tilde{\kappa}_2'(s)\right)ds^3 = \left(\kappa_1(t) - \kappa_2'(t)\right)\left(r(s)\,ds\right)^3 = \left(\kappa_1(t) - \kappa_2'(t)\right)dt^3.$$

The quartic form requires a bit more finesse to identify; for details, see the MAPLE worksheet for this chapter on the AMS webpage. But once we know the answer, verifying it is straightforward:

```
> Simf(Kappa0 - (1/2)*Kappa1_s + (1/5)*Kappa2_ss
  + (9/100)*Kappa2^2);
```

$$\frac{1}{100}r^4\left(100\,\kappa 0 + 9\,\kappa 2^2 + 20\,kappa2\_tt - 50\,kappa1\_t\right)$$

Now we move on to surfaces in $\mathbb{P}^3$. The initial setup is similar to that in Chapter 6, with a slightly different collection of indices:

```
> Form(omega[0,0], omega[0,1], omega[0,2], omega[0,3],
  omega[1,0], omega[1,1], omega[1,2], omega[1,3],
```

```
omega[2,0], omega[2,1], omega[2,2], omega[2,3],
omega[3,0], omega[3,1], omega[3,2], omega[3,3]);

for i from 0 to 3 do
  for j from 0 to 3 do
    d(omega[i,j]):= -add('omega[i,k] &^ omega[k,j]',
    k=0..3);
  end do;
end do;
```

It's a little bit clumsy to introduce the relation

$$\bar{\omega}_0^0 + \bar{\omega}_1^1 + \bar{\omega}_2^2 + \bar{\omega}_3^3 = 0$$

at this point because doing so would require solving for one of the forms in terms of the others, and it's not yet clear which form would be best to solve for. So for now, we'll simply keep the relation in mind, and we'll tell MAPLE about it when it becomes convenient to do so.

Next, set up the initial substitution for the Maurer-Cartan forms of a 0-adapted frame field:

```
> adaptedsub1:= [omega[3,0]=0,
  omega[3,1] = h[1,1]*omega[1,0] + h[1,2]*omega[2,0],
  omega[3,2] = h[1,2]*omega[1,0] + h[2,2]*omega[2,0]];
```

EXERCISE 7.34: Introduce new 1-forms to represent the transformed forms:

```
> Form(Omega[0,0], Omega[0,1], Omega[0,2], Omega[0,3],
  Omega[1,0], Omega[1,1], Omega[1,2], Omega[1,3],
  Omega[2,0], Omega[2,1], Omega[2,2], Omega[2,3],
  Omega[3,0], Omega[3,1], Omega[3,2], Omega[3,3]);
```

Ultimately, we're going to need to know how all of the Maurer-Cartan forms transform as we gradually reduce the transformation group. And while it would be quite a mess to write everything out explicitly, we know that under a transformation of the form (7.22), we have

$$\left[\tilde{\bar{\omega}}_\beta^\alpha\right] = A^{-1}dA + A^{-1}\left[\bar{\omega}_\beta^\alpha\right]A, \tag{7.44}$$

where A is the matrix in equation (7.22). We can go ahead and set up this substitution for all the Maurer-Cartan forms at once—although you will probably want to suppress the output so as not to have MAPLE spew out pages and pages of nasty expressions that we don't really need to see!

We can set up the matrix computation (7.44) as follows: First, introduce the necessary matrices:

```
> omegamatrix:= Matrix([
  [omega[0,0], omega[0,1], omega[0,2], omega[0,3]],
  [omega[1,0], omega[1,1], omega[1,2], omega[1,3]],
  [omega[2,0], omega[2,1], omega[2,2], omega[2,3]],
  [omega[3,0], omega[3,1], omega[3,2], omega[3,3]]]);
> Omegamatrix:= Matrix([
  [Omega[0,0], Omega[0,1], Omega[0,2], Omega[0,3]],
  [Omega[1,0], Omega[1,1], Omega[1,2], Omega[1,3]],
  [Omega[2,0], Omega[2,1], Omega[2,2], Omega[2,3]],
  [Omega[3,0], Omega[3,1], Omega[3,2], Omega[3,3]]]);
> groupmatrix:= Matrix([
  [lambda, r[1], r[2], s[0]],
  [0, b[1,1], b[1,2], s[1]],
  [0, b[2,1], b[2,2], s[2]],
  [0,0,0,1/(lambda*(b[1,1]*b[2,2] - b[1,2]*b[2,1]))]]);
```

Next, compute the right-hand side of equation (7.44) (and you'll definitely want to suppress the output):

```
> RHSmatrix:= map(Simf,
  MatrixInverse(groupmatrix).map(d, groupmatrix)
  + MatrixInverse(groupmatrix).omegamatrix.groupmatrix):
```

Now, here's the desired substitution:

```
> framechangesub:= [
  Omega[0,0] = RHSmatrix[1,1], Omega[0,1] = RHSmatrix[1,2],
  Omega[0,2] = RHSmatrix[1,3], Omega[0,3] = RHSmatrix[1,4],
  Omega[1,0] = RHSmatrix[2,1], Omega[1,1] = RHSmatrix[2,2],
  Omega[1,2] = RHSmatrix[2,3], Omega[1,3] = RHSmatrix[2,4],
  Omega[2,0] = RHSmatrix[3,1], Omega[2,1] = RHSmatrix[3,2],
  Omega[2,2] = RHSmatrix[3,3], Omega[2,3] = RHSmatrix[3,4],
  Omega[3,0] = RHSmatrix[4,1], Omega[3,1] = RHSmatrix[4,2],
  Omega[3,2] = RHSmatrix[4,3], Omega[3,3] = RHSmatrix[4,4]]:
```

We need to be careful when constructing the reverse substitution; the expressions on the right-hand sides in `framechangesub` involve exterior derivatives of the group parameters as well as the forms $(\bar{\omega}^\alpha_\beta)$, and this may prevent the `makebacksub` command from working as desired. It's safer to do the following instead:

```
> framechangebacksub:= [op(solve({op(framechangesub)},
  {seq(seq(omega[i,j], j=0..3), i=0..3)}))]:
```

We now proceed much as we did in Exercise 6.20: First, introduce another substitution describing the adaptations for the transformed frame:

```
> adaptedsub2:= [Omega[3,0]=0,
  Omega[3,1]  = H[1,1]*Omega[1,0]  + H[1,2]*Omega[2,0],
  Omega[3,2]  = H[1,2]*Omega[1,0]  + H[2,2]*Omega[2,0]];
```

Next, determine how the $(\tilde{h}_{ij})$ may be expressed in terms of the (h_{ij}). (Note that we need one extra substitution here that wasn't needed in Exercise 6.20; if you wonder why it's there, see what happens when you leave it out!)

```
> zero2:= Simf(subs(adaptedsub2, Omega[3,1])
  - Simf(subs(adaptedsub2, Simf(subs(framechangebacksub,
  Simf(subs(adaptedsub1, Simf(subs(framechangesub,
  Omega[3,1])))))))));
> eqns:= {op(ScalarForm(zero2))};
> zero3:= Simf(subs(adaptedsub2, Omega[3,2])
  - Simf(subs(adaptedsub2, Simf(subs(framechangebacksub,
  Simf(subs(adaptedsub1, Simf(subs(framechangesub,
  Omega[3,2])))))))));
> eqns:= eqns union {op(ScalarForm(zero3))};
> solve(eqns, {H[1,1], H[1,2], H[2,2]});
> assign(%);
```

You can now check that the result is precisely the same as that in Exercise 6.20.

EXERCISE 7.38: Now suppose that $[\Sigma]$ is an elliptic surface and that we have chosen a 1-adapted frame field, so that $[h_{ij}]$ is the identity matrix and $B \in SO(2)$. Since we now wish to explore transformations among 1-adapted frame fields, assign these conditions for both (h_{ij}) and $(\tilde{h}_{ij})$:

```
> h[1,1]:= 1;
  h[1,2]:= 0;
  h[2,2]:= 1;

  H[1,1]:= 1;
  H[1,2]:= 0;
  H[2,2]:= 1;

  b[1,1]:= cos(theta);
  b[1,2]:= -sin(theta);
```

```
b[2,1]:= sin(theta);
b[2,2]:= cos(theta);
```

Differentiate the first equation in (7.26):

```
> zero4:= Simf(subs(adaptedsub1,
  Simf(d(omega[3,1] - omega[1,0]))));
```

Use the pick command to write this expression in the form

$$\phi_1 \wedge \bar{\omega}_0^1 + \phi_2 \wedge \bar{\omega}_0^2$$

so that we can apply Cartan's lemma:

```
pick(zero4, omega[1,0]);
```

$$-\omega_{0,0} + 2\,\omega_{1,1} - \omega_{3,3}$$

```
pick(zero4, omega[2,0]);
```

$$\omega_{1,2} + \omega_{2,1}$$

Similarly for the second equation in (7.26):

```
> zero5:= Simf(subs(adaptedsub1,
  Simf(d(omega[3,2] - omega[2,0]))));
> pick(zero5, omega[1,0]);
```

$$\omega_{1,2} + \omega_{2,1}$$

```
> pick(zero5, omega[2,0]);
```

$$-\omega_{0,0} + 2\,\omega_{2,2} - \omega_{3,3}$$

Equation (7.27) then follows from Cartan's lemma. Now add these conditions to adaptedsub1:

```
> adaptedsub1:= [op(adaptedsub1),
  omega[1,1] = (1/2)*(omega[0,0] + omega[3,3]
  + h[1,1,1]*omega[1,0] + h[1,1,2]*omega[2,0]),
  omega[2,1] = -omega[1,2] + h[1,1,2]*omega[1,0]
  + h[1,2,2]*omega[2,0],
  omega[2,2] = (1/2)*(omega[0,0] + omega[3,3]
  + h[1,2,2]*omega[1,0] + h[2,2,2]*omega[2,0])];
```

EXERCISE 7.39: We may as well go ahead and determine how the (h_{ijk}) transform under the full 1-adapted group action. We can compute the functions $(\tilde{h}_{ijk})$ as follows:

```
> newform1:= Simf(subs(adaptedsub2,
  Simf(subs(framechangebacksub, Simf(subs(adaptedsub1,
```

```
  Simf(subs(framechangesub,
  2*Omega[1,1] - Omega[0,0] - Omega[3,3]))))))));
> H[1,1,1]:= pick(newform1, Omega[1,0]);
  H[1,1,2]:= pick(newform1, Omega[2,0]);
> newform2:= Simf(subs(adaptedsub2,
  Simf(subs(framechangebacksub, Simf(subs(adaptedsub1,
  Simf(subs(framechangesub,
  2*Omega[2,2] - Omega[0,0] - Omega[3,3]))))))));
> H[1,2,2]:= pick(newform2, Omega[1,0]);
  H[2,2,2]:= pick(newform2, Omega[2,0]);
```

The resulting expressions look fairly complicated, but they simplify considerably if we take $\theta = 0$, $\lambda = 1$:

```
> Simf(subs([theta=0, lambda=1], H[1,1,1]));
```

$$h_{1,1,1} - 3\, s_1 + 3\, r_1$$

```
> Simf(subs([theta=0, lambda=1], H[1,1,2]));
```

$$h_{1,1,2} - s_2 + r_2$$

```
> Simf(subs([theta=0, lambda=1], H[1,2,2]));
```

$$h_{1,2,2} - s_1 + r_1$$

```
> Simf(subs([theta=0, lambda=1], H[2,2,2]));
```

$$h_{2,2,2} - 3\, s_2 + 3\, r_2$$

EXERCISE 7.41: Now we assume that both frame fields are 2-adapted, so that the condition (7.28) holds for both sets of Maurer-Cartan forms. So, we tell MAPLE that it holds for the (h_{ijk}) and then determine what conditions must be satisfied by the group parameters in order for it to hold for the $(\tilde{h}_{ijk})$ as well:

```
> h[1,2,2]:= -h[1,1,1];
  h[1,1,2]:= -h[2,2,2];
> zero6:= Simf(H[1,1,1] + H[1,2,2]);
  zero7:= Simf(H[1,1,2] + H[2,2,2]);
> solve({zero6, zero7}, {r[1], r[2]});
```

$$\{r_1 = \sin(\theta)\lambda\, s_2 + \lambda\, s_1 \cos(\theta),\ r_2 = -\sin(\theta)\lambda\, s_1 + \lambda \cos(\theta) s_2\}$$

```
> assign(%);
```

Now compute how the functions h_{111}, h_{222} transform under this restricted group action:

```
> collect(Simf(H[1,1,1]), {h[1,1,1], h[2,2,2]});
> collect(Simf(H[2,2,2]), {h[1,1,1], h[2,2,2]});
```

If the output looks too complicated to recognize immediately, we can check that this does, in fact, yield the result in equation (7.30):

```
> Simf(Vector([H[1,1,1], H[2,2,2]])
  - (1/lambda)*B.B.B.Vector([h[1,1,1], h[2,2,2]]));
```

$$\begin{bmatrix} 0 \\ 0 \end{bmatrix}$$

The remainder of the adaptation process divides into cases based on whether or not the vector ${}^t\big[h_{111}\ h_{222}\big]$ vanishes. For MAPLE purposes, it becomes prudent to use substitutions rather than assignments from this point on, so that whatever information we learn about each case can be restricted to that case rather than applied globally.

In the non-umbilic case, we now restrict to 3-adapted frame fields, so we assume that $h_{111} = 2$, $h_{222} = 0$. In order to keep computations as simple as possible, we will restrict to those transformations in Exercise 7.42 for which $\lambda = 1$ and B is the identity matrix. We can collect this information into the following substitution:

```
> threeadaptedsub:= [h[1,1,1] = 2, h[2,2,2] = 0, theta = 0,
  lambda = 1];
```

EXERCISE 7.45: At this point, we have the following relation:

```
> Simf(subs(adaptedsub1, omega[1,1] + omega[2,2]));
```

$$\omega_{0,0} + \omega_{3,3}$$

Therefore, since $\bar{\omega}_0^0 + \bar{\omega}_1^1 + \bar{\omega}_2^2 + \bar{\omega}_3^3 = 0$, we must have

$$\bar{\omega}_0^0 + \bar{\omega}_3^3 = \bar{\omega}_1^1 + \bar{\omega}_2^2 = 0.$$

We need to add these relations to adaptedsub1 (and give it a new name since we're working with a subcase) via the same two-step process that we used in Chapter 6. Since the substitution already contains expressions for $\bar{\omega}_1^1$ and $\bar{\omega}_2^2$, we only need to add the relation $\bar{\omega}_0^0 + \bar{\omega}_3^3 = 0$:

```
> adaptedsub13:= Simf(subs([omega[3,3] = -omega[0,0]],
  Simf(adaptedsub1)));
> adaptedsub13:= [op(adaptedsub13), omega[3,3] = -omega[0,0]];
```

Now differentiate this relation:

```
> zero8:= Simf(subs(adaptedsub13,
  Simf(d(omega[0,0] + omega[3,3]))));
```

Identify the appropriate 1-forms so that we can apply Cartan's lemma:

```
> pick(zero8, omega[1,0]);
```

$$-\omega_{0,1} + \omega_{1,3}$$

```
> pick(zero8, omega[2,0]);
```

$$-\omega_{0,2} + \omega_{2,3}$$

It follows from Cartan's lemma that equations (7.33) must hold, and so we add them to our substitution:

```
> adaptedsub13:= [op(adaptedsub13),
  omega[1,3] = omega[0,1] + ell[1,1]*omega[1,0]
  + ell[1,2]*omega[2,0],
  omega[2,3] = omega[0,2] + ell[1,2]*omega[1,0]
  + ell[2,2]*omega[2,0]];
```

EXERCISE 7.46: Now we compute how the (ℓ_{ij}) transform under the transformation (7.32) between 3-adapted frame fields.

```
> newform4:= Simf(subs(threeadaptedsub, Simf(subs(adaptedsub2,
  Simf(subs(framechangebacksub, Simf(subs(adaptedsub13,
  Simf(subs(framechangesub, Omega[1,3] - Omega[0,1]))))))))));
> newform5:= Simf(subs(threeadaptedsub, Simf(subs(adaptedsub2,
  Simf(subs(framechangebacksub, Simf(subs(adaptedsub13,
  Simf(subs(framechangesub, Omega[2,3] - Omega[0,2]))))))))));
> LL[1,1]:= pick(newform4, Omega[1,0]);
  LL[1,2]:= pick(newform4, Omega[2,0]);
  LL[2,2]:= pick(newform5, Omega[2,0]);
```

In order to show that we can always find a 3-adapted frame for which the (ℓ_{ij}) are all equal to zero, it suffices to show that we can solve the equations

$$\tilde{\ell}_{11} = \tilde{\ell}_{12} = \tilde{\ell}_{22} = 0$$

for s_0, s_1, s_2:

```
> solve({LL[1,1], LL[1,2], LL[2,2]}, {s[0], s[1], s[2]});
```

$$\{s_0 = -\frac{1}{4}ell_{1,1} - \frac{1}{4}ell_{2,2} + \frac{1}{8}ell_{1,2}^2 + \frac{1}{32}ell_{2,2}^2 - \frac{1}{16}ell_{2,2}\,ell_{1,1} + \frac{1}{32}ell_{1,1}^2,$$
$$s_1 = \frac{1}{4}ell_{2,2} - \frac{1}{4}ell_{1,1},\ s_2 = \frac{1}{2}ell_{1,2}\}$$

EXERCISE 7.47: For a 4-adapted frame field, we have the following conditions:

```
> fouradaptedsub:= [h[1,1,1] = 2, h[2,2,2] = 0,
  ell[1,1] = 0, ell[1,2] = 0, ell[2,2] = 0];
```

We can refine our 3-adapted frame adaptation by incorporating these conditions as follows:

```
> adaptedsub14:= Simf(subs(fouradaptedsub, adaptedsub13));
```

We can now read off the relations (7.34) directly from `adaptedsub14`.

At this point, all the remaining Maurer-Cartan forms must be linear combinations of $(\bar{\omega}_0^1, \bar{\omega}_0^2)$. It follows from equations (7.34) that the first four equations in (7.35) must hold for some functions (p_i, q_{ij}). Before we continue, add these expressions to `adaptedsub14` in two steps:

```
> adaptedsub14:= Simf(subs([
  omega[1,2] = p[1]*omega[1,0] + (p[2] - 1)*omega[2,0],
  omega[0,1] = q[1,1]*omega[1,0] + q[1,2]*omega[2,0],
  omega[0,2] = q[2,1]*omega[1,0] + q[2,2]*omega[2,0]],
  adaptedsub14));
> adaptedsub14:= [op(adaptedsub14),
  omega[1,2] = p[1]*omega[1,0] + (p[2] - 1)*omega[2,0],
  omega[0,1] = q[1,1]*omega[1,0] + q[1,2]*omega[2,0],
  omega[0,2] = q[2,1]*omega[1,0] + q[2,2]*omega[2,0]];
```

Now differentiate the first two equations in (7.34):

```
> Simf(subs(adaptedsub14, Simf(d(omega[1,1] - omega[1,0]))));
> pick(%, omega[1,0]);
```

$$-\omega_{0,0} - 3\,p_1\,\omega_{2,0}$$

By Cartan's lemma, $(\bar{\omega}_0^0 + 3p_1\bar{\omega}_0^2)$ must be equal to a multiple of $\bar{\omega}_0^1$.

```
> Simf(subs(adaptedsub14,
  Simf(d(omega[1,2] + omega[2,1] + 2*omega[2,0]))));
> pick(%, omega[2,0]);
```

$$2\,\omega_{0,0} - 6\,p_2\,\omega_{1,0}$$

By Cartan's lemma, $(2\bar{\omega}_0^0 - 6p_2\bar{\omega}_0^1)$ must be equal to a multiple of $\bar{\omega}_0^2$. Together, these conditions imply that

$$\bar{\omega}_0^0 = 3p_2\bar{\omega}_0^1 - 3p_1\bar{\omega}_0^2.$$

Add this condition to adaptedsub14:

```
> adaptedsub14:= Simf(subs([
  omega[0,0] = 3*p[2]*omega[1,0] - 3*p[1]*omega[2,0]],
  adaptedsub14));
> adaptedsub14:= [op(adaptedsub14),
  omega[0,0] = 3*p[2]*omega[1,0] - 3*p[1]*omega[2,0]];
```

Finally, differentiating the last two equations in (7.34) and applying Cartan's lemma in this same fashion yields

$$\bar{\omega}_3^0 = (q_{11} - q_{22})\bar{\omega}_0^1 - (q_{12} + q_{21})\bar{\omega}_0^2,$$

as desired. Add this last condition to adaptedsub14:

```
> adaptedsub14:= [op(adaptedsub14),
  omega[0,3] = (q[1,1] - q[2,2])*omega[1,0]
  - (q[1,2] + q[2,1])*omega[2,0]];
```

The last step in the process is to check the remaining structure equations for any additional relations among the functions (p_i, q_{ij}). We can check them all at once with the following sequence of commands, which will produce any and all such relations:

```
> for i from 0 to 3 do
    for j from 0 to 3 do
      print(Simf(subs(adaptedsub14, Simf(d(omega[i,j])
      - d(Simf(subs(adaptedsub14, omega[i,j])))))));
    end do;
  end do;
```

The resulting equations all involve the exterior derivatives of the functions (p_i, q_{ij}); they are equivalent to a system of PDEs that these functions must satisfy. This system is analogous to the Gauss-Codazzi system for surfaces in $\mathbb{E}^3$; it determines the compatibility conditions for the invariants of elliptic projective surfaces without umbilic points.

The totally umbilic case is treated in the the MAPLE worksheet for this chapter on the AMS webpage.

Part 3

Applications of moving frames

Chapter 8

Minimal surfaces in $\mathbb{E}^3$ and $\mathbb{A}^3$

8.1. Introduction

The study of minimal surfaces goes back to 1760, when Lagrange posed the following problem: Given a closed curve in $\mathbb{E}^3$, can we find a surface of minimum area among all surfaces that have the given curve as their boundary? This question is called the *Plateau problem*; it is named after the physicist Joseph Plateau, who in the nineteenth century performed experiments with wire frames and soap films in order to study the properties of such surfaces. Despite the long history, rigorous existence results for area-minimizing surfaces with general boundary curves were only proved beginning in the 1930s [**Dou31**].

8.2. Minimal surfaces in $\mathbb{E}^3$

A regular surface in $\mathbb{E}^3$ is called *minimal* if it is "locally area minimizing". More precisely, $\Sigma \subset \mathbb{E}^3$ is minimal if for any sufficiently small open set $V \subset \Sigma$, the closure $\overline{V}$ has the minimum area of all surfaces in $\mathbb{E}^3$ with the same boundary as $\overline{V}$. Classical examples of minimal surfaces include the plane, the catenoid, and the helicoid.

8.2.1. Minimal surfaces and the calculus of variations. In order to study properties of minimal surfaces, we need to introduce some ideas from the calculus of variations. (For a more comprehensive introduction to the calculus of variations, see, e.g., [**Dac08**] or [**Olv00**].) Intuitively, we consider the set $\mathcal{S}$ of closed and bounded surfaces $\Sigma \subset \mathbb{E}^3$ to be an infinite-dimensional space. (Making this notion precise would require that we define a topology on this space, but we won't need this level of technicality.) We then define the *area functional* $\mathcal{A}$ on this space to be

$$\mathcal{A}(\Sigma) = \int_\Sigma dA;$$

i.e., $\mathcal{A}(\Sigma)$ is simply the area of Σ. The key idea is that minimal surfaces should be critical points of this functional.

Unfortunately, since the space $\mathcal{S}$ is infinite-dimensional (with a rather complicated topology!), it isn't entirely obvious how to go about finding critical points of the functional $\mathcal{A}$. The solution to this problem lies in the idea of a *variation*. We start with the observation that if

$$f : M \to \mathbb{R}$$

is a smooth function on a finite-dimensional manifold M and $\mathbf{q}_0 \in M$ is a critical point of f, then for any smooth curve $\alpha : I \to M$ with $\alpha(0) = \mathbf{q}_0$, we must have

$$\left.\frac{d}{dt}\right|_{t=0} (f \circ \alpha) = 0.$$

Conversely, if $\mathbf{q}_0 \in M$ is not a critical point of f, then there exists a smooth curve $\alpha : I \to M$ with $\alpha(0) = \mathbf{q}_0$ and $\left.\frac{d}{dt}\right|_{t=0} (f \circ \alpha) \neq 0$.

If we could come up with a good definition for a "smooth curve" in the space $\mathcal{S}$, then we could apply this same idea: In order for a surface Σ to be a critical point of $\mathcal{A}$, it should have the property that for any "smooth curve" $\alpha : I \to \mathcal{S}$ with $\alpha(0) = \Sigma$,

$$\left.\frac{d}{dt}\right|_{t=0} (\mathcal{A} \circ \alpha) = 0. \tag{8.1}$$

Conversely, if Σ is *not* a critical point of $\mathcal{A}$, then there should exist a "smooth curve" $\alpha : I \to \mathcal{S}$ with $\alpha(0) = \Sigma$ and $\left.\frac{d}{dt}\right|_{t=0} (\mathcal{A} \circ \alpha) \neq 0$.

Remark 8.1. It is customary to denote the surface $\alpha(t) \in \mathcal{S}$ by Σ_t, so that $\Sigma_0 = \Sigma$ and the condition (8.1) becomes

$$\left.\frac{d}{dt}\right|_{t=0} \mathcal{A}(\Sigma_t) = 0.$$

The intuitive idea of a "smooth curve" in $\mathcal{S}$ can be defined rigorously as follows.

Definition 8.2. Let $U \subset \mathbb{R}^2$ be an open set, and let $\mathbf{x} : U \to \mathbb{E}^3$ be a smooth immersion whose image is a surface $\Sigma = \mathbf{x}(U)$. A *variation* of $\mathbf{x}$ is a smooth map

$$\mathbf{X} : U \times (-\varepsilon, \varepsilon) \to \mathbb{E}^3$$

for some $\varepsilon > 0$, with the properties that:

(1) For each $t \in (-\varepsilon, \varepsilon)$, the set $\Sigma_t = \mathbf{X}(U, t)$ is a regular surface in $\mathbb{E}^3$.

(2) For each $\mathbf{u} \in U$, $\mathbf{X}(\mathbf{u}, 0) = \mathbf{x}(\mathbf{u})$. (In particular, $\Sigma_0 = \Sigma$.)

So intuitively, a variation defines a 1-parameter family of surfaces $\Sigma_t \subset \mathbb{E}^3$ that "vary smoothly" with t.

Definition 8.3. A variation $\mathbf{X}$ of an immersion $\mathbf{x} : U \to \mathbb{E}^3$ is called

(1) *compactly supported* if there exists a compact set $D \subset U$ such that for all $t \in (-\varepsilon, \varepsilon)$ and all $\mathbf{u} \in U \setminus D$,

$$\mathbf{X}(\mathbf{u}, t) = \mathbf{X}(\mathbf{u}, 0) = \mathbf{x}(\mathbf{u});$$

(2) *normal* if the vector $\frac{\partial \mathbf{X}}{\partial t}(\mathbf{u}, t)$ is parallel to the unit normal vector to the surface Σ_t for all $t \in (-\varepsilon, \varepsilon)$ and all $\mathbf{u} \in U$.

We will only consider compactly supported, normal variations. For surfaces with a given boundary, the restriction to compactly supported variations corresponds to considering only variations that leave the boundary fixed, which is a natural assumption in the context of the Plateau problem. (Moreover, it guarantees that the integrals that arise in the computations below remain finite.) The restriction to normal variations will simplify the analysis of minimal surfaces, and the following exercise shows that any compactly supported variation can be made normal via a reparametrization; hence, this assumption is not really a significant restriction.

***Exercise 8.4.** Let $\mathbf{x} : U \to \mathbb{E}^3$ be an immersion, and let $\mathbf{X} : U \times (-\varepsilon, \varepsilon) \to \mathbb{E}^3$ be a compactly supported variation of $\mathbf{x}$. For each $t \in (-\varepsilon, \varepsilon)$, the function $\mathbf{x}_t : U \to \mathbb{E}^3$ defined by

$$\mathbf{x}_t(\mathbf{u}) = \mathbf{X}(\mathbf{u}, t)$$

defines a parametrization of the surface Σ_t.

Consider a reparametrization $\bar{\mathbf{X}} : U' \times (-\varepsilon, \varepsilon) \to \mathbb{E}^3$ of $\mathbf{X}$ of the form

$$\bar{\mathbf{X}}(\bar{\mathbf{u}}, t) = \mathbf{X}(\mathbf{u}(\bar{\mathbf{u}}, t), t),$$

with $\mathbf{u}(\bar{\mathbf{u}}, 0) = \bar{\mathbf{u}}$ for all $\bar{\mathbf{u}} \in U'$. If we write $\mathbf{u} = (u^1, u^2)$, then we can write this reparametrization as

$$\bar{\mathbf{X}}\left(\bar{u}^1, \bar{u}^2, t\right) = \mathbf{X}\big(u^1(\bar{u}^1, \bar{u}^2, t),\ u^2(\bar{u}^1, \bar{u}^2,\ t),\ t\big).$$

(a) Show that for each $t \in (-\varepsilon, \varepsilon)$, the function

$$\bar{\mathbf{x}}_t(\bar{\mathbf{u}}) = \bar{\mathbf{X}}(\bar{\mathbf{u}}, t)$$

is a reparametrization of the surface Σ_t; i.e., the image of $\bar{\mathbf{x}}_t$ is the same as the image of $\mathbf{x}_t$. Moreover, $\bar{\mathbf{x}}_0 = \mathbf{x}_0 = \mathbf{x}$.

(b) Show that

$$\frac{\partial \bar{\mathbf{X}}}{\partial t} = \frac{\partial u^1}{\partial t}\frac{\partial \mathbf{X}}{\partial u^1} + \frac{\partial u^2}{\partial t}\frac{\partial \mathbf{X}}{\partial u^2} + \frac{\partial \mathbf{X}}{\partial t}.$$

(c) Let $\mathbf{n}_t(\mathbf{u})$ denote the unit normal vector to the surface Σ_t at the point $\mathbf{x}_t(\mathbf{u})$. Then we can decompose $\frac{\partial \mathbf{X}}{\partial t}$ as

$$\frac{\partial \mathbf{X}}{\partial t} = a(\mathbf{u}, t)\frac{\partial \mathbf{X}}{\partial u^1} + b(\mathbf{u}, t)\frac{\partial \mathbf{X}}{\partial u^2} + c(\mathbf{u}, t)\mathbf{n}_t$$

for some functions $a(\mathbf{u}, t), b(\mathbf{u}, t), c(\mathbf{u}, t)$. Show that the condition that $\frac{\partial \bar{\mathbf{X}}}{\partial t}$ is parallel to $\mathbf{n}_t(\mathbf{u})$ is equivalent to the system of differential equations

$$\frac{\partial u^1}{\partial t} = -a(u^1, u^2, t), \qquad \frac{\partial u^2}{\partial t} = -b(u^1, u^2, t) \tag{8.2}$$

for the functions $u^1(\bar{\mathbf{u}}, t), u^2(\bar{\mathbf{u}}, t)$. (These equations should be regarded as ordinary differential equations, with t as the independent variable and $\bar{\mathbf{u}} = (\bar{u}^1, \bar{u}^2)$ as parameters.) Moreover, the condition $\mathbf{u}(\bar{\mathbf{u}}, 0) = \bar{\mathbf{u}}$ is equivalent to the initial conditions

$$u^1(\bar{u}^1, \bar{u}^2, 0) = \bar{u}^1, \qquad u^2(\bar{u}^1, \bar{u}^2, 0) = \bar{u}^2. \tag{8.3}$$

(d) Conclude from the existence/uniqueness theorem for ordinary differential equations that for each choice of $(\bar{u}^1, \bar{u}^2)$, the system (8.2), (8.3) has a unique solution for t in some interval $(-\varepsilon, \varepsilon)$. Therefore, there exists a reparametrization $\bar{\mathbf{X}}$ of $\mathbf{X}$ (possibly defined for a smaller value of ε than the original variation) with the property that the variation defined by $\bar{\mathbf{X}}$ is normal.

Remark 8.5. You might worry that, because the value of ε in the previous exercise depends on $\bar{\mathbf{u}}$, it could happen that there is no single $\varepsilon > 0$ that suffices for all $\bar{\mathbf{u}} \in U'$. Fortunately, the hypothesis that the variation is compactly supported saves the day: Any positive-valued function on a compact set must have a positive lower bound; therefore there exists a real number $\varepsilon > 0$ such that the desired variation $\bar{\mathbf{X}}$ is well-defined on $U' \times (-\varepsilon, \varepsilon)$.

Now we can give a rigorous definition that corresponds to our intuitive idea of a critical point for the area functional $\mathcal{A}$ on $\mathcal{S}$:

Definition 8.6. Let $U \subset \mathbb{R}^2$ be an open set, and let $\mathbf{x} : U \to \mathbb{E}^3$ be an immersion whose image is a surface $\Sigma = \mathbf{x}(U)$. Σ is called a *minimal surface* if for every compactly supported variation $\mathbf{X} : U \times (-\varepsilon, \varepsilon)$ of $\mathbf{x}$, we have

$$\left.\frac{d}{dt}\right|_{t=0} \mathcal{A}(\Sigma_t) = 0. \tag{8.4}$$

Remark 8.7. Since the area functional is invariant under reparametrizations, minimality is a property of the surface Σ, independent of the choice of parametrization $\mathbf{x} : U \to \mathbb{E}^3$ of Σ. Moreover, in order to show that a given immersion $\mathbf{x}$ is minimal, it suffices to consider normal variations of $\mathbf{x}$.

Remark 8.8. Despite the name, minimal surfaces are not necessarily global minimizers for the area functional; for instance, there may be multiple minimal surfaces with different areas spanning a given boundary curve or curves. However, it can be shown that any minimal surface Σ is *locally* area-minimizing in the sense that, given any point $\mathbf{q} \in \Sigma$, there exists a neighborhood $V \subset \Sigma$ of $\mathbf{q}$ that is area-minimizing among all surfaces with the same boundary curve as V.

Now, it still isn't entirely obvious how to use this definition to *find* minimal surfaces. In order to find critical points $\mathbf{q}_0$ for a function $f : M \to \mathbb{R}$ on a finite-dimensional manifold M, it suffices to identify those points $\mathbf{q}_0 \in M$ where all the partial derivatives of f vanish; however, for the functional $\mathcal{A}$ on the infinite-dimensional space $\mathcal{S}$, there is no finite set of compactly supported variations that plays a role analogous to the partial derivatives of f. Instead, we will need to find a way to work directly with Definition 8.6.

Let $\mathbf{x} : U \to \mathbb{E}^3$ be an immersion whose image is a surface $\Sigma \subset \mathbb{E}^3$. We begin by choosing an adapted orthonormal frame field $(\mathbf{e}_1(\mathbf{u}), \mathbf{e}_2(\mathbf{u}), \mathbf{e}_3(\mathbf{u}))$ along Σ, so that $(\mathbf{e}_1(\mathbf{u}), \mathbf{e}_2(\mathbf{u}))$ span the tangent plane $T_{\mathbf{x}(\mathbf{u})}\Sigma$ and $\mathbf{e}_3(\mathbf{u})$ is orthogonal to $T_{\mathbf{x}(\mathbf{u})}\Sigma$. Recall that choosing an orthonormal frame field along Σ is equivalent to choosing a lifting $\tilde{\mathbf{x}} : U \to E(3)$ defined by

$$\tilde{\mathbf{x}}(\mathbf{u}) = (\mathbf{x}(\mathbf{u}); \mathbf{e}_1(\mathbf{u}), \mathbf{e}_2(\mathbf{u}), \mathbf{e}_3(\mathbf{u}))\,.$$

The pullbacks $(\bar{\omega}^i, \bar{\omega}^i_j)$ to U of the Maurer-Cartan forms (ω^i, ω^i_j) on $E(3)$ via the lifting $\tilde{\mathbf{x}}$ satisfy the conditions that $\bar{\omega}^3 = 0$ and

$$\begin{bmatrix} \bar{\omega}^3_1 \\ \bar{\omega}^3_2 \end{bmatrix} = \begin{bmatrix} h_{11} & h_{12} \\ h_{12} & h_{22} \end{bmatrix} \begin{bmatrix} \bar{\omega}^1 \\ \bar{\omega}^2 \end{bmatrix}$$

for some functions h_{11}, h_{12}, h_{22} on U, and the Gauss and mean curvature functions of Σ (cf. Definition 4.45) are given by

$$K = h_{11}h_{22} - h_{12}^2, \qquad H = \tfrac{1}{2}(h_{11} + h_{22}).$$

Now, consider a compactly supported, normal variation

$$\mathbf{X} : U \times (-\varepsilon, \varepsilon) \to \mathbb{E}^3$$

of $\mathbf{x}$. In order to compute $\frac{d}{dt}\big|_{t=0} \mathcal{A}(\Sigma_t)$, we will choose an orthonormal frame field *on the variation* $\mathbf{X}$, i.e., a lifting $\widetilde{\mathbf{X}} : U \times (-\varepsilon, \varepsilon) \to E(3)$—and consider the pullbacks $(\bar{\omega}^i, \bar{\omega}^i_j)$ of the Maurer-Cartan forms on $E(3)$ to $U \times (-\varepsilon, \varepsilon)$ via $\widetilde{\mathbf{X}}$.

We can define such a frame field as follows: For each $(\mathbf{u}, t) \in U \times (-\varepsilon, \varepsilon)$, let $(\mathbf{e}_1(\mathbf{u}, t), \mathbf{e}_2(\mathbf{u}, t), \mathbf{e}_3(\mathbf{u}, t))$ be an orthonormal frame for the surface Σ_t at the point $\mathbf{x}_t(\mathbf{u})$, with $\mathbf{e}_3(\mathbf{u}, t)$ normal to the tangent plane $T_{\mathbf{x}_t(\mathbf{u})}\Sigma_t$. Then the pullbacks $(\bar{\omega}^i)$ of the (ω^i) to U via $\widetilde{\mathbf{X}}$ must satisfy the equation

$$d\mathbf{X} = \mathbf{e}_i \bar{\omega}^i.$$

***Exercise 8.9.** (a) For each $t \in (-\varepsilon, \varepsilon)$, let $\imath_t : U \to U \times (-\varepsilon, \varepsilon)$ denote the inclusion map defined by $\imath_t(\mathbf{u}) = (\mathbf{u}, t)$. Show that the pullbacks of the 1-forms

$$\bar{\omega}^1 = \langle d\mathbf{X}, \mathbf{e}_1 \rangle, \qquad \bar{\omega}^2 = \langle d\mathbf{X}, \mathbf{e}_2 \rangle$$

to U via $\imath_t$ are the usual dual forms on the surface $\Sigma_t = \mathbf{x}_t(U)$. (Hint: This isn't as complicated as it sounds. It is an immediate consequence of the fact that the immersion $\mathbf{x}_t$ can be written as the composition

$$\mathbf{x}_t = \mathbf{X} \circ \imath_t$$

and the fact that pullbacks behave well with respect to composition—specifically,

$$(\mathbf{x}_t)^* = (\imath_t)^* \circ (\mathbf{X})^* .)$$

(b) Show that

$$\bar{\omega}^3 = \langle d\mathbf{X}, \mathbf{e}_3 \rangle = \pm \left| \frac{\partial \mathbf{X}}{\partial t} \right| dt.$$

(The sign is dependent on the choice of orientation for Σ, as determined by the choice of the vector field $\mathbf{e}_3(\mathbf{u})$ on Σ.)

In particular, note that $\bar{\omega}^3$ is no longer necessarily equal to zero as a 1-form on the variation $\mathbf{X}$; instead, it is a multiple of dt. Moreover, the 1-forms $(\bar{\omega}^1, \bar{\omega}^2, dt)$ are linearly independent and form a basis for the 1-forms on the 3-dimensional manifold $U \times (-\varepsilon, \varepsilon)$.

For convenience, let $\chi(\mathbf{u}, t) = \pm\left|\frac{\partial \mathbf{X}}{\partial t}(\mathbf{u}, t)\right|$, with the sign chosen so that $\bar{\omega}^3 = \chi(\mathbf{u}, t)\, dt$.

***Exercise 8.10.** (a) Differentiate the equation $\bar{\omega}^3 = \chi\, dt$ to obtain

$$\bar{\omega}^3_1 \wedge \bar{\omega}^1 + \bar{\omega}^3_2 \wedge \bar{\omega}^2 + d\chi \wedge dt = 0.$$

(b) Apply Cartan's lemma to conclude that

$$\begin{bmatrix} \bar{\omega}^3_1 \\ \bar{\omega}^3_2 \\ d\chi \end{bmatrix} = \begin{bmatrix} h_{11} & h_{12} & \chi_1 \\ h_{12} & h_{22} & \chi_2 \\ \chi_1 & \chi_2 & \chi_3 \end{bmatrix} \begin{bmatrix} \bar{\omega}^1 \\ \bar{\omega}^2 \\ dt \end{bmatrix} \tag{8.5}$$

for some functions $h_{11}, h_{12}, h_{22}, \chi_1, \chi_2, \chi_3$ on $U \times (-\varepsilon, \varepsilon)$.

(c) Show that the functions (h_{ij}) are precisely the coefficients of the second fundamental form of the surface Σ_t, while the functions (χ_i) are the directional derivatives of χ in the directions of the vectors $\mathbf{e}_i(\mathbf{u}, t)$.

Now, recall that for any adapted orthonormal frame field $\tilde{\mathbf{x}} : U \to E(3)$ on the surface $\Sigma = \mathbf{x}(U)$, the area of Σ is given by

$$\mathcal{A}(\Sigma) = \int_U \bar{\omega}^1 \wedge \bar{\omega}^2 = \int_U \tilde{\mathbf{x}}^*(\omega^1 \wedge \omega^2)$$

(cf. Exercise 4.47). Applying this formula to each surface Σ_t in the variation yields

$$\mathcal{A}(\Sigma_t) = \int_U \tilde{\mathbf{x}}_t^*(\omega^1 \wedge \omega^2).$$

The surface $\Sigma = \Sigma_0$ is minimal if and only if for every compactly supported normal variation, we have

$$\begin{aligned} 0 &= \left.\frac{d}{dt}\right|_{t=0} \mathcal{A}(\Sigma_t) \\ &= \left.\frac{d}{dt}\right|_{t=0} \int_U \tilde{\mathbf{x}}_t^*(\omega^1 \wedge \omega^2) \\ &= \int_U \left.\frac{d}{dt}\right|_{t=0} \left(\tilde{\mathbf{x}}_t^*(\omega^1 \wedge \omega^2)\right). \end{aligned} \tag{8.6}$$

***Exercise 8.11.** Show that

$$\tilde{\mathbf{x}}_t^*(\omega^1 \wedge \omega^2) = \imath_t^*(\bar{\omega}^1 \wedge \bar{\omega}^2),$$

where $(\bar{\omega}^1, \bar{\omega}^2)$ are the pullbacks of (ω^1, ω^2) to $U \times (-\varepsilon, \varepsilon)$ via $\widetilde{\mathbf{X}}$ (cf. Exercise 8.9). Therefore, the condition (8.6) can be written as

$$\int_U \left.\frac{d}{dt}\right|_{t=0} \left(\imath_t^*(\bar{\omega}^1 \wedge \bar{\omega}^2)\right) = 0. \tag{8.7}$$

In order to evaluate this integral, recall from §2.11 that the *Lie derivative* of a p-form Φ along a vector field $\mathbf{v}$ is the p-form

$$\mathcal{L}_{\mathbf{v}}\Phi = \lim_{t\to 0} \frac{\varphi_t^*\Phi - \Phi}{t},$$

where φ_t is the *flow* of the vector field $\mathbf{v}$. In other words,

$$\mathcal{L}_{\mathbf{v}}\Phi = \left.\frac{d}{dt}\right|_{t=0} \varphi_t^*\Phi.$$

***Exercise 8.12.** (a) Show that the inclusion map $\imath_t : U \to U \times (-\varepsilon, \varepsilon)$ may be regarded as the flow of the vector field $\frac{\partial}{\partial t}$ on $U \times (-\varepsilon, \varepsilon)$, restricted to the set $U \times \{0\} \subset U \times (-\varepsilon, \varepsilon)$.

(b) Use part (a) to show that

$$\left.\frac{d}{dt}\right|_{t=0} \left(\imath_t^*(\bar{\omega}^1 \wedge \bar{\omega}^2)\right) = \mathcal{L}_{\partial/\partial t}(\bar{\omega}^1 \wedge \bar{\omega}^2).$$

(See §2.11 for the relevant definitions.)

(c) Use equation (8.7) to conclude that if the immersion $\mathbf{x} : U \to \mathbb{E}^3$ is minimal, then for any compactly supported, normal variation of $\mathbf{x}$, we must have

$$\int_U \mathcal{L}_{\partial/\partial t}(\bar{\omega}^1 \wedge \bar{\omega}^2) = 0. \tag{8.8}$$

Now we're ready to put all the pieces together!

***Exercise 8.13.** Let $\mathbf{x} : U \to \mathbb{E}^3$ be an immersion, and let $\mathbf{X} : U \times (-\varepsilon, \varepsilon) \to \mathbb{E}^3$ be a compactly supported, normal variation of $\mathbf{x}$.

(a) Use Cartan's formula for the Lie derivative (cf. Theorem 2.55) to show that

$$\mathcal{L}_{\partial/\partial t}(\bar{\omega}^1 \wedge \bar{\omega}^2) = \frac{\partial}{\partial t} \lrcorner \left(d(\bar{\omega}^1 \wedge \bar{\omega}^2)\right).$$

(Hint: Since $\mathbf{X}$ is a normal variation of $\mathbf{x}$, the vector

$$\frac{\partial \mathbf{X}}{\partial t} = \mathbf{X}_*\left(\frac{\partial}{\partial t}\right)$$

is a multiple of the frame vector $\mathbf{e}_3(\mathbf{u}, t)$. Therefore,

$$\omega^1\left(\frac{\partial \mathbf{X}}{\partial t}\right) = \omega^2\left(\frac{\partial \mathbf{X}}{\partial t}\right) = 0,$$

and pulling these equations back via $\mathbf{X}^*$ yields

$$\bar{\omega}^1\left(\frac{\partial}{\partial t}\right) = \bar{\omega}^2\left(\frac{\partial}{\partial t}\right) = 0.$$

(b) Use the Cartan structure equations, the equation $\bar{\omega}^3 = \chi\, dt$, and equation (8.5) to show that

$$d(\bar{\omega}^1 \wedge \bar{\omega}^2) = -(h_{11} + h_{22})\chi\, \bar{\omega}^1 \wedge \bar{\omega}^2 \wedge dt = -2H\chi\, \bar{\omega}^1 \wedge \bar{\omega}^2 \wedge dt.$$

(c) Use parts (a) and (b) to show that

$$\int_U \mathcal{L}_{\partial/\partial t}(\bar{\omega}^1 \wedge \bar{\omega}^2) = \int_U -2H\chi\, \bar{\omega}^1 \wedge \bar{\omega}^2.$$

(d) Conclude that if the mean curvature H of Σ is identically equal to zero, then

$$\int_U \mathcal{L}_{\partial/\partial t}(\bar{\omega}^1 \wedge \bar{\omega}^2) = 0$$

for every compactly supported, normal variation of $\mathbf{x}$, and hence Σ is minimal.

Thus, we have proved the following proposition:

Proposition 8.14. *If a regular surface $\Sigma \subset \mathbb{E}^3$ has mean curvature H identically equal to zero, then Σ is minimal.*

In the following exercise, we will prove the converse of Propsition 8.14: If $\Sigma = \mathbf{x}(U)$ is a surface whose mean curvature is *not* identically zero, then Σ is not a critical point for the area functional $\mathcal{A}$, and hence Σ is not minimal.

***Exercise 8.15.** Let $\Sigma = \mathbf{x}(U)$ be a surface whose mean curvature is not identically zero.

(a) Let $\mathbf{u}_0 \in U$ be a point where $H(\mathbf{u}_0) \neq 0$. Show that there exists a neighborhood $U' \subset U$ of $\mathbf{u}_0$ whose closure $\bar{U}'$ is contained in U and such that H is nonzero and does not change sign on U'.

(b) Let $\mathbf{X} : U \times (-\varepsilon, \varepsilon)$ be a normal variation of $\mathbf{x}$ supported on $\bar{U}'$; i.e.,

$$\mathbf{X}(\mathbf{u}, t) = \mathbf{x}(\mathbf{u})$$

for all $\mathbf{u} \in U \setminus \bar{U}'$, chosen so that $\bar{\omega}^3 = \chi\, dt$ for some function χ that is nonzero in a neighborhood of $\{\mathbf{u}_0\} \times (-\varepsilon, \varepsilon)$ and has the same sign as H wherever it is nonzero. Show that for this variation,

$$\left.\frac{d}{dt}\right|_{t=0} \mathcal{A}(\Sigma_t) < 0.$$

Conclude that $\Sigma = \Sigma_0$ is not a critical point for $\mathcal{A}$.

Together with Proposition 8.14, this proves the following theorem:

Theorem 8.16. *A regular surface $\Sigma \subset \mathbb{E}^3$ is minimal if and only if its mean curvature H is identically equal to zero.*

Remark 8.17. Theorem 8.16 is often taken as a definition for minimal surfaces because it is much easier to work with than Definition 8.6.

In the following two exercises, we explore two classical minimal surfaces.

Exercise 8.18. The *catenoid* is the surface $\Sigma \subset \mathbb{E}^3$ obtained by rotating the curve $x = \cosh(z)$ about the z-axis. It can be parametrized by

$$\mathbf{x}(u, v) = {}^t[\cos(u) \cosh(v),\ \sin(u) \cosh(v),\ v].$$

(a) Show that the frame field

$$\mathbf{e}_1(u, v) = \frac{\mathbf{x}_u}{|\mathbf{x}_u|} = {}^t[-\sin(u),\ \cos(u),\ 0],$$

$$\mathbf{e}_2(u, v) = \frac{\mathbf{x}_v}{|\mathbf{x}_v|} = \frac{1}{\cosh(v)}\,{}^t[\cos(u) \sinh(v),\ \sin(u) \sinh(v),\ 1],$$

$$\mathbf{e}_3(u, v) = \mathbf{e}_1(u, v) \times \mathbf{e}_2(u, v) = \frac{1}{\cosh(v)}\,{}^t[\cos(u),\ \sin(u),\ -\sinh(v)]$$

is orthonormal and that $(\mathbf{e}_1(u, v), \mathbf{e}_2(u, v))$ span the tangent space to Σ at each point $\mathbf{x}(u, v) \in \Sigma$.

(b) Show that the dual forms of this frame field are

$$\bar{\omega}^1 = \cosh(v)\, du, \qquad \bar{\omega}^2 = \cosh(v)\, dv.$$

(c) Compute $d\mathbf{e}_3$, and show that

$$\bar{\omega}^3_1 = -\frac{1}{\cosh(v)}\, du, \qquad \bar{\omega}^3_2 = \frac{1}{\cosh(v)}\, dv.$$

(d) Use the results of parts (b) and (c) to compute the matrix $[h_{ij}]$, and show that the mean curvature of Σ is $H = 0$. Therefore, the catenoid is a minimal surface.

(e) (MAPLE recommended) Repeat the computations of parts (a)–(d) for an arbitrary non-planar surface of revolution, parametrized by

$$\mathbf{x}(u, v) = {}^t[\rho(v) \cos(u),\ \rho(v) \sin(u),\ v].$$

Show that the surface is minimal if and only if the function $\rho(v)$ satisfies the differential equation

$$\rho\rho'' = (\rho')^2 + 1. \tag{8.9}$$

(f) Show that the only solutions of equation (8.9) are

$$\rho(v) = \frac{1}{a}\cosh(av + b),$$

where a, b are constants. Conclude that catenoids are the only non-planar minimal surfaces of revolution.

Exercise 8.19. The *helicoid* is the ruled surface $\Sigma \subset \mathbb{E}^3$ parametrized by

$$\mathbf{x}(u, v) = {}^t[v\cos(u),\ v\sin(u),\ u].$$

(a) Show that the frame field

$$\mathbf{e}_1(u, v) = \frac{\mathbf{x}_u}{|\mathbf{x}_u|} = \frac{1}{\sqrt{v^2+1}}\,{}^t[-v\sin(u),\ v\cos(u),\ 1],$$

$$\mathbf{e}_2(u, v) = \frac{\mathbf{x}_v}{|\mathbf{x}_v|} = {}^t[\cos(u),\ \sin(u),\ 0],$$

$$\mathbf{e}_3(u, v) = \mathbf{e}_1(u, v) \times \mathbf{e}_2(u, v) = \frac{1}{\sqrt{v^2+1}}\,{}^t[-\sin(u),\ \cos(u),\ -v]$$

is orthonormal and that $(\mathbf{e}_1(u, v), \mathbf{e}_2(u, v))$ span the tangent space to Σ at each point $\mathbf{x}(u, v) \in \Sigma$.

(b) Show that the dual forms of this frame field are

$$\bar{\omega}^1 = \sqrt{v^2+1}\,du, \qquad \bar{\omega}^2 = dv.$$

(c) Compute $d\mathbf{e}_3$, and show that

$$\bar{\omega}^3_1 = \frac{1}{v^2+1}\,dv, \qquad \bar{\omega}^3_2 = \frac{1}{\sqrt{v^2+1}}\,du.$$

(d) Use the results of parts (b) and (c) to compute the matrix $[h_{ij}]$, and show that the mean curvature of Σ is $H = 0$. Therefore, the helicoid is a minimal surface.

8.2.2. The Weierstrass-Enneper representation for minimal surfaces. There is a beautiful connection between minimal surfaces in $\mathbb{E}^3$ and the theory of holomorphic functions of a complex variable, which was first described by Weierstrass and Enneper in the late nineteenth century [**Wei66**]. In this section, we will see how moving frames can be used to explore this relationship.

Let $\Sigma \subset \mathbb{E}^3$ be a regular surface with parametrization $\mathbf{x} : U \to \mathbb{E}^3$, and let $(\mathbf{e}_1(\mathbf{u}), \mathbf{e}_2(\mathbf{u}), \mathbf{e}_3(\mathbf{u}))$ be an orthonormal frame field on Σ, with $\mathbf{e}_3(\mathbf{u})$ normal to the tangent plane $T_{\mathbf{x}(\mathbf{u})}\Sigma$ at each point $\mathbf{x}(\mathbf{u}) \in \Sigma$. As we saw in the previous subsection, the associated Maurer-Cartan forms $(\bar{\omega}^i, \bar{\omega}^i_j)$ have the property that

$$\begin{bmatrix} \bar{\omega}^3_1 \\ \bar{\omega}^3_2 \end{bmatrix} = \begin{bmatrix} h_{11} & h_{12} \\ h_{12} & h_{22} \end{bmatrix} \begin{bmatrix} \bar{\omega}^1 \\ \bar{\omega}^2 \end{bmatrix}$$

for some functions (h_{ij}) on U, and Σ is minimal if and only if $h_{11} + h_{22} = 0$.

Consider the *complex*, vector-valued 1-form

$$\begin{aligned} \xi &= (\mathbf{e}_1 - i\mathbf{e}_2)(\bar{\omega}^1 + i\bar{\omega}^2) \\ &= (\mathbf{e}_1\bar{\omega}^1 + \mathbf{e}_2\bar{\omega}^2) + i(\mathbf{e}_1\bar{\omega}^2 - \mathbf{e}_2\bar{\omega}^1) \end{aligned}$$

on U, where $i = \sqrt{-1}$.

Remark 8.20. In keeping with the notation used throughout this book, we will continue to use $(\bar{\omega}^i, \bar{\omega}^i_j)$ to denote the (real-valued!) Maurer-Cartan forms associated to the orthonormal frame field $(\mathbf{e}_1(\mathbf{u}), \mathbf{e}_2(\mathbf{u}), \mathbf{e}_3(\mathbf{u}))$. In order to avoid confusion, we will use the notation z^* rather than $\bar{z}$ to denote complex conjugation.

***Exercise 8.21.** (a) Show that ξ is a well-defined 1-form on U, independent of the choice of orthonormal frame field $(\mathbf{e}_1(\mathbf{u}), \mathbf{e}_2(\mathbf{u}), \mathbf{e}_3(\mathbf{u}))$. (Hint: This is the same sort of computation that you used to show that the first fundamental form was well-defined in Exercise 4.24.)

(b) Show that

$$d\xi = i(h_{11} + h_{22})\, \mathbf{e}_3\, \bar{\omega}^1 \wedge \bar{\omega}^2.$$

Therefore, $d\xi = 0$ if and only if Σ is minimal.

Now suppose that Σ is minimal. Let $(\mathbf{e}_1(\mathbf{u}), \mathbf{e}_2(\mathbf{u}), \mathbf{e}_3(\mathbf{u}))$ be any adapted orthonormal frame field along Σ, with associated Maurer-Cartan forms $(\bar{\omega}^i, \bar{\omega}^i_j)$, and consider the complex, scalar-valued 1-form $\bar{\omega}^1 + i\bar{\omega}^2$ on U.

***Exercise 8.22.** (a) Use the Frobenius theorem (cf. Theorem 2.33) to show that every point $\mathbf{u} \in U$ has a neighborhood $V \subset U$ on which there exist complex-valued functions $z, \varphi : V \to \mathbb{C}$ such that

$$\bar{\omega}^1 + i\bar{\omega}^2 = \varphi\, dz. \tag{8.10}$$

(Hint: Show that the hypothesis of Theorem 2.33 is automatically satisfied for any 1-form on a 2-dimensional surface. Moreover, this theorem holds for complex-valued 1-forms Φ, in which case the functions f, g of Theorem 2.33 are complex-valued as well.)

(b) Show that the function φ must be nonzero at every point of V. (Hint: The 2-form

$$(\bar{\omega}^1 + i\bar{\omega}^2) \wedge (\bar{\omega}^1 + i\bar{\omega}^2)^* = (\bar{\omega}^1 + i\bar{\omega}^2) \wedge (\bar{\omega}^1 - i\bar{\omega}^2) = -2i\bar{\omega}^1 \wedge \bar{\omega}^2$$

is nonvanishing on Σ.)

(c) Write the function z as

$$z = u + iv,$$

where $u, v : V \to \mathbb{R}$ are real-valued functions on V, and let $z^* = u - iv$ denote the complex conjugate of z. Use your calculation from part (b) to show that

$$dz \wedge dz^* = -2i\, du \wedge dv \neq 0$$

at every point of V; therefore, the functions (u, v) can be used as a system of local (real) coordinates on V. (In fact, by a reparametrization of V, we can safely assume that $u = u^1$, $v = u^2$.)

The next exercise introduces some basic results from complex analysis that will be needed for the remainder of this section.

***Exercise 8.23.** Let $V \subset \mathbb{R}^2$ be a simply connected, open set with local coordinates (u, v). If we introduce the complex coordinate $z = u + iv$ on V, then z and its complex conjugate $z^* = u - iv$ form a complex-valued local coordinate system (z, z^*) on V.

(a) Show that we can write the original coordinates (u, v) as

$$u = \frac{1}{2}(z + z^*), \qquad v = \frac{1}{2i}(z - z^*).$$

Now, let $w : V \to \mathbb{C}$ be a complex-valued, differentiable function on V. The function w may be considered as a function of either the coordinates (u, v) or the coordinates (z, z^*). The function w is called *holomorphic* or *complex analytic* if, when considered as a function $w(z, z^*)$, it satisfies the condition

$$\frac{\partial w}{\partial z^*} = 0, \tag{8.11}$$

i.e., if w is a function of z alone.

(b) Show that equation (8.11) is equivalent to

$$w_u + i w_v = 0. \tag{8.12}$$

(c) Write

$$w(u,v) = x(u,v) + iy(u,v),$$

where $x, y : V \to \mathbb{R}$ are real-valued, differentiable functions on V. Show that equation (8.12) is equivalent to the *pair* of equations

$$x_u = y_v, \qquad x_v = -y_u. \tag{8.13}$$

Equations (8.13) are called the *Cauchy-Riemann equations.*

(d) Show that any differentiable functions $x, y : V \to \mathbb{R}$ satisfying equations (8.13) must also satisfy the equations

$$x_{uu} + x_{vv} = y_{uu} + y_{vv} = 0. \tag{8.14}$$

In other words, both x and y must be *harmonic* functions.

Now we return to the complex-valued function $z = u + iv$ of Exercise 8.22. Since this function satisfies

$$dz \wedge dz^* \neq 0,$$

it can be used as a local complex coordinate on V. This choice of complex coordinate z allows us to write the restriction of the parametrization $\mathbf{x}$ to the subset $V \subset U$ in the form $\mathbf{x}(z)$, thereby defining a *complex structure* on Σ; in other words, it gives Σ the structure of a 1-dimensional complex manifold.

Next, with φ as in equation (8.10), let $\mathbf{f} : V \to \mathbb{C}^3$ be the complex, vector-valued function

$$\mathbf{f}(z, z^*) = (\mathbf{e}_1(z, z^*) - i\mathbf{e}_2(z, z^*))\varphi(z, z^*).$$

Then we can write ξ as

$$\xi = \mathbf{f}(z, z^*)\, dz.$$

***Exercise 8.24.** (a) Show that the assumption that Σ is minimal, and hence $d\xi = 0$, is equivalent to

$$\frac{\partial \mathbf{f}}{\partial z^*} = 0.$$

Therefore, if Σ is minimal, then $\mathbf{f}(z, z^*) = \mathbf{f}(z)$ is a holomorphic function on V. For the remainder of this exercise, assume that this condition holds.

(b) Show that

$$\langle \mathbf{f}(z), \mathbf{f}(z) \rangle = 0.$$

(c) Apply the Poincaré lemma (cf. Theorem 2.31) to the closed 1-form $\xi = \mathbf{f}(z)\, dz$ to conclude that there exists a vector-valued, holomorphic function

$\mathbf{z} : V \to \mathbb{C}^3$ such that

$$\xi = d\mathbf{z},$$

and therefore, $\mathbf{f}(z) = \mathbf{z}'(z)$.

(d) Show that

$$\mathrm{Re}(d\mathbf{z}) = \mathrm{Re}(\xi) = \mathbf{e}_1\bar{\omega}^1 + \mathbf{e}_2\bar{\omega}^2 = d\mathbf{x},$$

where $\mathbf{x} : U \to \Sigma$ is the original parametrization of Σ. (Here Re denotes the "real part"; i.e., $\mathrm{Re}(a + ib) = a$.) Conclude that, up to a translation, we must have

$$\mathbf{x}(z) = \mathrm{Re}\left(\mathbf{z}(z)\right).$$

Therefore, we can write $\mathbf{z}(z)$ as

$$\mathbf{z}(z) = \mathbf{x}(z) + i\mathbf{y}(z)$$

for some function $\mathbf{y} : V \to \mathbb{R}^3$. Moreover, from part (b), we have

$$\langle \mathbf{z}'(z), \mathbf{z}'(z) \rangle = 0.$$

(Note, however, that the requirement that $\mathbf{x}$ be an immersion implies that the vector $\mathbf{z}'(z)$ is never zero for any $z \in V$.)

Exercise 8.24 shows that any minimal surface $\mathbf{x} : U \to \mathbb{E}^3$ can locally be written as the real part of a holomorphic function $\mathbf{z} : U \to \mathbb{C}^3$ with $\langle \mathbf{z}'(z), \mathbf{z}'(z) \rangle = 0$. The next exercise shows that the converse is true as well.

***Exercise 8.25.** Let $V \subset \mathbb{C}$ be an open set with local coordinates (u, v) and complex coordinate $z = u + iv$. Let $\mathbf{z} : V \to \mathbb{C}^3$ be a complex vector-valued, holomorphic function with the properties that $\mathbf{z}'(z)$ is never zero on V and $\langle \mathbf{z}'(z), \mathbf{z}'(z) \rangle = 0$. Write

$$\mathbf{z}(u, v) = \mathbf{x}(u, v) + i\mathbf{y}(u, v),$$

where $\mathbf{x}, \mathbf{y} : V \to \mathbb{R}^3$ are real vector-valued functions on V.

(a) Use the Cauchy-Riemann equations (8.13) to show that

$$d\mathbf{z} = (\mathbf{x}_u - i\mathbf{x}_v)(du + i\,dv) = (\mathbf{x}_u - i\mathbf{x}_v)\,dz.$$

Therefore, $\mathbf{z}'(z) = \mathbf{x}_u - i\mathbf{x}_v$.

(b) Use the condition $\langle \mathbf{z}'(z), \mathbf{z}'(z) \rangle = 0$ to show that the parametrization $\mathbf{x} : V \to \mathbb{R}^3$ is *conformal*; i.e.,

$$\langle \mathbf{x}_u, \mathbf{x}_u \rangle = \langle \mathbf{x}_v, \mathbf{x}_v \rangle, \qquad \langle \mathbf{x}_u, \mathbf{x}_v \rangle = 0.$$

In classical notation (cf. Exercise 4.24), this means that $E = G$ and $F = 0$.

(c) Use the fact that $\mathbf{x}_{uu} + \mathbf{x}_{vv} = 0$ (cf. Exercise 8.23) to show that the surface $\Sigma = \mathbf{x}(V)$ has mean curvature $H = 0$ and therefore Σ is minimal. (Hint: Exercise 4.27 may be helpful.)

Together, Exercises 8.24 and 8.25 imply the following proposition:

Proposition 8.26. *A regular surface $\Sigma \subset \mathbb{E}^3$ is minimal if and only if every point of Σ has a neighborhood that can be parametrized by a smooth immersion $\mathbf{x} : V \to \mathbb{E}^3$ that is the real part of a holomorphic function $\mathbf{z} : V \to \mathbb{C}^3$ with $\langle \mathbf{z}'(z), \mathbf{z}'(z) \rangle = 0$.*

This brings us to the *Weierstrass-Enneper representation* for minimal surfaces, which is defined as follows. Let $U \subset \mathbb{C}$ be open, and suppose that we are given

(1) a *meromorphic* function $g : U \to \mathbb{C}$, i.e., a holomorphic function that may have isolated singularities $z_0 \in U$ called *poles*, where $g(z) = \frac{1}{(z-z_0)^k} h(z)$ for some holomorphic function $h(z)$ and some integer $k \geq 1$, called the *order* of the pole at z_0;

(2) a holomorphic function $f : U \to \mathbb{C}$ with the properties that whenever g has a pole of order k at $z_0 \in U$, f has a zero of order exactly $2k$ at z_0, and f does not have a zero at any point that is not a pole of g.

Choose a base point $z_0 \in U$, and define $\mathbf{z} : U \to \mathbb{C}^3$ by

(8.15)

$$\mathbf{z}(z) = {}^t\left[\int_{z_0}^{z} \tfrac{1}{2} f(\zeta)(1 - g(\zeta)^2)\, d\zeta,\ \int_{z_0}^{z} \tfrac{i}{2} f(\zeta)(1 + g(\zeta)^2)\, d\zeta,\ \int_{z_0}^{z} f(\zeta) g(\zeta)\, d\zeta\right].$$

The following exercise shows that for any choice of functions $f, g : U \to \mathbb{C}$ as above, the real part of $\mathbf{z}(z)$ defines a parametrization for a minimal surface:

***Exercise 8.27.** (a) Show that under the given assumptions on f and g, the integrands in (8.15) are holomorphic and therefore the function $\mathbf{z}(z)$ is holomorphic.

(b) Show that $\mathbf{z}'(z) \neq 0$ for all $z \in U$ and that $\langle \mathbf{z}'(z), \mathbf{z}'(z) \rangle = 0$. Conclude that $\mathbf{x} = \mathrm{Re}(\mathbf{z}) : U \to \mathbb{R}^3$ is a parametrization for an immersed minimal surface in $\mathbb{E}^3$.

Conversely, the following exercise shows that every minimal surface has a local parametrization of this form.

***Exercise 8.28.** Let $U \subset \mathbb{R}^2$, and let $\mathbf{x} : U \to \mathbb{E}^3$ be a parametrization for an immersed minimal surface $\Sigma \subset \mathbb{R}^3$. Choose a complex coordinate z on U as in Exercise 8.22 and a holomorphic function $\mathbf{z} : U \to \mathbb{C}^3$ such that $\langle \mathbf{z}'(z), \mathbf{z}'(z) \rangle = 0$ and $\mathbf{x} = \mathrm{Re}(\mathbf{z})$.

(a) Write $\mathbf{z}'(z) = {}^t[\xi_1, \xi_2, \xi_3]$, where $\xi_1, \xi_2, \xi_3 : U \to \mathbb{C}^3$ are holomorphic functions. Show that the condition $\langle \mathbf{z}'(z), \mathbf{z}'(z) \rangle = 0$ is equivalent to

$$\xi_1^2 + \xi_2^2 + \xi_3^2 = 0.$$

(b) Define

$$f(z) = \xi_1(z) - i\xi_2(z), \qquad g(z) = \frac{\xi_3(z)}{\xi_1(z) - i\xi_2(z)}.$$

Show that f is holomorphic and that if g has a pole of order k at $z_0 \in U$, then f has a zero of order $2k$ at z_0. (Hint: Clearly, zeros of f and poles of g both occur at points $z_0 \in U$ where $\xi_1(z_0) - i\xi_2(z_0) = 0$. Use the relation in part (a) to relate the order of the pole of g at z_0 to the order of the zero of f at z_0.)

(c) Show that with f, g as in part (b), $\mathbf{z}$ has the form (8.15).

***Exercise 8.29** (The associated family of a minimal surface)**.** Let $\mathbf{x} : U \to \mathbb{E}^3$ be a parametrization for a minimal surface $\Sigma \subset \mathbb{E}^3$ with Weierstrass-Enneper representation $\mathbf{x} = \mathrm{Re}(\mathbf{z})$, with $\mathbf{z}$ as in (8.15).

(a) Show that the first fundamental form of Σ can be written as

$$\mathrm{I} = \tfrac{1}{2}\langle d\mathbf{z}, d\mathbf{z}^* \rangle.$$

(Hint: Recall that $d\mathbf{z} = \bar{\omega}^1 + i\bar{\omega}^2$.)

(b) For each $t \in [0, 2\pi]$, let

$$\mathbf{z}_t(z) = e^{it}\mathbf{z}(z).$$

Show that the function $\mathbf{z}_t$ is holomorphic, with $\langle \mathbf{z}_t'(z), \mathbf{z}_t'(z) \rangle = 0$, and therefore, the function $\mathbf{x}_t = \mathrm{Re}(\mathbf{z}_t)$ defines a parametrization for a minimal surface $\Sigma_t \subset \mathbb{E}^3$. The 1-parameter family of minimal surfaces Σ_t is called the *associated family* of Σ.

(c) Use part (a) to show that the first fundamental form of Σ_t is independent of t; therefore, the minimal surfaces Σ_t are all *isometric*. In particular, the surface $\Sigma_{-\pi/2}$ with parametrization

$$\mathbf{y} = \mathrm{Im}(\mathbf{z})$$

is isometric to Σ; this surface is called the *conjugate surface* of Σ.

Exercise 8.30 (Enneper's surface)**.** (a) In the Weierstrass-Enneper representation (8.15), let

$$f(z) = 2, \qquad g(z) = z.$$

Show that the resulting minimal surface Σ is parametrized by

$$\mathbf{x}(u,v) = {}^t\!\left[u - \tfrac{1}{3}u^3 + uv^2,\ -v + \tfrac{1}{3}v^3 - vu^2,\ u^2 - v^2\right],$$

where $z = u + iv$. This surface is called *Enneper's surface.*

(b) Use MAPLE to plot Enneper's surface over various ranges in u and v. Is it an embedded surface in $\mathbb{E}^3$?

(c) Compute the parametrizations $\mathbf{x}_t$ for the associated family of Enneper's surface. (Hint: The Weierstrass-Enneper representation for Σ_t can be obtained by taking

$$f(z) = 2e^{it}, \qquad g(z) = z.)$$

(d) Use MAPLE to create an animation of plots of the family of surfaces Σ_t, with $t \in [0, 2\pi]$ as the time parameter.

Exercise 8.31. Let $\Sigma \subset \mathbb{E}^3$ be the catenoid, parametrized as in Exercise 8.18.

(a) Show that the Weierstrass-Enneper representation for Σ is obtained (up to a translation) by taking

$$f(z) = -ie^{-iz}, \qquad g(z) = e^{iz}$$

and that the conjugate surface of the catenoid is the helicoid. (Hint: The formula that you find for the conjugate surface will require a slight reparametrization before it looks like the parametrization for the helicoid from Exercise 8.19.)

(b) Use MAPLE to create an animation of plots of the family of surfaces Σ_t in the associated family, with $t \in [0, 2\pi]$ as the time parameter.

8.3. Minimal surfaces in $\mathbb{A}^3$

Recall that in the process of constructing adapted frame fields for elliptic surfaces $\Sigma \subset \mathbb{A}^3$, we found two invariant quadratic forms that we referred to as the first and second *equi-affine fundamental forms.* These were defined in terms of the Maurer-Cartan forms $(\bar{\omega}^i, \bar{\omega}^i_j)$ associated to any 2-adapted frame field $(\mathbf{e}_1(\mathbf{u}), \mathbf{e}_2(\mathbf{u}), \mathbf{e}_3(\mathbf{u}))$ along Σ by

$$\begin{aligned} \mathrm{I} &= (\bar{\omega}^1)^2 + (\bar{\omega}^2)^2, \\ \mathrm{II} &= \bar{\omega}^1_3\, \bar{\omega}^1 + \bar{\omega}^2_3\, \bar{\omega}^2. \end{aligned}$$

These quadratic forms are equi-affine analogs of the Euclidean first and second fundamental forms for an elliptic surface Σ in $\mathbb{E}^3$, but despite the apparent similarities, they are quite different from their Euclidean counterparts. For instance, if $\mathbf{x} : U \to \mathbb{A}^3$ is a parametrization of Σ, then the coefficients of the equi-affine first fundamental form are defined in terms of second derivatives of $\mathbf{x}$, whereas the Euclidean first fundamental form is defined in terms of first derivatives of $\mathbf{x}$. (In fact, the equi-affine first fundamental form of a surface Σ is a scalar multiple of the Euclidean *second* fundamental form of Σ; cf. Exercise 6.40(c).) Nevertheless, we will see that an equi-affine analog of Theorem 8.16 is true: If we define an area functional for equi-affine surfaces based on the equi-affine first fundamental form I, then the critical points of this functional are precisely those surfaces for which the trace of II with respect to I is identically zero.

8.3.1. Variational calculations.

Let $\mathbf{x} : U \to \mathbb{A}^3$ be a smooth immersion whose image is an elliptic surface $\Sigma = \mathbf{x}(U)$. Let $(\mathbf{e}_1(\mathbf{u}), \mathbf{e}_2(\mathbf{u}), \mathbf{e}_3(\mathbf{u}))$ be a 2-adapted frame field along Σ, and let $(\bar{\omega}^i, \bar{\omega}^i_j)$ be the associated Maurer-Cartan forms.

***Exercise 8.32.** Show that the 2-form

$$dA = \bar{\omega}^1 \wedge \bar{\omega}^2$$

is well-defined, independent of the choice of 2-adapted frame field and associated Maurer-Cartan forms. (Hint: This computation makes use of Exercises 6.20 and 6.22.) This 2-form is called the *equi-affine area form* of Σ.

By analogy with the Euclidean case, we define the *equi-affine area functional* $\mathcal{A}$ on the set $\mathcal{S}$ of closed and bounded elliptic surfaces $\Sigma \subset \mathbb{A}^3$ to be

$$\mathcal{A}(\Sigma) = \int_\Sigma dA.$$

If $\mathbf{x} : U \to \mathbb{A}^3$ is a parametrization of Σ and $\tilde{\mathbf{x}} : U \to A(3)$ is a 2-adapted frame field along the surface $\Sigma = \mathbf{x}(U)$, then the equi-affine area of Σ is given by

$$\mathcal{A}(\Sigma) = \int_U \bar{\omega}^1 \wedge \bar{\omega}^2 = \int_U \tilde{\mathbf{x}}^*(\omega^1 \wedge \omega^2).$$

In order to look for critical points of the equi-affine area functional, we apply the same ideas that we developed in the Euclidean case. The notion of a compactly supported normal variation is precisely the same for surfaces in $\mathbb{A}^3$ as for surfaces in $\mathbb{E}^3$; the only difference is that "normal" refers to the equi-affine normal vector, which is well-defined for any 2-adapted frame field on Σ. Thus, the equi-affine analog of Definition 8.6 is as follows:

Definition 8.33. Let $U \subset \mathbb{R}^2$ be an open set, and let $\mathbf{x} : U \to \mathbb{A}^3$ be an immersion whose image is an elliptic surface $\Sigma = \mathbf{x}(U)$. Σ is called an *equi-affine minimal surface* if for every compactly supported variation $\mathbf{X} : U \times (-\varepsilon, \varepsilon)$ of $\mathbf{x}$, we have

$$\frac{d}{dt}\bigg|_{t=0} \mathcal{A}(\Sigma_t) = 0. \tag{8.16}$$

***Exercise 8.34.** Convince yourself that the results of Exercise 8.4 are equally valid in the equi-affine case; the only change is that the unit normal vector $\mathbf{n}_t(\mathbf{u})$ to the surface Σ_t should be replaced with the equi-affine normal vector to Σ_t at the point $\mathbf{x}_t(\mathbf{u})$. Therefore, as in the Euclidean case, it suffices to consider normal variations of $\mathbf{x}$.

Now, let $U \subset \mathbb{R}^2$ be an open set; let $\mathbf{x} : U \to \mathbb{A}^3$ be an immersion whose image is an elliptic surface $\Sigma = \mathbf{x}(U)$; and let $\mathbf{X} : U \times (-\varepsilon, \varepsilon) \to \mathbb{A}^3$ be a compactly supported, normal variation of $\mathbf{x}$. By analogy with the Euclidean case, define a 2-adapted frame field on the variation $\mathbf{X}$ as follows: For each $(\mathbf{u}, t) \in U \times (-\varepsilon, \varepsilon)$, let $(\mathbf{e}_1(\mathbf{u}, t), \mathbf{e}_2(\mathbf{u}, t), \mathbf{e}_3(\mathbf{u}, t))$ be a 2-adapted frame for the surface Σ_t at the point $\mathbf{x}_t(\mathbf{u})$, so that $\mathbf{e}_3(\mathbf{u}, t)$ is the equi-affine normal to the surface Σ_t at the point $\mathbf{x}_t(\mathbf{u})$. Let $(\bar{\omega}^i, \bar{\omega}^i_j)$ denote the pullbacks of the Maurer-Cartan forms on $A(3)$ to $U \times (-\varepsilon, \varepsilon)$ via $\widetilde{\mathbf{X}}$.

***Exercise 8.35** (Cf. Exercise 8.9)**.** (a) For each $t \in (-\varepsilon, \varepsilon)$, let $\imath_t : U \to U \times (-\varepsilon, \varepsilon)$ denote the inclusion map defined by $\imath_t(\mathbf{u}) = (\mathbf{u}, t)$. Show that the pullbacks of the 1-forms $(\bar{\omega}^1, \bar{\omega}^2)$ to U via $\imath_t$ are the usual dual forms on the surface $\Sigma_t = \mathbf{x}_t(U)$.

(b) Show that $\bar{\omega}^3 = \chi\, dt$, where the function $\chi(\mathbf{u}, t)$ is determined by the condition that

$$\mathbf{e}_3(\mathbf{u}, t)\, \bar{\omega}^3 = \frac{\partial \mathbf{X}}{\partial t}\, dt.$$

(c) Show that the pullbacks of the connection forms $(\bar{\omega}^i_j)$ to U via $\imath_t$ are the usual connection forms on the surface $\Sigma_t = \mathbf{x}_t(U)$. In particular, the relations

$$\begin{bmatrix} \bar{\omega}^1_3 \\ \bar{\omega}^2_3 \end{bmatrix} = \begin{bmatrix} \ell_{11} & \ell_{12} \\ \ell_{12} & \ell_{22} \end{bmatrix} \begin{bmatrix} \bar{\omega}^1 \\ \bar{\omega}^2 \end{bmatrix}, \tag{8.17}$$

which hold for a 2-adapted frame field on an elliptic equi-affine surface, still hold modulo dt for the corresponding forms on $U \times (-\varepsilon, \varepsilon)$. (This means, e.g., that the 1-form

$$\bar{\omega}^1_3 - (\ell_{11}\, \bar{\omega}^1 + \ell_{12}\, \bar{\omega}^2)$$

on $U \times (-\varepsilon, \varepsilon)$ is equal to a multiple of dt.)

As in the Euclidean case, the equi-affine area functional of the surface Σ_t is given by

$$\mathcal{A}(\Sigma_t) = \int_U \tilde{\mathbf{x}}_t^*(\omega^1 \wedge \omega^2),$$

and the surface $\Sigma = \Sigma_0$ is minimal if and only if for every compactly supported normal variation of $\mathbf{x}$, we have

$$0 = \left.\frac{d}{dt}\right|_{t=0} \mathcal{A}(\Sigma_t) = \int_U \left.\frac{d}{dt}\right|_{t=0} \left(\tilde{\mathbf{x}}_t^*(\omega^1 \wedge \omega^2)\right).$$

***Exercise 8.36.** Convince yourself that the results of Exercises 8.11 and 8.12 hold in the equi-affine case.

***Exercise 8.37** (Cf. Exercise 8.13)**.** Let $\mathbf{x} : U \to \mathbb{A}^3$ be an immersion, and let $\mathbf{X} : U \times (-\varepsilon, \varepsilon) \to \mathbb{A}^3$ be a compactly supported, normal variation of $\mathbf{x}$.

(a) Use Cartan's formula for the Lie derivative to show that

$$\mathcal{L}_{\partial/\partial t}(\bar{\omega}^1 \wedge \bar{\omega}^2) = \frac{\partial}{\partial t} \lrcorner \left(d(\bar{\omega}^1 \wedge \bar{\omega}^2)\right).$$

(b) Use the Cartan structure equations for a 2-adapted coframing, the equation $\bar{\omega}^3 = \chi\, dt$, and equation (8.17) to show that

$$d(\bar{\omega}^1 \wedge \bar{\omega}^2) = 2L\chi\bar{\omega}^1 \wedge \bar{\omega}^2 \wedge dt,$$

where $L = \frac{1}{2}(\ell_{11} + \ell_{22})$ is the equi-affine mean curvature of Σ. (Note that you only need to know that equation (8.17) holds modulo dt for this computation.)

(c) Use parts (a) and (b) to show that

$$\int_U \mathcal{L}_{\partial/\partial t}(\bar{\omega}^1 \wedge \bar{\omega}^2) = \int_U 2L\chi\, \bar{\omega}^1 \wedge \bar{\omega}^2.$$

(d) Conclude that if the equi-affine mean curvature L of Σ is identically equal to zero, then

$$\int_U \mathcal{L}_{\partial/\partial t}(\bar{\omega}^1 \wedge \bar{\omega}^2) = 0$$

for every compactly supported, normal variation of $\mathbf{x}$ and hence Σ is equi-affine minimal.

Thus, we have proved the following proposition:

Proposition 8.38. *If a regular elliptic surface $\Sigma \subset \mathbb{A}^3$ has mean curvature L identically equal to zero, then Σ is equi-affine minimal.*

***Exercise 8.39.** Adapt the argument of Exercise 8.15 to the equi-affine case to prove the converse of Proposition 8.38: If $\Sigma \subset \mathbb{A}^3$ is an elliptic surface whose equi-affine mean curvature is not identically zero, then Σ is not a critical point for the equi-affine area functional $\mathcal{A}$, and hence Σ is not equi-affine minimal.

Together with Proposition 8.38, this proves the following equi-affine analog of Theorem 8.16:

Theorem 8.40. *A regular elliptic surface $\Sigma \subset \mathbb{A}^3$ is equi-affine minimal if and only if its equi-affine mean curvature L is identically equal to zero.*

Exercise 8.41. (a) Show that for any values of a, b, c with $ac - b^2 > 0$, the elliptic paraboloid

$$z = ax^2 + bxy + cy^2$$

is equi-affinely equivalent to the paraboloid $\Sigma \subset \mathbb{A}^3$ defined by the equation

$$z = \tfrac{1}{2}(x^2 + y^2). \tag{8.18}$$

(b) Consider the parametrization $\mathbf{x} : \mathbb{R}^2 \to \mathbb{A}^3$ of Σ given by

$$\mathbf{x}(u, v) = {}^t\!\left[u,\ v,\ \tfrac{1}{2}\left(u^2 + v^2\right)\right].$$

Show that the equi-affine frame field

$$\begin{aligned} \mathbf{e}_1(u, v) &= \mathbf{x}_u = {}^t[1,\ 0,\ u], \\ \mathbf{e}_2(u, v) &= \mathbf{x}_v = {}^t[0,\ 1,\ v], \\ \mathbf{e}_3(u, v) &= {}^t[0,\ 0,\ 1] \end{aligned} \tag{8.19}$$

is a 2-adapted frame field on Σ by computing its dual and connection forms and showing that they satisfy the defining conditions

$$\bar{\omega}^3_1 = \bar{\omega}^1, \qquad \bar{\omega}^3_2 = \bar{\omega}^2, \qquad \bar{\omega}^3_3 = 0$$

for a 2-adapted frame field.

(c) Show that the equi-affine mean curvature of Σ is identically zero. Conclude that any elliptic paraboloid is equi-affine minimal.

Exercise 8.42 (MAPLE recommended)**.** In this exercise, we will derive the conditions that a function $f(x, y)$ must satisfy in order for the graph $z = f(x, y)$ to be an elliptic equi-affine minimal surface $\Sigma \subset \mathbb{A}^3$. Consider the parametrization $\mathbf{x} : \mathbb{R}^2 \to \mathbb{A}^3$ of Σ given by

$$\mathbf{x}(u, v) = {}^t[u,\ v,\ f(u, v)].$$

(a) Let $(\mathbf{e}_1(\mathbf{u}), \mathbf{e}_2(\mathbf{u}), \mathbf{e}_3(\mathbf{u}))$ be the 0-adapted equi-affine frame field on Σ given by

$$\begin{aligned}\mathbf{e}_1(u,v) &= \mathbf{x}_u = {}^t[1,\ 0,\ f_u]\,,\\ \mathbf{e}_2(u,v) &= \mathbf{x}_v = {}^t[0,\ 1,\ f_v]\,,\\ \mathbf{e}_3(u,v) &= {}^t[0,\ 0,\ 1]\,.\end{aligned}$$

Show that the dual forms associated to this frame field are

$$\bar{\omega}^1 = du, \qquad \bar{\omega}^2 = dv$$

and that the only nonzero connection forms are

$$\begin{aligned}\bar{\omega}^3_1 &= f_{uu}\,du + f_{uv}\,dv,\\ \bar{\omega}^3_2 &= f_{uv}\,du + f_{vv}\,dv.\end{aligned}$$

Thus, we have

$$\begin{bmatrix} h_{11} & h_{12} \\ h_{12} & h_{22}\end{bmatrix} = \begin{bmatrix} f_{uu} & f_{uv} \\ f_{uv} & f_{vv}\end{bmatrix}.$$

Assume that $f_{uu}f_{vv} - f_{uv}^2 > 0$, so that Σ is elliptic, and for simplicity assume that $f_{uu}, f_{vv} > 0$. In order to compute the equi-affine mean curvature of Σ, we need to construct a 2-adapted frame field on Σ and compute the associated Maurer-Cartan forms. Recall from Chapter 6 that any other 0-adapted frame field $(\tilde{\mathbf{e}}_1(\mathbf{u}), \tilde{\mathbf{e}}_2(\mathbf{u}), \tilde{\mathbf{e}}_3(\mathbf{u}))$ on Σ has the form

$$(8.20)\qquad \big[\tilde{\mathbf{e}}_1(\mathbf{u})\ \tilde{\mathbf{e}}_2(\mathbf{u})\ \tilde{\mathbf{e}}_3(\mathbf{u})\big] = \big[\mathbf{e}_1(\mathbf{u})\ \mathbf{e}_2(\mathbf{u})\ \mathbf{e}_3(\mathbf{u})\big]\begin{bmatrix} \multicolumn{2}{c}{\multirow{2}{*}{B}} & r_1 \\ & & r_2 \\ 0 & 0 & (\det B)^{-1}\end{bmatrix}$$

for some $GL(2)$-valued function B and real-valued functions r_1, r_2 on U and that the Maurer-Cartan forms $(\tilde{\bar{\omega}}^i, \tilde{\bar{\omega}}^i_j)$ associated to this frame field satisfy the relations

$$(8.21)\qquad \begin{bmatrix}\tilde{\bar{\omega}}^1 \\ \tilde{\bar{\omega}}^2\end{bmatrix} = B^{-1}\begin{bmatrix}\bar{\omega}^1 \\ \bar{\omega}^2\end{bmatrix}, \qquad \begin{bmatrix}\tilde{\bar{\omega}}^3_1 \\ \tilde{\bar{\omega}}^3_2\end{bmatrix} = (\det B)\,{}^tB\begin{bmatrix}\bar{\omega}^3_1 \\ \bar{\omega}^3_2\end{bmatrix}.$$

(b) Show that if we take

$$(8.22)\qquad B = \begin{bmatrix} \dfrac{(f_{uu}f_{vv} - f_{uv}^2)^{1/8}}{\sqrt{f_{uu}}} & \dfrac{-f_{uv}}{\sqrt{f_{uu}}(f_{uu}f_{vv} - f_{uv}^2)^{3/8}} \\ 0 & \dfrac{\sqrt{f_{uu}}}{(f_{uu}f_{vv} - f_{uv}^2)^{3/8}}\end{bmatrix},$$

then

$$\begin{bmatrix} \tilde{h}_{11} & \tilde{h}_{12} \\ \tilde{h}_{12} & \tilde{h}_{22} \end{bmatrix} = \begin{bmatrix} 1 & 0 \\ 0 & 1 \end{bmatrix},$$

and therefore $\tilde{\bar{\omega}}^3_1 = \tilde{\bar{\omega}}^1$, $\tilde{\bar{\omega}}^3_2 = \tilde{\bar{\omega}}^2$.

(c) Show that for this frame field (with r_1, r_2 still arbitrary), we have

$$\tilde{\bar{\omega}}^3_3 = \left(\frac{(f_{uu} r_1 + f_{uv} r_2)}{(f_{uu} f_{vv} - f_{uv}^2)^{1/4}} + \frac{(f_{uu} f_{uvv} - 2 f_{uv} f_{uuv} + f_{vv} f_{uuu})}{4(f_{uu} f_{vv} - f_{uv}^2)} \right) du$$

$$+ \left(\frac{(f_{uv} r_1 + f_{vv} r_2)}{(f_{uu} f_{vv} - f_{uv}^2)^{1/4}} + \frac{(f_{uu} f_{vvv} - 2 f_{uv} f_{uvv} + f_{vv} f_{uuv})}{4(f_{uu} f_{vv} - f_{uv}^2)} \right) dv.$$

Conclude that by choosing

$$\begin{bmatrix} r_1 \\ r_2 \end{bmatrix} = - \begin{bmatrix} f_{uu} & f_{uv} \\ f_{uv} & f_{vv} \end{bmatrix}^{-1} \begin{bmatrix} \dfrac{(f_{uu} f_{uvv} - 2 f_{uv} f_{uuv} + f_{vv} f_{uuu})}{4(f_{uu} f_{vv} - f_{uv}^2)^{3/4}} \\ \dfrac{(f_{uu} f_{vvv} - 2 f_{uv} f_{uvv} + f_{vv} f_{uuv})}{4(f_{uu} f_{vv} - f_{uv}^2)^{3/4}} \end{bmatrix}, \tag{8.23}$$

we can arrange that $\tilde{\bar{\omega}}^3_3 = 0$ and hence (together with the result of part (b)) that the resulting frame field $(\tilde{\mathbf{e}}_1(\mathbf{u}), \tilde{\mathbf{e}}_2(\mathbf{u}), \tilde{\mathbf{e}}_3(\mathbf{u}))$ is 2-adapted.

(d) Show that the dual forms associated to this frame field are

$$\tilde{\bar{\omega}}^1 = \frac{\sqrt{f_{uu}}}{(f_{uu} f_{vv} - f_{uv}^2)^{1/8}}\, du + \frac{f_{uv}}{\sqrt{f_{uu}}(f_{uu} f_{vv} - f_{uv}^2)^{1/8}}\, dv,$$

$$\tilde{\bar{\omega}}^2 = \frac{(f_{uu} f_{vv} - f_{uv}^2)^{3/8}}{\sqrt{f_{uu}}}\, dv.$$

Then compute $\tilde{\bar{\omega}}^1_3$ and $\tilde{\bar{\omega}}^2_3$, and find the functions (ℓ_{ij}) such that

$$\begin{bmatrix} \tilde{\bar{\omega}}^1_3 \\ \tilde{\bar{\omega}}^2_3 \end{bmatrix} = \begin{bmatrix} \ell_{11} & \ell_{12} \\ \ell_{12} & \ell_{22} \end{bmatrix} \begin{bmatrix} \tilde{\bar{\omega}}^1 \\ \tilde{\bar{\omega}}^2 \end{bmatrix}.$$

Finally, write the equi-affine mean curvature equation

$$L = \tfrac{1}{2}(\ell_{11} + \ell_{22}) = 0$$

as a (rather nasty!) fourth-order differential equation for f.

8.3.2. A Weierstrass-Enneper-type representation for elliptic equi-affine minimal surfaces in $\mathbb{A}^3$. In this section, we will derive a Weierstrass-Enneper-type representation for elliptic equi-affine minimal surfaces. This formula is originally due to Blaschke and may be found in [**Bla85**].

Let $\Sigma \subset \mathbb{A}^3$ be an elliptic surface with parametrization $\mathbf{x} : U \to \mathbb{A}^3$, and let $(\mathbf{e}_1(\mathbf{u}), \mathbf{e}_2(\mathbf{u}), \mathbf{e}_3(\mathbf{u}))$ be a 2-adapted frame field along Σ, so that the associated Maurer-Cartan forms $(\bar{\omega}^i, \bar{\omega}^i_j)$ satisfy

$$\bar{\omega}^3_1 = \bar{\omega}^1, \qquad \bar{\omega}^3_2 = \bar{\omega}^2, \qquad \bar{\omega}^3_3 = 0.$$

Recall that for such a frame field, we have

$$\begin{bmatrix} \bar{\omega}^1_3 \\ \bar{\omega}^2_3 \end{bmatrix} = \begin{bmatrix} \ell_{11} & \ell_{12} \\ \ell_{12} & \ell_{22} \end{bmatrix} \begin{bmatrix} \bar{\omega}^1 \\ \bar{\omega}^2 \end{bmatrix},$$

$$\begin{bmatrix} 2\bar{\omega}^1_1 \\ \bar{\omega}^1_2 + \bar{\omega}^2_1 \\ 2\bar{\omega}^2_2 \end{bmatrix} = \begin{bmatrix} h_1 & -h_2 \\ -h_2 & -h_1 \\ -h_1 & h_2 \end{bmatrix} \begin{bmatrix} \bar{\omega}^1 \\ \bar{\omega}^2 \end{bmatrix},$$

where we have set

$$h_1 = h_{111} = -h_{122}, \qquad h_2 = h_{222} = -h_{112},$$

and that Σ is equi-affine minimal if and only if $\ell_{11} + \ell_{22} = 0$.

Now, let $\mathbb{A}^3_{\mathbb{C}}$ denote the *complexified* equi-affine space $\mathbb{A}^3 \otimes \mathbb{C}$. (This is simply the vector space $\mathbb{C}^3$, but with an equi-affine structure rather than a Euclidean structure.) Consider the $\Lambda^2\mathbb{A}^3_{\mathbb{C}}$-valued 1-form (cf. Definition 2.11)

$$\begin{aligned} \xi &= \tfrac{1}{2}\mathbf{e}_3 \wedge (\mathbf{e}_1 - i\mathbf{e}_2)(\bar{\omega}^1 + i\bar{\omega}^2) \\ &= \tfrac{1}{2}\mathbf{e}_3 \wedge [(\mathbf{e}_1\,\bar{\omega}^1 + \mathbf{e}_2\,\bar{\omega}^2) + i(\mathbf{e}_1\,\bar{\omega}^2 - \mathbf{e}_2\,\bar{\omega}^1)] \end{aligned}$$

on U.

***Exercise 8.43.** (a) Show that ξ is a well-defined 1-form on U, independent of the choice of 2-adapted frame field $(\mathbf{e}_1(\mathbf{u}), \mathbf{e}_2(\mathbf{u}), \mathbf{e}_3(\mathbf{u}))$.

(b) Show that

$$d\xi = \tfrac{1}{2}(\ell_{11} + \ell_{22})(\mathbf{e}_1 \wedge \mathbf{e}_2)\,\bar{\omega}^1 \wedge \bar{\omega}^2.$$

Therefore, $d\xi = 0$ if and only if Σ is equi-affine minimal.

Now, suppose that Σ is equi-affine minimal. Let $(\mathbf{e}_1(\mathbf{u}), \mathbf{e}_2(\mathbf{u}), \mathbf{e}_3(\mathbf{u}))$ be any 2-adapted frame field on Σ, with associated Maurer-Cartan forms $(\bar{\omega}^i, \bar{\omega}^i_j)$, and consider the scalar-valued, complex 1-form $\bar{\omega}^1 + i\bar{\omega}^2$ on U.

***Exercise 8.44.** Convince yourself that the results of Exercise 8.22 hold in the equi-affine case. Thus, every point $\mathbf{u} \in U$ has a neighborhood $V \subset U$ on which there exist complex-valued functions $z, \varphi : V \to \mathbb{C}$ such that

$$\bar{\omega}^1 + i\bar{\omega}^2 = \varphi\, dz.$$

Without loss of generality, we can assume that

$$z = u + iv,$$

where (u, v) are local coordinates on V, and we can write the restriction of the parametrization $\mathbf{x} : U \to \mathbb{A}^3$ of Σ to $V \subset U$ in the form $\mathbf{x}(z)$, thereby defining a complex structure on Σ.

Let $\mathbf{F} : V \to \Lambda^2\mathbb{A}^3_{\mathbb{C}}$ be the function

$$\mathbf{F}(z, z^*) = \tfrac{1}{2}\mathbf{e}_3 \wedge (\mathbf{e}_1 - i\mathbf{e}_2)\varphi,$$

so that

$$\xi = \mathbf{F}(z, z^*)\, dz.$$

***Exercise 8.45.** (a) Show that the assumption that Σ is equi-affine minimal, and hence $d\xi = 0$, is equivalent to

$$\frac{\partial \mathbf{F}}{\partial z^*} = 0.$$

Therefore, $\mathbf{F}(z, z^*) = \mathbf{F}(z)$ is a holomorphic, $\Lambda^2\mathbb{A}^3_{\mathbb{C}}$-valued function on V. For the remainder of this exercise, assume that this condition holds.

(b) Apply the Poincaré lemma (cf. Theorem 2.31) to the closed 1-form $\xi = \mathbf{F}(z)\, dz$ to conclude that there exists a $\Lambda^2\mathbb{A}^3_{\mathbb{C}}$-valued holomorphic function $\mathbf{Z} : V \to \Lambda^2\mathbb{A}^3_{\mathbb{C}}$ such that

$$\xi = d\mathbf{Z},$$

and therefore, $\mathbf{F}(z) = \mathbf{Z}'(z)$.

For ease of notation, let

$$\mathbf{e} = \tfrac{1}{2}(\mathbf{e}_1 - i\mathbf{e}_2), \qquad \bar{\omega} = \bar{\omega}^1 + i\bar{\omega}^2,$$

so that

$$\xi = d\mathbf{Z} = \mathbf{e}_3 \wedge \mathbf{e}\,\bar{\omega}.$$

Then the complex conjugate of ξ is

$$\xi^* = d\mathbf{Z}^* = \mathbf{e}_3 \wedge \mathbf{e}^*\,\bar{\omega}^*.$$

(c) Show that

$$d(\mathbf{e} \wedge \mathbf{e}^*) = \tfrac{1}{2}(\xi^* - \xi) = \tfrac{1}{2}(d\mathbf{Z}^* - d\mathbf{Z}).$$

Conclude that

$$\mathbf{Z}^* - \mathbf{Z} = 2\mathbf{e} \wedge \mathbf{e}^* + 2i\mathbf{C}$$

for some real-valued constant $\mathbf{C} \in \Lambda^2\mathbb{A}^3$. (Why must $\mathbf{C}$ be real-valued?)

(d) Show that by adding an imaginary constant to $\mathbf{Z}$, we can arrange that $\mathbf{C} = \mathbf{0}$. Moreover, this will have no effect on the condition that $\xi = d\mathbf{Z}$.

At this point, we need to introduce an operation called the *special linear cross product*. This is somewhat different from the usual cross product on $\mathbb{R}^3$, in that it operates on a pair of elements of $\Lambda^2\mathbb{A}^3$ and produces an element of $\mathbb{A}^3$.

Definition 8.46. The *special linear cross product* is the unique skew-symmetric, bilinear map

$$\times_{sl} : \Lambda^2\mathbb{A}^3 \times \Lambda^2\mathbb{A}^3 \to \mathbb{A}^3$$

that is $SL(3)$-equivariant and satisfies

$$(\mathbf{e}_1 \wedge \mathbf{e}_2) \times_{sl} (\mathbf{e}_1 \wedge \mathbf{e}_3) = \mathbf{e}_1 \tag{8.24}$$

for any unimodular basis $(\mathbf{e}_1, \mathbf{e}_2, \mathbf{e}_3)$ of $\mathbb{A}^3$.

A few comments on this definition are in order:

(1) "$SL(3)$-equivariant" means that for any matrix $A \in SL(3)$, we have

$$(A\mathbf{e}_1 \wedge A\mathbf{e}_2) \times_{sl} (A\mathbf{e}_1 \wedge A\mathbf{e}_3) = A\mathbf{e}_1.$$

(2) Let $(\underline{\mathbf{e}}_1, \underline{\mathbf{e}}_2, \underline{\mathbf{e}}_3)$ denote the standard basis for $\mathbb{A}^3$. Since any other unimodular basis $(\mathbf{e}_1, \mathbf{e}_2, \mathbf{e}_3)$ can be expressed as

$$(\mathbf{e}_1, \mathbf{e}_2, \mathbf{e}_3) = (A\underline{\mathbf{e}}_1, A\underline{\mathbf{e}}_2, A\underline{\mathbf{e}}_3)$$

for some matrix $A \in SL(3)$, requiring that (8.24) hold for every unimodular basis is equivalent to requiring that it hold only for the standard basis and that it be $SL(3)$-equivariant.

(3) This cross product extends via bilinearity in the usual way to elements of the complexified space $\Lambda^2\mathbb{A}^3_{\mathbb{C}}$.

For a geometric interpretation of this cross product, think of an element $\mathbf{v}_1 \wedge \mathbf{v}_2 \in \Lambda^2\mathbb{A}^3$ as representing the oriented plane spanned by the vectors $(\mathbf{v}_1, \mathbf{v}_2)$ in $\mathbb{A}^3$. The cross product of two such planes $(\mathbf{v}_1 \wedge \mathbf{v}_2,\ \mathbf{w}_1 \wedge \mathbf{w}_2)$ is a vector that spans the line of intersection of the two planes.

Exercise 8.47. Show that the special linear cross product can be expressed in terms of the usual cross product on $\mathbb{R}^3$ (or $\mathbb{C}^3$) as

$$(\mathbf{v}_1 \wedge \mathbf{v}_2) \times_{sl} (\mathbf{w}_1 \wedge \mathbf{w}_2) = (\mathbf{v}_1 \times \mathbf{v}_2) \times (\mathbf{w}_1 \times \mathbf{w}_2).$$

(Hint: Since both sides are skew-symmetric and bilinear, it suffices to show that the equation holds in the case where

$$\mathbf{v}_1 = \mathbf{w}_1 = \underline{\mathbf{e}}_1, \qquad \mathbf{v}_2 = \underline{\mathbf{e}}_2, \qquad \mathbf{w}_2 = \underline{\mathbf{e}}_3$$

and that the right-hand side is $SL(3)$-equivariant.)

We are now ready to derive Blaschke's formula.

***Exercise 8.48.** (a) Use the results of Exercise 8.45 to show that

$$(\mathbf{Z}^* - \mathbf{Z}) \times_{sl} d(\mathbf{Z}^* + \mathbf{Z}) = -i(\mathbf{e}\,\bar{\omega} + \mathbf{e}^*\,\bar{\omega}^*) = -i\,d\mathbf{x},$$

where $\mathbf{x} : V \to \mathbb{A}^3$ is the given parametrization of Σ.

(b) Conclude that

$$\begin{aligned} d\mathbf{x} &= i[(\mathbf{Z}^* - \mathbf{Z}) \times_{sl} d(\mathbf{Z}^* + \mathbf{Z})] \\ &= i[\mathbf{Z}^* \times_{sl} d\mathbf{Z}^* - \mathbf{Z} \times_{sl} d\mathbf{Z} + d(\mathbf{Z}^* \times_{sl} \mathbf{Z})]. \end{aligned}$$

Therefore, up to translation, the parametrization $\mathbf{x} : V \to \mathbb{A}^3$ of the Σ is given in terms of the complex coordinate z on V by

(8.25)
$$\mathbf{x}(z) = i\left(\mathbf{Z}(z)^* \times_{sl} \mathbf{Z}(z) + \int_{z_0}^{z} (\mathbf{Z}(\zeta)^* \times_{sl} \mathbf{Z}'(\zeta)^* - \mathbf{Z}(\zeta) \times_{sl} \mathbf{Z}'(\zeta))\, d\zeta\right)$$

for any choice of base point $z_0 \in V$. This formula is called the *Blaschke representation* for Σ.

Therefore, for any equi-affine minimal surface $\mathbf{x} : U \to \mathbb{A}^3$ and any point $\mathbf{u} \in U$, there exists a neighborhood $V \subset U$ of $\mathbf{u}$ for which the restriction of $\mathbf{x}$ to V can be written in the form (8.25) for some holomorphic function $\mathbf{Z} : V \to \Lambda^2\mathbb{A}^3_{\mathbb{C}}$.

***Exercise 8.49.** Write

$$\mathbf{Z} = \tfrac{1}{2}(\mathbf{X} + i\mathbf{Y})$$

for some smooth functions $\mathbf{X}, \mathbf{Y} : V \to \Lambda^2\mathbb{A}^3_{\mathbb{R}}$. Recall that the functions $\mathbf{X}, \mathbf{Y}$ must satisfy the Cauchy-Riemann equations

$$\mathbf{X}_u = \mathbf{Y}_v, \qquad \mathbf{X}_v = -\mathbf{Y}_u, \tag{8.26}$$

which in turn imply that $\mathbf{X}$ and $\mathbf{Y}$ are harmonic; i.e.,

$$\mathbf{X}_{uu} + \mathbf{X}_{vv} = \mathbf{Y}_{uu} + \mathbf{Y}_{vv} = 0 \tag{8.27}$$

(cf. Exercise 8.23).

(a) Show that

$$d\mathbf{x} = (\mathbf{Y} \times_{sl} \mathbf{Y}_v)\, du - (\mathbf{Y} \times_{sl} \mathbf{Y}_u)\, dv.$$

(b) Conclude that in order for $\mathbf{x}$ to be an immersion, the vectors $(\mathbf{Y}(z), \mathbf{Y}_u(z), \mathbf{Y}_v(z))$ must be linearly independent elements of $\Lambda^2\mathbb{A}^3_{\mathbb{R}}$ for each $z \in V$.

Conversely, let $V \subset \mathbb{C}$ be an open set, and let $\mathbf{Z} = \frac{1}{2}(\mathbf{X} + i\mathbf{Y}) : V \to \Lambda^2\mathbb{A}^3_{\mathbb{C}}$ be a holomorphic function such that the vectors $(\mathbf{Y}(z),\ \mathbf{Y}_u(z),\ \mathbf{Y}_v(z))$ are linearly independent for each $z \in V$. Define an immersion $\mathbf{x} : V \to \mathbb{A}^3$ by (8.25).

***Exercise 8.50.** Let $\lambda : V \to \mathbb{R}$ be a smooth function, and consider the frame field along $\Sigma = \mathbf{x}(V)$ defined by

$$\begin{aligned} \mathbf{e}_1(\mathbf{u}) &= \lambda \mathbf{x}_u = \lambda(\mathbf{Y} \times_{sl} \mathbf{Y}_v), \\ \mathbf{e}_2(\mathbf{u}) &= \pm\lambda \mathbf{x}_v = \mp\lambda(\mathbf{Y} \times_{sl} \mathbf{Y}_u), \\ \mathbf{e}_3(\mathbf{u}) &= \lambda^2(\mathbf{Y}_u \times_{sl} \mathbf{Y}_v), \end{aligned} \tag{8.28}$$

with the sign for $\mathbf{e}_2(\mathbf{u})$ chosen so that the matrix $\left[\mathbf{e}_1(\mathbf{u})\ \mathbf{e}_2(\mathbf{u})\ \mathbf{e}_3(\mathbf{u})\right]$ has positive determinant.

(a) Show that there exists a unique choice for the function λ (up to sign) for which $(\mathbf{e}_1(\mathbf{u}), \mathbf{e}_2(\mathbf{u}), \mathbf{e}_3(\mathbf{u}))$ is a unimodular frame field. For the remainder of this exercise, assume that λ has been chosen accordingly.

(b) Show that the dual forms associated to the frame field (8.28) are

$$\bar{\omega}^1 = \lambda^{-1}\,du, \qquad \bar{\omega}^2 = \pm\lambda^{-1}\,dv,$$

with the sign of $\bar{\omega}^2$ chosen according to the sign of $\mathbf{e}_2(\mathbf{u})$.

In order to compute the connection forms associated to the frame field (8.28), observe that, since $(\mathbf{Y}, \mathbf{Y}_u, \mathbf{Y}_v)$ are linearly independent, the second derivatives of $\mathbf{Y}$ can be written as

$$\begin{aligned} \mathbf{Y}_{uu} &= h_0\mathbf{Y} + h_1\mathbf{Y}_u + h_2\mathbf{Y}_v, \\ \mathbf{Y}_{uv} &= k_0\mathbf{Y} + k_1\mathbf{Y}_u + k_2\mathbf{Y}_v, \\ \mathbf{Y}_{vv} &= -h_0\mathbf{Y} - h_1\mathbf{Y}_u - h_2\mathbf{Y}_v \end{aligned}$$

for some functions $h_i, k_i : V \to \mathbb{R}$.

(c) Differentiate equations (8.28) and show that the connection forms associated to the frame field (8.28) satisfy the conditions

$$\bar{\omega}^3_1 = \bar{\omega}^1, \qquad \bar{\omega}^3_2 = \bar{\omega}^2, \qquad \bar{\omega}^3_3 = 0.$$

Conclude that the frame field (8.28) is 2-adapted. (Hint: For the last condition, first show that

$$\bar{\omega}^1_1 + \bar{\omega}^2_2 = \bar{\omega}^3_3 = 2\lambda^{-1}\,d\lambda + \lambda(h_1 + k_2)\bar{\omega}^1 \mp \lambda(h_2 + k_1)\bar{\omega}^2.$$

Then use the assumption that the frame field is unimodular to conclude that $\bar{\omega}^3_3 = 0$.)

(d) Show that

$$\begin{bmatrix} \bar{\omega}^1_3 \\ \bar{\omega}^2_3 \end{bmatrix} = \lambda^2 \begin{bmatrix} h_0 & k_0 \\ k_0 & -h_0 \end{bmatrix} \begin{bmatrix} \bar{\omega}^1 \\ \bar{\omega}^2 \end{bmatrix}.$$

Conclude that Σ is equi-affine minimal.

Together, Exercises 8.48 and 8.50 imply the following proposition:

Proposition 8.51. *A regular elliptic surface $\Sigma \subset \mathbb{A}^3$ is equi-affine minimal if and only if every point of Σ has a neighborhood that can be parametrized by a smooth immersion $\mathbf{x} : V \to \mathbb{A}^3$ of the form* (8.25) *for some holomorphic function $\mathbf{Z} : V \to \Lambda^2\mathbb{A}^3_{\mathbb{C}}$.*

***Exercise 8.52.** Let Σ be the elliptic paraboloid (8.18) of Exercise 8.41, with the 2-adapted frame field (8.19).

Show that the Blaschke representation for Σ may be obtained by taking

$$\mathbf{Z}(z) = \tfrac{1}{2}\left(-i\underline{\mathbf{e}}_1 \wedge \underline{\mathbf{e}}_2 + iz\,\underline{\mathbf{e}}_2 \wedge \underline{\mathbf{e}}_3 + z\,\underline{\mathbf{e}}_3 \wedge \underline{\mathbf{e}}_1\right),$$

where $(\underline{\mathbf{e}}_1, \underline{\mathbf{e}}_2, \underline{\mathbf{e}}_3)$ represents the standard basis of $\mathbb{A}^3$. (Hint: Write the frame vectors $(\mathbf{e}_1(u,v), \mathbf{e}_2(u,v), \mathbf{e}_3(u,v))$ as

$$\begin{aligned} \mathbf{e}_1(u,v) &= \underline{\mathbf{e}}_1 + u\underline{\mathbf{e}}_3, \\ \mathbf{e}_2(u,v) &= \underline{\mathbf{e}}_2 + v\underline{\mathbf{e}}_3, \\ \mathbf{e}_3(u,v) &= \underline{\mathbf{e}}_3, \end{aligned}$$

and note that, since $(\underline{\mathbf{e}}_1, \underline{\mathbf{e}}_2, \underline{\mathbf{e}}_3)$ is a unimodular basis, we have

$$\begin{aligned} (\underline{\mathbf{e}}_1 \wedge \underline{\mathbf{e}}_2) \times (\underline{\mathbf{e}}_1 \wedge \underline{\mathbf{e}}_3) &= \underline{\mathbf{e}}_1, \\ (\underline{\mathbf{e}}_2 \wedge \underline{\mathbf{e}}_3) \times (\underline{\mathbf{e}}_2 \wedge \underline{\mathbf{e}}_1) &= \underline{\mathbf{e}}_2, \\ (\underline{\mathbf{e}}_3 \wedge \underline{\mathbf{e}}_1) \times (\underline{\mathbf{e}}_3 \wedge \underline{\mathbf{e}}_2) &= \underline{\mathbf{e}}_3.) \end{aligned}$$

8.4. Maple computations

As usual, begin by loading the `Cartan` and `LinearAlgebra` packages into Maple.

Exercise 8.18 (e): Define the parametrization for Σ and the orthonormal frame field $(\mathbf{e}_1(\mathbf{u}), \mathbf{e}_2(\mathbf{u}), \mathbf{e}_3(\mathbf{u}))$:

```
> PDETools[declare](rho(v));
  X:= Vector([rho(v)*cos(u), rho(v)*sin(u), v]);
> Xu:= map(diff, X, u);
  Xv:= map(diff, X, v);
> e1:= Xu/simplify(Norm(Xu, Euclidean, conjugate=false),
```

```
  symbolic);
  e2:= Xv/simplify(Norm(Xv, Euclidean, conjugate=false),
  symbolic);
  e3:= simplify(CrossProduct(e1, e2));
```

We can use the following substitution to go back and forth between the (du, dv) basis and the $(\bar{\omega}^1, \bar{\omega}^2)$ basis as necessary (but first we need to declare the 1-forms $(\bar{\omega}^1, \bar{\omega}^2)$):

```
> Form(omega[1], omega[2]);
> framesub:= [
  omega[1] = simplify(Norm(Xu, Euclidean, conjugate=false),
  symbolic)*d(u),
  omega[2] = simplify(Norm(Xv, Euclidean, conjugate=false),
  symbolic)*d(v)];
> framebacksub:= makebacksub(framesub);
```

In order to compute the (h_{ij}), we need to express $d\mathbf{e}_3$ as a linear combination of $(\mathbf{e}_1(\mathbf{u}), \mathbf{e}_2(\mathbf{u}))$. We know from the Cartan structure equations (3.1) that the coefficients of $(\mathbf{e}_1(\mathbf{u}), \mathbf{e}_2(\mathbf{u}))$ will be the 1-forms $\bar{\omega}_3^1 = -\bar{\omega}_1^3$, $\bar{\omega}_3^2 = -\bar{\omega}_2^3$, respectively.

```
> de3:= map(Simf, subs(framebacksub, map(d, e3)));
> zero1:= Simf(de3 + (omega[3,1]*e1 + omega[3,2]*e2));
> Simf(solve({zero1[1], zero1[2], zero1[3]},
  {omega[3,1], omega[3,2]}));
```

$$\left\{ \omega_{3,1} = -\frac{\omega_1}{\sqrt{1+\rho_v^2}\,\rho},\ \omega_{3,2} = \frac{\rho_{v,v}\,\omega_2\,\sqrt{1+\rho_v^2}}{1+2\,\rho_v^2+\rho_v^4} \right\}$$

```
> assign(%);
```

We can read off the (h_{ij}) directly from $\bar{\omega}_1^3$ and $\bar{\omega}_2^3$:

```
> h:= Matrix([
  [pick(omega[3,1], omega[1]), pick(omega[3,1], omega[2])],
  [pick(omega[3,2], omega[1]), pick(omega[3,2], omega[2])]]);
```

Finally, Σ is minimal if and only if $H = \frac{1}{2}(h_{11} + h_{22}) = 0$:

```
> minsurfeq:= numer(simplify(Trace(h)));
```

$$minsurfeq := -\rho_v^2 - 1 + \rho_{v,v}\,\rho$$

Now, MAPLE can solve this equation with the `dsolve` command. Unfortunately, MAPLE can be rather clumsy about simplifying exponentials and trigonometric functions, so it requires a bit of manipulation to get the solution into a nice form.

```
> soln:= dsolve(minsurfeq, rho(v));
```

$$soln := \rho = \frac{1}{2} \_C1 \left(\frac{1}{\left(e^{\frac{v}{\_C1}}\right)^2 \left(e^{\frac{\_C2}{\_C1}}\right)^2} + 1 \right) e^{\frac{v}{\_C1}} e^{\frac{\_C2}{\_C1}},$$

$$f = \frac{1}{2} \frac{\_C1 \left(\left(e^{\frac{v}{\_C1}}\right)^2 \left(e^{\frac{\_C2}{\_C1}}\right)^2 + 1 \right)}{e^{\frac{v}{\_C1}} e^{\frac{\_C2}{\_C1}}}$$

MAPLE apparently doesn't even realize that these two solutions are, in fact, the same! But we can verify that our eyes are not deceiving us on this point:

```
> Simf(subs(soln[1], rho(v)) - subs(soln[2], rho(v)));
```

$$0$$

So, we can choose either solution and perform some gymnastics to force MAPLE to put it into a nicer form. (Try unpacking this command and applying these operations one at a time to see the intermediate steps.)

```
> Simf(convert(combine(expand(Simf(subs(soln[1], rho(v)))),
  exp), trig));
```

$$\_C1 \cosh\left(\frac{v + \_C2}{\_C1}\right)$$

EXERCISE 8.30: Define the Weierstrass parametrization associated to functions $f(z)$, $g(z)$:

```
> PDETools[declare](f(z), g(z));
> dZ:= Vector([(1/2)*f(z)*(1 - g(z)^2),
  (I/2)*f(z)*(1 + g(z)^2), f(z)*g(z)]);
> Z:= map(int, dZ, z);
```

Now consider the case $f(z) = 2$, $g(z) = z$:

```
> Z0:= Simf(subs([f(z) = 2, g(z) = z], Z));
```

In order to compute the parametrization $\mathbf{x}(u, v)$, we need to introduce the real and imaginary parts of z and tell Maple that the components are real:

```
> assume(u, real);
  assume(v, real);
> X0:= map(Re, Simf(subs([z = u + I*v], Z0)));
```

The `plot3d` command can be used to plot the surface over various parameter ranges; e.g.,

```
> plot3d(X0, u=-2..2, v=-2..2, axes = normal,
  scaling=constrained);
```

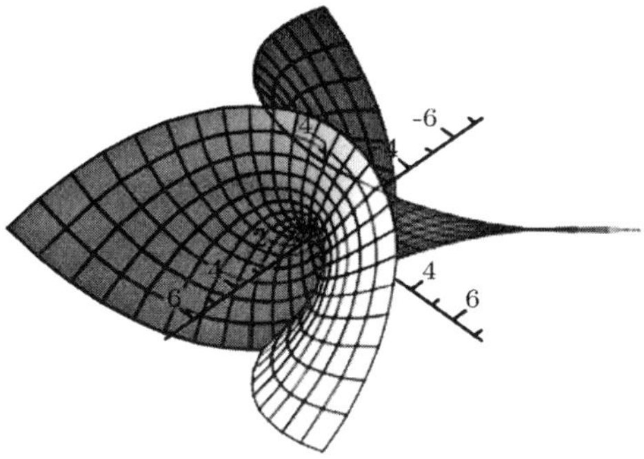

Next, compute the associated family of surfaces:

```
> assume(t, real);
> Zt:= Simf(subs([f(z) = 2*exp(I*t), g(z) = z], Z));
> Xt:= map(Re, Simf(subs([z = u + I*v], Zt)));
```

In order to animate the associated family, we need to load the `plots` package, and then we can use the `animate3d` command:

```
> with(plots);
> animate3d(Xt, u=-2..2, v=-2..2, t=0..2*Pi, axes=normal,
  scaling=constrained, frames=50);
```

To view the animation, click on the plot. Then, from the `Plot` menu, select `Animation` → `Play`.

Exactly the same procedure can be used for Exercise 8.31 to animate the associated family that interpolates between the helicoid and the catenoid.

EXERICSE 8.41: First of all, let's remove the assumptions on (u, v), just because the trailing tildes in the output are distracting:

```
> unassign('u', 'v');
```

Define the parametrization for Σ and the unimodular frame field $(\mathbf{e}_1(\mathbf{u}), \mathbf{e}_2(\mathbf{u}), \mathbf{e}_3(\mathbf{u}))$:

```
> X:= Vector([u, v, (1/2)*(u^2 + v^2)]);
> e1:= map(diff, X, u);
  e2:= map(diff, X, v);
  e3:= Vector([0,0,1]);
```

Since we have $\mathbf{e}_1(\mathbf{u}) = \mathbf{x}_u(\mathbf{u})$, $\mathbf{e}_2(\mathbf{u}) = \mathbf{x}_v(\mathbf{u})$, it follows that the associated Maurer-Cartan forms $\bar{\omega}^1, \bar{\omega}^2$ are equal to

$$\bar{\omega}^1 = du, \qquad \bar{\omega}^2 = dv.$$

Set this up as a substitution:

```
> framesub:= [omega[1] = d(u), omega[2] = d(v)];
  framebacksub:= makebacksub(framesub);
```

In order to compute the connection forms $(\bar{\omega}^i_j)$, differentiate the frame fields:

```
> de1:= map(Simf, subs(framebacksub, map(d, e1)));
  de2:= map(Simf, subs(framebacksub, map(d, e2)));
  de3:= map(Simf, subs(framebacksub, map(d, e3)));
```

From the output, it is easy to read off that

$$d\mathbf{e}_1 = \mathbf{e}_3\,\bar{\omega}^1, \qquad d\mathbf{e}_2 = \mathbf{e}_3\bar{\omega}^2, \qquad d\mathbf{e}_3 = 0.$$

Therefore, we have

$$\bar{\omega}^3_1 = \bar{\omega}^1, \qquad \bar{\omega}^3_2 = \bar{\omega}^2,$$

and the remaining connection forms $(\bar{\omega}^i_j)$ are equal to zero. It follows that this frame field is 2-adapted. Moreover, since $\bar{\omega}^1_3 = \bar{\omega}^2_3 = 0$, Σ is an equi-affine minimal surface.

Exercise 8.42: Define the parametrization for Σ and the 0-adapted frame field of part (a) (we'll call this frame field `(e10, e20, e30)` because it will be refined later):

```
> PDETools[declare](f(u,v));
> X:= Vector([u, v, f(u,v)]);
> e10:= map(diff, X, u);
  e20:= map(diff, X, v);
  e30:= Vector([0,0,1]);
```

As in the previous exercise, for this frame field we have $\bar{\omega}^1 = du$, $\bar{\omega}^2 = dv$:

```
> framesub0:= [omega[1] = d(u), omega[2] = d(v)];
> framebacksub0:= makebacksub(framesub0);
```

In order to compute the connection forms $(\bar{\omega}^i_j)$, differentiate the frame fields:

```
> de10:= map(Simf, subs(framebacksub0, map(d, e10)));
  de20:= map(Simf, subs(framebacksub0, map(d, e20)));
  de30:= map(Simf, subs(framebacksub0, map(d, e30)));
```

From the output, we see that

$$\bar{\omega}_1^3 = f_{uu}\,\bar{\omega}^1 + f_{uv}\,\bar{\omega}^2, \qquad \bar{\omega}_2^3 = f_{uv}\,\bar{\omega}^1 + f_{vv}\,\bar{\omega}^2,$$

and the remaining connection forms $(\bar{\omega}_j^i)$ are equal to zero.

Now we need to find a matrix-valued function B and real-valued functions r_1, r_2 on U that will make the frame field (8.20) 2-adapted. Rather than simply verifying that the expressions given in equations (8.22) and (8.23) will do the trick, let's see if we can figure out where they came from. First, look for a matrix-valued function B such that the transformed Maurer-Cartan forms in equation (8.21) will satisfy the 1-adapted condition $\tilde{\bar{\omega}}_1^3 = \tilde{\bar{\omega}}^1$, $\tilde{\bar{\omega}}_2^3 = \tilde{\bar{\omega}}^2$. We can make this problem slightly simpler by observing that, since B is only determined up to multiplication by a rotation matrix, we should be able to find a suitable matrix B whose lower left-hand entry is zero. With this assumption, define the new frame field and the new Maurer-Cartan forms as follows:

```
> B:= Matrix([[b[1,1], b[1,2]], [0, b[2,2]]]);
> e1:= Simf(B[1,1]*e10 + B[2,1]*e20);
  e2:= Simf(B[1,2]*e10 + B[2,2]*e20);
  e3:= Simf(r1*e10 + r2*e20 + (1/Determinant(B))*e30);
> dualformsvec:= Simf(MatrixInverse(B).
  Vector([Simf(subs(framesub0, omega[1])),
  Simf(subs(framesub0, omega[2]))]));
> connformsvec1:= Simf(Determinant(B)*Transpose(B).
  Vector([Simf(subs(framesub0, omega[3,1])),
  Simf(subs(framesub0, omega[3,2]))]));
```

The 1-adapted condition is simply the condition that these last two vectors of 1-forms are equal. Thus, collecting all the scalar coefficients of the 1-forms in their difference gives a system of equations that can be solved for the entries of B:

```
> zero2:= map(Simf,connformsvec1 - dualformsvec);
> eqns:= {op(ScalarForm(zero2[1])), op(ScalarForm(zero2[2]))};
> solve(eqns, {b[1,1], b[1,2], b[2,2]});
```

The resulting output gives complicated expressions involving `RootOf`, but it is straightforward to check that the solution yields the matrix B in equation (8.22). Make these assignments so that we can go on to the next step:

```
> b[1,1]:= (diff(f(u,v),u,u)*diff(f(u,v),v,v)
  - diff(f(u,v),u,v)^2)^(1/8)/sqrt(diff(f(u,v),u,u));
  b[1,2]:= -diff(f(u,v),u,v)/(sqrt(diff(f(u,v),u,u))*
  (diff(f(u,v),u,u)*diff(f(u,v),v,v)
```

```
  - diff(f(u,v),u,v)^2)^(3/8));
  b[2,2]:= sqrt(diff(f(u,v),u,u))/
  (diff(f(u,v),u,u)*diff(f(u,v),v,v)
  - diff(f(u,v),u,v)^2)^(3/8);
```

Set up a substitution for the Maurer-Cartan forms for this new frame field:

```
> framesub:= [omega[1] = Simf(dualformsvec[1]),
  omega[2] = Simf(dualformsvec[2])];
> framebacksub:= makebacksub(framesub);
```

We still need to solve for r_1, r_2. This requires computing $d\tilde{\mathbf{e}}_3$ so that we can find $\tilde{\omega}^3_3$ and set it equal to zero. We can do this fairly compactly as follows: If we let A be the matrix

$$A = \left[\tilde{\mathbf{e}}_1(\mathbf{u})\ \tilde{\mathbf{e}}_2(\mathbf{u})\ \tilde{\mathbf{e}}_3(\mathbf{u})\right],$$

then we have

$$d\tilde{\mathbf{e}}_3 = A \begin{bmatrix} \tilde{\omega}^1_3 \\ \tilde{\omega}^2_3 \\ \tilde{\omega}^3_3 \end{bmatrix}.$$

Therefore, $\tilde{\omega}^3_3$ is the last entry of the vector $A^{-1}d\tilde{\mathbf{e}}_3$, and we can solve for r_1, r_2 by setting the scalar coefficients of $\tilde{\omega}^3_3$ equal to zero.

```
> A:= Matrix([e1, e2, e3]);
> de3:= map(Simf, subs(framebacksub, map(d, e3)));
> connformsvec2:= map(Simf, MatrixInverse(A).de3);
> zero3:= connformsvec2[3];
> solve({op(ScalarForm(zero3))}, {r1, r2});
> assign(%);
```

You should check that the resulting expressions for r_1, r_2 agree with equation (8.23).

Finally, the equi-affine mean curvature L of Σ can be computed from $\tilde{\omega}^1_3$ and $\tilde{\omega}^2_3$, which are the first two entries of `connformsvec2`. First, we need to use the substitution to express these forms as linear combinations of $(\tilde{\omega}^1, \tilde{\omega}^2)$ and then compute the trace of the associated coefficient matrix:

```
> omega[1,3]:= Simf(subs(framebacksub, connformsvec2[1]));
  omega[2,3]:= Simf(subs(framebacksub, connformsvec2[2]));
> LL:= (1/2)*(Simf(pick(omega[1,3], omega[1])
  + pick(omega[2,3], omega[2])));
```

Now, aren't you glad that you didn't have to compute L by hand?

Chapter 9

Pseudospherical surfaces and Bäcklund's theorem

9.1. Introduction

In this chapter, we will show how moving frames may be used to prove *Bäcklund's theorem.* Bäcklund's theorem concerns surfaces of constant negative Gauss curvature, also known as *pseudospherical* surfaces. The best-known pseudospherical surface is, of course, the *pseudosphere*, which is the surface of revolution obtained by revolving the tractrix

$$\alpha(t) = {}^t[t - \tanh t,\ \operatorname{sech} t,\ 0]$$

about the x-axis. But there are infinitely many other pseudospherical surfaces as well.

Bäcklund's theorem is based on a beautiful geometric construction that starts with a given pseudospherical surface and produces from it a 2-parameter family of new pseudospherical surfaces. (See [**RS82**] for a discussion of Bäcklund's original construction.) The construction can be iterated, thereby producing an arbitrary number of increasingly complicated families of pseudospherical surfaces from a single starting surface. For example, we might take the pseudosphere as our initial surface and use Bäcklund's construction to generate new families of surfaces.

In §9.4, we will see how pseudospherical surfaces are intimately connected with solutions $\phi(x, y)$ of the partial differential equation known as the *sine-Gordon equation*:

$$\phi_{xy} = \sin(\phi). \tag{9.1}$$

This is a nonlinear partial differential equation, and it is one of a number of nonlinear PDEs known as "integrable systems". PDEs in this class share a number of important features, including special families of solutions known as "soliton" solutions. As we will see, Bäcklund's geometric construction for pseudospherical surfaces gives rise to a corresponding transformation between solutions of equation (9.1). This transformation is called a *Bäcklund transformation*, and, just as Bäcklund's construction generates new pseudospherical surfaces from a known pseudospherical surface, the Bäcklund transformation for equation (9.1) can be used to generate new solutions from any known solution (cf. Exercise 9.17). Applying this transformation, even starting from the trivial solution $\phi(x, y) = 0$, produces nontrivial new solutions; in fact, the soliton solutions of (9.1) can be constructed in precisely this way.

9.2. Line congruences

Bäcklund's construction begins with the notion of a *line congruence*. Roughly, a line congruence in $\mathbb{E}^3$ is simply a 2-parameter family of lines in $\mathbb{E}^3$. A more formal definition requires that the set of lines in $\mathbb{E}^3$ be given a topological structure so that it becomes a manifold, called the *affine Grassmannian* $G_1(\mathbb{E}^3)$. This can be accomplished as follows: Any line ℓ in $\mathbb{E}^3$ is determined by a pair $(\mathbf{x}, \mathbf{e})$, where $\mathbf{x}$ is a point on ℓ and $\mathbf{e}$ is a unit vector parallel to ℓ. The set of all such pairs is equal to the product $\mathbb{E}^3 \times \mathbb{S}^2$. Now define an equivalence relation $\sim$ on $\mathbb{E}^3 \times \mathbb{S}^2$ by the condition that $(\mathbf{x}, \mathbf{e}) \sim (\mathbf{y}, \mathbf{f})$ if and only if the pairs $(\mathbf{x}, \mathbf{e})$ and $(\mathbf{y}, \mathbf{f})$ determine the same line.

Exercise 9.1. Show that $(\mathbf{x}, \mathbf{e}) \sim (\mathbf{y}, \mathbf{f})$ if and only if $\mathbf{f} = \pm\mathbf{e}$ and $\mathbf{y} = \mathbf{x} + t\mathbf{e}$ for some $t \in \mathbb{R}$.

The affine Grassmannian $G_1(\mathbb{E}^3)$ is then defined to be the set of equivalence classes

$$G_1(\mathbb{E}^3) = \left(\mathbb{E}^3 \times \mathbb{S}^2\right) / \sim,$$

with the line $\ell \in G_1(\mathbb{E}^3)$ determined by the pair $(\mathbf{x}, \mathbf{e})$ denoted by $[(\mathbf{x}, \mathbf{e})]$. The set $G_1(\mathbb{E}^3)$ can be given the structure of a smooth manifold of dimension 4 in a natural way. With this definition in hand, the formal definition of a line congruence is as follows:

Definition 9.2. A *line congruence* in $\mathbb{E}^3$ is an immersed surface in $G_1(\mathbb{E}^3)$.

Line congruences were the object of much study in the nineteenth century; for a thorough treatment, see [**Eis60**].

Example 9.3. Let U be an open set in $\mathbb{R}^2$, and let $\mathbf{x} : U \to \mathbb{E}^3$ be an immersion whose image is a regular surface $\Sigma \subset \mathbb{E}^3$. For each $\mathbf{u} \in U$, let $\ell(\mathbf{u})$ denote the line in $\mathbb{E}^3$ passing through the point $\mathbf{x}(\mathbf{u})$ and parallel to the normal vector $\mathbf{e}_3(\mathbf{u})$. Then the collection

$$\{\ell(\mathbf{u}) \mid \mathbf{u} \in U\}$$

is a line congruence in $\mathbb{E}^3$. A line congruence of this type is called a *normal congruence.*

Given an open set $U \subset \mathbb{R}^2$ and a line congruence $\ell : U \to G_1(\mathbb{E}^3)$, we can express the congruence—in infinitely many different ways—as

$$\ell(\mathbf{u}) = [(\mathbf{x}(\mathbf{u}), \mathbf{e}(\mathbf{u})],$$

where $\mathbf{x}(\mathbf{u})$ is a point on the line $\ell(\mathbf{u})$ and $\mathbf{e}(\mathbf{u})$ is a unit vector parallel to $\ell(\mathbf{u})$. If $\mathbf{x}(\mathbf{u})$ is chosen to be a smooth function of $\mathbf{u}$, then the surface $\Sigma = \mathbf{x}(U)$ is called a *surface of reference* for the line congruence. Any other surface of reference $\widetilde{\Sigma} = \tilde{\mathbf{x}}(U)$ for the congruence can then be parametrized as

$$\tilde{\mathbf{x}}(\mathbf{u}) = \mathbf{x}(\mathbf{u}) + \lambda(\mathbf{u})\mathbf{e}(\mathbf{u})$$

for some smooth, real-valued function λ on U.

For a generic line congruence $\ell : U \to G_1(\mathbb{E}^3)$, there are two distinguished surfaces of reference $\Sigma = \mathbf{x}(U)$, $\widetilde{\Sigma} = \tilde{\mathbf{x}}(U)$, called *focal surfaces* of the congruence. The definition and construction of the focal surfaces are rather involved and will be omitted here (for details, see [**Eis60**]); for our purposes, the key property of these surfaces is that each line in the congruence is tangent to both focal surfaces. We will assume that both focal surfaces are parametrized as above, so that for each $\mathbf{u} \in U$, the points $\mathbf{x}(\mathbf{u}), \tilde{\mathbf{x}}(\mathbf{u})$ are contained in the line $\ell(\mathbf{u})$ and the line $\ell(\mathbf{u})$ is tangent to the surfaces $\Sigma, \widetilde{\Sigma}$ at the points $\mathbf{x}(\mathbf{u}), \tilde{\mathbf{x}}(\mathbf{u})$, respectively.

9.3. Bäcklund's theorem

Bäcklund's theorem concerns a special category of line congruences known as *pseudospherical* congruences.

Definition 9.4. Let $U \subset \mathbb{R}^2$, and let $\ell : U \to G_1(\mathbb{E}^3)$ be a line congruence in $\mathbb{E}^3$ with focal surfaces $\Sigma, \widetilde{\Sigma}$, parametrized as above. The congruence is called *pseudospherical* if the following two conditions hold:

(1) The distance $r = |\tilde{\mathbf{x}}(\mathbf{u}) - \mathbf{x}(\mathbf{u})|$ is a constant, independent of $\mathbf{u}$.

(2) The angle α (assumed to be nonzero) between the surface normal vectors $\mathbf{e}_3(\mathbf{u}), \tilde{\mathbf{e}}_3(\mathbf{u})$ to the surfaces $\Sigma, \tilde{\Sigma}$ at the points $\mathbf{x}(\mathbf{u}), \tilde{\mathbf{x}}(\mathbf{u})$, respectively, is a constant, independent of $\mathbf{u}$.

Theorem 9.5 (Bäcklund).

(1) *Let $U \subset \mathbb{R}^2$, and suppose that $\ell : U \to G_1(\mathbb{E}^3)$ is a pseudospherical line congruence in $\mathbb{E}^3$ with focal surfaces $\Sigma, \tilde{\Sigma}$. Then both Σ and $\tilde{\Sigma}$ have constant negative Gauss curvature $K = -\frac{\sin^2(\alpha)}{r^2}$, where r, α are as in Definition 9.4.*

(2) *If $\Sigma \subset \mathbb{E}^3$ is any surface of constant negative Gauss curvature $K = -\frac{\sin^2(\alpha)}{r^2}$, then given any point $\mathbf{x}_0 \in \Sigma$ and any unit tangent vector $\mathbf{e}_0 \in T_{\mathbf{x}_0}\Sigma$ that is not a principal direction at $\mathbf{x}_0$, there exists a unique surface $\tilde{\Sigma} \subset \mathbb{E}^3$ and a pseudospherical line congruence with focal surfaces $\Sigma, \tilde{\Sigma}$ such that if $\tilde{\mathbf{x}}_0$ is the point in $\tilde{\Sigma}$ corresponding to $\mathbf{x}_0$, then $\tilde{\mathbf{x}}_0 - \mathbf{x}_0 = r\mathbf{e}_0$ and the angle between the surface normal vectors to $\Sigma, \tilde{\Sigma}$ at the points $\mathbf{x}_0, \tilde{\mathbf{x}}_0$, respectively, is α.*

The construction in part (2) of Theorem 9.5 is due to Bianchi and Bäcklund ([**B8̈3**], [**Bia79**]) and is called a *Bäcklund transformation*; the terminology refers to the idea that the new surface $\tilde{\Sigma}$ is a "transformation" of the original surface Σ.

Bäcklund's theorem can be proved using local coordinates on the surfaces $\Sigma, \tilde{\Sigma}$, but the computations are rather ugly. The proof can be greatly simplified by using the method of moving frames because frame fields can be adapted to the geometry of the problem in a way that local coordinates cannot. Whereas in previous chapters we have adapted our frames according to the geometry of a single surface, here we have to consider *two* surfaces and the geometric conditions relating them. We will use these considerations to guide our choices of orthonormal frame fields on the surfaces $\Sigma, \tilde{\Sigma}$. (This proof is taken from [**CT80**].)

Proof. (1) Let $\mathbf{x}, \tilde{\mathbf{x}} : U \to \mathbb{E}^3$ be the parametrizations of $\Sigma, \tilde{\Sigma}$, respectively, induced from the line congruence $\ell : U \to G_1(\mathbb{E}^3)$ as above. We can choose orthonormal frame fields $(\underline{\mathbf{e}}_1(\mathbf{u}), \underline{\mathbf{e}}_2(\mathbf{u}), \underline{\mathbf{e}}_3(\mathbf{u}))$ along Σ and $(\underline{\tilde{\mathbf{e}}}_1(\mathbf{u}), \underline{\tilde{\mathbf{e}}}_2(\mathbf{u}), \underline{\tilde{\mathbf{e}}}_3(\mathbf{u}))$ along $\tilde{\Sigma}$ such that:

(1) $\underline{\mathbf{e}}_3(\mathbf{u})$ is the unit normal vector to Σ at $\mathbf{x}(\mathbf{u})$ and $\underline{\tilde{\mathbf{e}}}_3(\mathbf{u})$ is the unit normal vector to $\tilde{\Sigma}$ at $\tilde{\mathbf{x}}(\mathbf{u})$; therefore, $(\underline{\mathbf{e}}_1(\mathbf{u}), \underline{\mathbf{e}}_2(\mathbf{u}))$ span the tangent space $T_{\mathbf{x}(\mathbf{u})}\Sigma$ and $(\underline{\tilde{\mathbf{e}}}_1(\mathbf{u}), \underline{\tilde{\mathbf{e}}}_2(\mathbf{u}))$ span the tangent space $T_{\tilde{\mathbf{x}}(\mathbf{u})}\tilde{\Sigma}$.

(2) $\underline{\mathbf{e}}_1(\mathbf{u}) = \tilde{\underline{\mathbf{e}}}_1(\mathbf{u})$ is the common unit tangent vector to both surfaces in the direction of $\tilde{\mathbf{x}}(\mathbf{u}) - \mathbf{x}(\mathbf{u})$.

Remark 9.6. The notation $\underline{\mathbf{e}}_i(\mathbf{u})$ is intended to distinguish these orthonormal frame vectors from the *principal* orthonormal frame vectors $\mathbf{e}_i(\mathbf{u})$ that will be introduced in §9.4. This notational distinction will become important in §9.5, when we need to consider both of these orthonormal frame fields simultaneously! Similarly, we will denote the Maurer-Cartan forms associated to the frame field $(\underline{\mathbf{e}}_1(\mathbf{u}),\ \underline{\mathbf{e}}_2(\mathbf{u}),\ \underline{\mathbf{e}}_3(\mathbf{u}))$ by $(\underline{\bar{\omega}}^i, \underline{\bar{\omega}}^i_j)$, and those associated to the frame field $(\tilde{\underline{\mathbf{e}}}_1(\mathbf{u}),\ \tilde{\underline{\mathbf{e}}}_2(\mathbf{u}),\ \tilde{\underline{\mathbf{e}}}_3(\mathbf{u}))$ by $(\underline{\tilde{\bar{\omega}}}^i, \underline{\tilde{\bar{\omega}}}^i_j)$.

***Exercise 9.7.** Show that it follows from these conditions (and the definition of α as the angle between $\underline{\mathbf{e}}_3(\mathbf{u})$, $\tilde{\underline{\mathbf{e}}}_3(\mathbf{u})$) that

$$\begin{aligned}\tilde{\underline{\mathbf{e}}}_1(\mathbf{u}) &= \underline{\mathbf{e}}_1(\mathbf{u}),\\ \tilde{\underline{\mathbf{e}}}_2(\mathbf{u}) &= \cos(\alpha)\underline{\mathbf{e}}_2(\mathbf{u}) + \sin(\alpha)\underline{\mathbf{e}}_3(\mathbf{u}),\\ \tilde{\underline{\mathbf{e}}}_3(\mathbf{u}) &= -\sin(\alpha)\underline{\mathbf{e}}_2(\mathbf{u}) + \cos(\alpha)\underline{\mathbf{e}}_3(\mathbf{u})\end{aligned} \tag{9.2}$$

and that

$$\tilde{\mathbf{x}}(\mathbf{u}) = \mathbf{x}(\mathbf{u}) + r\underline{\mathbf{e}}_1(\mathbf{u}). \tag{9.3}$$

Let $(\underline{\bar{\omega}}^i, \underline{\bar{\omega}}^i_j)$ denote the pullbacks of the Maurer-Cartan forms (ω^i, ω^i_j) on $E(3)$ to U via the frame field $(\mathbf{x}(\mathbf{u}); \underline{\mathbf{e}}_1(\mathbf{u}), \underline{\mathbf{e}}_2(\mathbf{u}), \underline{\mathbf{e}}_3(\mathbf{u}))$ for Σ, and let $(\underline{\tilde{\bar{\omega}}}^i, \underline{\tilde{\bar{\omega}}}^i_j)$ denote the pullbacks of (ω^i, ω^i_j) to U via the frame field $(\tilde{\mathbf{x}}(\mathbf{u}); \tilde{\underline{\mathbf{e}}}_1(\mathbf{u}), \tilde{\underline{\mathbf{e}}}_2(\mathbf{u}), \tilde{\underline{\mathbf{e}}}_3(\mathbf{u}))$ for $\tilde{\Sigma}$.

***Exercise 9.8.** (a) Show that taking the exterior derivative of equation (9.3) and applying the Cartan structure equations (3.1) yields

$$\tilde{\underline{\mathbf{e}}}_1\, \underline{\tilde{\bar{\omega}}}^1 + \tilde{\underline{\mathbf{e}}}_2\, \underline{\tilde{\bar{\omega}}}^2 = \underline{\mathbf{e}}_1\, \underline{\bar{\omega}}^1 + \underline{\mathbf{e}}_2\, \underline{\bar{\omega}}^2 + r(\underline{\mathbf{e}}_2\, \underline{\bar{\omega}}^2_1 + \underline{\mathbf{e}}_3\, \underline{\bar{\omega}}^3_1). \tag{9.4}$$

Then substitute the expressions (9.2) for $(\tilde{\underline{\mathbf{e}}}_1(\mathbf{u}), \tilde{\underline{\mathbf{e}}}_2(\mathbf{u}))$ into equation (9.4) to obtain

$$\underline{\mathbf{e}}_1\, \underline{\tilde{\bar{\omega}}}^1 + \underline{\mathbf{e}}_2 \cos(\alpha)\, \underline{\tilde{\bar{\omega}}}^2 + \underline{\mathbf{e}}_3 \sin(\alpha)\, \underline{\tilde{\bar{\omega}}}^2 = \underline{\mathbf{e}}_1\, \underline{\bar{\omega}}^1 + \underline{\mathbf{e}}_2(\underline{\bar{\omega}}^2 + r\, \underline{\bar{\omega}}^2_1) + \underline{\mathbf{e}}_3(r\, \underline{\bar{\omega}}^3_1).$$

Conclude that we have the following relationships between the Maurer-Cartan forms on Σ and those on $\tilde{\Sigma}$:

$$\begin{aligned}\underline{\tilde{\bar{\omega}}}^1 &= \underline{\bar{\omega}}^1,\\ \cos(\alpha)\, \underline{\tilde{\bar{\omega}}}^2 &= \underline{\bar{\omega}}^2 + r\, \underline{\bar{\omega}}^2_1,\\ \sin(\alpha)\, \underline{\tilde{\bar{\omega}}}^2 &= r\, \underline{\bar{\omega}}^3_1.\end{aligned} \tag{9.5}$$

(b) Show that the last two equations in (9.5) imply that

$$\underline{\bar{\omega}}^2 + r\, \underline{\bar{\omega}}^2_1 = r\cot(\alpha)\, \underline{\bar{\omega}}^3_1. \tag{9.6}$$

(c) Use the fact that $\bar{\omega}^i_j = \langle d\underline{\mathbf{e}}_j, \underline{\mathbf{e}}_i \rangle$ (and similarly for $\tilde{\bar{\omega}}^i_j$), the expressions (9.2), and equation (9.6) to show that

$$\tilde{\bar{\omega}}^3_1 = \frac{\sin(\alpha)}{r}\bar{\omega}^2, \qquad \tilde{\bar{\omega}}^3_2 = \bar{\omega}^3_2. \tag{9.7}$$

Next, recall that the coefficients (h_{ij}) of the second fundamental form of Σ are defined by the equations

$$\begin{aligned} \bar{\omega}^3_1 &= h_{11}\bar{\omega}^1 + h_{12}\bar{\omega}^2, \\ \bar{\omega}^3_2 &= h_{12}\bar{\omega}^1 + h_{22}\bar{\omega}^2. \end{aligned} \tag{9.8}$$

Because we have *not* made any attempt to arrange for $\underline{\mathbf{e}}_1(\mathbf{u})$ and $\underline{\mathbf{e}}_2(\mathbf{u})$ to be principal vector fields, we should not expect to have $h_{12} = 0$; in fact, the following exercise shows that h_{12} *cannot* be equal to zero.

***Exercise 9.9.** Use the first and third equations in (9.5), the first equation in (9.8), and the fact that $(\tilde{\bar{\omega}}^1, \tilde{\bar{\omega}}^2)$ are linearly independent 1-forms on U to conclude that $h_{12} \neq 0$.

Finally, consider the Gauss curvature $\tilde{K}$ of $\tilde{\Sigma}$. Recall that

$$\tilde{\bar{\omega}}^3_1 \wedge \tilde{\bar{\omega}}^3_2 = \tilde{K}\,\tilde{\bar{\omega}}^1 \wedge \tilde{\bar{\omega}}^2 \tag{9.9}$$

(cf. Exericse 4.47).

***Exercise 9.10.** (a) Use equations (9.7) and (9.8) to show that

$$\tilde{\bar{\omega}}^3_1 \wedge \tilde{\bar{\omega}}^3_2 = -\frac{\sin(\alpha)}{r} h_{12}\,\bar{\omega}^1 \wedge \bar{\omega}^2.$$

(b) Use equation (9.9), the first and third equations of (9.5), and (9.8) to show that

$$\tilde{\bar{\omega}}^3_1 \wedge \tilde{\bar{\omega}}^3_2 = \tilde{K}\frac{r}{\sin(\alpha)} h_{12}\,\bar{\omega}^1 \wedge \bar{\omega}^2.$$

(c) Conclude from parts (a) and (b) and the fact that $h_{12} \neq 0$ that

$$\tilde{K} = -\frac{\sin^2(\alpha)}{r^2}.$$

An analogous argument shows that $K = -\frac{\sin^2(\alpha)}{r^2}$ as well.

The proof of part (2) of Theorem 9.5 involves concepts from the theory of exterior differential systems similar to those needed for the proof of Lemma 4.12. The key step involves constructing an adapted orthonormal frame field $(\underline{\mathbf{e}}_1(\mathbf{u}), \underline{\mathbf{e}}_2(\mathbf{u}), \underline{\mathbf{e}}_3(\mathbf{u}))$ along Σ with the property that a parametrization $\tilde{\mathbf{x}} : U \to \mathbb{E}^3$ for the desired surface $\tilde{\Sigma}$ will be given by

$$\tilde{\mathbf{x}}(\mathbf{u}) = \mathbf{x}(\mathbf{u}) + r\underline{\mathbf{e}}_1(\mathbf{u}). \tag{9.10}$$

The Maurer-Cartan forms for such a frame field must satisfy equation (9.6); in fact (as we will see in §9.5), equation (9.6) is equivalent to an overdetermined system of partial differential equations for the frame field $(\underline{\mathbf{e}}_1(\mathbf{u}), \underline{\mathbf{e}}_2(\mathbf{u}), \underline{\mathbf{e}}_3(\mathbf{u}))$, and the compatibility condition for this system is precisely the condition that Σ has Gauss curvature $K = -\frac{\sin^2(\alpha)}{r^2}$. The initial condition $\underline{\mathbf{e}}_1(\mathbf{u}_0) = \mathbf{e}_0 \in T_{\mathbf{x}_0}\Sigma$ then determines the desired frame field uniquely, and the remainder of the proof consists of showing that the surface $\widetilde{\Sigma}$ defined by (9.10) satisfies all the desired conditions. □

The proof of part (2) of Theorem 9.5 shows how, given a surface of constant negative Gauss curvature K, one can construct new pseudospherical surfaces $\widetilde{\Sigma}$. The 2-parameter family of such surfaces alluded to earlier arises as follows: One parameter comes from the choice of constants r, α such that $K = -\frac{\sin^2(\alpha)}{r^2}$, and one comes from the choice of a non-principal unit vector $\mathbf{e}_0 \in T_{\mathbf{x}_0}\Sigma$. The choice of r, α determines the coefficients of the PDE system (9.6), while the choice of $\mathbf{e}_0$ determines the initial conditions that give rise to a particular solution of this PDE system.

9.4. Pseudospherical surfaces and the sine-Gordon equation

Let Σ be a pseudospherical surface, and for simplicity assume that its Gauss curvature is $K = -1$. Since the Gauss curvature of Σ is negative, Σ cannot have any umbilic points; consequently, it can be shown (for a proof, see [**dC76**]) that every point $\mathbf{x} \in \Sigma$ has a neighborhood for which there exists a local parametrization $\mathbf{x} : U \to \mathbb{E}^3$ of Σ whose coordinate curves are principal curves in Σ (cf. Exercises 4.24 and 4.27). As in Exercise 4.24, we can then choose the adapted orthonormal frame field

$$\mathbf{e}_1(\mathbf{u}) = \frac{1}{\sqrt{E}}\mathbf{x}_u, \qquad \mathbf{e}_2(\mathbf{u}) = \frac{1}{\sqrt{G}}\mathbf{x}_v, \qquad \mathbf{e}_3(\mathbf{u}) = \mathbf{e}_1(\mathbf{u}) \times \mathbf{e}_2(\mathbf{u})$$

along Σ. (Note that, unlike in §9.3, we are only considering a single pseudospherical surface, so there is no line congruence to take into consideration when choosing an adapted frame field.) Then, as we saw in Exercises 4.24 and 4.27, the coefficients of the first and second fundamental forms satisfy $F = f = 0$, and the associated Maurer-Cartan forms $(\bar{\omega}^i, \bar{\omega}^i_j)$ are given by

$$\begin{aligned} &\bar{\omega}^1 = \sqrt{E}\,du, \qquad \bar{\omega}^2 = \sqrt{G}\,dv, \\ &\bar{\omega}^3_1 = \frac{e}{\sqrt{E}}\,du = \kappa_1\sqrt{E}\,du, \qquad \bar{\omega}^3_2 = \frac{g}{\sqrt{G}}\,dv = \kappa_2\sqrt{G}\,dv, \\ &\bar{\omega}^1_2 = \frac{1}{2\sqrt{EG}}\,(E_v\,du - G_u\,dv), \end{aligned} \tag{9.11}$$

where $\kappa_1 = \frac{e}{E}$, $\kappa_2 = \frac{g}{G}$ are the principal curvatures of Σ.

The following exercises will show how the surface $\Sigma = \mathbf{x}(U)$ gives rise to a solution of the sine-Gordon equation (9.1). First, we investigate the Gauss and Codazzi equations for Σ and show that the parametrization $\mathbf{x} : U \to \mathbb{E}^3$ can be fine-tuned to arrange that the first and second fundamental forms of Σ can be expressed nicely in terms of a single function $\psi : U \to \mathbb{R}$.

***Exercise 9.11.** (a) Show that the Codazzi equations of Exercise 4.41(f) can be written in the form

$$\frac{\partial \kappa_1}{\partial v} = (\kappa_2 - \kappa_1)\frac{\partial}{\partial v}\left(\ln(\sqrt{E})\right), \qquad \frac{\partial \kappa_2}{\partial u} = (\kappa_1 - \kappa_2)\frac{\partial}{\partial u}\left(\ln(\sqrt{G})\right). \tag{9.12}$$

(b) Divide equations (9.12) by $(\kappa_1 - \kappa_2)$, multiply the left-hand sides by $\frac{\kappa_1}{\kappa_1}$ and $\frac{\kappa_2}{\kappa_2}$, respectively, and use the Gauss equation $\kappa_1\kappa_2 = -1$ to show that

$$\begin{aligned} \frac{\partial}{\partial v}\left(\ln(\kappa_1^2 + 1)\right) &= -2\frac{\partial}{\partial v}\left(\ln(\sqrt{E})\right), \\ \frac{\partial}{\partial u}\left(\ln(\kappa_2^2 + 1)\right) &= -2\frac{\partial}{\partial u}\left(\ln(\sqrt{G})\right). \end{aligned} \tag{9.13}$$

(c) Integrate equations (9.13) and conclude that

$$\kappa_1^2 + 1 = \frac{c_1(u)}{E}, \qquad \kappa_2^2 + 1 = \frac{c_2(v)}{G} \tag{9.14}$$

for some functions $c_1(u), c_2(v) > 0$.

(d) Show that under a change of coordinates of the form

$$\tilde{u} = h(u), \qquad \tilde{v} = k(v), \tag{9.15}$$

the coefficients of the first fundamental form with respect to the coordinates $(\tilde{u}, \tilde{v})$ become

$$\tilde{E} = \frac{1}{(h'(u))^2}E, \qquad \tilde{G} = \frac{1}{(k'(v))^2}G.$$

(The functions κ_1, κ_2, however, are invariants and are unchanged by the coordinate transformation.) Conclude that the functions $h(u), k(v)$ can be chosen so as to arrange that

$$\tilde{c}_1(\tilde{u}) = \tilde{c}_2(\tilde{v}) = 1.$$

(e) Now, assume that the coordinate functions (u, v) (without the tildes) have been chosen so that $c_1(u) = c_2(v) = 1$. Show that there exists a function $\psi(u, v)$ such that

$$\begin{aligned} \kappa_1 &= \tan(\psi), & \kappa_2 &= -\cot(\psi), \\ E &= \cos^2(\psi), & G &= \sin^2(\psi). \end{aligned}$$

Thus, the first and second fundamental forms of Σ are

$$\mathrm{I} = \cos^2(\psi)\, du^2 + \sin^2(\psi)\, dv^2,$$
$$\mathrm{II} = \sin(\psi)\cos(\psi)\,(du^2 - dv^2).$$

(Hint: Recall that $\kappa_1\kappa_2 = -1$ and that the tangent and cotangent functions are surjective onto $\mathbb{R}$.)

Next, we give a geometric interpretation of the function ψ and show that the function $\phi = 2\psi$ is a solution of the sine-Gordon equation (9.1).

***Exercise 9.12.** (a) Show that the angle between the asymptotic directions at any point $\mathbf{x}(u,v)$ is equal to $2\psi(u,v)$. (Hint: The asymptotic directions are the null directions for the second fundamental form; in this case, they are represented by the tangent vectors $\mathbf{x}_u \pm \mathbf{x}_v$.)

(b) Show that $\bar{\omega}_2^1$ (cf. equation (9.11)) is given by

$$\bar{\omega}_2^1 = -\psi_v\, du - \psi_u\, dv.$$

(c) Use the Gauss equation

$$d\bar{\omega}_2^1 = K\bar{\omega}^1 \wedge \bar{\omega}^2$$

to show that the function ψ satisfes the PDE

(9.16) $$\psi_{uu} - \psi_{vv} = \sin(\psi)\cos(\psi).$$

(d) Let $\phi = 2\psi$, so that $\phi(u,v)$ is the angle between the asymptotic directions at the point $\mathbf{x}(u,v)$. Show that equation (9.16) is equivalent to the PDE

(9.17) $$\phi_{uu} - \phi_{vv} = \sin(\phi)$$

for the function ϕ.

(e) Consider the change of coordinates

$$x = \frac{1}{2}(u+v), \qquad y = \frac{1}{2}(u-v).$$

Show that in terms of the (x,y)-coordinates, the first and second fundamental forms of Σ are given by

(9.18) $$\mathrm{I} = dx^2 + 2\cos(2\psi)\, dx\, dy + dy^2,$$
$$\mathrm{II} = 2\sin(2\psi)\, dx\, dy.$$

Note that the x- and y-coordinate directions are now the asymptotic directions at each point of Σ; for this reason, (x,y) are called *asymptotic coordinates* on Σ, and the corresponding parametrization is called an *asymptotic parametrization* of Σ.

(f) Show that equation (9.17) is equivalent to equation (9.1).

Remark 9.13. Equations (9.17) and (9.1) are both referred to as the *sine-Gordon equation.* The local coordinates (u, v) of equation (9.17) are called *space-time* coordinates because the left-hand side has the same form as the wave equation

$$\phi_{uu} - \phi_{vv} = 0,$$

where u is often thought of as a time coordinate and v as a spatial coordinate. The local coordinates (x, y) of equation (9.1), on the other hand, are called *null* or *characteristic* coordinates. The term "characteristic" comes from the fact that the x- and y-coordinate curves are the characteristic curves for the PDE (9.1), while the term "null" arises from thinking of the (u, v)-plane as the Minkowski space $\mathbb{M}^{1,1}$ with its standard metric, for which the x- and y-coordinate curves are the null lines.

We have shown that, at least locally, any pseudospherical surface Σ determines a solution ϕ of the sine-Gordon equation (9.1). The following exercise shows that the converse is true as well.

***Exercise 9.14.** Let $\phi : U \to \mathbb{R}$ be any solution of the sine-Gordon equation (9.1), and let $\psi(x, y) = \frac{1}{2}\phi(x, y)$. Define 1-forms $(\bar{\omega}^1, \bar{\omega}^2, \bar{\omega}^3_1, \bar{\omega}^3_2, \bar{\omega}^1_2)$ by

$$\begin{aligned} \bar{\omega}^1 &= \cos(\psi)(dx + dy), & \bar{\omega}^2 &= \sin(\psi)(dx - dy), \\ \bar{\omega}^3_1 &= \sin(\psi)(dx + dy), & \bar{\omega}^3_2 &= -\cos(\psi)(dx - dy), \\ & & \bar{\omega}^1_2 &= -\psi_x \, dx + \psi_y \, dy. \end{aligned} \tag{9.19}$$

(a) Check that the first and second fundamental forms

$$\begin{aligned} \mathrm{I} &= (\bar{\omega}^1)^2 + (\bar{\omega}^2)^2, \\ \mathrm{II} &= \bar{\omega}^3_1 \, \bar{\omega}^1 + \bar{\omega}^3_2 \, \bar{\omega}^2 \end{aligned}$$

agree with those in equations (9.18).

(b) Check that these 1-forms (together with $\bar{\omega}^3 = 0$) satisfy the Cartan structure equations (3.8). (Hint: This result depends on the fact that ϕ is a solution of equation (9.1).)

(c) Conclude from Bonnet's theorem (cf. Theorem 4.39) that there exists an immersed surface $\mathbf{x} : U \to \mathbb{E}^3$ whose first and second fundamental forms are given by (9.18). In particular, the Gauss curvature of the surface $\Sigma = \mathbf{x}(U)$ is $K = -1$, and the angle between the asymptotic directions at the point $\mathbf{x}(x, y)$ is equal to $\phi(x, y)$.

Remark 9.15. In recent years, many other integrable systems have been shown to be connected with pseudospherical geometry, and this connection provides an important tool for studying the space of solutions to these equations. This connection was first considered by Chern and Tenenblat [**CT86**] and further explored by Reyes [**Rey98**] and others.

9.5. The Bäcklund transformation for the sine-Gordon equation

In this section, we will see how the geometric Bäcklund transformation between pseudospherical surfaces gives rise to a corresponding analytic transformation between solutions of the sine-Gordon equation (9.1).

Suppose that we have a Bäcklund transformation between two pseudospherical surfaces $\Sigma, \widetilde{\Sigma}$ of Gauss curvature $K = -1$. (Note that the condition $K = -1$ implies that $r = \sin(\alpha)$.) Let $(\underline{\mathbf{e}}_1(\mathbf{u}), \underline{\mathbf{e}}_2(\mathbf{u}), \underline{\mathbf{e}}_3(\mathbf{u}))$ be the orthonormal frame field on Σ adapted to the Bäcklund transformation as in §9.3, and let $(\mathbf{e}_1(\mathbf{u}), \mathbf{e}_2(\mathbf{u}), \mathbf{e}_3(\mathbf{u}))$ be the principal adapted frame field on Σ as in §9.4.

Let $\eta(\mathbf{u})$ denote the angle between $\mathbf{e}_1(\mathbf{u})$ and $\underline{\mathbf{e}}_1(\mathbf{u})$. The following exercise shows how the function η is related to the function ψ of §9.4.

***Exercise 9.16.** (a) Show that

$$[\underline{\mathbf{e}}_1(\mathbf{u})\ \underline{\mathbf{e}}_2(\mathbf{u})] = [\mathbf{e}_1(\mathbf{u})\ \mathbf{e}_2(\mathbf{u})] \begin{bmatrix} \cos(\eta(\mathbf{u})) & -\sin(\eta(\mathbf{u})) \\ \sin(\eta(\mathbf{u})) & \cos(\eta(\mathbf{u})) \end{bmatrix}. \tag{9.20}$$

(b) Show that

$$\begin{bmatrix} \underline{\bar{\omega}}^1 \\ \underline{\bar{\omega}}^2 \end{bmatrix} = \begin{bmatrix} \cos(\eta) & \sin(\eta) \\ -\sin(\eta) & \cos(\eta) \end{bmatrix} \begin{bmatrix} \bar{\omega}^1 \\ \bar{\omega}^2 \end{bmatrix} = \begin{bmatrix} \cos(\psi - \eta) & \cos(\psi + \eta) \\ \sin(\psi - \eta) & -\sin(\psi + \eta) \end{bmatrix} \begin{bmatrix} dx \\ dy \end{bmatrix},$$

$$\begin{bmatrix} \underline{\bar{\omega}}^3_1 \\ \underline{\bar{\omega}}^3_2 \end{bmatrix} = \begin{bmatrix} \cos(\eta) & \sin(\eta) \\ -\sin(\eta) & \cos(\eta) \end{bmatrix} \begin{bmatrix} \bar{\omega}^3_1 \\ \bar{\omega}^3_2 \end{bmatrix} = \begin{bmatrix} \sin(\psi - \eta) & \sin(\psi + \eta) \\ -\cos(\psi - \eta) & \cos(\psi + \eta) \end{bmatrix} \begin{bmatrix} dx \\ dy \end{bmatrix}, \tag{9.21}$$

$$\underline{\bar{\omega}}^1_2 = \bar{\omega}^1_2 - d\eta = -(\psi_x + \eta_x)\, dx + (\psi_y - \eta_y)\, dy.$$

(Hint: Cf. Exercises 4.28 and 4.47 and equations (9.19).)

(c) Show that the Bäcklund equation (9.6) is equivalent to the first-order system of partial differential equations

$$\begin{aligned} \psi_x + \eta_x &= \lambda \sin(\psi - \eta), \\ \psi_y - \eta_y &= \frac{1}{\lambda} \sin(\psi + \eta), \end{aligned} \tag{9.22}$$

where $\lambda = \cot(\alpha) - \csc(\alpha)$. (Hint: Recall that, since $K = -1$, we have $r = \sin(\alpha)$, and you will need the trigonometric identity

$$-(\cot(\alpha) + \csc(\alpha)) = \frac{1}{(\cot(\alpha) - \csc(\alpha))}.)$$

The system (9.22) is a coupled system of partial differential equations for the pair of functions $(\psi(x, y), \eta(x, y))$. The following exercise explores some properties of solutions of this system.

***Exercise 9.17.** (a) Suppose that the pair of functions $(\psi(x, y), \eta(x, y))$ satisfies the PDE system (9.22), where λ is any nonzero constant. Show that the functions $2\psi, 2\eta$ must *each* be solutions of the sine-Gordon equation (9.1). (Hint: Differentiate the first equation in (9.22) with respect to y and the second equation with respect to x; then add and subtract the resulting equations and apply trigonometric identities to simplify the right-hand sides.)

(b) Now, suppose that 2ψ is any known solution of (9.1). Then the system (9.22) can be regarded as the overdetermined PDE system

$$\begin{aligned} \eta_x &= -\psi_x + \lambda \sin(\psi - \eta), \\ \eta_y &= \psi_y - \frac{1}{\lambda} \sin(\psi + \eta) \end{aligned}$$

for the unknown function $\eta(x, y)$. Show that this system is *compatible*, i.e., that $(\eta_x)_y = (\eta_y)_x$—precisely because 2ψ satisfies (9.1). It follows that this system has a 1-parameter family of solutions $\eta(x, y)$ and that these solutions can be constructed using only techniques of *ordinary* differential equations.

The PDE system (9.22) is called a *Bäcklund transformation* for the sine-Gordon equation (9.1); the construction in Exercise 9.17 is the analog of part (2) of Theorem 9.5. It shows how, given one solution 2ψ of the sine-Gordon equation (9.1), the PDE system (9.22) can be used to construct a 2-parameter family of new solutions 2η: One parameter comes from the choice of the constant $\lambda \neq 0$, and one comes from the 1-parameter family of solutions η to the system (9.22).

At this point, we have established several relationships involving pseudospherical surfaces and solutions of the sine-Gordon equation:

(1) Every pseudospherical surface Σ of Gauss curvature $K = -1$ is associated with a solution 2ψ to the sine-Gordon equation (9.1) that describes the angle between its asymptotic directions at each point (and vice versa).

(2) Every pseudospherical surface Σ of Gauss curvature $K = -1$ has a 2-parameter family of associated pseudospherical surfaces $\widetilde{\Sigma}$ generated by the construction in part (2) of Theorem 9.5.

(3) Every solution 2ψ to the sine-Gordon equation (9.1) has a 2-parameter family of associated solutions 2η to (9.1) generated by the PDE system (9.22).

It seems natural to ask: Given a pseudospherical surface Σ with associated solution 2ψ to (9.1), what is the relationship between the new surfaces $\widetilde{\Sigma}$ associated to Σ and the new solutions 2η to (9.1) associated to 2ψ? The following exercise answers this question.

***Exercise 9.18.** Let $K = -1$ (so that $r = \sin(\alpha)$), and let $(\bar{\underline{\omega}}^i, \bar{\underline{\omega}}^i_j)$ be as in Exercise 9.16.

(a) Use equations (9.5), (9.7) to show that the Maurer-Cartan forms ($\tilde{\bar{\underline{\omega}}}^1$, $\tilde{\bar{\underline{\omega}}}^2$, $\tilde{\bar{\underline{\omega}}}^3_1$, $\tilde{\bar{\underline{\omega}}}^3_2$) on $\widetilde{\Sigma}$ are given by

$$
\begin{aligned}
\tilde{\bar{\underline{\omega}}}^1 &= \cos(\psi - \eta)\, dx + \cos(\psi + \eta)\, dy, \\
\tilde{\bar{\underline{\omega}}}^2 &= \sin(\psi - \eta)\, dx + \sin(\psi + \eta)\, dy, \\
\tilde{\bar{\underline{\omega}}}^3_1 &= \sin(\psi - \eta)\, dx - \sin(\psi + \eta)\, dy, \\
\tilde{\bar{\underline{\omega}}}^3_2 &= -\cos(\psi - \eta)\, dx + \cos(\psi + \eta)\, dy.
\end{aligned} \tag{9.23}
$$

(b) Show that the first and second fundamental forms of $\widetilde{\Sigma}$ are

$$
\begin{aligned}
\mathrm{I} &= (\tilde{\bar{\underline{\omega}}}^1)^2 + (\tilde{\bar{\underline{\omega}}}^2)^2 = dx^2 + 2\cos(2\eta)\, dx\, dy + dy^2, \\
\mathrm{II} &= \tilde{\bar{\underline{\omega}}}^3_1\, \tilde{\bar{\underline{\omega}}}^1 + \tilde{\bar{\underline{\omega}}}^3_2\, \tilde{\bar{\underline{\omega}}}^2 = -2\sin(2\eta)\, dx\, dy.
\end{aligned} \tag{9.24}
$$

(c) Conclude that the angle between the asymptotic directions of $\widetilde{\Sigma} = \tilde{\mathbf{x}}(U)$ at the point $\tilde{\mathbf{x}}(x, y)$ is equal to $2\eta(x, y)$. (The sign of the second fundamental form is unimportant, as it can be reversed by reversing the orientation of $\widetilde{\Sigma}$.)

The result of Exercise 9.18 completes the picture: The analytic Bäcklund transformation (9.22) between solutions of the sine-Gordon equation (9.1) is the precise analog of the geometric Bäcklund transformation between pseudospherical surfaces of Gauss curvature $K = -1$.

Exercise 9.19. Let $2\psi(x, y) \equiv 0$ be the trivial solution of the sine-Gordon equation (9.1). (This solution corresponds to the degenerate "surface" consisting of a straight line in $\mathbb{E}^3$.) Show that the corresponding solutions $2\eta(x, y)$ generated by the Bäcklund transformation (9.22) are given by

$$\eta(x, y) = 2\tan^{-1}(Ce^{-(\lambda x + \frac{1}{\lambda} y)}),$$

where $C \neq 0$ is constant. (Hint: You may find the trig identity $\csc(\eta) + \cot(\eta) = \cot(\frac{1}{2}\eta)$ useful.) The functions

$$2\eta = 4\tan^{-1}(Ce^{-(\lambda x + \frac{1}{\lambda}y)}) \tag{9.25}$$

are called the 1-*soliton solutions* of the sine-Gordon equation. Iterating this procedure gives the 2-solitons, etc.

Exercise 9.20. Let Σ be the pseudosphere, with parametrization

$$\mathbf{x}(x,y) = \begin{bmatrix} x + y - \tanh(x+y) \\ \operatorname{sech}(x+y)\cos(x-y) \\ \operatorname{sech}(x+y)\sin(x-y) \end{bmatrix}.$$

(a) Show that the first and second fundamental forms of Σ are given by

$$\mathrm{I} = dx^2 + \frac{2(\cosh^2(x+y) - 2)}{\cosh^2(x+y)}\, dx\, dy + dy^2,$$

$$\mathrm{II} = \frac{4\sinh(x+y)}{\cosh^2(x+y)}\, dx\, dy.$$

(b) Conclude that:

(1) (x, y) are asymptotic coordinates for Σ.

(2) The Gauss curvature of Σ is $K = -1$.

(3) The angle $2\psi(x, y)$ between the asymptotic directions of Σ at the point $\mathbf{x}(x, y)$ satisfies the conditions

$$\cos(2\psi(x,y)) = \frac{\cosh^2(x+y) - 2}{\cosh^2(x+y)}, \quad \sin(2\psi(x,y)) = \frac{2\sinh(x+y)}{\cosh^2(x+y)}. \tag{9.26}$$

(In particular, check that

$$\left(\frac{\cosh^2(x+y) - 2}{\cosh^2(x+y)}\right)^2 + \left(\frac{2\sinh(x+y)}{\cosh^2(x+y)}\right)^2 = 1.)$$

(c) Show that the function $\psi(x, y)$ defined by equation (9.26) is

$$\psi(x,y) = 2\tan^{-1}(e^{x+y}).$$

Thus, the pseudosphere is one of the family of pseudospherical surfaces that arise from the 1-soliton solutions (9.25) of the sine-Gordon equation (9.1). Surfaces in this family are sometimes referred to as "1-soliton" pseudospherical surfaces.

Exercise 9.21. In this exercise, we will give a proof of *Hilbert's theorem*, which states that there is no isometric immersion of the complete hyperbolic plane $\mathbb{H}^2$ into the Euclidean space $\mathbb{E}^3$. (For purposes of this exercise, all you need to know about $\mathbb{H}^2$ is that it is a complete surface of infinite area, diffeomorphic to $\mathbb{R}^2$, with constant Gauss curvature $K = -1$. We will explore hyperbolic spaces in more detail in Chapter 11.)

Suppose that $\mathbf{x} : \mathbb{H}^2 \to \mathbb{E}^3$ is an isometric immersion. Recall that $\mathbb{H}^2$ has Gauss curvature $K = -1$; therefore the image $\Sigma = \mathbf{x}(\mathbb{H}^2)$ is an immersed pseudospherical surface. ("Immersed" allows for the possibility of self-intersection; an immersion $\mathbf{x} : \mathbb{H}^2 \to \mathbb{E}^3$ must be a regular mapping at every point of the domain $\mathbb{H}^2$, but it need not necessarily be one-to-one.) Moreover, since $\mathbb{H}^2$ is a complete surface of infinite area, the same must be true of Σ.

(a) Show that, because $\mathbf{x}$ is an immersion, there exist global coordinates (x, y) on $\mathbb{H}^2$ such that $\mathbf{x}(x, y)$ is an asymptotic parametrization of Σ and for which the first and second fundamental forms of Σ are as in equations (9.18). (Hint: We already know that local coordinates (x, y) satisfying these conditions exist in a neighborhood of each point in $\mathbb{H}^2$. Moreover, these local coordinates are determined only up to additive constants. Show that for any two such overlapping coordinate patches with local coordinates (x, y) and $(\tilde{x}, \tilde{y})$, respectively, we must have

$$\tilde{x} = x + x_0, \qquad \tilde{y} = y + y_0$$

for some constants x_0, y_0. Since each pair of coordinates is only determined up to additive constants, we can set $\tilde{x} = x, \tilde{y} = y$. A topological argument shows that this patching construction works globally because $\mathbb{H}^2$ is simply connected.)

(b) Let $2\psi : \mathbb{H}^2 \to \mathbb{R}$ be the solution of the sine-Gordon equation (9.1) associated to Σ. Show that, because $\mathbf{x}$ is an immersion, the function ψ must satisfy

$$0 < \psi(x, y) < \frac{\pi}{2}$$

for all $(x, y) \in \mathbb{H}^2$. (Hint: An immersed pseudospherical surface must have linearly independent asymptotic directions at each point.)

(c) Let $R \subset \mathbb{H}^2$ be a rectangle of the form

$$a \le x \le b, \qquad c \le y \le d.$$

Use the sine-Gordon equation (9.1) and the Fundamental Theorem of Calculus to show that the area of $\mathbf{x}(R) \subset \Sigma$ is given by

$$\begin{aligned}\mathcal{A}(\mathbf{x}(R)) &= \int_R \bar{\omega}^1 \wedge \bar{\omega}^2 \\ &= 2\left(\psi(b,d) - \psi(b,c) - \psi(a,d) + \psi(a,c)\right).\end{aligned}$$

Conclude from part (b) that $\mathcal{A}(\mathbf{x}(R)) < 2\pi$.

(d) Observe that, because $\mathbb{H}^2$ is complete and has infinite area, the area of R can be made arbitrarily large by choosing a, b, c, d appropriately. Therefore, since $\mathbf{x}$ is an isometric immersion, the area of $\mathbf{x}(R)$ can be made arbitrarily large as well. This contradicts the result of part (c); thus, no such isometric immersion can exist.

Exercise 9.22. While the Bäcklund transformation (9.22) relates two different solutions of the same PDE (9.1), it is also possible for a Bäcklund transformation to relate solutions of two *different* PDEs. For example, consider the first-order system of partial differential equations

$$\begin{aligned}\psi_x + \eta_x &= 2\lambda e^{\frac{(\psi-\eta)}{2}}, \\ \psi_y - \eta_y &= \frac{1}{\lambda} e^{\frac{(\psi+\eta)}{2}}.\end{aligned} \tag{9.27}$$

(a) Suppose that the pair of functions $(\psi(x,y),\ \eta(x,y))$ satisfies the PDE system (9.27), where λ is any nonzero constant. Show that $\eta(x,y)$ must be a solution of the wave equation (in characteristic coordinates)

$$\eta_{xy} = 0, \tag{9.28}$$

while $\psi(x,y)$ must be a solution of *Liouville's equation*

$$\psi_{xy} = e^{\psi}. \tag{9.29}$$

(b) Show by integration that the general solution of the wave equation (9.28) is

$$\eta(x,y) = \rho(x) + \sigma(y), \tag{9.30}$$

where $\rho(x), \sigma(y)$ are arbitrary functions of a single variable.

(c) Substitute equation (9.30) into the system (9.27) to obtain a system of two first-order PDEs for the function $\psi(x,y)$. By treating each of these equations as a separable ODE in the appropriate variable, show that

$$e^{-\frac{1}{2}\psi} = -e^{\frac{1}{2}(\rho(x)-\sigma(y))}\left(\lambda \int e^{-\rho(x)}\, dx + \frac{1}{2\lambda}\int e^{\sigma(y)}\, dy\right). \tag{9.31}$$

(Hint: Write the right-hand sides of equations (9.27) as

$$2\lambda e^{\frac{1}{2}(\psi+\rho(x)-2\rho(x)-\sigma(y))} \quad \text{and} \quad \frac{1}{\lambda}e^{\frac{1}{2}(\psi-\sigma(y)+\rho(x)+2\sigma(y))}.)$$

(d) Set

$$X(x) = -\lambda \int e^{-\rho(x)}\, dx, \qquad Y(y) = -\frac{1}{2\lambda}\int e^{\sigma(y)}\, dy.$$

Solve equation (9.31) for ψ to obtain the general solution of Liouville's equation:

$$\psi(x,y) = \ln\left(\frac{2X'(x)Y'(y)}{(X(x)+Y(y))^2}\right), \tag{9.32}$$

where $X(x)$ and $Y(y)$ are arbitrary functions of a single variable, with the property that $X'(x)$ and $Y'(y)$ are nonzero and have the same sign.

9.6. MAPLE computations

As usual, begin by loading the **Cartan** and **LinearAlgebra** packages into MAPLE. Since we have three different adapted frame fields in play in this chapter (one frame field on each of the surfaces $\Sigma, \tilde{\Sigma}$ adapted to the Bäcklund transformation and one principal adapted frame field on Σ), we'll need to keep track of three different sets of Maurer-Cartan forms. In order to distinguish them a bit, we'll use `(theta1[i], theta1[i,j])` and `(theta2[i], theta2[i,j])` for the Maurer-Cartan forms on $\Sigma, \tilde{\Sigma}$, respectively, corresponding to the Bäcklund-adapted frame fields, and `(omega[i], omega[i,j])` for the Maurer-Cartan forms corresponding to the principal frame field on Σ. Start by declaring the necessary 1-forms and telling MAPLE about their symmetries and structure equations:

```
> Form(omega[1], omega[2], omega[3]);
  Form(omega[1,2], omega[3,1], omega[3,2]);

  Form(theta1[1], theta1[2], theta1[3]);
  Form(theta1[1,2], theta1[3,1], theta1[3,2]);

  Form(theta2[1], theta2[2], theta2[3]);
  Form(theta2[1,2], theta2[3,1], theta2[3,2]);

> omega[1,1]:= 0;
  omega[2,2]:= 0;
  omega[3,3]:= 0;
  omega[2,1]:= -omega[1,2];
  omega[1,3]:= -omega[3,1];
  omega[2,3]:= -omega[3,2];
```

```
  theta1[1,1]:= 0;
  theta1[2,2]:= 0;
  theta1[3,3]:= 0;
  theta1[2,1]:= -theta1[1,2];
  theta1[1,3]:= -theta1[3,1];
  theta1[2,3]:= -theta1[3,2];

  theta2[1,1]:= 0;
  theta2[2,2]:= 0;
  theta2[3,3]:= 0;
  theta2[2,1]:= -theta2[1,2];
  theta2[1,3]:= -theta2[3,1];
  theta2[2,3]:= -theta2[3,2];

> for i from 1 to 3 do
    d(omega[i]):= -add('omega[i,j] &^ omega[j]', j=1..3);
  end do;
  d(omega[1,2]):= -add('omega[1,k] &^ omega[k,2]', k=1..3);
  d(omega[3,1]):= -add('omega[3,k] &^ omega[k,1]', k=1..3);
  d(omega[3,2]):= -add('omega[3,k] &^ omega[k,2]', k=1..3);

  for i from 1 to 3 do
    d(theta1[i]):= -add('theta1[i,j] &^ theta1[j]', j=1..3);
  end do;
  d(theta1[1,2]):= -add('theta1[1,k] &^ theta1[k,2]', k=1..3);
  d(theta1[3,1]):= -add('theta1[3,k] &^ theta1[k,1]', k=1..3);
  d(theta1[3,2]):= -add('theta1[3,k] &^ theta1[k,2]', k=1..3);

  for i from 1 to 3 do
    d(theta2[i]):= -add('theta2[i,j] &^ theta2[j]', j=1..3);
  end do;
  d(theta2[1,2]):= -add('theta2[1,k] &^ theta2[k,2]', k=1..3);
  d(theta2[3,1]):= -add('theta2[3,k] &^ theta2[k,1]', k=1..3);
  d(theta2[3,2]):= -add('theta2[3,k] &^ theta2[k,2]', k=1..3);
```

We'll also need the vector form of the structure equations; we'll use `x1` and `x2` for the parametrizations of Σ and $\widetilde{\Sigma}$, respectively, and `(e11, e12, e13)`, `(e21, e22, e23)` for the frame fields on the two surfaces that are adapted to the Bäcklund transformation.

```
> d(x1):= e11*theta1[1] + e12*theta1[2];
  d(e11):= e12*theta1[2,1] + e13*theta1[3,1];
  d(e12):= e11*theta1[1,2] + e13*theta1[3,2];
  d(e13):= e11*theta1[1,3] + e12*theta1[2,3];
```

```
d(x2):= e21*theta2[1] + e22*theta2[2];
d(e21):= e22*theta2[2,1] + e23*theta2[3,1];
d(e22):= e21*theta2[1,2] + e23*theta2[3,2];
d(e23):= e21*theta2[1,3] + e22*theta2[2,3];
```

EXERCISE 9.8: Now, assume that we have chosen adapted frame fields $(\underline{\mathbf{e}}_1(\mathbf{u}), \underline{\mathbf{e}}_2(\mathbf{u}), \underline{\mathbf{e}}_3(\mathbf{u}))$ on Σ and $(\underline{\tilde{\mathbf{e}}}_1(\mathbf{u}), \underline{\tilde{\mathbf{e}}}_2(\mathbf{u}), \underline{\tilde{\mathbf{e}}}_3(\mathbf{u}))$ on $\widetilde{\Sigma}$ as in equations (9.2) and that equation (9.3) holds. First, declare α and r to be constants, and set up a substitution to move from one frame field to the other:

```
> Form(r=-1, alpha=-1);
> frame2to1sub:= [e21 = e11,
  e22 = cos(alpha)*e12 + sin(alpha)*e13,
  e23 = -sin(alpha)*e12 + cos(alpha)*e13];
```

Now differentiate equation (9.3), and write the results in terms of the first adapted frame field:

```
> zero1:= Simf(subs(frame2to1sub, d(x2 - x1 - r*e11)));
```

Collect terms, and then use the fact that $(\underline{\mathbf{e}}_1(\mathbf{u}), \underline{\mathbf{e}}_2(\mathbf{u}), \underline{\mathbf{e}}_3(\mathbf{u}))$ are linearly independent to conclude that each of their coefficients must be zero:

```
> collect(zero1, {e11, e12, e13});
> zero1a:= coeff(zero1, e11);
  zero1b:= coeff(zero1, e12);
  zero1c:= coeff(zero1, e13);
```

This computation yields equations (9.5), and we can combine the last two equations to obtain a relation that only involves the Maurer-Cartan forms on Σ:

```
> zero1d:= Simf(zero1b - cot(alpha)*zero1c);
```

The result yields equation (9.6).

Next, we need to compute the 1-forms

$$\underline{\tilde{\omega}}_1^3 = \langle d\underline{\tilde{\mathbf{e}}}_1, \underline{\tilde{\mathbf{e}}}_3 \rangle, \qquad \underline{\tilde{\omega}}_2^3 = \langle d\underline{\tilde{\mathbf{e}}}_2, \underline{\tilde{\mathbf{e}}}_3 \rangle.$$

Since we haven't told MAPLE that (`e11, e12, e13`) are vectors, we'll have to define our own procedure to compute these inner products. We can do this as follows:

```
> innprod1:= proc(exp1, exp2)
    RETURN(coeff(exp1, e11)*coeff(exp2, e11)
    + coeff(exp1, e12)*coeff(exp2, e12)
```

```
    + coeff(exp1, e13)*coeff(exp2, e13));
  end proc;
```

Now we can compute $\tilde{\underline{\omega}}_1^3$ and $\tilde{\underline{\omega}}_2^3$ via the formulas above. First, compute $\tilde{\underline{\omega}}_1^3$:

```
> Simf(innprod1(d(Simf(subs(frame2to1sub, e21))),
  Simf(subs(frame2to1sub, e23))));
```

$$\theta 1_{1,2} \sin(\alpha) + \theta 1_{3,1} \cos(\alpha)$$

Apply the relation (9.6):

```
> Simf(subs(solve({zero1d}, {theta1[3,1]}), %));
```

This yields the expression in the first equation of (9.7). Similarly, the computation

```
> Simf(innprod1(d(Simf(subs(frame2to1sub, e22))),
  Simf(subs(frame2to1sub, e23))));
```

yields the expression in the second equation of (9.7).

Exercise 9.14: Set up two substitutions: one to tell Maple that the function $\phi(x, y) = 2\psi(x, y)$ satisfies the sine-Gordon equation (9.1) and one to define the desired Maurer-Cartan forms associated to the principal adapted frame field $(\mathbf{e}_1(\mathbf{u}), \mathbf{e}_2(\mathbf{u}), \mathbf{e}_3(\mathbf{u}))$:

```
> PDETools[declare](psi(x,y));
> SGEsub:= [diff(psi(x,y), x, y) = (1/2)*sin(2*psi(x,y))];
> SGEformsub:= [omega[1] = cos(psi(x,y))*(d(x) + d(y)),
  omega[2] = sin(psi(x,y))*(d(x) - d(y)), omega[3] = 0,
  omega[3,1] = sin(psi(x,y))*(d(x) + d(y)),
  omega[3,2] = -cos(psi(x,y))*(d(x) - d(y)),
  omega[1,2] = -diff(psi(x,y), x)*d(x)
  + diff(psi(x,y), y)*d(y)];
```

Maple doesn't really know how to compute symmetric products of differential forms, but the following commands will work for computing the first and second fundamental forms:

```
> collect(simplify(Simf(subs(SGEformsub, omega[1]))^2
  + Simf(subs(SGEformsub, omega[2]))^2), {d(x), d(y)});
> collect(simplify(Simf(subs(SGEformsub, omega[3,1]))*
  Simf(subs(SGEformsub, omega[1]))
  + Simf(subs(SGEformsub, omega[3,2]))*
  Simf(subs(SGEformsub, omega[2]))), {d(x), d(y)});
```

In order to verify that the structure equations are satisfied, check that computations such as the following all yield zero:

```
> Simf(d(Simf(subs(SGEformsub, omega[1]))))
  - Simf(subs(SGEformsub, Simf(d(omega[1]))));
```

The only one that doesn't immediately reduce to zero is the one for $\bar{\omega}_2^1$, which requires an application of `SGEsub` in order to see that it vanishes.

EXERCISE 9.16: Equation (9.20) is immediate from the definition of $\eta(\mathbf{u})$, and the results of Chapter 4 yield the expressions for $(\underline{\bar{\omega}}^1, \underline{\bar{\omega}}^2, \underline{\bar{\omega}}_1^3, \underline{\bar{\omega}}_2^3)$ in equations (9.21). Set up a substitution to go back and forth between the two sets of Maurer-Cartan forms:

```
> PDETools[declare](eta(x,y));
> rotationsub:= [
  theta1[1] = cos(eta(x,y))*omega[1] + sin(eta(x,y))*omega[2],
  theta1[2] = -sin(eta(x,y))*omega[1]
  + cos(eta(x,y))*omega[2],
  theta1[3,1] = cos(eta(x,y))*omega[3,1]
  + sin(eta(x,y))*omega[3,2],
  theta1[3,2] = -sin(eta(x,y))*omega[3,1]
  + cos(eta(x,y))*omega[3,2]];
> rotationbacksub:= makebacksub(rotationsub);
```

We can then compute $\underline{\bar{\omega}}_2^1$ in terms of $\bar{\omega}_2^1$ as follows: Compute $d\underline{\bar{\omega}}^1$ and $d\underline{\bar{\omega}}^2$ by differentiating the expressions for $\underline{\bar{\omega}}^1$ and $\underline{\bar{\omega}}^2$ in equation (9.21), and then use the reverse substitution to express the result in terms of $(\underline{\bar{\omega}}^1, \underline{\bar{\omega}}^2)$:

```
> dtheta1[1]:= Simf(subs([omega[3]=0],
  Simf(subs(rotationbacksub, Simf(d(Simf(subs(rotationsub,
  theta1[1]))))))));
> pick(dtheta1[1], theta1[2]);
```

$$-\omega_{1,2} + \eta_x\, d(x) + \eta_y\, d(y)$$

```
> dtheta1[2]:= Simf(subs([omega[3]=0],
  Simf(subs(rotationbacksub, Simf(d(Simf(subs(rotationsub,
  theta1[2]))))))));
> pick(dtheta1[2], theta1[1]);
```

$$\omega_{1,2} - \eta_x\, d(x) - \eta_y\, d(y)$$

It follows from the structure equations for $d\underline{\bar{\omega}}^1$ and $d\underline{\bar{\omega}}^2$ that $\underline{\bar{\omega}}_2^1 = \bar{\omega}_2^1 - d\eta$. Add this to our substitution:

```
> rotationsub:= [op(rotationsub),
  theta1[1,2] = omega[1,2] - d(eta(x,y))];
```

We can combine this with the expressions for $(\bar{\omega}^i, \bar{\omega}^i_j)$ in SGEformsub in order to write $(\underline{\bar{\omega}}^i, \underline{\bar{\omega}}^i_j)$ in terms of (dx, dy):

```
> SGErotationsub:= Simf(subs(SGEformsub, rotationsub));
```

Since $K = -1$, we have $r = \sin(\alpha)$:

```
> r:= sin(alpha);
```

Now consider the Bäcklund equation (9.6), which says that the following expression is zero:

```
> Backlundeq:= theta1[2] + r*theta1[2,1]
   - r*cot(alpha)*theta1[3,1];
```

We can express this equation in terms of (dx, dy) via SGErotationsub:

```
> Backlundzero:= Simf(subs(SGErotationsub, Backlundeq));
```

The coefficients of dx and dy in this expression are PDEs involving the functions ψ and η:

```
> PDE1:= pick(Backlundzero, d(x));
  PDE2:= pick(Backlundzero, d(y));
```

These PDEs contain some coefficients involving trigonometric functions of α, and they will look nicer if we rename them. In the first equation, divide by $\sin(\alpha)$ and set $\lambda = \cot(\alpha) - \csc(\alpha)$ via the substitution:

```
> lambdasub:= [cos(alpha) = lambda*sin(alpha) + 1];
> Simf(subs(lambdasub, PDE1/sin(alpha)));
```

$$-\lambda \sin(\psi) \cos(\eta) + \lambda \sin(\eta) \cos(\psi) + \psi_x + \eta_x$$

```
> combine(%, trig);
```

$$\lambda \sin(\eta - \psi) + \psi_x + \eta_x$$

Similarly, for the second equation divide by $\sin(\alpha)$ and set $\mu = -(\cot(\alpha) + \csc(\alpha))$:

```
> musub:= [cos(alpha) = -mu*sin(alpha) - 1];
> Simf(subs(musub, PDE2/sin(alpha)));
```

$$\mu \sin(\psi) \cos(\eta) + \mu \sin(\eta) \cos(\psi) - \psi_y + \eta_y$$

```
> combine(%, trig);
```

$$\mu \sin(\eta + \psi) - \psi_y + \eta_y$$

All that remains is to show that $\lambda\mu = 1$, so that $\mu = \frac{1}{\lambda}$:

```
> Simf(subs(solve({op(lambdasub), op(musub)}, {lambda, mu}),
  lambda*mu));
```

$$1$$

EXERCISE 9.18: This exercise is now a simple matter of combining things that we have already computed in order to write the forms $(\tilde{\bar{\underline{\omega}}}^i, \tilde{\bar{\underline{\omega}}}^i_j)$ in terms of (dx, dy). Equations (9.7) and the first and third equations in (9.5) allow us to write $(\tilde{\bar{\underline{\omega}}}^1, \tilde{\bar{\underline{\omega}}}^2, \tilde{\bar{\underline{\omega}}}^3_1, \tilde{\bar{\underline{\omega}}}^3_2)$ in terms of $(\bar{\underline{\omega}}^1, \bar{\underline{\omega}}^2, \bar{\underline{\omega}}^3_1, \bar{\underline{\omega}}^3_2)$ as follows:

```
> theta2sub:= [op(solve({zero1a, zero1c},
  {theta2[1], theta2[2]})),
  theta2[3,1] = theta1[2], theta2[3,2] = theta1[3,2]];
```

Then we apply `SGErotationsub` to express these forms in terms of (dx, dy):

```
> theta2coordsub:= Simf(subs(SGErotationsub, theta2sub));
```

We can compactify these expressions a bit as follows:

```
> map(combine, theta2coordsub, trig);
```

This should yield the expressions in equation (9.23). We can compute the first and second fundamental forms in equation (9.24) as follows:

```
> collect(simplify(Simf(subs(theta2coordsub, theta2[1]))^2
  + Simf(subs(theta2coordsub, theta2[2]))^2), {d(x), d(y)});
> collect(simplify(Simf(subs(theta2coordsub, theta2[3,1]))*
  Simf(subs(theta2coordsub, theta2[1]))
  + Simf(subs(theta2coordsub, theta2[3,2]))*
  Simf(subs(theta2coordsub, theta2[2]))), {d(x), d(y)});
```

Chapter 10

Two classical theorems

In this chapter, we will see how moving frames may be used to prove two classical results in differential geometry: the classification of doubly ruled surfaces in $\mathbb{R}^3$ and the Cauchy-Crofton formula for the length of a curve in the Euclidean plane $\mathbb{E}^2$.

10.1. Doubly ruled surfaces in $\mathbb{R}^3$

A regular surface $\Sigma \subset \mathbb{R}^3$ is called *ruled* if Σ contains a straight line segment passing through each point $\mathbf{x} \in \Sigma$. Ruled surfaces have been the subject of much study in classical differential geometry; see, e.g., [**Eis60**] or [**Wil62**]. A surface Σ is called *doubly ruled* if it can be realized as a ruled surface in two distinct ways, i.e., if Σ contains two linearly independent line segments passing through each point $\mathbf{x} \in \Sigma$.

There are two well-known examples of non-planar doubly ruled surfaces in $\mathbb{R}^3$: the hyperboloid of one sheet

$$\Sigma = \left\{ {}^t[x,\, y,\, z] \in \mathbb{R}^3 \mid x^2 + y^2 - z^2 = r^2 \right\}, \tag{10.1}$$

where $r > 0$ is a positive constant, and the hyperbolic paraboloid

$$\Sigma = \left\{ {}^t[x,\, y,\, z] \in \mathbb{R}^3 \mid z = xy \right\}. \tag{10.2}$$

***Exercise 10.1.** Recall that a regular surface $\Sigma \subset \mathbb{R}^3$ is ruled if and only if every point $\mathbf{x} \in \Sigma$ has a neighborhood that can be given a parametrization of the form

$$\mathbf{x}(u, v) = \alpha(u) + v\beta(u),$$

where α is a regular curve in $\mathbb{R}^3$ and $\beta(u) \neq 0$.

(a) Show that the following maps $\mathbf{x}_1, \mathbf{x}_2 : \mathbb{R}^2 \to \mathbb{R}^3$ are parametrizations of the hyperboloid (10.1):

$$\mathbf{x}_1(u,v) = {}^t[r\cos(u),\, r\sin(u),\, 0] + v\, {}^t[-\sin(u),\, \cos(u),\, 1]\,,$$

$$\mathbf{x}_2(u,v) = {}^t[r\cos(u),\, r\sin(u),\, 0] + v\, {}^t[-\sin(u),\, \cos(u),\, -1]\,.$$

Conclude that the hyperboloid (10.1) is doubly ruled. (The rulings described by these parametrizations are shown in Figure 10.1.)

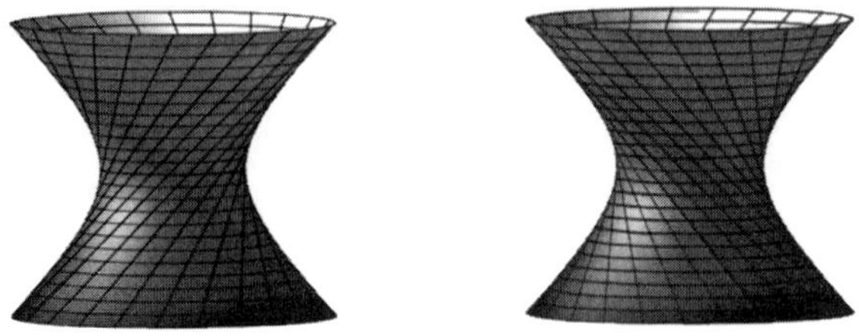

Figure 10.1. Two families of rulings on the hyperboloid of one sheet

(b) Show that the following map $\mathbf{x} : \mathbb{R}^2 \to \mathbb{R}^3$ is a parametrization of the hyperbolic paraboloid (10.2) that simultaneously realizes both families of rulings:

$$\mathbf{x}(u,v) = {}^t[u,\, v,\, uv]\,.$$

(Hint: Find curves $\alpha(u), \beta(u), \gamma(v), \delta(v)$ in $\mathbb{R}^3$ such that

$$\mathbf{x}(u,v) = \alpha(u) + v\beta(u) = \gamma(v) + u\delta(v).)$$

Conclude that the hyperbolic paraboloid (10.2) is doubly ruled. (The rulings described by this parametrization are shown in Figure 10.2.)

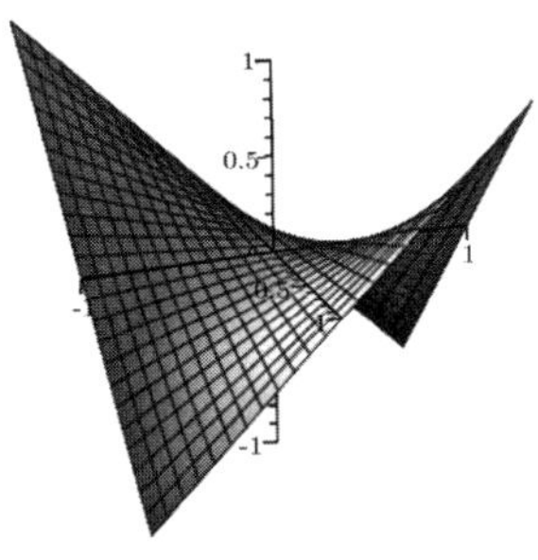

Figure 10.2. Two families of rulings on the hyperbolic paraboloid

You may have noticed that so far in this chapter, we have been describing surfaces as subsets of $\mathbb{R}^3$ without specifying any particular homogeneous space structure on $\mathbb{R}^3$. The following exercise suggests such a structure that might be well-suited to our purposes:

***Exercise 10.2.** Show that the set of doubly ruled surfaces is invariant under the action of the equi-affine group $A(3)$. Specifically, if $\Sigma \subset \mathbb{R}^3$ is a doubly ruled surface and $g \in A(3)$ is an equi-affine transformation, then the surface $g \cdot \Sigma \subset \mathbb{R}^3$ is also doubly ruled.

Based on the result of Exercise 10.2, for the remainder of this section, we will regard $\mathbb{R}^3$ as the equi-affine space $\mathbb{A}^3$.

The goal of this section is to prove the following theorem [**HCV52**]:

Theorem 10.3. *Any non-planar doubly ruled regular surface $\Sigma \subset \mathbb{A}^3$ is equivalent via an equi-affine transformation to (an open subset of) either a hyperboloid of one sheet as in* (10.1) *or a hyperbolic paraboloid as in* (10.2).

Remark 10.4. We will use the theory developed in Chapter 6 to prove Theorem 10.3 as stated, but in fact more is true:

(1) The set of doubly ruled surfaces is invariant under the action of the full affine group without the equi-affine restriction. Under this group action, the hyperboloids of one sheet (10.1) for all positive values of r are all equivalent, whereas different values of r give surfaces that are inequivalent under the action of the equi-affine group.

(2) If we regard $\mathbb{R}^3$ as an open subset of $\mathbb{P}^3$ via the identification
$$(x^1, x^2, x^3) \leftrightarrow \left[1 : x^1 : x^2 : x^3\right],$$
the set of doubly ruled surfaces is also invariant under the action of the group of projective transformations on $\mathbb{P}^3$. Under this group action, the surfaces (10.1) and (10.2) are all equivalent to each other.

To begin the proof, let U be an open set in $\mathbb{R}^2$, and let $\mathbf{x} : U \to \mathbb{A}^3$ be an immersion whose image is a regular surface $\Sigma \subset \mathbb{A}^3$ that is doubly ruled. In order to apply the method of moving frames, it seems natural to choose a frame field $(\mathbf{e}_1(\mathbf{u}), \mathbf{e}_2(\mathbf{u}), \mathbf{e}_3(\mathbf{u}))$ along Σ with the property that for each $\mathbf{u} \in U$, the vectors $\mathbf{e}_1(\mathbf{u})$ and $\mathbf{e}_2(\mathbf{u})$ are each tangent to one of the rulings passing through the point $\mathbf{x}(\mathbf{u})$. Here we see another reason to regard $\mathbb{R}^3$ as $\mathbb{A}^3$ (as opposed to, say, $\mathbb{E}^3$): Having the freedom to choose a unimodular frame field on Σ (as opposed to, say, an orthonormal frame field) allows us sufficient flexibility to adapt our choice of frame field to the surface in precisely this way. So, in this context, a unimodular frame field

$(\mathbf{e}_1(\mathbf{u}), \mathbf{e}_2(\mathbf{u}), \mathbf{e}_3(\mathbf{u}))$ along Σ will be called 0-*adapted* if for each $\mathbf{u} \in U$, the vectors $\mathbf{e}_1(\mathbf{u})$ and $\mathbf{e}_2(\mathbf{u})$ are each tangent to one of the rulings passing through the point $\mathbf{x}(\mathbf{u})$. (Note that this notion of "0-adapted" is more restrictive than that in Chapter 6 because here we have additional geometric information to guide our initial choice for an adapted frame field.)

Exercise 10.5. Explain why orthonormal frame fields would generally be too restrictive to allow for an adapted frame field satisfying this tangency condition.

Recall that choosing such a unimodular frame field along Σ is equivalent to choosing a lifting $\tilde{\mathbf{x}} : U \to A(3)$ defined by

$$\tilde{\mathbf{x}}(\mathbf{u}) = (\mathbf{x}(\mathbf{u}); \mathbf{e}_1(\mathbf{u}), \mathbf{e}_2(\mathbf{u}), \mathbf{e}_3(\mathbf{u}))$$

and that the pullbacks $(\bar{\omega}^i, \bar{\omega}^i_j)$ to U of the Maurer-Cartan forms (ω^i, ω^i_j) on $A(3)$ via the lifting $\tilde{\mathbf{x}}$ satisfy the conditions that $\bar{\omega}^3 = 0$ and

$$\begin{aligned} d\mathbf{x} &= \mathbf{e}_1\,\bar{\omega}^1 + \mathbf{e}_2\,\bar{\omega}^2, \\ d\mathbf{e}_i &= \mathbf{e}_1\,\bar{\omega}^1_i + \mathbf{e}_2\,\bar{\omega}^2_i + \mathbf{e}_3\,\bar{\omega}^3_i, \qquad i = 1, 2, 3. \end{aligned} \tag{10.3}$$

***Exercise 10.6.** (a) Show that the condition that the vector fields $(\mathbf{e}_1(\mathbf{u}), \mathbf{e}_2(\mathbf{u}))$ are tangent to the rulings implies that

$$d\mathbf{e}_1(\mathbf{e}_1) \equiv 0 \mod \mathbf{e}_1, \qquad d\mathbf{e}_2(\mathbf{e}_2) \equiv 0 \mod \mathbf{e}_2. \tag{10.4}$$

(Hint: Recall that $d\mathbf{e}_1(\mathbf{e}_1)$ computes the directional derivative of the vector field $\mathbf{e}_1(\mathbf{u})$ in the direction of $\mathbf{e}_1(\mathbf{u})$. What limitations are imposed on this derivative by the condition that the vector field $\mathbf{e}_1(\mathbf{u})$ is tangent to a straight line along this direction?)

(b) Show that the condition (10.4) implies that

$$d\mathbf{e}_1 \equiv 0 \mod (\mathbf{e}_1, \bar{\omega}^2), \qquad d\mathbf{e}_2 \equiv 0 \mod (\mathbf{e}_2, \bar{\omega}^1). \tag{10.5}$$

Conclude from equations (10.3) and (10.5) that

$$\begin{aligned} \bar{\omega}^2_1 &= a^2_{12}\,\bar{\omega}^2, & \bar{\omega}^3_1 &= a^3_{12}\,\bar{\omega}^2, \\ \bar{\omega}^1_2 &= a^1_{21}\,\bar{\omega}^1, & \bar{\omega}^3_2 &= a^3_{21}\,\bar{\omega}^1 \end{aligned} \tag{10.6}$$

for some functions $a^1_{21}, a^2_{12}, a^3_{12}, a^3_{21}$ on U.

(c) Use the Cartan structure equation

$$0 = d\bar{\omega}^3 = -\bar{\omega}^3_1 \wedge \bar{\omega}^1 - \bar{\omega}^3_2 \wedge \bar{\omega}^2$$

to show that $a^3_{21} = a^3_{12}$.

In order to minimize notational clutter, we will rename the (a^k_{ij}) and write

$$\begin{aligned} \bar{\omega}^2_1 &= a\,\bar{\omega}^2, & \bar{\omega}^3_1 &= c\,\bar{\omega}^2, \\ \bar{\omega}^1_2 &= b\,\bar{\omega}^1, & \bar{\omega}^3_2 &= c\,\bar{\omega}^1. \end{aligned} \tag{10.7}$$

Now we will investigate how the functions a, b, c transform if we vary our choice of 0-adapted frame field on Σ.

***Exercise 10.7.** Let $(\mathbf{e}_1(\mathbf{u}), \mathbf{e}_2(\mathbf{u}), \mathbf{e}_3(\mathbf{u}))$, $(\tilde{\mathbf{e}}_1(\mathbf{u}), \tilde{\mathbf{e}}_2(\mathbf{u}), \tilde{\mathbf{e}}_3(\mathbf{u}))$ be any two 0-adapted frame fields on Σ, with associated Maurer-Cartan forms $(\bar{\omega}^i, \bar{\omega}^i_j)$ and $(\tilde{\bar{\omega}}^i, \tilde{\bar{\omega}}^i_j)$, respectively. For simplicity, assume that the vector field $\tilde{\mathbf{e}}_1(\mathbf{u})$ is tangent to the same family of rulings as $\mathbf{e}_1(\mathbf{u})$, and similarly for the vector fields $\mathbf{e}_2(\mathbf{u})$ and $\tilde{\mathbf{e}}_2(\mathbf{u})$.

(a) Show that

$$\begin{bmatrix}\tilde{\mathbf{e}}_1(\mathbf{u}) & \tilde{\mathbf{e}}_2(\mathbf{u}) & \tilde{\mathbf{e}}_3(\mathbf{u})\end{bmatrix} = \begin{bmatrix}\mathbf{e}_1(\mathbf{u}) & \mathbf{e}_2(\mathbf{u}) & \mathbf{e}_3(\mathbf{u})\end{bmatrix} \begin{bmatrix} \lambda_1 & 0 & r_1 \\ 0 & \lambda_2 & r_2 \\ 0 & 0 & \frac{1}{\lambda_1\lambda_2} \end{bmatrix} \tag{10.8}$$

for some functions $\lambda_1, \lambda_2, r_1, r_2$ on U, with $\lambda_1, \lambda_2 \neq 0$.

(b) Show that under a transformation of the form (10.8), we have

$$\begin{bmatrix}\tilde{\bar{\omega}}^1 \\ \tilde{\bar{\omega}}^2\end{bmatrix} = \begin{bmatrix}\frac{1}{\lambda_1}\bar{\omega}^1 \\ \frac{1}{\lambda_2}\bar{\omega}^2\end{bmatrix}, \qquad \begin{bmatrix}\tilde{\bar{\omega}}^3_1 \\ \tilde{\bar{\omega}}^3_2\end{bmatrix} = \begin{bmatrix}\lambda_1^2\lambda_2\,\bar{\omega}^3_1 \\ \lambda_1\lambda_2^2\,\bar{\omega}^3_2\end{bmatrix}. \tag{10.9}$$

Conclude that the transformed function $\tilde{c}$ defined by the conditions

$$\tilde{\bar{\omega}}^3_1 = \tilde{c}\,\tilde{\bar{\omega}}^2, \qquad \tilde{\bar{\omega}}^3_2 = \tilde{c}\,\tilde{\bar{\omega}}^1$$

is given by

$$\tilde{c} = \lambda_1^2\lambda_2^2\,c. \tag{10.10}$$

Thus, the function c is a *relative invariant*: If it vanishes for any 0-adapted frame based at a point $\mathbf{u} \in U$, then it vanishes for every 0-adapted frame based at $\mathbf{u}$. At this point, we need to consider separately the cases where c is zero or nonzero. First, we examine the case where c vanishes identically on U.

***Exercise 10.8.** Suppose that for any 0-adapted frame field $(\mathbf{e}_1(\mathbf{u}), \mathbf{e}_2(\mathbf{u}), \mathbf{e}_3(\mathbf{u}))$ along Σ, the function c is identically equal to zero on U. Show that this assumption implies that

$$d\mathbf{e}_1 \equiv d\mathbf{e}_2 \equiv 0 \mod (\mathbf{e}_1, \mathbf{e}_2).$$

Conclude that the plane spanned by the vectors $(\mathbf{e}_1(\mathbf{u}), \mathbf{e}_2(\mathbf{u}))$ is constant on U and hence that the surface Σ is contained in this plane.

As a consequence of Exercise 10.8, the function c associated to any 0-adapted frame field $(\mathbf{e}_1(\mathbf{u}), \mathbf{e}_2(\mathbf{u}), \mathbf{e}_3(\mathbf{u}))$ on a non-planar doubly ruled surface cannot vanish identically. By shrinking the domain U of our parametrization $\mathbf{x} : U \to \mathbb{A}^3$ of Σ if necessary, we can assume that $c \neq 0$ at every point of U.

Remark 10.9. The assumption that the function c is either identically zero or never zero on U is an example of a *constant type* assumption; it may be thought of as somewhat analogous to the assumption that a regular surface in $\mathbb{E}^3$ either contains no umbilic points or is totally umbilic. In principle, there could exist doubly ruled surfaces for which this assumption does not hold, i.e., for which $c = 0$ only on a proper, nonempty subset of U. But once we prove Theorem 10.3 for the open subset of U on which $c \neq 0$, a patching argument based on the assumption that Σ is smooth can be used to show that, in fact, this cannot happen.

According to equation (10.10), there exists a choice of 0-adapted frame field $(\mathbf{e}_1(\mathbf{u}), \mathbf{e}_2(\mathbf{u}), \mathbf{e}_3(\mathbf{u}))$ for which $c = \pm 1$. If $c = -1$, we can exchange $\mathbf{e}_1(\mathbf{u})$ and $\mathbf{e}_2(\mathbf{u})$ and replace $\mathbf{e}_3(\mathbf{u})$ by $-\mathbf{e}_3(\mathbf{u})$ to arrive at a new 0-adapted frame field for which $c = 1$; thus, without loss of generality, we will assume that we can choose a 0-adapted frame field for which $c = 1$. Such a frame field will be called 1-*adapted.*

***Exercise 10.10.** Let $(\mathbf{e}_1(\mathbf{u}), \mathbf{e}_2(\mathbf{u}), \mathbf{e}_3(\mathbf{u}))$ be any 1-adapted frame field along Σ.

(a) Show that any other 1-adapted frame field $(\tilde{\mathbf{e}}_1(\mathbf{u}), \tilde{\mathbf{e}}_2(\mathbf{u}), \tilde{\mathbf{e}}_3(\mathbf{u}))$ along Σ must have the form

$$\begin{bmatrix}\tilde{\mathbf{e}}_1(\mathbf{u}) & \tilde{\mathbf{e}}_2(\mathbf{u}) & \tilde{\mathbf{e}}_3(\mathbf{u})\end{bmatrix} = \begin{bmatrix}\mathbf{e}_1(\mathbf{u}) & \mathbf{e}_2(\mathbf{u}) & \mathbf{e}_3(\mathbf{u})\end{bmatrix} \begin{bmatrix} \lambda & 0 & r_1 \\ 0 & \pm\frac{1}{\lambda} & r_2 \\ 0 & 0 & \pm 1 \end{bmatrix} \tag{10.11}$$

for some functions λ, r_1, r_2 on U, with $\lambda \neq 0$.

(b) Note that the condition $c = 1$ is equivalent to the relations

$$\bar{\omega}^3_1 = \bar{\omega}^2, \qquad \bar{\omega}^3_2 = \bar{\omega}^1 \tag{10.12}$$

among the associated Maurer-Cartan forms. Compare with Exercise 6.42(a), and conclude that any 1-adapted frame field along Σ satisfies the defining conditions for what we called a "1-adapted null frame" there. In particular, Σ must be a *hyperbolic* equi-affine surface; otherwise no such frame field could exist along Σ.

(c) Apply the result of Exercise 6.42(c) to show that there exists a choice of 1-adapted frame field along Σ for which $\bar{\omega}^3_3 = 0$. (This condition corresponds to choosing $\mathbf{e}_3(\mathbf{u})$ to be parallel to the equi-affine normal direction.) Such a frame field will be called 2-*adapted.*

(d) Show that any two 2-adapted frame fields $(\mathbf{e}_1(\mathbf{u}), \mathbf{e}_2(\mathbf{u}), \mathbf{e}_3(\mathbf{u}))$, $(\tilde{\mathbf{e}}_1(\mathbf{u}), \tilde{\mathbf{e}}_2(\mathbf{u}), \tilde{\mathbf{e}}_3(\mathbf{u}))$ along Σ must be related by a transformation of the form

$$\left[\tilde{\mathbf{e}}_1(\mathbf{u})\ \tilde{\mathbf{e}}_2(\mathbf{u})\ \tilde{\mathbf{e}}_3(\mathbf{u})\right] = \left[\mathbf{e}_1(\mathbf{u})\ \mathbf{e}_2(\mathbf{u})\ \mathbf{e}_3(\mathbf{u})\right] \begin{bmatrix} \lambda & 0 & 0 \\ 0 & \pm\frac{1}{\lambda} & 0 \\ 0 & 0 & \pm 1 \end{bmatrix} \tag{10.13}$$

for some function $\lambda \neq 0$ on U.

***Exercise 10.11.** Let $(\mathbf{e}_1(\mathbf{u}), \mathbf{e}_2(\mathbf{u}), \mathbf{e}_3(\mathbf{u}))$ be any 2-adapted frame field along Σ.

(a) Apply the results of Exercises 6.42(e) and 6.42(f) and equations (10.7) to show that

$$a = b = 0.$$

(Alternatively, differentiate equations (10.12) to arrive at the same result.) Consequently, the Maurer-Cartan forms associated to any 2-adapted frame field on Σ must satisfy the conditions

$$\bar{\omega}^1_2 = \bar{\omega}^2_1 = 0. \tag{10.14}$$

Conclude from equation (10.14) that the Fubini-Pick form (6.26) of Σ is identically zero.

(b) Conversely, show that if Σ is a hyperbolic equi-affine surface whose Fubini-Pick form is identically zero, then Σ is doubly ruled. (Hint: It suffices to show that the condition (10.5) holds.)

(c) Differentiate equations (10.14) and apply Cartan's lemma to conclude that there exist functions f, g on U such that

$$\bar{\omega}^1_3 = f\,\bar{\omega}^1, \qquad \bar{\omega}^2_3 = g\,\bar{\omega}^2. \tag{10.15}$$

(d) Differentiate the equation $\bar{\omega}^3_3 = 0$ and conclude that $g = f$.

(e) Substitute equations (10.15) (with $g = f$) into the Cartan structure equations (3.8) for $d\bar{\omega}^1_3$ and $d\bar{\omega}^2_3$ to obtain

$$df \wedge \bar{\omega}^1 = df \wedge \bar{\omega}^2 = 0.$$

Conclude that $df = 0$, and hence f is equal to a constant $C \in \mathbb{R}$ on U.

Once again, we need to divide into cases based on whether or not C is zero.

Case 1: $C = 0$.

***Exercise 10.12.** Suppose that $C = 0$.

(a) Show that $d\mathbf{e}_3 = 0$, and conclude that the vector field $\mathbf{e}_3(\mathbf{u})$ is constant on U. (Note that this implies that Σ is an improper equi-affine sphere; cf. Exercise 6.41.)

(b) Recall that the Maurer-Cartan forms associated to a 2-adapted frame field on Σ satisfy

$$\bar{\omega}_1^1 + \bar{\omega}_2^2 = 0.$$

Show that $d\bar{\omega}_1^1 = d\bar{\omega}_2^2 = 0$, and use the Poincaré lemma (cf. Theorem 2.31) to conclude that there exists a function μ on U such that

$$\bar{\omega}_1^1 = d\mu = -\bar{\omega}_2^2. \tag{10.16}$$

Remark 10.13. Technically, this result only holds if U is homeomorphic to an open disk in $\mathbb{R}^2$. If this is not the case, then U can be covered by such open sets, and the result of Theorem 10.3 can be applied to the restriction of the parametrization $\mathbf{x} : U \to \mathbb{A}^3$ to each of these open subsets of U. The theorem can then be obtained for the entire surface $\Sigma = \mathbf{x}(U)$ via a patching argument.

(c) Use equations (10.14) and (10.16) to show that

$$d\bar{\omega}^1 = -d\mu \wedge \bar{\omega}^1, \qquad d\bar{\omega}^2 = d\mu \wedge \bar{\omega}^2. \tag{10.17}$$

(d) Show that

$$d\left(e^{\mu}\bar{\omega}^1\right) = d\left(e^{-\mu}\bar{\omega}^2\right) = 0.$$

Conclude that by a transformation of the form (10.13) with $\lambda = e^{-\mu}$, we can arrange that

$$d\bar{\omega}^1 = d\bar{\omega}^2 = 0. \tag{10.18}$$

Then use equation (10.17) to conclude that $d\mu = 0$, and hence

$$\bar{\omega}_1^1 = \bar{\omega}_2^2 = 0. \tag{10.19}$$

To summarize, we have shown that when $C = 0$, there exists a 2-adapted frame field along Σ for which the associated Maurer-Cartan forms $(\bar{\omega}^i, \bar{\omega}_j^i)$ satisfy the following conditions:

$$\begin{gathered} d\bar{\omega}^1 = d\bar{\omega}^2 = 0, \\ \bar{\omega}_1^1 = \bar{\omega}_2^1 = \bar{\omega}_1^2 = \bar{\omega}_2^2 = \bar{\omega}_3^1 = \bar{\omega}_3^2 = \bar{\omega}_3^3 = 0, \\ \bar{\omega}_1^3 = \bar{\omega}^2, \qquad \bar{\omega}_2^3 = \bar{\omega}^1. \end{gathered} \tag{10.20}$$

The following exercise shows how these equations can be integrated in order to determine the surface Σ.

***Exercise 10.14.** (a) Apply the Poincaré lemma to the first equations in (10.20) to show that there exist functions u, v on U such that

$$\bar{\omega}^1 = du, \qquad \bar{\omega}^2 = dv \tag{10.21}$$

(cf. Remark 10.13). Note that the condition $\bar{\omega}^1 \wedge \bar{\omega}^2 \neq 0$ implies that the functions (u, v) form a local coordinate system on U.

(b) From equations (10.21) and the remainder of equations (10.20), conclude that the structure equations (10.3) now take the form

$$\begin{aligned} d\mathbf{x} &= \mathbf{e}_1\, du + \mathbf{e}_2\, dv, \\ d\mathbf{e}_1 &= \mathbf{e}_3\, dv, \\ d\mathbf{e}_2 &= \mathbf{e}_3\, du, \\ d\mathbf{e}_3 &= 0. \end{aligned} \tag{10.22}$$

(c) Integrate equations (10.22) (beginning with the equation for $d\mathbf{e}_3$ and working backwards) to show that there exist *constant* vectors $(\bar{\mathbf{x}}, \bar{\mathbf{e}}_1, \bar{\mathbf{e}}_2, \bar{\mathbf{e}}_3)$ such that

$$\begin{aligned} \mathbf{e}_3(u, v) &= \bar{\mathbf{e}}_3, \\ \mathbf{e}_1(u, v) &= \bar{\mathbf{e}}_1 + v\bar{\mathbf{e}}_3, \\ \mathbf{e}_2(u, v) &= \bar{\mathbf{e}}_2 + u\bar{\mathbf{e}}_3, \\ \mathbf{x}(u, v) &= \bar{\mathbf{x}} + u\bar{\mathbf{e}}_1 + v\bar{\mathbf{e}}_2 + uv\bar{\mathbf{e}}_3. \end{aligned} \tag{10.23}$$

(d) Use equations (10.23) and the fact that $(\mathbf{e}_1(\mathbf{u}), \mathbf{e}_2(\mathbf{u}), \mathbf{e}_3(\mathbf{u}))$ is a unimodular frame field to show that

$$\det \begin{bmatrix} \bar{\mathbf{e}}_1 & \bar{\mathbf{e}}_2 & \bar{\mathbf{e}}_3 \end{bmatrix} = 1.$$

Conclude that via an equi-affine transformation, we can arrange that $\bar{\mathbf{x}} = {}^t[0,\ 0,\ 0]$ and

$$\bar{\mathbf{e}}_1 = {}^t[1,\ 0,\ 0], \qquad \bar{\mathbf{e}}_2 = {}^t[0,\ 1,\ 0], \qquad \bar{\mathbf{e}}_3 = {}^t[0,\ 0,\ 1],$$

and hence

$$\mathbf{x}(u, v) = {}^t[u,\ v,\ uv].$$

Therefore, up to equi-affine equivalence, Σ must be an open subset of the hyperbolic paraboloid (10.2).

Case 2: $C \neq 0$.

Suppose that $C \neq 0$. Without loss of generality, we may suppose that $C > 0$: If instead we have $C < 0$, an equi-affine transformation of the form

$$\mathbf{x} = {}^t[x^1, x^2, x^3] \to {}^t[-x^1, x^2, -x^3]$$

will reverse the sign of C.

***Exercise 10.15.** (a) Suppose that $C > 0$. Show that

$$d\bar{\omega}^1 = -\bar{\omega}^1_1 \wedge \bar{\omega}^1, \qquad d\bar{\omega}^2 = -\bar{\omega}^2_2 \wedge \bar{\omega}^2, \tag{10.24}$$

and use the Frobenius theorem (cf. Theorem 2.33) to conclude that every point $\mathbf{u} \in U$ has a neighborhood $V \subset U$ on which there exist functions u, v, g_1, g_2 such that

$$\bar{\omega}^1 = e^{g_1}\, du, \qquad \bar{\omega}^2 = e^{g_2}\, dv. \tag{10.25}$$

It suffices to prove Theorem 10.3 on each of these restricted neighborhoods V (cf. Remark 10.13); for simplicity, we will assume that $V = U$. Note that the condition $\bar{\omega}^1 \wedge \bar{\omega}^2 \neq 0$ implies that the functions (u, v) form a local coordinate system on U.

(b) Show that by a transformation of the form (10.13) with $\lambda = e^{\frac{1}{2}(g_1 - g_2)}$, we can arrange that

$$\bar{\omega}^1 = e^h\, du, \qquad \bar{\omega}^2 = e^h\, dv, \tag{10.26}$$

where $h = \frac{1}{2}(g_1 + g_2)$.

(c) Use equations (10.24) (together with the fact that $\bar{\omega}^1_1 + \bar{\omega}^2_2 = 0$) to show that

$$\bar{\omega}^1_1 = -\bar{\omega}^2_2 = h_u\, du - h_v\, dv. \tag{10.27}$$

Then substitute this expression into the structure equation for $d\bar{\omega}^1_1$ to show that the function h must satisfy the PDE

$$2h_{uv} = Ce^{2h}. \tag{10.28}$$

Conclude that the function $z(u, v) = 2h(u, v) + \ln C$ must be a solution of Liouville's equation

$$z_{uv} = e^z. \tag{10.29}$$

***Exercise 10.16.** (a) The general solution to Liouville's equation (10.29) is

$$z(u, v) = \ln\left(\frac{2U'(u)V'(v)}{(U(u) + V(v))^2}\right), \tag{10.30}$$

where $U(u)$ and $V(v)$ are arbitrary functions of a single variable with the property that $U'(u)$ and $V'(v)$ are nonzero and have the same sign (cf. Exercise 9.22). Show that by making the change of coordinates

$$\tilde{u} = U(u), \qquad \tilde{v} = V(v),$$

we can arrange that

$$z(\tilde{u}, \tilde{v}) = \ln\left(\frac{2}{(\tilde{u} + \tilde{v})^2}\right).$$

Dropping the tildes, this yields

$$h(u, v) = \frac{1}{2}(z(u, v) - \ln C) = \ln\left(\frac{\sqrt{2}}{\sqrt{C}(u + v)}\right). \tag{10.31}$$

(b) Use equations (10.12), (10.14), (10.15), (10.26), (10.27), and (10.31) to show that

$$\begin{gathered}
\bar{\omega}^1 = \frac{\sqrt{2}}{\sqrt{C}(u+v)}\, du, \qquad \bar{\omega}^2 = \frac{\sqrt{2}}{\sqrt{C}(u+v)}\, dv, \\
\bar{\omega}^1_2 = \bar{\omega}^2_1 = \bar{\omega}^3_3 = 0, \\
\bar{\omega}^1_1 = -\bar{\omega}^2_2 = \frac{1}{(u+v)}(-du + dv), \\
\bar{\omega}^3_1 = \frac{\sqrt{2}}{\sqrt{C}(u+v)}\, dv, \qquad \bar{\omega}^3_2 = \frac{\sqrt{2}}{\sqrt{C}(u+v)}\, du, \\
\bar{\omega}^1_3 = \frac{\sqrt{2C}}{(u+v)}\, du, \qquad \bar{\omega}^2_3 = \frac{\sqrt{2C}}{(u+v)}\, dv.
\end{gathered} \tag{10.32}$$

(c) From equations (10.32), conclude that the structure equations (10.3) now take the form

$$\begin{aligned}
d\mathbf{x} &= \mathbf{e}_1\left(\frac{\sqrt{2}}{\sqrt{C}(u+v)}\, du\right) + \mathbf{e}_2\left(\frac{\sqrt{2}}{\sqrt{C}(u+v)}\, dv\right), \\
d\mathbf{e}_1 &= \mathbf{e}_1\left(\frac{1}{(u+v)}(-du + dv)\right) + \mathbf{e}_3\left(\frac{\sqrt{2}}{\sqrt{C}(u+v)}\, dv\right), \\
d\mathbf{e}_2 &= \mathbf{e}_2\left(\frac{1}{(u+v)}(du - dv)\right) + \mathbf{e}_3\left(\frac{\sqrt{2}}{\sqrt{C}(u+v)}\, du\right), \\
d\mathbf{e}_3 &= \mathbf{e}_1\left(\frac{\sqrt{2C}}{(u+v)}\, du\right) + \mathbf{e}_2\left(\frac{\sqrt{2C}}{(u+v)}\, dv\right).
\end{aligned} \tag{10.33}$$

***Exercise 10.17.** (a) Show that equations (10.33) are equivalent to the PDE system

(10.34)

$$\mathbf{x}_u = \frac{\sqrt{2}}{\sqrt{C}(u+v)}\mathbf{e}_1, \qquad \mathbf{x}_v = \frac{\sqrt{2}}{\sqrt{C}(u+v)}\mathbf{e}_2,$$

$$(\mathbf{e}_1)_u = -\frac{1}{(u+v)}\mathbf{e}_1, \qquad (\mathbf{e}_1)_v = \frac{1}{(u+v)}\mathbf{e}_1 + \frac{\sqrt{2}}{\sqrt{C}(u+v)}\mathbf{e}_3,$$

$$(\mathbf{e}_2)_u = \frac{1}{(u+v)}\mathbf{e}_2 + \frac{\sqrt{2}}{\sqrt{C}(u+v)}\mathbf{e}_3, \quad (\mathbf{e}_2)_v = -\frac{1}{(u+v)}\mathbf{e}_2,$$

$$(\mathbf{e}_3)_u = \frac{\sqrt{2C}}{(u+v)}\mathbf{e}_1, \qquad (\mathbf{e}_3)_v = \frac{\sqrt{2C}}{(u+v)}\mathbf{e}_2.$$

(b) Integrate the equations for $(\mathbf{e}_1)_u$ and $(\mathbf{e}_2)_v$ to show that there exist $\mathbb{A}^3$-valued functions $\mathbf{f}(v), \mathbf{g}(u)$ such that

$$\mathbf{e}_1(u,v) = \frac{1}{(u+v)}\mathbf{f}(v), \qquad \mathbf{e}_2(u,v) = \frac{1}{(u+v)}\mathbf{g}(u). \tag{10.35}$$

(c) Use the equations for $(\mathbf{e}_1)_v$ and $(\mathbf{e}_2)_u$ to write

$$\mathbf{e}_3(u,v) = \frac{\sqrt{C}}{\sqrt{2}}\left((u+v)(\mathbf{e}_2)_u - \mathbf{e}_2\right) = \frac{\sqrt{C}}{\sqrt{2}}\left((u+v)(\mathbf{e}_1)_v - \mathbf{e}_1\right). \tag{10.36}$$

Substitute equations (10.35) into this equation to show that

$$(u+v)\mathbf{f}'(v) - 2\mathbf{f}(v) = (u+v)\mathbf{g}'(u) - 2\mathbf{g}(u). \tag{10.37}$$

(d) Differentiate equation (10.37) with respect to u and v, and use the result to show that there exist *constant* vectors $\mathbf{c}_{10}, \mathbf{c}_{20}, \mathbf{c}_{11}, \mathbf{c}_{21}, \mathbf{c}_2$ such that

$$\mathbf{f}(v) = \mathbf{c}_2 v^2 + \mathbf{c}_{11}v + \mathbf{c}_{10}, \qquad \mathbf{g}(u) = \mathbf{c}_2 u^2 + \mathbf{c}_{21}u + \mathbf{c}_{20}. \tag{10.38}$$

Substitute these expressions into equation (10.37), and by comparing like powers of u and v, conclude that $\mathbf{c}_{21} = -\mathbf{c}_{11}$, $\mathbf{c}_{20} = \mathbf{c}_{10}$. Set

$$\mathbf{c}_1 = \mathbf{c}_{11} = -\mathbf{c}_{21}, \qquad \mathbf{c}_0 = \mathbf{c}_{10} = \mathbf{c}_{20},$$

so that (10.38) becomes

$$\mathbf{f}(v) = v^2\mathbf{c}_2 + v\mathbf{c}_1 + \mathbf{c}_0, \qquad \mathbf{g}(u) = u^2\mathbf{c}_2 - u\mathbf{c}_1 + \mathbf{c}_0. \tag{10.39}$$

(e) Substitute (10.39) into equations (10.35), (10.36) to obtain

$$
\begin{aligned}
\mathbf{e}_1(u,v) &= \frac{1}{(u+v)}(v^2\mathbf{c}_2 + v\mathbf{c}_1 + \mathbf{c}_0),\\
\mathbf{e}_2(u,v) &= \frac{1}{(u+v)}(u^2\mathbf{c}_2 - u\mathbf{c}_1 + \mathbf{c}_0),\\
\mathbf{e}_3(u,v) &= \frac{\sqrt{C}}{\sqrt{2}(u+v)}(2uv\mathbf{c}_2 + (u-v)\mathbf{c}_1 - 2\mathbf{c}_0).
\end{aligned}
\tag{10.40}
$$

Observe from equations (10.33) that $d(\mathbf{x} - \frac{1}{C}\mathbf{e}_3) = 0$; this implies that $\mathbf{x} - \frac{1}{C}\mathbf{e}_3$ is constant along Σ (in particular, Σ is a proper equi-affine sphere; cf. Exercise 6.41), and by a translation we can arrange that $\mathbf{x} - \frac{1}{C}\mathbf{e}_3 = \mathbf{0}$. Then we have

$$
\mathbf{x}(u,v) = \frac{1}{C}\mathbf{e}_3(u,v) = \frac{1}{\sqrt{2C}(u+v)}(2uv\mathbf{c}_2 + (u-v)\mathbf{c}_1 - 2\mathbf{c}_0).
\tag{10.41}
$$

(f) Verify that the functions in equations (10.40), (10.41) satisfy the PDE system (10.34).

(g) Use equations (10.40) to show that

$$
\det\left[\mathbf{e}_1(u,v)\ \mathbf{e}_2(u,v)\ \mathbf{e}_3(u,v)\right] = \frac{\sqrt{C}}{\sqrt{2}}\det\left[\mathbf{c}_2\ \mathbf{c}_1\ \mathbf{c}_0\right].
$$

Since $(\mathbf{e}_1, \mathbf{e}_2, \mathbf{e}_3)$ must be a unimodular frame, conclude that

$$
\det\left[\mathbf{c}_2\ \mathbf{c}_1\ \mathbf{c}_0\right] = \frac{\sqrt{2}}{\sqrt{C}}.
\tag{10.42}
$$

(h) By an equi-affine transformation, we can arrange that

$$
\mathbf{c}_2 = \frac{1}{\sqrt{2}\sqrt[6]{C}}\,{}^t[1,\,0,\,1], \qquad \mathbf{c}_1 = \frac{\sqrt{2}}{\sqrt[6]{C}}\,{}^t[0,\,1,\,0], \qquad \mathbf{c}_0 = \frac{1}{\sqrt{2}\sqrt[6]{C}}\,{}^t[-1,\,0,\,1].
$$

Check that these vectors satisfy the determinant condition (10.42) and that with this choice, we have

$$
\mathbf{x}(u,v) = \frac{1}{C^{\frac{2}{3}}(u+v)}\,{}^t[uv+1,\, u-v,\, uv-1].
$$

Finally, check that the coordinates of $\mathbf{x}(u,v)$ satisfy the defining equation (10.1) for the hyperboloid, with $r = C^{-\frac{2}{3}}$. Therefore, up to equi-affine equivalence, Σ must be an open subset of a hyperboloid of one sheet.

This completes the proof of Theorem 10.3.

10.2. The Cauchy-Crofton formula

The Cauchy-Crofton formula comes from the subject of *integral geometry.* It relates the length of a curve in the Euclidean plane to the "size" of the set of lines in the plane that intersect the curve (counted with multiplicity). A classical approach to this formula may be found in [**dC76**]; the moving frames approach described here is outlined in [**Che42**].

In order to make sense of the idea of the "size" of a set of lines, we need to introduce the notion of a *measure* on the set of lines in $\mathbb{E}^2$, a.k.a. the *affine Grassmannian* $G_1(\mathbb{E}^2)$ (cf. §9.2). In general, a measure on a manifold M is a differential form $d\mu$ on M (which, despite the notation, is not necessarily an exact form) that is used to define the size of any subset of M via integration. Specifically, the measure of a subset $\Omega \subset M$ is given by

$$\mu(\Omega) = \int_\Omega d\mu.$$

One familiar example is the *area measure* on the Euclidean plane $\mathbb{E}^2$: The area measure is the 2-form $dA = dx \wedge dy$, and the area of any (measurable) subset $\Omega \subset \mathbb{E}^2$ is given by

$$A(\Omega) = \int_\Omega dA.$$

An important feature of the area measure is that it is invariant under the action of the Euclidean group $E(2)$: For any element $g \in E(2)$ and any measurable subset $\Omega \subset \mathbb{E}^2$, we have

$$A(g \cdot \Omega) = A(\Omega).$$

We can see how this invariance comes about as follows: Recall that we can regard $\mathbb{E}^2$ as the homogeneous space $\mathbb{E}^2 \cong E(2)/SO(2)$; i.e., $\mathbb{E}^2$ is the set of left cosets of the subgroup $SO(2) \subset E(2)$:

$$\begin{array}{ccc} SO(2) & \longrightarrow & E(2) \\ & & \big\downarrow \pi \\ & & \mathbb{E}^2 \cong E(2)/SO(2). \end{array}$$

The Maurer-Cartan forms $(\omega^1, \omega^2, \omega^1_2)$ form a basis for the left-invariant 1-forms on the Lie group $E(2)$, and of these, ω^1 and ω^2 are *semi-basic* for the projection $\pi : E(2) \to \mathbb{E}^2$ (cf. Exercise 3.19). The area measure on the quotient space $\mathbb{E}^2$ is given by the wedge product of these semi-basic forms:

$$dA = \omega^1 \wedge \omega^2.$$

(While the 1-forms (ω^1, ω^2) are not individually well-defined on $\mathbb{E}^2$, it turns out that their wedge product is; cf. Exercise 4.47.) So the invariance of the area measure follows immediately from the left-invariance of the Maurer-Cartan forms: Given any measurable subset $\Omega \subset \mathbb{E}^2$ and any element $g \in E(2)$, we have

$$\begin{aligned} A(g \cdot \Omega) &= \int_{g\cdot\Omega} \omega^1 \wedge \omega^2 \\ &= \int_\Omega L_g^*(\omega^1 \wedge \omega^2) \\ &= \int_\Omega \omega^1 \wedge \omega^2 \\ &= A(\Omega). \end{aligned}$$

Now consider the space of lines in $\mathbb{E}^2$. As we saw in §9.2 when we constructed the affine Grassmannian $G_1(\mathbb{E}^3)$, we can specify a line ℓ in the plane $\mathbb{E}^2$ by choosing a point $\mathbf{x}$ on ℓ and a unit vector $\mathbf{e}_1$ parallel to ℓ. Two pairs $(\mathbf{x}, \mathbf{e}_1)$ and $(\mathbf{y}, \mathbf{f}_1)$ determine the same line if and only if $\mathbf{f}_1 = \pm\mathbf{e}_1$ and $\mathbf{y} = \mathbf{x} + t\mathbf{e}_1$ for some $t \in \mathbb{R}$ (cf. Exercise 9.1). In this case, we write $(\mathbf{x}, \mathbf{e}_1) \sim (\mathbf{y}, \mathbf{f}_1)$, and the affine Grassmannian is defined to be the quotient space

$$G_1(\mathbb{E}^2) = \left(\mathbb{E}^2 \times \mathbb{S}^1\right) / \sim .$$

***Exercise 10.18.** (a) Show that the product $\mathbb{E}^2 \times \mathbb{S}^1$ is diffeomorphic to the oriented, orthonormal frame bundle $\mathcal{F}(\mathbb{E}^2)$, which in turn is diffeomorphic to the Lie group $E(2)$. (Hint: Choosing one unit vector $\mathbf{e}_1 \in T_\mathbf{x}\mathbb{E}^2$ uniquely determines a second unit vector $\mathbf{e}_2 \in T_\mathbf{x}\mathbb{E}^2$ such that $(\mathbf{e}_1, \mathbf{e}_2)$ is an oriented, orthonormal frame for $T_\mathbf{x}\mathbb{E}^2$. This is a feature that is particular to $\mathbb{E}^2$ and not true for $\mathbb{E}^n$ when $n \geq 3$.) Consequently, we can write

$$G_1(\mathbb{E}^2) = E(2)/ \sim .$$

(b) Show that the equivalence relation $(\mathbf{x}, \mathbf{e}_1) \sim (\mathbf{y}, \mathbf{f}_1)$ on $\mathbb{E}^2 \times \mathbb{S}^1$ corresponds to the following equivalence relation on $E(2)$: If

$$\tilde{\mathbf{x}} = \begin{bmatrix} 1 & 0 & 0 \\ \mathbf{x} & \mathbf{e}_1 & \mathbf{e}_2 \end{bmatrix}, \quad \tilde{\mathbf{y}} = \begin{bmatrix} 1 & 0 & 0 \\ \mathbf{y} & \mathbf{f}_1 & \mathbf{f}_2 \end{bmatrix} \in E(2),$$

then $\tilde{\mathbf{x}} \sim \tilde{\mathbf{y}}$ if and only if

$$\tilde{\mathbf{y}} = \tilde{\mathbf{x}} \begin{bmatrix} 1 & 0 & 0 \\ t & \pm 1 & 0 \\ 0 & 0 & \pm 1 \end{bmatrix}$$

for some $t \in \mathbb{R}$, where the choice of sign is the same in both diagonal entries. Therefore, $G_1(\mathbb{E}^2)$ may be regarded as the set of left cosets of the subgroup

$$H = \left\{ \begin{bmatrix} 1 & 0 & 0 \\ t & \pm 1 & 0 \\ 0 & 0 & \pm 1 \end{bmatrix} : t \in \mathbb{R} \right\} \subset E(2),$$

and $E(2)$ may be regarded as a principal bundle over $G_1(\mathbb{E}^2)$ with fiber group H:

$$\begin{array}{ccc} H & \longrightarrow & E(2) \\ & & \downarrow \pi \\ & & G_1(\mathbb{E}^2) \cong E(2)/H. \end{array}$$

(c) Show that the Maurer-Cartan forms (ω^2, ω^1_2) on $E(2)$ are semi-basic for the projection $\pi : E(2) \to G_1(\mathbb{E}^2)$.

This suggests that a reasonable way to define a measure $d\mu$ on $G_1(\mathbb{E}^2)$ might be to set

$$d\mu = |\omega^2 \wedge \omega^1_2|.$$

Like the area measure $dA = \omega^1 \wedge \omega^2$, the 2-form $d\mu$ is well-defined on the quotient $G_1(\mathbb{E}^2)$, even though the individual 1-forms (ω^2, ω^1_2) are not. (The absolute value sign is necessary because we have defined $G_1(\mathbb{E}^2)$ to be the space of unoriented lines, but reversing the orientation changes the sign of the 2-form $\omega^2 \wedge \omega^1_2$; therefore, this 2-form is only well-defined up to sign on $G_1(\mathbb{E}^2)$.) Then, given any (measurable) subset $\Omega \subset G_1(\mathbb{E}^2)$, we define the measure $\mu(\Omega)$ to be

$$\mu(\Omega) = \int_\Omega d\mu = \int_\Omega |\omega^2 \wedge \omega^1_2|.$$

Exercise 10.19. Another way to determine a line ℓ in $\mathbb{E}^2$ is to specify two parameters: the angle θ with $0 \leq \theta < \pi$ such that the vector $\mathbf{e}_1 = {}^t[\cos(\theta), \sin(\theta)]$ is parallel to ℓ and the shortest distance ρ from the origin to ℓ. This identification gives rise to a diffeomorphism

$$G_1(\mathbb{E}^2) \cong \mathbb{R} \times \mathbb{S}^1$$

(where $\mathbb{S}^1$ represents $\mathbb{R}/\pi\mathbb{Z}$ rather than the more typical $\mathbb{R}/2\pi\mathbb{Z}$), with coordinates

$$\{(\rho, \theta) \mid \rho \in \mathbb{R}, 0 \leq \theta < \pi\}.$$

Show that the measure $d\mu$ on $G_1(\mathbb{E}^2)$ is then given by the 2-form

$$d\mu = |d\rho \wedge d\theta|$$

on $\mathbb{R} \times \mathbb{S}^1$. (Hint: How are the Maurer-Cartan forms (ω^2, ω^1_2) related to $(d\rho, d\theta)$?)

Before we state the Cauchy-Crofton formula, we need to introduce the notion of *incidence*:

Definition 10.20. A point $\mathbf{x} \in \mathbb{E}^2$ and a line $\ell \in G_1(\mathbb{E}^2)$ are called *incident* if the point $\mathbf{x}$ lies on the line ℓ. A curve α in $\mathbb{E}^2$ and a line $\ell \in G_1(\mathbb{E}^2)$ are called *incident* if any point of α is incident with ℓ.

***Exercise 10.21.** Let $\pi_1 : E(2) \to \mathbb{E}^2$ denote the projection from $E(2)$ to $\mathbb{E}_2$ and let $\pi_2 : E(2) \to G_1(\mathbb{E}^2)$ denote the projection from $E(2)$ to $G_1(\mathbb{E}^2)$, as in the following diagram:

$$\begin{array}{ccccc} SO(2) & \longrightarrow & E(2) & \longleftarrow & H \\ & \swarrow^{\pi_1} & & \searrow^{\pi_2} & \\ \mathbb{E}^2 & & & & G_1(\mathbb{E}^2). \end{array}$$

Show that a point $\mathbf{x} \in \mathbb{E}^2$ and a line $\ell \in G_1(\mathbb{E}^2)$ are incident if and only if the left cosets $\pi_1^{-1}(\mathbf{x}), \pi_2^{-1}(\ell) \subset E(2)$ have nonempty intersection.

We are now ready to state the Cauchy-Crofton formula.

Theorem 10.22. *Let $\alpha : [a, b] \to \mathbb{E}^2$ be a curve in the Euclidean plane, and let $\Omega \subset G_1(\mathbb{E}^2)$ be the set of lines incident with α, counted with multiplicity (e.g., if a line ℓ intersects α at two distinct points, then it counts twice). Then*

$$\mu(\Omega) = 2L,$$

where L is the length of α.

To begin the proof, let

$$\widetilde{\Omega} = \pi_1^{-1}\left(\alpha([a, b])\right) \subset E(2).$$

The set $\widetilde{\Omega}$ consists of all oriented, orthonormal frames $(\mathbf{x}; \mathbf{e}_1, \mathbf{e}_2)$ based at all points $\mathbf{x} \in \alpha$, each of which corresponds to a line ℓ passing through the point $\mathbf{x}$ and parallel to the frame vector $\mathbf{e}_1$. Moreover, any line ℓ that intersects α in k distinct points is represented by $2k$ distinct points in $\widetilde{\Omega}$: Each line appears twice in $\widetilde{\Omega}$ for each point $\mathbf{x}$ where it intersects α, represented by frames of the form $(\mathbf{x}; \mathbf{e}_1, \mathbf{e}_2)$ and $(\mathbf{x}; -\mathbf{e}_1, -\mathbf{e}_2)$.

***Exercise 10.23.** Convince yourself that $\widetilde{\Omega}$ is a 2-dimensional submanifold of $E(2)$.

Now, let

$$\Omega = \pi_2(\widetilde{\Omega}) \subset G_1(\mathbb{E}^2),$$

and observe that Ω consists of all lines incident with α. Moreover, for any $\ell \in \Omega$, the cardinality of the set $\pi_2^{-1}(\ell) \cap \widetilde{\Omega}$ is precisely twice the number of distinct points in which ℓ intersects α. So the set $\widetilde{\Omega}$ represents all the lines in $\mathbb{E}^2$ that are incident with α, counted with double multiplicity.

The restriction of the projection $\pi_2 : E(2) \to G_1(\mathbb{E}^2)$ to the set $\widetilde{\Omega}$ is an immersion $\pi_2 : \widetilde{\Omega} \to \Omega$; in fact, this map realizes $\widetilde{\Omega}$ as a double cover of Ω. Because the measure $d\mu = |\omega^2 \wedge \omega^1_2|$ is semi-basic for the projection π_2, it follows that

$$\mu(\Omega) = \int_{\Omega} d\mu = \tfrac{1}{2} \int_{\widetilde{\Omega}} \pi_2^*(d\mu) = \tfrac{1}{2} \int_{\widetilde{\Omega}} d\mu = \tfrac{1}{2}\mu(\widetilde{\Omega}),$$

where the integral on the left must be counted with multiplicity. (We are abusing notation slightly here by writing $d\mu$ for both the 2-form $|\omega^2 \wedge \omega^1_2|$ on $E(2)$ and the well-defined 2-form $d\mu$ on $G_1(\mathbb{E}^2)$ whose pullback to $E(2)$ via π_2 is equal to $|\omega^2 \wedge \omega^1_2|$.)

***Exercise 10.24.** Let $\alpha : [a, b] \to \mathbb{E}^2$ be an arc-length parametrization of α, given by $\alpha(s) = {}^t[x(s), y(s)]$.

(a) Show that the vectors

$$\bar{\mathbf{e}}_1(s) = {}^t\big[x'(s),\, y'(s)\big], \qquad \bar{\mathbf{e}}_2(s) = {}^t\big[-y'(s),\, x'(s)\big]$$

form an oriented, orthonormal frame at the point $\alpha(s)$.

(b) Show that the set $\widetilde{\Omega}$ can be parametrized by the map

$$\tilde{\alpha} : [a, b] \times [0, 2\pi] \to E(2)$$

defined by

$$\begin{aligned}
\tilde{\alpha}(s, \theta) &= (\mathbf{x}(s, \theta);\, \mathbf{e}_1(s, \theta),\, \mathbf{e}_2(s, \theta)) \\
&= (\alpha(s);\, \cos(\theta)\, \bar{\mathbf{e}}_1(s) + \sin(\theta)\, \bar{\mathbf{e}}_2(s),\, -\sin(\theta)\bar{\mathbf{e}}_1(s) + \cos(\theta)\, \bar{\mathbf{e}}_2(s)) \\
&= \left(\begin{bmatrix} x(s) \\ y(s) \end{bmatrix} ; \begin{bmatrix} x'(s)\cos(\theta) - y'(s)\sin(\theta) \\ y'(s)\cos(\theta) + x'(s)\sin(\theta) \end{bmatrix}, \begin{bmatrix} -x'(s)\sin(\theta) - y'(s)\cos(\theta) \\ -y'(s)\sin(\theta) + x'(s)\cos(\theta) \end{bmatrix} \right).
\end{aligned}$$

(c) Pull back the Cartan structure equation

$$d\mathbf{x} = \mathbf{e}_1\omega^1 + \mathbf{e}_2\omega^2$$

on $E(2)$ via $\tilde{\alpha}$ to show that the 1-forms $\bar{\omega}^1 = \tilde{\alpha}^*\omega^1$ and $\bar{\omega}^2 = \tilde{\alpha}^*\omega^2$ are given by

$$\bar{\omega}^1 = \cos(\theta)\, ds, \qquad \bar{\omega}^2 = -\sin(\theta)\, ds. \tag{10.43}$$

Then use the structure equations for $d\bar{\omega}^1$ and $d\bar{\omega}^2$ to show that

$$\bar{\omega}^1_2 = -d\theta + \lambda\, ds \tag{10.44}$$

for some function $\lambda(s,\theta)$. (In fact, the structure equation for $d\bar{\omega}^1_2$ implies that λ is a function of s alone.)

(d) Use equations (10.43) and (10.44), together with the fact that

$$\int_{\widetilde{\Omega}} |\omega^2 \wedge \omega^1_2| = \int_{\tilde{\alpha}([a,b]\times[0,2\pi])} |\omega^2 \wedge \omega^1_2| = \int_{[a,b]\times[0,2\pi]} \tilde{\alpha}^*(|\omega^2 \wedge \omega^1_2|),$$

to show that

$$\mu(\widetilde{\Omega}) = \int_0^{2\pi}\int_a^b |\sin(\theta)|\, ds\, d\theta = 4L,$$

where $L = (b-a)$ is the length of α. Conclude that $\mu(\Omega) = 2L$.

This completes the proof of Theorem 10.22.

10.3. MAPLE computations

Begin by setting up exactly as we did for Chapter 6: Load the `Cartan` and `LinearAlgebra` packages into MAPLE, declare the Maurer-Cartan forms $(\bar{\omega}^i, \bar{\omega}^i_j)$ associated to a unimodular frame field on Σ, and tell MAPLE about the structure equations and the relation $\bar{\omega}^1_1 + \bar{\omega}^2_2 + \bar{\omega}^3_3 = 0$.

Now, suppose that Σ is doubly ruled and that we have chosen an adapted frame field with $(\mathbf{e}_1(\mathbf{u}), \mathbf{e}_2(\mathbf{u}))$ tangent to the rulings. Set up a substitution that encodes the conditions (10.7) for the associated Maurer-Cartan forms:

```
> ruledsub:= [omega[3]=0, omega[2,1] = a*omega[2],
  omega[1,2] = b*omega[1], omega[3,1] = c*omega[2],
  omega[3,2] = c*omega[1]];
```

EXERCISE 10.7: Begin by introducing new 1-forms $(\tilde{\bar{\omega}}^i, \tilde{\bar{\omega}}^i_j)$ to represent the transformed forms:

```
> Form(Omega[1], Omega[2], Omega[3]);
  Form(Omega[1,1], Omega[1,2], Omega[1,3], Omega[2,1],
```

```
  Omega[2,2], Omega[2,3], Omega[3,1], Omega[3,2], Omega[3,3]);
  Omega[3,3]:= -(Omega[1,1] + Omega[2,2]);
```

Under a transformation of the form (10.8), we have the relations (10.9). Set this up as a substitution:

```
> framechangesub:= [Omega[1] = (1/lambda1)*omega[1],
  Omega[2] = (1/lambda2)*omega[2],
  Omega[3,1] = lambda1^2*lambda2*omega[3,1],
  Omega[3,2] = lambda1*lambda2^2*omega[3,2]];
> framechangebacksub:= makebacksub(framechangesub);
```

In order to find how the function c transforms, combine the substitutions `framechangesub` and `ruledsub` to see how $\tilde{\bar{\omega}}_1^3$ and $\tilde{\bar{\omega}}_2^3$ are related to $\tilde{\bar{\omega}}^2$ and $\tilde{\bar{\omega}}^1$, respectively:

```
> Simf(subs(framechangebacksub, Simf(subs(ruledsub,
  Simf(subs(framechangesub, Omega[3,1]))))));
```

$$\lambda 1^2\, \lambda 2^2\, c\, \Omega_2$$

```
> Simf(subs(framechangebacksub, Simf(subs(ruledsub,
  Simf(subs(framechangesub, Omega[3,2]))))));
```

$$\lambda 1^2\, \lambda 2^2\, c\, \Omega_1$$

So we have $\tilde{c} = \lambda_1^2\lambda_2^2 c$, as in equation (10.10).

EXERCISE 10.11: Since $(\mathbf{e}_1(\mathbf{u}), \mathbf{e}_2(\mathbf{u}), \mathbf{e}_3(\mathbf{u}))$ is a 2-adapted frame field, we have $c = 1$, and since we have already set $\bar{\omega}_3^3 = -(\bar{\omega}_1^1 + \bar{\omega}_2^2)$, the condition $\bar{\omega}_3^3 = 0$ is equivalent to $\bar{\omega}_1^1 + \bar{\omega}_2^2 = 0$. Add these conditions to our substitution:

```
> c:= 1;
> ruledsub:= [op(ruledsub), omega[2,2] = -omega[1,1]];
```

Now differentiate equations (10.12):

```
> Simf(subs(ruledsub, Simf(d(omega[3,1] - omega[2]))));
```

$$-2\, a\, (\omega_1)\ \&\hat{}\ (\omega_2)$$

```
> Simf(subs(ruledsub, Simf(d(omega[3,2] - omega[1]))));
```

$$2\, b\, (\omega_1)\ \&\hat{}\ (\omega_2)$$

Since $\bar{\omega}^1 \wedge \bar{\omega}^2 \neq 0$, it follows that $a = b = 0$.

```
> a:= 0;
  b:= 0;
```

Now differentiate equations (10.14):

```
> Simf(subs(ruledsub, Simf(d(omega[1,2]))));
```

$$(\omega_1)\ \&\char`\^\ (\omega_{1,3})$$

```
> Simf(subs(ruledsub, Simf(d(omega[2,1]))));
```

$$(\omega_2)\ \&\char`\^\ (\omega_{2,3})$$

By Cartan's lemma, equations (10.15) must hold. Add these conditions to our substitution:

```
> ruledsub:= [op(ruledsub), omega[1,3] = f*omega[1],
  omega[2,3] = g*omega[2]];
```

Now differentiate the equation $\bar{\omega}_1^1 + \bar{\omega}_2^2 = 0$:

```
> Simf(subs(ruledsub, Simf(d(omega[1,1] + omega[2,2]))));
```

$$(-f+g)\,(\omega_1)\ \&\char`\^\ (\omega_2)$$

Therefore, $g = f$.

```
> g:= f;
```

Finally, differentiate equations (10.15):

```
> Simf(subs(ruledsub, Simf(d(omega[1,3]))
  - d(Simf(subs(ruledsub, omega[1,3])))));
```

$$(\omega_1)\ \&\char`\^\ (d(f))$$

```
> Simf(subs(ruledsub, Simf(d(omega[2,3]))
  - d(Simf(subs(ruledsub, omega[2,3])))));
```

$$(\omega_2)\ \&\char`\^\ (d(f))$$

It follows from Cartan's lemma that $df = 0$, and hence f is equal to a constant $C \in \mathbb{R}$.

Details for the case $C = 0$ may be found in the MAPLE worksheet for this chapter on the AMS webpage; here we will explore the case $C > 0$.

EXERCISE 10.15: First, declare C to be constant and set $f = C$. (It is useful to set up separate substitutions for the cases $C = 0, C > 0$, so we give our substitution a new name when we divide into cases.)

```
> Form(C=-1);
> ruledsubcase2:= Simf(subs([f=C], ruledsub));
```

Compute $d\bar{\omega}^1$ and $d\bar{\omega}^2$:

```
> Simf(subs(ruledsubcase2, Simf(d(omega[1]))));
> Simf(subs(ruledsubcase2, Simf(d(omega[2]))));
```

This should yield equations (10.24). By an application of the Frobenius theorem followed by a transformation of the form (10.13), we can arrange that $(\bar{\omega}^1, \bar{\omega}^2)$ have the form (10.26). Add these conditions to our substitution, and note that since $\bar{\omega}^1$ and $\bar{\omega}^2$ already appear on the right-hand sides of some of the equations in `ruledsubcase2`, we need to do this via the two-step process that we introduced in Chapter 6:

```
> PDETools[declare](h(u,v));
> ruledsubcase2:= Simf(subs([omega[1] = exp(h(u,v))*d(u),
  omega[2] = exp(h(u,v))*d(v)], ruledsubcase2));
> ruledsubcase2:= [op(ruledsubcase2),
  omega[1] = exp(h(u,v))*d(u), omega[2] = exp(h(u,v))*d(v)];
```

Now we can use the equations (10.24) to determine the 1-form $\bar{\omega}^1_1 = -\bar{\omega}^2_2$:

```
> zero1:= Simf(subs(ruledsubcase2, Simf(d(omega[1]))
  - d(Simf(subs(ruledsubcase2, omega[1])))));
> factor(pick(zero1, d(u)));
```

$$-e^h\,(\omega^1_1 + h_v\, d(v))$$

```
> zero2:= Simf(subs(ruledsubcase2, Simf(d(omega[2]))
  - d(Simf(subs(ruledsubcase2, omega[2])))));
> factor(pick(zero1, d(v)));
```

$$-e^h\,(-\omega^1_1 + h_u\, d(u))$$

Equation (10.27) then follows from Cartan's lemma. Add this condition to our substitution:

```
> ruledsubcase2:= Simf(subs([omega[1,1] = diff(h(u,v), u)*d(u)
  - diff(h(u,v), v)*d(v)], ruledsubcase2));
> ruledsubcase2:= [op(ruledsubcase2),
  omega[1,1] = diff(h(u,v), u)*d(u) - diff(h(u,v), v)*d(v)];
```

Finally, consider the structure equation for $d\bar{\omega}^1_1$:

```
> Simf(subs(ruledsubcase2, Simf(d(omega[1,1]))
  - d(Simf(subs(ruledsubcase2, omega[1,1])))));
```

$$(C\,e^{2\,h} - 2\,h_{u,v})\,(d(v))\ \&\hat{}\ (d(u))$$

Therefore, h must be a solution of equation (10.28).

EXERCISE 10.16: Set h equal to the function given in equation (10.31):

```
> ruledsubcase2:= Simf(subs([
  h(u,v) = ln(sqrt(2)/(sqrt(C)*(u+v)))], ruledsubcase2));
```

Equations (10.32) can then be read off directly from the equations in `ruledsubcase2`.

EXERCISE 10.17: In order to construct the PDE system (10.34), declare the variables (`X, e1, e2, e3`) to be functions of u and v; then apply our substitution to the structure equations (10.3):

```
> PDETools[declare](X(u,v), e1(u,v), e2(u,v), e3(u,v));
> zero1:= Simf(subs(ruledsubcase2,
  Simf(d(X(u,v)) - (e1(u,v)*omega[1] + e2(u,v)*omega[2]))));
  zero2:= Simf(subs(ruledsubcase2,
  Simf(d(e1(u,v)) - (e1(u,v)*omega[1,1] + e2(u,v)*omega[2,1]
  + e3(u,v)*omega[3,1]))));
  zero3:= Simf(subs(ruledsubcase2,
  Simf(d(e2(u,v)) - (e1(u,v)*omega[1,2] + e2(u,v)*omega[2,2]
  + e3(u,v)*omega[3,2]))));
  zero4:= Simf(subs(ruledsubcase2,
  Simf(d(e3(u,v)) - (e1(u,v)*omega[1,3] + e2(u,v)*omega[2,3]
  + e3(u,v)*omega[3,3]))));
```

The PDE system (10.34) now consists of the scalar coefficients of du and dv in each of these expressions:

```
> pde1a:= pick(zero1, d(u));
  pde1b:= pick(zero1, d(v));
  pde2a:= pick(zero2, d(u));
  pde2b:= pick(zero2, d(v));
  pde3a:= pick(zero3, d(u));
  pde3b:= pick(zero3, d(v));
  pde4a:= pick(zero4, d(u));
  pde4b:= pick(zero4, d(v));
```

In this case, it turns out that MAPLE can solve the entire system in one step. (This often isn't the case, especially with overdetermined systems, so the `pdsolve` command should generally be used with caution!)

```
> pdsolve({pde1a, pde1b, pde2a, pde2b, pde3a, pde3b,
  pde4a, pde4b}, {X(u,v), e1(u,v), e2(u,v), e3(u,v)});
```

(The constants in the resulting solution should be interpreted as being vector-valued.) Up to renaming the constants and shifting $\mathbf{x}$ by a translation to eliminate one of them, this gives the expressions in equations (10.40), (10.41).

EXERCISE 10.24: Since this exercise requires different Maurer-Cartan forms with different structure equations from the previous exercises, it's probably best to restart MAPLE and reload the `Cartan` and `LinearAlgebra` packages.

Declare the Maurer-Cartan forms on $\mathcal{F}(\mathbb{E}^2)$ and tell MAPLE about their symmetries and structure equations:

```
> Form(omega[1], omega[2], omega[1,2]);
> omega[2,1]:= -omega[1,2];
> d(omega[1]):= -omega[1,2] &^ omega[2];
  d(omega[2]):= -omega[2,1] &^ omega[1];
  d(omega[1,2]):= 0;
```

Declare the functions $(x(s), y(s))$, and define the vectors $(\mathbf{x}(s,\theta)$, $\mathbf{e}_1(s,\theta)$, $\mathbf{e}_2(s,\theta))$:

```
> PDETools[declare](x(s), y(s));
> X:= Vector([x(s), y(s)]);
  e10:= Vector([diff(x(s), s), diff(y(s), s)]);
  e20:= Vector([-diff(y(s), s), diff(x(s), s)]);

  e1:= cos(theta)*e10 + sin(theta)*e20;
  e2:= -sin(theta)*e10 + cos(theta)*e20;
```

Now, express $d\mathbf{x}$ as a linear combination of $(\mathbf{e}_1(s,\theta)$, $\mathbf{e}_2(s,\theta))$ in order to determine $(\bar{\omega}^1, \bar{\omega}^2)$ via the structure equation

$$d\mathbf{x} = \mathbf{e}_1\,\bar{\omega}^1 + \mathbf{e}_2\,\bar{\omega}^2 :$$

```
> zero1:= map(d, X) - e1*omega[1] - e2*omega[2];
> coordsub:= [op(Simf(solve({zero1[1], zero1[2]},
  {omega[1], omega[2]})))];
```

This should yield the expressions (10.43) for $(\bar{\omega}^1, \bar{\omega}^2)$. Now use the structure equations for $d\bar{\omega}^1$ and $d\bar{\omega}^2$ to determine $\bar{\omega}^1_2$:

```
> zero2:= Simf(subs(coordsub, Simf(d(omega[1]))
  - d(Simf(subs(coordsub, omega[1])))));
> factor(pick(zero2, d(s)));
```

$$\sin(\theta)\,(\omega_{1,2} + d(\theta))$$

```
> zero3:= Simf(subs(coordsub, Simf(d(omega[2]))
  - d(Simf(subs(coordsub, omega[2])))));
```

```
> factor(pick(zero3, d(s)));
```

$$\cos(\theta)\,(\omega_{1,2} + d(\theta))$$

Since $\sin(\theta)$ and $\cos(\theta)$ cannot vanish simultaneously, Cartan's lemma implies that $\bar{\omega}_2^1$ has the form (10.44). Add this expression to `coordsub`:

```
> coordsub:= [op(coordsub),
  omega[1,2] = -d(theta) + lambda*d(s)];
```

Now we can compute $\tilde{\alpha}^*(\omega^2 \wedge \omega_2^1)$ as follows:

```
> Simf(subs(coordsub, omega[2] &^ omega[1,2]));
```

$$-\sin(\theta)\,(d(\theta)\ \&\hat{}\ d(s))$$

Part 4

Beyond the flat case: Moving frames on Riemannian manifolds

Chapter 11

Curves and surfaces in elliptic and hyperbolic spaces

11.1. Introduction

Until now, all the homogeneous spaces that we have encountered have been modeled on the vector space $\mathbb{R}^n$, and we have relied extensively on the fact that all the various structures that we have defined (Euclidean, Minkowski, equi-affine, projective) are *flat*. This property is encoded in the Cartan structure equations (3.8): According to the second equation in (3.8), the matrix of connection forms $\omega_c = [\omega^i_j]$ satisfies the structure equation

$$d\omega_c + \omega_c \wedge \omega_c = 0. \tag{11.1}$$

(Note that this is not *quite* the same thing as the Maurer-Cartan equation (3.13) because ω_c is a submatrix of the matrix-valued Maurer-Cartan form ω.)

This might not seem like such a big deal, but in fact flatness is a necessary condition for the existence of *canonical* isomorphisms $T_{\mathbf{x}}\mathbb{R}^n \cong \mathbb{R}^n$ (cf. Remark 3.13 and the discussion in §3.3.2). It is these canonical isomorphisms that allow us to think of the components $(\mathbf{e}_1(\mathbf{x}), \ldots, \mathbf{e}_n(\mathbf{x}))$ of a frame field on $\mathbb{R}^n$ as functions from $\mathbb{R}^n$ to $\mathbb{R}^n$ rather than as sections of the tangent bundle $T\mathbb{R}^n$, which in turn allows us to define their exterior derivatives in a straightforward way.

In this chapter, we will explore moving frames on two homogeneous spaces that are "curved" versions of $\mathbb{E}^n$: *elliptic space* $\mathbb{S}^n$ (which is diffeomorphic to an n-dimensional sphere) and *hyperbolic space* $\mathbb{H}^n$ (which is diffeomorphic to $\mathbb{R}^n$, but with a different metric structure than that of $\mathbb{E}^n$). We will see that for these spaces, the flatness condition (11.1) no longer holds; rather, the matrix of 2-forms

$$\Omega = d\omega_c + \omega_c \wedge \omega_c, \tag{11.2}$$

called the *curvature* of the connection matrix ω_c, is nonzero and is an important feature of the associated geometry.

11.2. The homogeneous spaces $\mathbb{S}^n$ and $\mathbb{H}^n$

In this section, we will introduce Riemannian homogeneous space structures on $\mathbb{S}^n$ and $\mathbb{H}^n$, much as we did for flat homogeneous spaces in Chapter 3. Fortunately, $\mathbb{S}^n$ and $\mathbb{H}^n$ may naturally be regarded as submanifolds of the flat spaces $\mathbb{E}^{n+1}$ and $\mathbb{M}^{1,n}$, respectively, so we can still use the tools developed in Chapter 3 to get started.

11.2.1. Elliptic space $\mathbb{S}^n$.

Definition 11.1. The n-dimensional *elliptic space* $\mathbb{S}^n$ is the unit sphere in $\mathbb{E}^{n+1}$; i.e.,

$$\mathbb{S}^n = \{\mathbf{x} \in \mathbb{E}^{n+1} \mid \langle \mathbf{x}, \mathbf{x} \rangle = 1\}.$$

The symmetry group of $\mathbb{S}^n$ is defined to be the subgroup of $E(n+1)$ that preserves the set $\mathbb{S}^n$.

***Exercise 11.2.** Show that the symmetry group of $\mathbb{S}^n$ consists of all matrices $\hat{A} \in GL(n+2)$ of the form

$$\hat{A} = \begin{bmatrix} 1 & {}^t\mathbf{0} \\ \mathbf{0} & A \end{bmatrix}, \tag{11.3}$$

where $A \in SO(n+1)$. Therefore, the symmetry group of $\mathbb{S}^n$ is isomorphic to $SO^+(n+1)$, and we will generally identify the element $\hat{A}$ in equation (11.3) with the corresponding element $A \in SO(n+1)$.

In order to describe $\mathbb{S}^n$ as a homogeneous space of the Lie group $SO(n+1)$, we need to compute the isotropy group of a point $\mathbf{x} \in \mathbb{S}^n$.

***Exercise 11.3.** Let $\mathbf{x}_0 = {}^t[1, 0, \ldots, 0] \in \mathbb{S}^n$. Show that:

(a) The isotropy group $H_{\mathbf{x}_0}$ of $\mathbf{x}_0$ in $SO(n+1)$ is

$$H_{\mathbf{x}_0} = \left\{ \begin{bmatrix} 1 & {}^t\mathbf{0} \\ \mathbf{0} & \bar{A} \end{bmatrix} : \bar{A} \in SO(n) \right\}. \tag{11.4}$$

(b) The isotropy group $H_{\mathbf{x}}$ of any other point $\mathbf{x} \in \mathbb{S}^n$ is

$$H_{\mathbf{x}} = t_{\mathbf{x}} H_{\mathbf{x}_0} t_{\mathbf{x}}^{-1},$$

where $t_{\mathbf{x}}$ is any matrix in $SO(n+1)$ whose first column is $\mathbf{x}$. (The transformation $t_{\mathbf{x}}$ will then have the property that $t_{\mathbf{x}}(\mathbf{x}_0) = \mathbf{x}$.)

Orthonormal frames on $\mathbb{S}^n$ are defined as follows:

Definition 11.4. An *orthonormal frame* $\mathbf{f}$ on $\mathbb{S}^n$ is a list of vectors $\mathbf{f} = (\mathbf{e}_0, \ldots, \mathbf{e}_n)$, where $\mathbf{e}_0 \in \mathbb{S}^n \subset \mathbb{E}^{n+1}$, $\mathbf{e}_1, \ldots, \mathbf{e}_n \in \mathbb{E}^{n+1}$, and $\begin{bmatrix} \mathbf{e}_0 & \cdots & \mathbf{e}_n \end{bmatrix} \in SO(n+1)$. We identify $\mathbf{e}_0$ with the position vector $\mathbf{x} \in \mathbb{S}^n$, and the condition that $\begin{bmatrix} \mathbf{e}_0 & \cdots & \mathbf{e}_n \end{bmatrix} \in SO(n+1)$ implies that the vectors $(\mathbf{e}_1, \ldots, \mathbf{e}_n)$ may be regarded as an orthonormal basis for the tangent space $T_{\mathbf{e}_0}\mathbb{S}^n \subset T_{\mathbf{e}_0}\mathbb{E}^{n+1} \cong \mathbb{E}^{n+1}$. (We may also say that $(\mathbf{e}_1, \ldots, \mathbf{e}_n)$ is an *orthonormal frame based at* $\mathbf{e}_0$.) The collection of all orthonormal frames on $\mathbb{S}^n$ is called the *orthonormal frame bundle* of $\mathbb{S}^n$, denoted $\mathcal{F}(\mathbb{S}^n)$.

The same reasoning as in prior cases shows that the orthonormal frame bundle $\mathcal{F}(\mathbb{S}^n)$ may be regarded as the group $SO(n+1)$ via the one-to-one correspondence

$$g(\mathbf{e}_0, \ldots, \mathbf{e}_n) = \begin{bmatrix} \mathbf{e}_0 & \cdots & \mathbf{e}_n \end{bmatrix}.$$

The projection map $\pi : SO(n+1) \to \mathbb{S}^n$ defined by

$$\pi([\mathbf{e}_0 \ \ldots \ \mathbf{e}_n]) = \mathbf{e}_0$$

describes $SO(n+1)$ as a principal bundle over $\mathbb{S}^n$ with fiber group $SO(n)$; therefore, we have a natural correspondence $\mathbb{S}^n \cong SO(n+1)/SO(n)$.

Because we have defined the components $(\mathbf{e}_0, \ldots, \mathbf{e}_n)$ of a frame as elements of the vector space $\mathbb{E}^{n+1}$, we can regard them as functions $\mathbf{e}_\alpha : \mathcal{F}(\mathbb{S}^n) \to \mathbb{E}^{n+1}$ and define the Maurer-Cartan forms (ω^α_β) on $SO(n+1)$ as usual by the equations

$$d\mathbf{e}_\alpha = \mathbf{e}_\beta \omega^\beta_\alpha, \tag{11.5}$$

where $0 \leq \alpha, \beta \leq n$ and $\omega^\beta_\alpha = -\omega^\alpha_\beta$.

***Exercise 11.5.** (a) Show that the forms (ω^i_0) for $1 \leq i \leq n$ are semi-basic for the projection $\pi : SO(n+1) \to \mathbb{S}^n$ and so may be regarded as the dual forms of any orthonormal frame $(\mathbf{e}_1, \ldots, \mathbf{e}_n)$ based at $\mathbf{e}_0 \in \mathbb{S}^n$. The forms (ω^i_j) for $1 \leq i, j \leq n$ may then be regarded as the connection forms.

(b) Set $\omega^i = \omega^i_0$. Show that the dual forms (ω^i) and the connection forms (ω^i_j) satisfy the structure equations

$$\begin{aligned} d\omega^i &= -\omega^i_j \wedge \omega^j, \\ d\omega^i_j &= -\omega^i_k \wedge \omega^k_j + \omega^i \wedge \omega^j. \end{aligned} \tag{11.6}$$

The second equation in (11.6) implies that the curvature (11.2) of the connection matrix $\omega_c = [\omega^i_j]$ is given by

$$\Omega = [\Omega^i_j] = [\omega^i \wedge \omega^j].$$

The nonzero entries of Ω reflect the fact that all sectional curvatures of $\mathbb{S}^n$ are identically equal to 1 (see, e.g., [**dC92**]).

(c) Show that the structure equations (11.6) are equivalent to the Maurer-Cartan equation

$$d\omega = -\omega \wedge \omega,$$

where

$$\omega = \begin{bmatrix} 0 & -\omega^1 & \cdots & -\omega^n \\ \omega^1 & \omega^1_1 & \cdots & \omega^1_n \\ \vdots & \vdots & & \vdots \\ \omega^n & \omega^n_1 & \cdots & \omega^n_n \end{bmatrix}$$

is the $\mathfrak{so}(n+1)$-valued Maurer-Cartan form on $SO(n+1)$.

11.2.2. Hyperbolic space $\mathbb{H}^n$.

Definition 11.6. The n-dimensional *hyperbolic space* $\mathbb{H}^n$ is the upper sheet of the two-sheeted hyperboloid

$$(x^0)^2 - (x^1)^2 - \cdots - (x^n)^2 = 1$$

in $\mathbb{M}^{1,n}$; i.e.,

$$\mathbb{H}^n = \{\mathbf{x} \in \mathbb{M}^{1,n} \mid \langle \mathbf{x}, \mathbf{x} \rangle = 1, \quad x^0 > 0\}.$$

$\mathbb{H}^n$ is a spacelike hypersurface in $\mathbb{M}^{1,n}$, so the restriction of the Minkowski metric on $\mathbb{M}^{1,n}$ defines a Riemannian metric on $\mathbb{H}^n$. (Technically, this requires reversing signs in the definition of the Minkowski inner product so that its restriction to each tangent space $T_{\mathbf{x}}\mathbb{H}^n$ yields a quadratic form that is positive definite rather than negative definite, but for the sake of consistency we will retain the sign conventions of Chapter 5.) The symmetry group of $\mathbb{H}^n$ is defined to be the subgroup of $M(1,n)$ that preserves the set $\mathbb{H}^n$.

***Exercise 11.7.** Show that the symmetry group of $\mathbb{H}^n$ consists of all matrices $\hat{A} \in GL(n+2)$ of the form

$$\hat{A} = \begin{bmatrix} 1 & {}^t\mathbf{0} \\ \mathbf{0} & A \end{bmatrix}, \tag{11.7}$$

where $A \in SO^+(1,n)$. Therefore, the symmetry group of $\mathbb{H}^n$ is isomorphic to $SO^+(1,n)$, and we will generally identify the element $\hat{A}$ in equation (11.7) with the corresponding element $A \in SO^+(1,n)$.

In order to describe $\mathbb{H}^n$ as a homogeneous space of the Lie group $SO^+(1,n)$, we need to compute the isotropy group of a point $\mathbf{x} \in \mathbb{H}^n$.

***Exercise 11.8.** Let $\mathbf{x}_0 = {}^t[1, 0, \ldots, 0] \in \mathbb{H}^n$. Show that:

(a) The isotropy group $H_{\mathbf{x}_0}$ of $\mathbf{x}_0$ in $SO^+(1,n)$ is

$$H_{\mathbf{x}_0} = \left\{ \begin{bmatrix} 1 & {}^t\mathbf{0} \\ \mathbf{0} & \bar{A} \end{bmatrix} : \bar{A} \in SO(n) \right\}. \tag{11.8}$$

(b) The isotropy group $H_{\mathbf{x}}$ of any other point $\mathbf{x} \in \mathbb{H}^n$ is

$$H_{\mathbf{x}} = t_{\mathbf{x}} H_{\mathbf{x}_0} t_{\mathbf{x}}^{-1},$$

where $t_{\mathbf{x}}$ is any matrix in $SO^+(1,n)$ whose first column is $\mathbf{x}$. (The transformation $t_{\mathbf{x}}$ will then have the property that $t_{\mathbf{x}}(\mathbf{x}_0) = \mathbf{x}$.)

Orthonormal frames on $\mathbb{H}^n$ are defined as follows:

Definition 11.9. An *orthonormal frame* $\mathbf{f}$ on $\mathbb{H}^n$ is a list of vectors $\mathbf{f} = (\mathbf{e}_0, \ldots, \mathbf{e}_n)$, where $\mathbf{e}_0 \in \mathbb{H}^n \subset \mathbb{M}^{1,n}$, $\mathbf{e}_1, \ldots, \mathbf{e}_n \in \mathbb{M}^{1,n}$, and $\begin{bmatrix} \mathbf{e}_0 & \cdots & \mathbf{e}_n \end{bmatrix} \in SO^+(1,n)$. We identify $\mathbf{e}_0$ with the position vector $\mathbf{x} \in \mathbb{H}^n$, and the condition that $\begin{bmatrix} \mathbf{e}_0 & \cdots & \mathbf{e}_n \end{bmatrix} \in SO^+(1,n)$ implies that the vectors $(\mathbf{e}_1, \ldots, \mathbf{e}_n)$ may be regarded as an orthonormal basis for the tangent space $T_{\mathbf{e}_0}\mathbb{H}^n$. (We may also say that $(\mathbf{e}_1, \ldots, \mathbf{e}_n)$ is an *orthonormal frame based at* $\mathbf{e}_0$.) The collection of all orthonormal frames on $\mathbb{H}^n$ is called the *orthonormal frame bundle* of $\mathbb{H}^n$, denoted $\mathcal{F}(\mathbb{H}^n)$.

Remark 11.10. For an orthonormal frame $\mathbf{f} = (\mathbf{e}_0, \ldots, \mathbf{e}_n)$ on $\mathbb{S}^n$, each of the vectors $\mathbf{e}_\alpha \in \mathbb{E}^{n+1}$ satisfies $\langle \mathbf{e}_\alpha, \mathbf{e}_\alpha \rangle = 1$, and so could, if desired, be identified with a point of $\mathbb{S}^n$. But for an orthonormal frame on $\mathbb{H}^n$, only the vector $\mathbf{e}_0$ satisfies the defining condition $\langle \mathbf{e}_0, \mathbf{e}_0 \rangle = 1$ for points in $\mathbb{H}^n$, while the vectors $(\mathbf{e}_1, \ldots, \mathbf{e}_n)$ each satisfy $\langle \mathbf{e}_i, \mathbf{e}_i \rangle = -1$. This illustrates the fact that for non-flat homogeneous spaces (and for Riemannian manifolds in general), we will no longer be able to regard the frame vectors $(\mathbf{e}_1, \ldots, \mathbf{e}_n)$ as taking values in the same space as the position vector $\mathbf{e}_0$, as we did for frames on flat homogeneous spaces. Instead, we must regard each of the vectors $(\mathbf{e}_1, \ldots, \mathbf{e}_n)$ as taking values in the tangent bundle of the underlying manifold.

The same reasoning as in prior cases shows that the orthonormal frame bundle $\mathcal{F}(\mathbb{H}^n)$ may be regarded as the group $SO^+(1,n)$ via the one-to-one correspondence

$$g(\mathbf{e}_0, \ldots, \mathbf{e}_n) = \begin{bmatrix} \mathbf{e}_0 & \cdots & \mathbf{e}_n \end{bmatrix}.$$

The projection map $\pi : SO^+(1,n) \to \mathbb{H}^n$ defined by

$$\pi([\mathbf{e}_0 \ \ldots \ \mathbf{e}_n]) = \mathbf{e}_0$$

describes $SO^+(1,n)$ as a principal bundle over $\mathbb{H}^n$ with fiber group $SO(n)$; therefore, we have a natural correspondence $\mathbb{H}^n \cong SO^+(1,n)/SO(n)$.

Because we have defined the components $(\mathbf{e}_0, \ldots, \mathbf{e}_n)$ of a frame as elements of the vector space $\mathbb{M}^{1,n}$, we can regard them as functions $\mathbf{e}_\alpha : \mathcal{F}(\mathbb{H}^n) \to \mathbb{M}^{1,n}$ and define the Maurer-Cartan forms (ω^α_β) on $SO^+(1,n)$ as usual by the equations

$$d\mathbf{e}_\alpha = \mathbf{e}_\beta \omega^\beta_\alpha, \tag{11.9}$$

where $0 \le \alpha, \beta \le n$. Recall from Exercise 3.50 that these forms satisfy the relations

$$\omega^\beta_\alpha = \begin{cases} 0, & \alpha = \beta, \\ \omega^\alpha_\beta, & \alpha = 0 \text{ or } \beta = 0, \\ -\omega^\alpha_\beta, & \alpha, \beta \ge 1. \end{cases}$$

***Exercise 11.11.** (a) Show that the forms (ω^i_0) for $1 \le i \le n$ are semi-basic for the projection $\pi : SO^+(1,n) \to \mathbb{H}^n$ and so may be regarded as the dual forms of any orthonormal frame $(\mathbf{e}_1, \ldots, \mathbf{e}_n)$ based at $\mathbf{e}_0 \in \mathbb{H}^n$. The forms (ω^i_j) for $1 \le i, j \le n$ may then be regarded as the connection forms.

(b) Set $\omega^i = \omega^i_0$. Show that the dual forms (ω^i) and the connection forms (ω^i_j) satisfy the structure equations

$$\begin{aligned} d\omega^i &= -\omega^i_j \wedge \omega^j, \\ d\omega^i_j &= -\omega^i_k \wedge \omega^k_j - \omega^i \wedge \omega^j. \end{aligned} \tag{11.10}$$

(Compare with equations (11.6).) The second equation in (11.10) implies that the curvature (11.2) of the connection matrix $\omega_c = [\omega^i_j]$ is given by

$$\Omega = [\Omega^i_j] = [-\omega^i \wedge \omega^j].$$

The nonzero entries of Ω reflect the fact that all sectional curvatures of $\mathbb{H}^n$ are identically equal to -1.

(c) Show that the structure equations (11.10) are equivalent to the Maurer-Cartan equation

$$d\omega = -\omega \wedge \omega,$$

where

$$\omega = \begin{bmatrix} 0 & \omega^1 & \cdots & \omega^n \\ \omega^1 & \omega^1_1 & \cdots & \omega^1_n \\ \vdots & \vdots & & \vdots \\ \omega^n & \omega^n_1 & \cdots & \omega^n_n \end{bmatrix}$$

is the $\mathfrak{so}^+(1,n)$-valued Maurer-Cartan form on $SO^+(1,n)$.

11.3. A more intrinsic view of $\mathbb{S}^n$ and $\mathbb{H}^n$

While it is certainly useful to regard $\mathbb{S}^n$ and $\mathbb{H}^n$ as submanifolds of the flat metric spaces $\mathbb{E}^{n+1}$ and $\mathbb{M}^{1,n}$, it is also somewhat artificial. These manifolds are perfectly well-defined as intrinsic n-dimensional homogeneous spaces of the Lie groups $SO(n+1)$ and $SO^+(1,n)$, respectively, and they shouldn't need to be extrinsically embedded as submanifolds of larger spaces in order to study their geometry.

For ease of exposition, let $\mathbb{X}^n$ denote either $\mathbb{S}^n$ or $\mathbb{H}^n$; let $\mathbb{V}^{n+1}$ denote the flat homogeneous space in which $\mathbb{X}^n$ is defined as an embedded hypersurface ($\mathbb{E}^{n+1}$ or $\mathbb{M}^{1,n}$), and let G denote the symmetry group of $\mathbb{X}^n$ ($SO(n+1)$ or $SO^+(1,n)$). Each tangent space $T_{\mathbf{x}}\mathbb{X}^n \subset T_{\mathbf{x}}\mathbb{V}^{n+1} \cong \mathbb{V}^{n+1}$ inherits an inner product $\langle\cdot,\cdot\rangle$ from the inner product (either Euclidean or Minkowski) on $\mathbb{V}^{n+1}$, thereby giving $\mathbb{X}^n$ the structure of a Riemannian manifold.

We can give an intrinsic definition for orthonormal frames on $\mathbb{X}^n$ that is equivalent to Definitions 11.4 and 11.9, as follows. (Note that the condition $\det\begin{bmatrix}\mathbf{e}_0 & \cdots & \mathbf{e}_n\end{bmatrix} = 1$ implied by these definitions induces an orientation on each tangent space $T_{\mathbf{e}_0}\mathbb{X}^n$.)

Definition 11.12. An (oriented) *orthonormal frame* $\mathbf{f}$ on $\mathbb{X}^n$ is a list $\mathbf{f} = (\mathbf{e}_0, \ldots, \mathbf{e}_n)$, where $\mathbf{e}_0 \in \mathbb{X}^n$ and $(\mathbf{e}_1, \ldots, \mathbf{e}_n)$ is an oriented, orthonormal basis for the tangent space $T_{\mathbf{e}_0}\mathbb{X}^n$. Alternatively, we may say that $(\mathbf{e}_1, \ldots, \mathbf{e}_n)$ is an *orthonormal frame based at* $\mathbf{e}_0$. The collection of all orthonormal frames on $\mathbb{X}^n$ is called the *orthonormal frame bundle* of $\mathbb{X}^n$ and is denoted $\mathcal{F}(\mathbb{X}^n)$; it may be identified with the Lie group G.

The Maurer-Cartan form ω and its components (ω^i, ω^i_j) are defined as usual on G, and the structure equations (11.6) and (11.10) remain valid in the intrinsic setting as consequences of the Maurer-Cartan equation on G. The primary issue that must be addressed is how to make sense of the idea of differentiating vector fields on $\mathbb{X}^n$. We can no longer think of the components $(\mathbf{e}_0, \ldots, \mathbf{e}_n)$ of an orthonormal frame as functions from $\mathcal{F}(\mathbb{X}^n)$ to $\mathbb{V}^{n+1}$; rather, $\mathbf{e}_0$ is a function from $\mathcal{F}(\mathbb{X}^n)$ to $\mathbb{X}^n$, while $(\mathbf{e}_1, \ldots, \mathbf{e}_n)$ are functions from $\mathcal{F}(\mathbb{X}^n)$ to the tangent bundle $T\mathbb{X}^n$, with the property that for any frame $\mathbf{f} = (\mathbf{e}_0, \ldots, \mathbf{e}_n) \in \mathcal{F}(\mathbb{X}^n)$, we have

$$\mathbf{e}_i(\mathbf{f}) \in T_{\mathbf{e}_0(\mathbf{f})}\mathbb{X}^n$$

for $1 \le i \le n$. Moreover, the derivatives of the functions $\mathbf{e}_i : \mathcal{F}(\mathbb{X}^n) \to T\mathbb{X}^n$ at a point $\mathbf{f} \in \mathcal{F}(\mathbb{X}^n)$ must also take values in the tangent space $T_{\mathbf{e}_0(\mathbf{f})}\mathbb{X}^n$ and hence must be linear combinations of the vectors $(\mathbf{e}_1(\mathbf{f}), \ldots, \mathbf{e}_n(\mathbf{f}))$. This is manifestly not the case for the extrinsically defined exterior derivatives of equations (11.5) and (11.9), as each of the exterior derivatives $d\mathbf{e}_1, \ldots, d\mathbf{e}_n$ contains a nonzero $\mathbf{e}_0$ term.

It turns out that the way to solve this problem is simply to take the orthogonal projection of these extrinsically defined exterior derivatives onto the tangent plane $T_{\mathbf{e}_0}\mathbb{X}^n$. This idea leads to the notion of the *covariant derivative* for vector fields on $\mathbb{X}^n$:

Definition 11.13. Let $\mathbf{e}_0 \in \mathbb{X}^n \subset \mathbb{V}^{n+1}$, and let $\mathbf{w} \in T_{\mathbf{e}_0}\mathbb{X}^n \subset T_{\mathbf{e}_0}\mathbb{V}^{n+1} \cong \mathbb{V}^{n+1}$. Let $\mathbf{v}$ be a (tangent) vector field on $\mathbb{X}^n$, regarded as a function $\mathbf{v} : \mathbb{X}^n \to \mathbb{V}^{n+1}$ with the property that for every $\mathbf{x} \in \mathbb{X}^n$, we have $\mathbf{v}(\mathbf{x}) \in T_{\mathbf{x}}\mathbb{X}^n \subset \mathbb{V}^{n+1}$. Let $\pi_{\mathbf{e}_0} : \mathbb{V}^{n+1} \to T_{\mathbf{e}_0}\mathbb{X}^n$ denote orthogonal projection with respect to the (Euclidean or Minkowski) metric on $\mathbb{V}^{n+1}$. The *covariant derivative* of $\mathbf{v}$ with respect to $\mathbf{w}$ is the vector $\nabla_{\mathbf{w}}\mathbf{v} \in T_{\mathbf{e}_0}\mathbb{X}^n$ defined by

$$\nabla_{\mathbf{w}}\mathbf{v} = \pi_{\mathbf{e}_0}(d\mathbf{v}(\mathbf{w})). \tag{11.11}$$

In particular, if we take $\mathbf{v}(\mathbf{x}) = \mathbf{e}_i(\mathbf{x})$ $(1 \leq i \leq n)$ for some orthonormal frame field $(\mathbf{e}_1(\mathbf{x}), \ldots, \mathbf{e}_n(\mathbf{x}))$ on $\mathbb{X}^n$, then we have

$$\nabla_{\mathbf{w}}\mathbf{e}_i = \pi_{\mathbf{e}_0}(d\mathbf{e}_i(\mathbf{w})) = \mathbf{e}_j\bar{\omega}^j_i(\mathbf{w}),$$

where the repeated index j is summed from 1 to n and $\bar{\omega}^i_j$ represents the pullback of the Maurer-Cartan form ω^i_j on $\mathcal{F}(\mathbb{X}^n)$ to $\mathbb{X}^n$ via the frame field $(\mathbf{e}_1(\mathbf{x}), \ldots, \mathbf{e}_n(\mathbf{x}))$. Moreover, the Leibniz rule for exterior derivatives implies that for any vector field $\mathbf{v}(\mathbf{x}) = v^i(\mathbf{x})\mathbf{e}_i(\mathbf{x})$ on $\mathbb{X}^n$, we have

$$\begin{aligned} \nabla_{\mathbf{w}}\mathbf{v} &= \mathbf{w}(v^i)\mathbf{e}_i(\mathbf{x}) + v^i\nabla_{\mathbf{w}}\mathbf{e}_i(\mathbf{x}) \\ &= \mathbf{w}(v^i)\mathbf{e}_i(\mathbf{x}) + v^i\mathbf{e}_j(\mathbf{x})\bar{\omega}^j_i(\mathbf{w}) \\ &= \mathbf{e}_i(\mathbf{x})\left(\mathbf{w}(v^i) + v^j\bar{\omega}^i_j(\mathbf{w})\right). \end{aligned} \tag{11.12}$$

The remarkable fact about equation (11.12) is that, even though we used extrinsic objects to define the covariant derivative, the result is described entirely in intrinsic terms: The tangent vector fields $(\mathbf{e}_i(\mathbf{x}))$ and the pulled-back Maurer-Cartan forms $(\bar{\omega}^i_j)$ are well-defined on $\mathbb{X}^n$ as a homogeneous space of the Lie group G, without regard to its embedding as a submanifold of $\mathbb{V}^{n+1}$. So, unlike the exterior derivative of equations (11.5) and (11.9), the covariant derivative is intrinsically defined on $\mathbb{X}^n$.

The following exercise shows how the covariant derivative may be thought of as an analog to the exterior derivative for vector fields on $\mathbb{X}^n$.

***Exercise 11.14.** Given a vector field $\mathbf{v}$ on $\mathbb{X}^n$, consider the map

$$\nabla\mathbf{v} : T\mathbb{X}^n \to T\mathbb{X}^n$$

defined by

$$\nabla\mathbf{v}(\mathbf{w}) = \nabla_{\mathbf{w}}\mathbf{v}. \tag{11.13}$$

Use the explicit formula (11.12) to show that $\nabla\mathbf{v}$ is a $T\mathbb{X}^n$-valued 1-form on $\mathbb{X}^n$ (cf. Definition 2.18). Specifically, if $(\mathbf{e}_1(\mathbf{x}), \ldots, \mathbf{e}_n(\mathbf{x}))$ is an orthonormal frame field on $\mathbb{X}^n$ with associated connection forms $(\bar{\omega}^i_j)$, then

$$\nabla\mathbf{v} = \mathbf{e}_i\left(dv^i + v^j\bar{\omega}^i_j\right).$$

In particular, if $\mathbf{v}(\mathbf{x}) = \mathbf{e}_i(\mathbf{x})$, then we have

$$\nabla\mathbf{e}_i(\mathbf{x}) = \mathbf{e}_j(\mathbf{x})\,\bar{\omega}^j_i. \tag{11.14}$$

Thus, we can think of ∇ as a generalization of the exterior derivative d that appears in the structure equations (3.1).

An important property of the covariant derivative is that it is *compatible with the metric* on $\mathbb{X}^n$. This means that for any vector fields $\mathbf{v}_1, \mathbf{v}_2$ on $\mathbb{X}^n$ and any vector $\mathbf{w} \in T_{\mathbf{x}}\mathbb{X}^n$, we have

$$\mathbf{w}(\langle\mathbf{v}_1, \mathbf{v}_2\rangle) = \langle\nabla_{\mathbf{w}}\mathbf{v}_1, \mathbf{v}_2\rangle + \langle\mathbf{v}_1, \nabla_{\mathbf{w}}\mathbf{v}_2\rangle. \tag{11.15}$$

(You can think of this as a Leibniz rule for computing the directional derivative of the real-valued function $\langle\mathbf{v}_1, \mathbf{v}_2\rangle$ in the direction of $\mathbf{w}$.)

***Exercise 11.15.** Use equation (11.12) to show that equation (11.15) holds. (Hint: You will need to use the fact that $\bar{\omega}^j_i = -\bar{\omega}^i_j$ for $1 \le i, j \le n$, which is true for the connection forms on both $\mathbb{S}^n$ and $\mathbb{H}^n$.)

Definition 11.16. Let $\Gamma(T\mathbb{X}^n)$ denote the space of smooth local sections of $T\mathbb{X}^n$ (i.e., smooth vector fields on open sets in $\mathbb{X}^n$). The map

$$\nabla : \Gamma(T\mathbb{X}^n) \times \Gamma(T\mathbb{X}^n) \to \Gamma(T\mathbb{X}^n)$$

defined by

$$\nabla(\mathbf{v}, \mathbf{w})(\mathbf{x}) = \nabla_{\mathbf{w}(\mathbf{x})}\mathbf{v} \in T_{\mathbf{x}}\mathbb{X}^n$$

is called the *Levi-Civita connection* on $\mathbb{X}^n$. The 1-forms $(\bar{\omega}^i_j)$ determined by the frame field $(\mathbf{e}_1(\mathbf{x}), \ldots, \mathbf{e}_n(\mathbf{x}))$ and equation (11.14) are called the *connection forms* associated to ∇ and the frame field $(\mathbf{e}_1(\mathbf{x}), \ldots, \mathbf{e}_n(\mathbf{x}))$. (Note that this terminology is consistent with their definition as the pullbacks of the connection forms on the frame bundle $G \to \mathbb{X}^n$ via the frame field.)

Remark 11.17. Although the Levi-Civita connection is defined as an operator on vector fields on $\mathbb{X}^n$, equation (11.14) suggests—correctly!—that there should be a related operator (also denoted ∇) on an appropriate class of smooth maps from $\mathcal{F}(\mathbb{X}^n)$ to $T\mathbb{X}^n$, determined by the condition

$$\nabla\mathbf{e}_i = \mathbf{e}_j\omega^j_i \tag{11.16}$$

together with an extension to appropriate maps $\tilde{\mathbf{v}} : \mathcal{F}(\mathbb{X}^n) \to T\mathbb{X}^n$ via linearity and a Leibniz rule akin to equation (11.12). Once all the details are worked out, equation (11.16) is the natural analog of the structure equations (3.1). We will develop this idea more fully in Chapter 12.

One of the most important differences between the Levi-Civita operator ∇ and the exterior derivative d is that there is no analog of the identity $d \circ d = 0$ for ∇, and so it is not immediately clear how to differentiate equation (11.16) in order to obtain structure equations for the derivatives $(d\omega^i_j)$. Fortunately, we have already derived these for the cases at hand via extrinsic techniques (cf. equations (11.6) and (11.10)). In Chapter 12, we will see how to derive these equations more intrinsically and in more general scenarios.

For the remainder of this chapter, we will restrict our consideration to the case $n = 3$. We will continue to use the notation $(\mathbb{X}^3, \mathbb{V}^4, G)$ to denote either $(\mathbb{S}^3, \mathbb{E}^4, SO(4))$ or $(\mathbb{H}^3, \mathbb{M}^{1,3}, SO^+(1,3))$.

11.4. Moving frames for curves in $\mathbb{S}^3$ and $\mathbb{H}^3$

Consider a smooth, parametrized curve $\alpha : I \to \mathbb{X}^3$ that maps some open interval $I \subset \mathbb{R}$ into $\mathbb{X}^3$. Since $\mathbb{X}^3$ has the structure of the homogeneous space $G/SO(3)$, an adapted frame field along α should be a lifting $\tilde{\alpha} : I \to G$. Any such lifting can be written as

$$\tilde{\alpha}(t) = (\mathbf{e}_0(t), \mathbf{e}_1(t), \mathbf{e}_2(t), \mathbf{e}_3(t)),$$

where for each $t \in I$, $\mathbf{e}_0(t) = \alpha(t)$ and $(\mathbf{e}_1(t), \mathbf{e}_2(t), \mathbf{e}_3(t))$ is an oriented, orthonormal basis for the tangent space $T_{\alpha(t)}\mathbb{X}^3$. Such an adapted frame field is usually called an *orthonormal frame field* along α.

As in the Euclidean case, we say that α is *regular* if $\alpha'(t) \neq \mathbf{0}$ for every $t \in I$; as usual, we will only consider regular curves. We begin our construction of an adapted orthonormal frame field along α by setting

$$\mathbf{e}_1(t) = \frac{\alpha'(t)}{|\alpha'(t)|};$$

i.e., we require that $\mathbf{e}_1(t)$ be the unit tangent vector to the curve at $\alpha(t)$. The same argument as in the Euclidean case shows that α can be smoothly reparametrized by its arc length function $s(t)$, so henceforth we will assume that $\alpha = \alpha(s)$ is parametrized by arc length and that

$$\mathbf{e}_1(s) = \alpha'(s).$$

Note that this makes sense even in the intrinsic setting: Since $\alpha(s) \in \mathbb{X}^3$, the derivative $\mathbf{e}_1(s) = \alpha'(s)$ is an element of $T_{\alpha(s)}\mathbb{X}^3$, which is where we expect the frame vectors to live.

The next step is where things start to look a bit different from the Euclidean case: Since $\mathbf{e}_1(s)$ is a vector field along α, we have to use the covariant derivative to differentiate it. In particular, differentiating any vector field along the curve α means taking its covariant derivative with respect to the unit tangent vector field along α. So the natural analog for the Euclidean derivative $\mathbf{e}_1'(s)$ is the covariant derivative $\nabla_{\mathbf{e}_1(s)}\mathbf{e}_1(s)$. We will say that α is *nondegenerate* if α is regular and, in addition, $\nabla_{\mathbf{e}_1(s)}\mathbf{e}_1(s) \neq \mathbf{0}$ for all $s \in I$.

***Exercise 11.18.** In the Euclidean case, any regular curve $\alpha : I \to \mathbb{E}^3$ with $\mathbf{e}_1'(s) = \alpha''(s) = 0$ for all $s \in I$ is contained in a straight line in $\mathbb{E}^3$. Consider the analogous condition for curves in $\mathbb{X}^3$: Let $\alpha : I \to \mathbb{X}^3$ be a regular curve parametrized by arc length; let $\mathbf{e}_0(s) = \alpha(s)$, $\mathbf{e}_1(s) = \alpha'(s)$, and suppose that $\nabla_{\mathbf{e}_1(s)}\mathbf{e}_1(s) = \mathbf{0}$ for all $s \in I$.

(a) Use the extrinsic definition (11.11) together with the structure equations (11.5) and (11.9) to show that, when regarded as functions $\mathbf{e}_0, \mathbf{e}_1 : I \to \mathbb{V}^4$, we have

$$\mathbf{e}_0'(s) = \mathbf{e}_1(s), \qquad \mathbf{e}_1'(s) = k(s)\mathbf{e}_0(s), \tag{11.17}$$

where $k(s) = \omega_1^0(\mathbf{e}_1(s))$.

(b) Use the fact that $\omega_1^0 = \pm\omega_0^1 = \pm\omega^1$ (with the sign depending on whether $\mathbb{X}^3 = \mathbb{S}^3$ or $\mathbb{H}^3$) to show that

$$k(s) = \begin{cases} -1, & \mathbb{X}^3 = \mathbb{S}^3, \\ 1, & \mathbb{X}^3 = \mathbb{H}^3. \end{cases}$$

(c) Solve the differential equations (11.17) and show that:

(1) If $\mathbb{X}^3 = \mathbb{S}^3$, then there exist orthogonal unit vectors $\bar{\mathbf{e}}_0, \bar{\mathbf{e}}_1 \in \mathbb{E}^4$ such that

$$\alpha(s) = \mathbf{e}_0(s) = \cos(s)\,\bar{\mathbf{e}}_0 + \sin(s)\,\bar{\mathbf{e}}_1. \tag{11.18}$$

In particular, α is contained in the great circle determined by the intersection of $\mathbb{S}^3$ with the plane spanned by $(\bar{\mathbf{e}}_0, \bar{\mathbf{e}}_1)$.

(2) If $\mathbb{X}^3 = \mathbb{H}^3$, then there exist orthogonal unit vectors $\bar{\mathbf{e}}_0, \bar{\mathbf{e}}_1 \in \mathbb{M}^{1,3}$, with $\bar{\mathbf{e}}_0$ timelike and $\bar{\mathbf{e}}_1$ spacelike, such that

$$\alpha(s) = \mathbf{e}_0(s) = \cosh(s)\,\bar{\mathbf{e}}_0 + \sinh(s)\,\bar{\mathbf{e}}_1. \tag{11.19}$$

In particular, α is contained in the "great hyperbola" determined by the intersection of $\mathbb{H}^3$ with the timelike plane spanned by $(\bar{\mathbf{e}}_0, \bar{\mathbf{e}}_1)$.

The differential equation $\nabla_{\alpha'(s)}\alpha'(s) = 0$ is called the *geodesic equation*, and the curves α in equations (11.18) and (11.19) are the *geodesics* in $\mathbb{S}^3$ and $\mathbb{H}^3$, respectively.

Now, suppose that $\alpha : I \to \mathbb{X}^3$ is a nondegenerate curve parametrized by arc length. According to equation (11.15), differentiating the equation

$$\langle \mathbf{e}_1(s), \mathbf{e}_1(s) \rangle = 1$$

along α yields

$$\langle \nabla_{\mathbf{e}_1(s)} \mathbf{e}_1(s),\ \mathbf{e}_1(s) \rangle = 0.$$

Thus, $\nabla_{\mathbf{e}_1(s)} \mathbf{e}_1(s)$ is orthogonal to $\mathbf{e}_1(s)$, and we define

$$\mathbf{e}_2(s) = \frac{\nabla_{\mathbf{e}_1(s)} \mathbf{e}_1(s)}{|\nabla_{\mathbf{e}_1(s)} \mathbf{e}_1(s)|}.$$

This vector will be called the *unit normal vector* to the curve at $\alpha(s)$.

The adapted frame field is now uniquely determined: Because the frame must be oriented and orthonormal, $\mathbf{e}_3(s)$ is uniquely determined by the condition that

$$\mathbf{e}_3(s) = \mathbf{e}_1(s) \times \mathbf{e}_2(s).$$

This makes sense because $\mathbf{e}_1(s)$ and $\mathbf{e}_2(s)$ are elements of the oriented, 3-dimensional Euclidean vector space $T_{\alpha(s)}\mathbb{X}^3$, where the cross product is well-defined. The vector $\mathbf{e}_3(s)$ is called the *binormal vector* to the curve at $\alpha(s)$. The adapted frame field $(\mathbf{e}_1(s), \mathbf{e}_2(s), \mathbf{e}_3(s))$ is called the *Frenet frame* of the curve $\alpha(s)$; it determines a canonical, left-invariant lifting $\tilde{\alpha} : I \to G$ given by

$$\tilde{\alpha}(s) = (\mathbf{e}_0(s), \mathbf{e}_1(s), \mathbf{e}_2(s), \mathbf{e}_3(s)),$$

where $\mathbf{e}_0(s) = \alpha(s)$, for any nondegenerate curve α in $\mathbb{X}^3$ parametrized by arc length.

***Exercise 11.19.** Show that we have the following analog of the Frenet equations for nondegenerate curves $\alpha : I \to \mathbb{X}^3$ parametrized by arc length:

$$(11.20)\quad \begin{bmatrix} \alpha'(s) & \nabla_{\mathbf{e}_1(s)} \mathbf{e}_1(s) & \nabla_{\mathbf{e}_1(s)} \mathbf{e}_2(s) & \nabla_{\mathbf{e}_1(s)} \mathbf{e}_3(s) \end{bmatrix}$$
$$= \begin{bmatrix} \alpha(s) & \mathbf{e}_1(s) & \mathbf{e}_2(s) & \mathbf{e}_3(s) \end{bmatrix} \begin{bmatrix} 0 & 0 & 0 & 0 \\ 1 & 0 & -\kappa(s) & 0 \\ 0 & \kappa(s) & 0 & -\tau(s) \\ 0 & 0 & \tau(s) & 0 \end{bmatrix},$$

where $\kappa, \tau : I \to \mathbb{R}$ are smooth functions along α with

$$\kappa(s) = |\nabla_{\mathbf{e}_1(s)} \mathbf{e}_1(s)| > 0.$$

(Hint: Equation (11.14) might be helpful.) As in the Euclidean case, the functions $\kappa(s), \tau(s)$ are called the *curvature* and *torsion*, respectively, of α.

Exercise 11.20. Let $\alpha : I \to \mathbb{X}^3 \subset \mathbb{V}^4$ be a nondegenerate curve parametrized by arc length.

(a) Show that, when regarded as functions $\mathbf{e}_0, \mathbf{e}_1, \mathbf{e}_2, \mathbf{e}_3 : I \to \mathbb{V}^4$, we have

$$\left[\mathbf{e}_0'(s)\ \mathbf{e}_1'(s)\ \mathbf{e}_2'(s)\ \mathbf{e}_3'(s)\right] = \left[\mathbf{e}_0(s)\ \mathbf{e}_1(s)\ \mathbf{e}_2(s)\ \mathbf{e}_3(s)\right] \begin{bmatrix} 0 & \pm 1 & 0 & 0 \\ 1 & 0 & -\kappa(s) & 0 \\ 0 & \kappa(s) & 0 & -\tau(s) \\ 0 & 0 & \tau(s) & 0 \end{bmatrix},$$

with the sign depending on whether $\mathbb{X}^3 = \mathbb{S}^3$ or $\mathbb{H}^3$.

(b) Show that there exist orthogonal unit vectors $\bar{\mathbf{e}}_0, \bar{\mathbf{e}}_1, \bar{\mathbf{e}}_2 \in \mathbb{V}^4$ such that α is contained in the intersection of $\mathbb{X}^3$ with the plane spanned by $(\bar{\mathbf{e}}_0, \bar{\mathbf{e}}_1, \bar{\mathbf{e}}_2)$ if and only if the torsion $\tau(s)$ is identically zero. (This intersection is a "great sphere" when $\mathbb{X}^3 = \mathbb{S}^3$ and a "great hyperboloid" when $\mathbb{X}^3 = \mathbb{H}^3$.) This is the analog of the fact that a nondegenerate curve in $\mathbb{E}^3$ is contained in a plane if and only if its torsion $\tau(s)$ is identically zero.

11.5. Moving frames for surfaces in $\mathbb{S}^3$ and $\mathbb{H}^3$

Now, let U be an open set in $\mathbb{R}^2$, and let $\mathbf{x} : U \to \mathbb{X}^3$ be an immersion whose image is a surface $\Sigma = \mathbf{x}(U)$. Just as for curves, an adapted frame field along Σ is a lifting $\tilde{\mathbf{x}} : U \to G$ of the form

$$\tilde{\mathbf{x}}(\mathbf{u}) = (\mathbf{e}_0(\mathbf{u}), \mathbf{e}_1(\mathbf{u}), \mathbf{e}_2(\mathbf{u}), \mathbf{e}_3(\mathbf{u})),$$

where for each $\mathbf{u} \in U$, $\mathbf{e}_0(\mathbf{u}) = \mathbf{x}(\mathbf{u})$ and $(\mathbf{e}_1(\mathbf{u}), \mathbf{e}_2(\mathbf{u}), \mathbf{e}_3(\mathbf{u}))$ is an oriented, orthonormal basis for the tangent space $T_{\mathbf{x}(\mathbf{u})}\mathbb{X}^3$.

***Exercise 11.21.** Convince yourself that the following statements, which we proved for surfaces in $\mathbb{E}^3$ in Chapter 4, remain true for a surface $\Sigma = \mathbf{x}(U) \subset \mathbb{X}^3$. (In particular, this means that none of the following constructions relied on the flatness of $\mathbb{E}^3$ or involved differentiating vector fields on $\mathbb{E}^3$.)

(a) We can choose an orthonormal frame $(\mathbf{e}_1(\mathbf{u}), \mathbf{e}_2(\mathbf{u}), \mathbf{e}_3(\mathbf{u}))$ for each tangent space $T_{\mathbf{x}(\mathbf{u})}\mathbb{X}^3$ so that that $\mathbf{e}_3(\mathbf{u})$ is orthogonal to $T_{\mathbf{x}(\mathbf{u})}\Sigma$ and $(\mathbf{e}_1(\mathbf{u}), \mathbf{e}_2(\mathbf{u}))$ span $T_{\mathbf{x}(\mathbf{u})}\Sigma$. Any such choice defines a lifting $\tilde{\mathbf{x}} : U \to G$ by

$$\tilde{\mathbf{x}}(\mathbf{u}) = (\mathbf{e}_0(\mathbf{u}), \mathbf{e}_1(\mathbf{u}), \mathbf{e}_2(\mathbf{u}), \mathbf{e}_3(\mathbf{u})),$$

where $\mathbf{e}_0(\mathbf{u}) = \mathbf{x}(\mathbf{u})$.

(b) Let $(\bar{\omega}^i, \bar{\omega}^i_j)$ denote the pullbacks of the Maurer-Cartan forms (ω^i, ω^i_j) on G to U via $\tilde{\mathbf{x}}$. Then we have $\bar{\omega}^3 = 0$, and the 1-forms $(\bar{\omega}^1, \bar{\omega}^2)$ form a basis for the 1-forms on U.

(c) The metric on $\mathbb{X}^3$ naturally induces a metric on $\Sigma = \mathbf{x}(U) \subset \mathbb{X}^3$, given by the *first fundamental form*

$$\mathrm{I} = (\bar{\omega}^1)^2 + (\bar{\omega}^2)^2.$$

(d) Differentiating the equation $\bar{\omega}^3 = 0$ and applying Cartan's lemma implies that there exist functions h_{11}, h_{12}, h_{22} on U such that

$$\begin{bmatrix} \bar{\omega}_1^3 \\ \bar{\omega}_2^3 \end{bmatrix} = \begin{bmatrix} h_{11} & h_{12} \\ h_{12} & h_{22} \end{bmatrix} \begin{bmatrix} \bar{\omega}^1 \\ \bar{\omega}^2 \end{bmatrix}.$$

Once again, the functions (h_{ij}) are related to the derivative of the Gauss map on Σ. However, there are two important differences:

(1) Since there are no canonical isomorphisms between the individual tangent spaces $T_{\mathbf{x}(\mathbf{u})}\mathbb{X}^3$, we cannot view the Gauss map as a map from Σ to the unit sphere in $\mathbb{E}^3$; rather, it is a map from Σ to the tangent bundle $T\mathbb{X}^3$.

(2) In order to define the second fundamental form, we must use the covariant derivative to differentiate the Gauss map.

Definition 11.22. Let $U \subset \mathbb{R}^2$ be an open set, and let $\mathbf{x} : U \to \mathbb{X}^3$ be an immersion with image $\Sigma = \mathbf{x}(U)$. The *Gauss map* of $\Sigma = \mathbf{x}(U)$ is the map $N : \Sigma \to T\mathbb{X}^3$ defined by

$$N(\mathbf{x}(\mathbf{u})) = \mathbf{e}_3(\mathbf{u}) \in T_{\mathbf{x}(\mathbf{u})}\mathbb{X}^3,$$

where $(\mathbf{e}_1(\mathbf{u}), \mathbf{e}_2(\mathbf{u}), \mathbf{e}_3(\mathbf{u}))$ is any adapted frame field on $\Sigma = \mathbf{x}(U)$.

Definition 11.23. Let $U \subset \mathbb{R}^2$ be an open set, and let $\mathbf{x} : U \to \mathbb{X}^3$ be an immersion. The *second fundamental form* of $\Sigma = \mathbf{x}(U)$ is the quadratic form II on TU defined by

$$\mathrm{II}(\mathbf{v}) = -\langle \nabla_{\mathbf{v}} \mathbf{e}_3, d\mathbf{x}(\mathbf{v}) \rangle$$

for $\mathbf{v} \in T_{\mathbf{u}}U$, where $(\mathbf{e}_1(\mathbf{u}), \mathbf{e}_2(\mathbf{u}), \mathbf{e}_3(\mathbf{u}))$ is any adapted frame field on $\Sigma = \mathbf{x}(U)$.

***Exercise 11.24.** Show that

$$\mathrm{II} = \bar{\omega}_1^3 \, \bar{\omega}^1 + \bar{\omega}_2^3 \, \bar{\omega}^2 = h_{11} \, (\bar{\omega}^1)^2 + 2h_{12} \, \bar{\omega}^1 \, \bar{\omega}^2 + h_{22} \, (\bar{\omega}^2)^2.$$

As in the Euclidean case, the eigenvalues $\kappa_1(\mathbf{u}), \kappa_2(\mathbf{u})$ of the matrix $[h_{ij}(\mathbf{u})]$ are called the *principal curvatures* of Σ at the point $\mathbf{x}(\mathbf{u})$, and the eigenvectors of the self-adjoint map $-dN_{\mathbf{x}(\mathbf{u})} : T_{\mathbf{x}(\mathbf{u})}\Sigma \to T_{\mathbf{x}(\mathbf{u})}\Sigma$ are called the *principal vectors* or *principal directions* of Σ at the point $\mathbf{x}(\mathbf{u})$.

Definition 11.25. The functions

$$\bar{K} = \kappa_1\kappa_2, \qquad H = \tfrac{1}{2}(\kappa_1 + \kappa_2) \tag{11.21}$$

on U are called the *extrinsic curvature* and the *mean curvature*, respectively, of Σ. The *Gauss curvature* K of Σ (also called the *intrinsic curvature* of Σ) is the function on U defined by the condition that

$$d\bar{\omega}^1_2 = K\,\bar{\omega}^1 \wedge \bar{\omega}^2,$$

where $(\bar{\omega}^i, \bar{\omega}^i_j)$ are the Maurer-Cartan forms associated to any adapted frame field on Σ.

***Exercise 11.26.** (a) Show that the functions K, $\bar{K}$, and H are all well-defined on U, independent of the choice of adapted frame field.

(b) Use the structure equations (11.6) and (11.10) to show that:

(1) If $\mathbb{X}^3 = \mathbb{S}^3$, then $K = \bar{K} + 1$.

(2) If $\mathbb{X}^3 = \mathbb{H}^3$, then $K = \bar{K} - 1$.

So, unlike in the Euclidean case where the Gauss equation implies that $K = \bar{K}$, here these two notions of curvature differ by the sectional curvature of the underlying homogeneous space.

As in the Euclidean case, the pullbacks to U of the structure equations for the derivatives $(d\bar{\omega}^i_j)$ in equations (11.6) and (11.10) are called the *Gauss* and *Codazzi equations.*

***Exercise 11.27.** Show that the Gauss and Codazzi equations take the form

$$\begin{aligned} d\bar{\omega}^1_2 &= \bar{\omega}^3_1 \wedge \bar{\omega}^3_2 \pm \bar{\omega}^1 \wedge \bar{\omega}^2, \\ d\bar{\omega}^3_1 &= \bar{\omega}^3_2 \wedge \bar{\omega}^1_2, \\ d\bar{\omega}^3_2 &= -\bar{\omega}^3_1 \wedge \bar{\omega}^1_2, \end{aligned} \tag{11.22}$$

with the sign in the first equation depending on whether $\mathbb{X}^3 = \mathbb{S}^3$ or $\mathbb{H}^3$.

***Exercise 11.28.** Suppose that $\mathbf{x} : U \to \mathbb{X}^3$ is an immersion whose coordinate curves are all principal curves. Then the first and second fundamental forms may be written as

$$\begin{aligned} \mathrm{I} &= E\,du^2 + G\,dv^2, \\ \mathrm{II} &= e\,du^2 + g\,dv^2. \end{aligned}$$

(a) Show that the principal adapted frame field

$$\mathbf{e}_1(\mathbf{u}) = \frac{1}{\sqrt{E}}\mathbf{x}_u, \qquad \mathbf{e}_2(\mathbf{u}) = \frac{1}{\sqrt{G}}\mathbf{x}_v, \qquad \mathbf{e}_3(\mathbf{u}) = \mathbf{e}_1(\mathbf{u}) \times \mathbf{e}_2(\mathbf{u})$$

has associated Maurer-Cartan forms

$$\bar{\omega}^1 = \sqrt{E}\,du, \qquad \bar{\omega}^2 = \sqrt{G}\,dv, \qquad \bar{\omega}^3 = 0,$$
$$\bar{\omega}^1_2 = \frac{1}{2\sqrt{EG}}\,(E_v\,du - G_u\,dv),$$
$$\bar{\omega}^3_1 = \frac{e}{E}\,\bar{\omega}^1 = \frac{e}{\sqrt{E}}\,du,$$
$$\bar{\omega}^3_2 = \frac{g}{G}\,\bar{\omega}^2 = \frac{g}{\sqrt{G}}\,dv.$$

(b) Show that the Gauss equation is equivalent to

$$\frac{eg}{EG} \pm 1 = -\frac{1}{2\sqrt{EG}}\left[\left(\frac{E_v}{\sqrt{EG}}\right)_v + \left(\frac{G_u}{\sqrt{EG}}\right)_u\right], \tag{11.23}$$

with the sign on the left-hand side depending on whether $\mathbb{X}^3$ is equal to $\mathbb{S}^3$ or $\mathbb{H}^3$.

(c) Show that the Codazzi equations are equivalent to

$$\begin{aligned} e_v &= \frac{1}{2}E_v\left(\frac{e}{E} + \frac{g}{G}\right), \\ g_u &= \frac{1}{2}G_u\left(\frac{e}{E} + \frac{g}{G}\right) \end{aligned} \tag{11.24}$$

(cf. Exercises 4.24 and 4.41).

As for surfaces in $\mathbb{E}^3$, the first and second fundamental forms are *invariants* of the surface. Consequently, if two surfaces have different first and second fundamental forms, then they cannot be equivalent via an isometry of M. Lemmas 4.2 and 4.12 imply that the converse is true as well, and we have the following analog of Bonnet's theorem:

Theorem 11.29. *Let $(\bar{\omega}^1, \bar{\omega}^2, \bar{\omega}^3_1, \bar{\omega}^3_2)$ be 1-forms on a connected and simply connected open set $U \subset \mathbb{R}^2$ satisfying the conditions that $(\bar{\omega}^1, \bar{\omega}^2)$ are linearly independent at each point of U and that $\bar{\omega}^3_i$ is a scalar multiple of $\bar{\omega}^i$ for $i = 1, 2$. Suppose that, together with the Levi-Civita connection form $\bar{\omega}^1_2$ determined by $\bar{\omega}^1$ and $\bar{\omega}^2$, these forms satisfy the Gauss and Codazzi equations (11.22). Then there exists an immersed surface $\mathbf{x} : U \to \mathbb{X}^3$, unique up to transformation by an element of G, whose first and second fundamental forms are*

$$\mathrm{I} = (\bar{\omega}^1)^2 + (\bar{\omega}^2)^2,$$
$$\mathrm{II} = \bar{\omega}^3_1\,\bar{\omega}^1 + \bar{\omega}^3_2\,\bar{\omega}^2.$$

We will conclude this chapter by exploring some special families of surfaces in $\mathbb{S}^3$ and $\mathbb{H}^3$.

Definition 11.30. A regular surface $\Sigma = \mathbf{x}(U) \subset \mathbb{X}^3$ is called *totally geodesic* if every geodesic in Σ is also a geodesic in $\mathbb{X}^3$.

It turns out that Σ is totally geodesic if and only if

$$\nabla_{\mathbf{w}}\mathbf{e}_3 = \mathbf{0} \quad \text{for all } \mathbf{w} \in T\Sigma, \tag{11.25}$$

where $\mathbf{e}_3$ is a unit normal vector field along Σ.

Remark 11.31. By way of comparison, for a surface $\Sigma \subset \mathbb{E}^3$, the totally geodesic condition says that every geodesic in Σ is a straight line in $\mathbb{E}^3$, which implies that Σ is contained in a plane in $\mathbb{E}^3$. Meanwhile, the condition (11.25) says that the normal vector field $\mathbf{e}_3$ to Σ is constant along Σ, which also implies that Σ is contained in a plane in $\mathbb{E}^3$.

Exercise 11.32. Suppose that $\Sigma = \mathbf{x}(U) \subset \mathbb{X}^3$ is a totally geodesic surface in $\mathbb{X}^3$.

(a) Show that the condition (11.25) is equivalent to the condition that the Maurer-Cartan forms associated to any adapted frame field along Σ satisfy

$$\bar{\omega}^3_1 = \bar{\omega}^3_2 = 0. \tag{11.26}$$

(b) Use the extrinsic definition (11.11) together with the structure equations (11.5) and (11.9) to show that, when regarded as functions $\mathbf{e}_0, \mathbf{e}_1, \mathbf{e}_2, \mathbf{e}_3 : U \to \mathbb{V}^4$, equation (11.26) implies that

$$d\mathbf{e}_3 = 0.$$

Conclude that there exist orthogonal unit vectors $\bar{\mathbf{e}}_0, \bar{\mathbf{e}}_1, \bar{\mathbf{e}}_2 \in \mathbb{V}^4$ such that $\Sigma = \mathbf{x}(U)$ is contained in the "great sphere" of $\mathbb{S}^3$ or the "great hyperboloid" of $\mathbb{H}^3$ determined by the intersection of $\mathbb{X}^3$ with the plane spanned by $(\bar{\mathbf{e}}_0, \bar{\mathbf{e}}_1, \bar{\mathbf{e}}_2)$ (cf. Exercise 11.20). This is the analog of the fact that any totally geodesic surface in $\mathbb{E}^3$ is contained in a plane.

Definition 11.33. A surface $\Sigma = \mathbf{x}(U) \subset \mathbb{X}^3$ is called *flat* if its Gauss curvature K is identically zero.

Exercise 11.34. Let Σ be a flat surface in $\mathbb{S}^3$.

(a) Show that every point $\mathbf{x}_0 \in \Sigma$ has a neighborhood for which there exists an asymptotic parametrization $\mathbf{x} : U \to \mathbb{S}^3$ of Σ such that the first and second fundamental forms of $\Sigma = \mathbf{x}(U)$ are given by

$$\begin{aligned} \mathrm{I} &= dx^2 + 2\cos(2\psi)\, dx\, dy + dy^2, \\ \mathrm{II} &= 2\sin(2\psi)\, dx\, dy, \end{aligned} \tag{11.27}$$

where the function $\psi : U \to \mathbb{R}$, called the *angle function*, satisfies the wave equation $\psi_{xy} = 0$. (Hint: Observe that $\kappa_1\kappa_2 = -1$, and adapt the construction of §9.4.) Conversely, Theorem 11.29 implies that any solution $\psi(x, y)$

of the wave equation gives rise to a flat surface in $\mathbb{S}^3$ whose first and second fundamental forms are given by (11.27).

(b) Let $a, b > 0$ with $a^2 + b^2 = 1$, and consider the map $\mathbf{x} : \mathbb{S}^1 \times \mathbb{S}^1 \to \mathbb{S}^3 \subset \mathbb{E}^4$ defined by

$$\mathbf{x}(\theta, \varphi) = {}^t[a\cos(\theta),\ a\sin(\theta),\ -b\cos(\varphi),\ b\sin(\varphi)]\,,$$

where θ and φ denote angle coordinates on the two copies of $\mathbb{S}^1$. Show that $\Sigma = \mathbf{x}(\mathbb{S}^1 \times \mathbb{S}^1)$ is a flat torus in $\mathbb{S}^3$, with angle function equal to a constant ψ_0 such that

$$\cos(\psi_0) = a, \qquad \sin(\psi_0) = b.$$

(Hint: Begin by considering the principal adapted frame field

$$\mathbf{e}_0(\theta, \varphi) = \mathbf{x}(\theta, \varphi) = {}^t[a\cos(\theta),\ a\sin(\theta),\ -b\cos(\varphi),\ b\sin(\varphi)],$$

$$\mathbf{e}_1(\theta, \varphi) = \frac{\mathbf{x}_\theta(\theta, \varphi)}{|\mathbf{x}_\theta(\theta, \varphi)|} = {}^t[-\sin(\theta),\ \cos(\theta),\ 0,\ 0],$$

$$\mathbf{e}_2(\theta, \varphi) = \frac{\mathbf{x}_\varphi(\theta, \varphi)}{|\mathbf{x}_\varphi(\theta, \varphi)|} = {}^t[0,\ 0,\ \sin(\varphi),\ \cos(\varphi)],$$

$$\mathbf{e}_3(\theta, \varphi) = {}^t[-b\cos(\theta),\ -b\sin(\theta),\ -a\cos(\varphi),\ a\sin(\varphi)]$$

and its associated Maurer-Cartan forms.)

Exercise 11.35. Let Σ be a flat surface in $\mathbb{H}^3$, and assume that Σ has no umbilic points (i.e., points where $\kappa_1 = \kappa_2$).

(a) Show that every point $\mathbf{x} \in \Sigma$ has a neighborhood for which there exists a principal parametrization $\mathbf{x} : U \to \mathbb{H}^3$ of Σ such that the first and second fundamental forms of $\Sigma = \mathbf{x}(U)$ are given by

$$\begin{aligned} \mathrm{I} &= \cosh^2(\psi)\, du^2 + \sinh^2(\psi)\, dv^2, \\ \mathrm{II} &= \sinh(\psi)\cosh(\psi)(du^2 + dv^2), \end{aligned} \tag{11.28}$$

where the angle function $\psi : U \to \mathbb{R}$ satisfies Laplace's equation $\psi_{uu} + \psi_{vv} = 0$. (Hint: Observe that $\kappa_1\kappa_2 = 1$, and adapt the construction of §9.4.) Conversely, Theorem 11.29 implies that any solution $\psi(u, v)$ of Laplace's equation gives rise to a flat surface in $\mathbb{H}^3$ whose first and second fundamental forms are given by (11.28).

(b) Let $a, b > 0$ with $a^2 - b^2 = 1$, and consider the map $\mathbf{x} : \mathbb{R} \times \mathbb{S}^1 \to \mathbb{H}^3 \subset \mathbb{M}^{1,3}$ defined by

$$\mathbf{x}(t, \varphi) = {}^t[a\cosh(t),\ a\sinh(t),\ b\cos(\varphi),\ b\sin(\varphi)]\,,$$

where t denotes a standard coordinate on $\mathbb{R}$ and φ denotes an angle coordinate on $\mathbb{S}^1$. Show that $\Sigma = \mathbf{x}(\mathbb{R} \times \mathbb{S}^1)$ is a flat cylinder in $\mathbb{H}^3$, with angle function equal to a constant ψ_0 such that

$$\cosh(\psi_0) = a, \qquad \sinh(\psi_0) = b.$$

(Hint: Begin by considering the principal adapted frame field

$$\begin{aligned}
\mathbf{e}_0(t,\varphi) &= \mathbf{x}(t,\varphi) = {}^t[a\cosh(t),\, a\sinh(t),\, b\cos(\varphi),\, b\sin(\varphi)],\\
\mathbf{e}_1(t,\varphi) &= \frac{\mathbf{x}_t(t,\varphi)}{|\mathbf{x}_t(t,\varphi)|} = {}^t[\sinh(t),\, \cosh(t),\, 0,\, 0],\\
\mathbf{e}_2(t,\varphi) &= \frac{\mathbf{x}_\varphi(t,\varphi)}{|\mathbf{x}_\varphi(t,\varphi)|} = {}^t[0,\, 0,\, -\sin(\varphi),\, \cos(\varphi)],\\
\mathbf{e}_3(t,\varphi) &= {}^t[-b\cosh(t),\, -b\sinh(t),\, -a\cos(\varphi),\, -a\sin(\varphi)]
\end{aligned}$$

and its associated Maurer-Cartan forms.)

11.6. MAPLE computations

We will need to set up separately for computations in $\mathbb{S}^3$ and $\mathbb{H}^3$ because the structure equations for their Maurer-Cartan forms are different due to the curvature terms. Here we will work through Exercise 11.34 regarding flat surfaces in $\mathbb{S}^3$; only minor modifications are required for the computations for Exercise 11.35 regarding flat surfaces in $\mathbb{H}^3$. (See the MAPLE worksheet for this chapter on the AMS webpage for details.)

EXERCISE 11.34: After loading the `Cartan` and `LinearAlgebra` packages into MAPLE, declare the Maurer-Cartan forms on $\mathbb{S}^3$, and tell MAPLE about their symmetries and structure equations:

```
> Form(omega[1], omega[2], omega[3]);
  Form(omega[1,2], omega[3,1], omega[3,2]);
> omega[1,1]:= 0;
  omega[2,2]:= 0;
  omega[3,3]:= 0;
  omega[2,1]:= -omega[1,2];
  omega[1,3]:= -omega[3,1];
  omega[2,3]:= -omega[3,2];
> for i from 1 to 3 do
    d(omega[i]):= -add('omega[i,j] &^ omega[j]', j=1..3);
  end do;
```

```
> d(omega[1,2]):= -omega[1,3] &^ omega[3,2]
  + omega[1] &^ omega[2];
  d(omega[3,1]):= -omega[3,2] &^ omega[2,1]
  + omega[3] &^ omega[1];
  d(omega[3,2]):= -omega[3,1] &^ omega[1,2]
  + omega[3] &^ omega[2];
```

Now, suppose that $\mathbf{x} : U \to \mathbb{S}^3$ is a principal parametrization of a flat surface $\Sigma \subset \mathbb{S}^3$ and that $(\mathbf{e}_1(\mathbf{u}), \mathbf{e}_2(\mathbf{u}), \mathbf{e}_3(\mathbf{u}))$ is a principal adapted frame field along Σ. Since Σ is flat, the principal curvatures $\kappa_1(\mathbf{u}), \kappa_2(\mathbf{u})$ satisfy the condition

$$\kappa_1(\mathbf{u})\kappa_2(\mathbf{u}) = -1.$$

The same argument as that in §9.4 shows that the coordinates (u, v) can be chosen in such a way that the associated dual forms are given by the following expressions for some function $\psi_0(u, v)$:

```
> PDETools[declare](psi0(u,v));
> adaptedsub:= [
  omega[1]  = cos(psi0(u,v))*d(u),
  omega[2]  = sin(psi0(u,v))*d(v),
  omega[3]  = 0,
  omega[3,1]  = sin(psi0(u,v))*d(u),
  omega[3,2]  = -cos(psi0(u,v))*d(v),
  omega[1,2]  = -diff(psi0(u,v), v)*d(u)
  - diff(psi0(u,v), u)*d(v)];
```

These forms satisfy the structure equations for the dual forms, as well as the Codazzi equations for $d\bar{\omega}_1^3$ and $d\bar{\omega}_2^3$. Now check the Gauss equation:

```
> Simf(d(Simf(subs(adaptedsub, omega[1,2])))
  + Simf(subs(adaptedsub, omega[1,3] &^ omega[3,2]
  - omega[1] &^ omega[2])));
```

$$(-\psi 0_{v,v} + \psi 0_{u,u})\ (d(v))\ \&\hat{}\ (d(u))$$

So the Gauss equation is satisfied if and only if the function $\psi_0(u, v)$ satisfies the wave equation $(\psi_0)_{uu} - (\psi_0)_{vv} = 0$.

Now make the change to asymptotic coordinates

$$x = \frac{(u+v)}{2}, \qquad y = \frac{(u-v)}{2},$$

or, equivalently,

$$u = x + y, \qquad v = x - y.$$

We'll need to remove $\bar{\omega}_2^1$ from the list in `adaptedsub` because MAPLE won't like this substitution in the expressions `diff(psi0(u,v), u)` and diff(psi0(u,v), v). We'll also introduce a new name for the function

$$\psi(x, y) = \psi_0(x + y, x - y).$$

```
> PDETools[declare](psi(x,y));
> adaptedsub_asymp:= Simf(subs([u = x + y, v = x - y],
  Simf(subs([psi0(u,v) = psi(x,y)],
  [seq(adaptedsub[i], i=1..5)]))));
```

MAPLE doesn't really know how to compute symmetric products of differential forms, but the following commands will work for computing the first and second fundamental forms:

```
> collect(simplify(Simf(subs(adaptedsub_asymp, omega[1]))^2
  + Simf(subs(adaptedsub_asymp, omega[2]))^2), {d(x), d(y)});
> collect(simplify(Simf(subs(adaptedsub_asymp, omega[3,1]))*
  Simf(subs(adaptedsub_asymp, omega[1]))
  + Simf(subs(adaptedsub_asymp, omega[3,2]))*
  Simf(subs(adaptedsub_asymp, omega[2]))), {d(x), d(y)});
```

For Exercise 11.34(b), we can compute the Maurer-Cartan forms associated to the given frame field, as follows. First, declare a, b to be constants and define the frame vectors $(\mathbf{e}_0, \mathbf{e}_1, \mathbf{e}_2, \mathbf{e}_3)$:

```
> Form(a=-1, b=-1);
> e0:= Vector([a*cos(theta), a*sin(theta), -b*cos(phi),
  b*sin(phi)]);
  e1:= Vector([-sin(theta), cos(theta), 0,0]);
  e2:= Vector([0,0, sin(phi), cos(phi)]);
  e3:= Vector([-b*cos(theta), -b*sin(theta), -a*cos(phi),
  a*sin(phi)]);
```

The fastest way to compute the Maurer-Cartan forms is to define the corresponding group element

$$g = \begin{bmatrix} \mathbf{e}_0 & \mathbf{e}_1 & \mathbf{e}_2 & \mathbf{e}_3 \end{bmatrix} \in SO(4)$$

and compute the matrix-valued Maurer-Cartan form

$$\bar{\omega} = g^{-1} dg,$$

as follows:

```
> g:= Matrix([e0, e1, e2, e3]);
  connection_matrix:= simplify(MatrixInverse(g).map(d, g));
```

$$connection\_matrix = \begin{bmatrix} 0 & -\frac{d(\theta)a}{a^2+b^2} & -b\frac{d(\phi)}{a^2+b^2} & 0 \\ d(\theta)a & 0 & 0 & -b\,d(\theta) \\ b\,d(\phi) & 0 & 0 & a\,d(\phi) \\ 0 & b\frac{d(\theta)}{a^2+b^2} & -\frac{d(\phi)a}{a^2+b^2} & 0 \end{bmatrix}$$

Since $a^2 + b^2 = 1$, we see that we have the following Maurer-Cartan forms:

```
> examplesub:= [omega[1] = a*d(theta), omega[2] = b*d(phi),
  omega[3] = 0, omega[1,2] = 0, omega[3,1] = b*d(theta),
  omega[3,2] = -a*d(phi)];
```

These forms agree with the forms in `adaptedsub`, with $\cos(\psi_0) = a$, $\sin(\psi_0) = b$, as desired.

Chapter 12

The nonhomogeneous case: Moving frames on Riemannian manifolds

12.1. Introduction

So far, we have been using moving frames to study the geometry of curves and surfaces as submanifolds Σ of homogeneous spaces G/H. In this context, the geometry of Σ is determined by the geometry of the ambient homogeneous space G/H and the particular way that Σ is embedded in G/H as a submanifold. But there are many interesting geometric problems for which this scenario is too restrictive. For instance, we may be interested in the geometric structure of a nonhomogeneous manifold that is defined intrinsically and not as a submanifold of some larger ambient homogeneous space. Or, even in the study of submanifolds, we might be interested in submanifolds Σ of some manifold M that is not homogeneous.

Recall that for a homogeneous space G/H, the natural projection map

$$\pi : G \to G/H$$

leads to a description of G as the *frame bundle* $\mathcal{F}(G/H)$ of the space G/H and that the set of frames over a given point $\mathbf{x} \in G/H$ is in one-to-one correspondence with the subgroup H of G. The fundamental property of homogeneous spaces is that for any two points $\mathbf{x}, \mathbf{y} \in G/H$ and any frames

$\mathbf{f_x}, \mathbf{f_y}$ based at the points $\mathbf{x}$ and $\mathbf{y}$, respectively, there is a symmetry of the manifold G/H that takes $\mathbf{x}$ to $\mathbf{y}$ and $\mathbf{f_x}$ to $\mathbf{f_y}$. In particular, the entire frame bundle $\mathcal{F}(G/H)$ is diffeomorphic to G, and G acts freely and transitively on $\mathcal{F}(G/H)$.

When we replace the homogeneous space G/H with a more general n-dimensional manifold M, the situation is a bit different. We can still associate to each point $\mathbf{x} \in M$ a collection of frames $(\mathbf{x}; \mathbf{e}_1, \dots, \mathbf{e}_n)$ for the tangent space $T_{\mathbf{x}}M$, and the specific collection will depend on what sort of geometric structure we want to consider on M. For instance, if M is a Riemannian manifold, we might consider the collection of frames that are orthonormal with respect to the Riemannian metric on M. Moreover, we still have a group action on the set of frames *based at each point* $\mathbf{x} \in M$, simply because any two frames for the tangent space $T_{\mathbf{x}}M$ are related by an element of $GL(n)$. In the case of a Riemannian manifold M, any two orthonormal frames based at a point $\mathbf{x} \in M$ are related by an element of the orthogonal group $O(n)$; in fact, the set of orthonormal frames over each point $\mathbf{x}$ is in one-to-one correspondence with the Lie group $H = O(n)$. What is different is that if $\mathbf{x}, \mathbf{y}$ are distinct points of M, then there is no obvious relationship between the orthonormal frames based at $\mathbf{x}$ and those based at $\mathbf{y}$ and no obvious group action that can be used to transform a frame based at $\mathbf{x}$ to one based at $\mathbf{y}$. The collection of all orthonormal frames based at all points of M forms a *principal bundle* $\mathcal{F}(M)$ with fiber group $H = O(n)$ (cf. §1.5), but there may be no larger Lie group G that acts transitively on the entire frame bundle $\mathcal{F}(M)$.

In this chapter, we will illustrate this more general scenario by exploring how moving frames can be applied to study the geometry of a Riemannian manifold M and submanifolds Σ of M. Similar constructions can be applied to manifolds with other types of geometry—e.g., Lorentzian, equi-affine, or projective manifolds.

12.2. Orthonormal frames and connections on Riemannian manifolds

Definition 12.1. A *Riemannian manifold* of dimension n is a smooth manifold of dimension n, together with a smoothly varying metric g on M (cf. Exercise 1.15). The metric g determines an inner product $\langle \cdot, \cdot \rangle$ on each tangent space $T_{\mathbf{x}}M$ defined by

$$\langle \mathbf{v}, \mathbf{w} \rangle = g(\mathbf{v}, \mathbf{w})$$

for $\mathbf{v}, \mathbf{w} \in T_{\mathbf{x}}M$.

Exercise 12.2. The phrase "smoothly varying" in Definition 12.1 means that for any local coordinate system $\mathbf{x} = (x^1, \dots, x^n)$ on M, the functions $\left\{g_{ij}(\mathbf{x}) = g\left(\frac{\partial}{\partial x^i}, \frac{\partial}{\partial x^j}\right) \mid 1 \leq i, j, \leq n\right\}$ are smooth functions of $\mathbf{x}$. Show that this condition is independent of the choice of local coordinates on M: If $\bar{\mathbf{x}} = (\bar{x}^1, \dots, \bar{x}^n)$ is another local coordinate system on M with smooth local coordinate transformation functions $x^i = x^i(\bar{x}^1, \dots, \bar{x}^n)$, then the functions

$$\bar{g}_{k\ell}(\bar{\mathbf{x}}) = g\left(\frac{\partial}{\partial \bar{x}^k}, \frac{\partial}{\partial \bar{x}^\ell}\right)$$

are smooth if and only if the original functions (g_{ij}) are smooth. (Hint: The results of Exercise 1.22 should be helpful.)

For simplicity, we will assume that the manifold M is oriented. Orthonormal frames on M are defined almost exactly as they were on $\mathbb{E}^n$ (cf. Definition 3.12); the only difference is that now $\mathbf{x}$ is simply a point of M and, in general, not an element of a vector space.

Definition 12.3. An (oriented) *orthonormal frame* $\mathbf{f}$ on an oriented Riemannian manifold M is a list $\mathbf{f} = (\mathbf{x}; \mathbf{e}_1, \dots, \mathbf{e}_n)$ where $\mathbf{x} \in M$ and $(\mathbf{e}_1, \dots, \mathbf{e}_n)$ is an oriented, orthonormal basis for the tangent space $T_{\mathbf{x}}M$. Alternatively, we may say that $(\mathbf{e}_1, \dots, \mathbf{e}_n)$ is an *orthonormal frame based at* $\mathbf{x}$.

The set of orthonormal frames at each point is in one-to-one correspondence with the Lie group $SO(n)$, and the set of orthonormal frames on M forms a principal bundle over M with fiber $SO(n)$, called the (oriented) *orthonormal frame bundle* of M and denoted $\mathcal{F}(M)$:

$$\begin{array}{ccc} SO(n) & \longrightarrow & \mathcal{F}(M) \\ & & \downarrow{\scriptstyle \pi} \\ & & M. \end{array}$$

There are several important differences between the orthonormal frame bundle of the homogeneous space $\mathbb{E}^n$ and that of a general Riemannian manifold M. First, while each fiber of the frame bundle is acted on freely and transitively by $SO(n)$, there is no larger group that acts transitively on the entire frame bundle $\mathcal{F}(M)$. But the most significant change is that, given an orthonormal frame field $(\mathbf{e}_1(\mathbf{x}), \dots, \mathbf{e}_n(\mathbf{x}))$ on an open set in M, there is no natural way of thinking of the frame vector fields $(\mathbf{e}_i(\mathbf{x}))$ as functions from M to a fixed vector space—not even by regarding M as a submanifold of some larger Euclidean space, as we did for $\mathbb{S}^n$ and $\mathbb{H}^n$ in Chapter 11. Rather, they are sections of the tangent bundle TM, which means that for each point $\mathbf{x} \in M$, the vectors $(\mathbf{e}_1(\mathbf{x}), \dots, \mathbf{e}_n(\mathbf{x}))$ take values in the vector space $T_\mathbf{x}M$.

In the case of $\mathbb{E}^n$, we were able to differentiate vector fields by using the fact that each tangent space $T_\mathbf{x}\mathbb{E}^n$ is *canonically* isomorphic to $\mathbb{E}^n$ and regarding the vector fields $(\mathbf{e}_1(\mathbf{x}), \dots, \mathbf{e}_n(\mathbf{x}))$ as functions into this fixed vector space (cf. Remark 3.13). And even for $\mathbb{S}^n$ and $\mathbb{H}^n$, we were able to make use of these canonical isomorphisms for the ambient spaces $\mathbb{E}^{n+1}$ and $\mathbb{M}^{1,n}$ in order to define the covariant derivatives of vector fields. But for a more general Riemannian manifold M, there is no such canonical isomorphism between each tangent space $T_\mathbf{x}M$ and a fixed vector space $\mathbb{E}^n$, or even a canonical embedding of $T_\mathbf{x}M$ into a larger fixed vector space; indeed, there are *infinitely* many ways of identifying each tangent space $T_\mathbf{x}M$ with a fixed vector space $\mathbb{E}^n$, all of which are equally valid. The following exercise illustrates some of the complications that may arise as a result of this ambiguity.

***Exercise 12.4.** To any local orthonormal frame field $(\underline{\mathbf{e}}_1(\mathbf{x}), \dots, \underline{\mathbf{e}}_n(\mathbf{x}))$ on an open set $U \subset M$, we can associate a *local trivialization* (cf. §1.5) $\phi : TU \to U \times \mathbb{E}^n$ of the tangent bundle TM by defining

$$\phi\left(\mathbf{x}, a^i \underline{\mathbf{e}}_i(\mathbf{x})\right) = \left(\mathbf{x}, {}^t[a^1, \dots, a^n]\right). \tag{12.1}$$

(a) Show that the map ϕ defines an isometry between each tangent space $T_\mathbf{x}M$ and the vector space $\mathbb{E}^n$.

(b) We may regard a local trivialization ϕ as a local choice of basis vector fields $(\underline{\mathbf{e}}_1(\mathbf{x}), \dots, \underline{\mathbf{e}}_n(\mathbf{x}))$ for sections of TM, and the identification (12.1) makes it tempting to think that we might be able to regard these vector fields as "constant" for purposes of differentiation. But what happens when we choose a different basis? Let $(\tilde{\underline{\mathbf{e}}}_1(\mathbf{x}), \dots, \tilde{\underline{\mathbf{e}}}_n(\mathbf{x}))$ be any other orthonormal frame field on U, related to the original frame field by

$$\left[\tilde{\underline{\mathbf{e}}}_1(\mathbf{x}) \ \dots \ \tilde{\underline{\mathbf{e}}}_n(\mathbf{x})\right] = \left[\underline{\mathbf{e}}_1(\mathbf{x}) \ \dots \ \underline{\mathbf{e}}_n(\mathbf{x})\right] \bar{A}(\mathbf{x}), \tag{12.2}$$

where $\bar{A}(\mathbf{x})$ is an $SO(n)$-valued function on U. Show that under the analogous local trivialization $\tilde{\phi}$ associated to the orthonormal frame field $(\tilde{\underline{\mathbf{e}}}_1(\mathbf{x}), \dots, \tilde{\underline{\mathbf{e}}}_n(\mathbf{x}))$, the vector fields $(\underline{\mathbf{e}}_1(\mathbf{x}), \dots, \underline{\mathbf{e}}_n(\mathbf{x}))$ are identified with the columns of the matrix $\left(\bar{A}(\mathbf{x})\right)^{-1}$. In particular, vector fields that appear "constant" with respect to one trivialization do not necessarily remain "constant" with respect to a different trivialization.

This exercises raises a crucial question: If there is no canonical trivialization of TM, and hence no consistent notion of a "constant" vector field on M, then how can we possibly *differentiate* vector fields on M? (This is a special case of the more general question of how to differentiate sections of a vector bundle.) It turns out that, in order to make sense of the notion of differentiation for vector fields on M, we need to introduce an additional structure, called a *connection*, on the tangent bundle $T_{\mathbf{x}}M$.

Definition 12.5. Let $\Gamma(TM)$ denote the space of smooth local sections of TM (i.e., smooth vector fields on open sets in M). An *affine connection* (or, more succinctly, a *connection*) ∇ on TM is a map

$$\nabla : \Gamma(TM) \times \Gamma(TM) \to \Gamma(TM),$$

with $\nabla(\mathbf{w}, \mathbf{v})$ denoted by $\nabla_{\mathbf{w}}\mathbf{v}$, such that for any vector fields $\mathbf{v}$, $\mathbf{v}_1$, $\mathbf{v}_2$, $\mathbf{w}$, $\mathbf{w}_1$, $\mathbf{w}_2 \in \Gamma(TM)$, any smooth, real-valued functions f, g on M, and any real numbers $a, b \in \mathbb{R}$, we have

(1) $\nabla_{f\mathbf{w}_1+g\mathbf{w}_2}\mathbf{v} = f\nabla_{\mathbf{w}_1}\mathbf{v} + g\nabla_{\mathbf{w}_2}\mathbf{v}$;

(2) $\nabla_{\mathbf{w}}(a\mathbf{v}_1 + b\mathbf{v}_2) = a\nabla_{\mathbf{w}}\mathbf{v}_1 + b\nabla_{\mathbf{w}}\mathbf{v}_2$;

(3) $\nabla_{\mathbf{w}}(f\mathbf{v}) = \mathbf{w}(f)\mathbf{v} + f\nabla_{\mathbf{w}}\mathbf{v}$.

Remark 12.6. The Levi-Civita connection of Definition 11.16 is, of course, a connection according to this definition. It also has certain additional properties, which we will discuss shortly.

Conditions (1) and (2) are *linearity* properties: They say that the map ∇ is linear over smooth functions in its first input and linear over real numbers in its second input. Condition (3) is an analog of the Leibniz rule that describes how ∇ behaves when its second input is multiplied by a smooth function. The vector field $\nabla_{\mathbf{w}}\mathbf{v}$ should be regarded as defining a sort of "directional derivative" of the vector field $\mathbf{v}$ in the direction of $\mathbf{w}$, and conditions (1)–(3) are precisely the conditions that such a directional derivative must satisfy.

Remark 12.7. This definition applies more generally to any smooth vector bundle B over a manifold M: If $\pi : B \to M$ is a vector bundle with fibers isomorphic to a fixed k-dimensional vector space V (cf. §1.5), then an affine connection ∇ on B is a map

$$\nabla : \Gamma(TM) \times \Gamma(B) \to \Gamma(B)$$

satisfying the properties of Definition 12.5. This more general setting illustrates the fact that the two copies of $\Gamma(TM)$ used for the inputs of ∇ in Definition 12.5 play significantly different roles: The first input is the direction along which differentiation should occur (which must be a tangent vector to M), and the second input is the object to be differentiated. The output is the resulting differentiated object, and it should live in the same space as the second input.

***Exercise 12.8.** Let $M = \mathbb{E}^n$, and let $\mathbf{v}, \mathbf{w}$ be smooth vector fields on $\mathbb{E}^n$. By taking advantage of the usual canonical identification of each tangent space $T_{\mathbf{x}}\mathbb{E}^n$ with the vector space $\mathbb{E}^n$, we may regard these vector fields as smooth functions $\mathbf{v}, \mathbf{w} : \mathbb{E}^n \to \mathbb{E}^n$. Having done so, define

$$\nabla_{\mathbf{w}}\mathbf{v} = d\mathbf{v}(\mathbf{w}).$$

(a) Show that ∇ is a connection on $T\mathbb{E}^n$. (Hint: It might be helpful to write everything out in terms of the canonical local coordinates on $T\mathbb{E}^n$; i.e.,

$$\mathbf{v}(\mathbf{x}) = {}^t\big[v^1(\mathbf{x}), \ldots, v^n(\mathbf{x})\big],$$

etc. This should make it more obvious that $\nabla_{\mathbf{w}}\mathbf{v}$ is a vector field on $\mathbb{E}^n$—in fact, the components of $\nabla_{\mathbf{w}}\mathbf{v}$ are obtained by computing the directional derivatives of the components of $\mathbf{v}$ in the direction $\mathbf{w}$.)

(b) The canonical isomorphism $T_{\mathbf{x}}\mathbb{E}^n \cong \mathbb{E}^n$ allows us to write the tangent bundle $T\mathbb{E}^n$ as the product manifold

$$T\mathbb{E}^n \cong \mathbb{E}^n \times \mathbb{E}^n, \tag{12.3}$$

where the first factor represents the base manifold $\mathbb{E}^n$ and the second factor represents the fibers $T_{\mathbf{x}}\mathbb{E}^n$. (In other words, we have a canonical *global* trivialization of the tangent bundle $T\mathbb{E}^n$.) This, in turn, allows us to write the tangent bundle $T(T\mathbb{E}^n)$ as the product manifold

$$T(T\mathbb{E}^n) \cong T\mathbb{E}^n \times T\mathbb{E}^n \tag{12.4}$$

in the obvious way. The vector field $\mathbf{v}$ is a section of $T\mathbb{E}^n$, and via the identification (12.3), we can write it as a function $\sigma : \mathbb{E}^n \to \mathbb{E}^n \times \mathbb{E}^n$ defined by

$$\sigma(\mathbf{x}) = (\mathbf{x}, \mathbf{v}(\mathbf{x})).$$

Show that $\nabla_{\mathbf{w}}\mathbf{v}$ is given by the composition

$$\nabla_{\mathbf{w}}\mathbf{v} = \pi_2 \circ d\sigma(\mathbf{w}), \tag{12.5}$$

where $d\sigma : T\mathbb{E}^n \to T(T\mathbb{E}^n)$ is the differential of the map σ (cf. §1.3) and $\pi_2 : T\mathbb{E}^n \times T\mathbb{E}^n \to T\mathbb{E}^n$ is the projection onto the second factor. This connection is called the *flat connection* on $\mathbb{E}^n$.

Remark 12.9. For a general Riemannian manifold M, there is generally no global trivialization analogous to equation (12.3) for TM. However, for each point $(\mathbf{x}, \mathbf{v}) \in TM$, the $2n$-dimensional tangent space $T_{(\mathbf{x},\mathbf{v})}(TM)$ can be decomposed in a manner analogous to equation (12.4) in many different ways. The n-dimensional subspace

$$V_{(\mathbf{x},\mathbf{v})} = T_{(\mathbf{x},\mathbf{v})}(T_{\mathbf{x}}M) \subset T_{(\mathbf{x},\mathbf{v})}(TM)$$

(corresponding to the second factor in (12.4)) is canonically defined: It is the tangent space to the fiber $T_{\mathbf{x}}M \subset TM$ and is called the *vertical* tangent space; moreover, $V_{(\mathbf{x},\mathbf{v})}$ is canonically isomorphic to $T_{\mathbf{x}}M$. But there is no single canonical choice for the complementary n-dimensional subspace corresponding to the first factor in (12.4) (called the *horizontal* tangent space); in fact, any n-dimensional subspace $H_{(\mathbf{x},\mathbf{v})} \subset T_{(\mathbf{x},\mathbf{v})}(TM)$ for which $H_{(\mathbf{x},\mathbf{v})} \cap V_{(\mathbf{x},\mathbf{v})} = \{\mathbf{0}\}$ is, a priori, an equally valid choice for the horizontal tangent space at $(\mathbf{x},\mathbf{v})$. But the choice of a connection on TM can resolve this issue: A connection ∇ on TM determines a projection map

$$\pi_2 : T_{(\mathbf{x},\mathbf{v})}(TM) \to V_{(\mathbf{x},\mathbf{v})}$$

as in equation (12.5), and this in turn determines the n-dimensional horizontal subspace $H_{(\mathbf{x},\mathbf{v})} = \ker(\pi_2) \subset T_{(\mathbf{x},\mathbf{v})}(TM)$ and the corresponding decomposition

$$T_{(\mathbf{x},\mathbf{v})}(TM) = H_{(\mathbf{x},\mathbf{v})} \oplus V_{(\mathbf{x},\mathbf{v})}.$$

Conversely, a choice of such a decomposition for each $(\mathbf{x},\mathbf{v}) \in TM$ (subject to certain consistency conditions implied by Definition 12.5) determines a connection on TM via a projection formula analogous to (12.5). In the case of $M = \mathbb{E}^n$, the canonical decomposition (12.4) for $T(T\mathbb{E}^n)$ is precisely equivalent to the flat connection ∇ on $T\mathbb{E}^n$.

As a consequence of properties (1)–(3) of Definition 12.5, a connection ∇ is completely determined by its action on any given orthonormal frame field on M. Given an orthonormal frame field $(\underline{\mathbf{e}}_1(\mathbf{x}), \ldots, \underline{\mathbf{e}}_n(\mathbf{x}))$ on M, the connection ∇ determines scalar-valued 1-forms $(\bar{\omega}^i_j)$, with $1 \le i, j \le n$, on M, called the *connection forms* associated to this frame field, defined by the condition that for any $\mathbf{w} \in T_{\mathbf{x}}M$,

$$(\nabla_{\mathbf{w}}\underline{\mathbf{e}}_i)(\mathbf{x}) = \underline{\mathbf{e}}_j(\mathbf{x})\,\bar{\omega}^j_i(\mathbf{w}). \tag{12.6}$$

The following exercise shows how a connection on TM may be thought of as an analog to the exterior derivative for vector fields on M (cf. Exercise 11.14).

***Exercise 12.10.** Given a connection ∇ on TM and a vector field $\mathbf{v}$ on M, consider the map

$$\nabla\mathbf{v} : TM \to TM$$

defined by

$$\nabla\mathbf{v}(\mathbf{w}) = \nabla_{\mathbf{w}}\mathbf{v}. \tag{12.7}$$

(a) Use the linearity properties of Definition 12.5 to show that $\nabla\mathbf{v}$ is a TM-valued 1-form on M (cf. Definition 2.18).

(b) Show that if $(\underline{\mathbf{e}}_1(\mathbf{x}), \ldots, \underline{\mathbf{e}}_n(\mathbf{x}))$ is an orthonormal frame field on an open set $U \subset M$, then $\nabla \underline{\mathbf{e}}_i$ is the TM-valued 1-form

$$\nabla \underline{\mathbf{e}}_i = \underline{\mathbf{e}}_j \bar{\omega}_i^j, \tag{12.8}$$

where $(\bar{\omega}_i^j)$ are the scalar-valued 1-forms on M defined by equation (12.6). Thus, we can think of ∇ as a generalization of the exterior derivative d that appears in the structure equations (3.1).

***Exercise 12.11.** Let ∇ be a connection on TM, and let $(\underline{\mathbf{e}}_1(\mathbf{x}), \ldots, \underline{\mathbf{e}}_n(\mathbf{x}))$ be an orthonormal frame field on an open set $U \subset M$, with associated connection forms $(\bar{\omega}_j^i)$ on U. (Think of the frame field as determining a local trivialization of TM, as in Exercise 12.4.)

(a) Show that for any vector field $\mathbf{v}(\mathbf{x}) = \sum v^i(\mathbf{x})\underline{\mathbf{e}}_i(\mathbf{x})$ on U,

$$(\nabla \mathbf{v})(\mathbf{x}) = \underline{\mathbf{e}}_i(\mathbf{x})(dv^i + v^j \bar{\omega}_j^i). \tag{12.9}$$

(Note that, while the exterior derivative of a vector field on M is not well-defined, it still makes perfect sense to compute the exterior derivative of real-valued functions, such as v^i, on M.) For this reason, a connection is sometimes expressed *with respect to a given trivialization of* TM as

$$\nabla = d + \bar{\omega},$$

where $\bar{\omega}$ is the matrix of 1-forms $\bar{\omega} = [\bar{\omega}_j^i]$. The notation means that if the vector field $\mathbf{v}(\mathbf{x}) = \sum v^i(\mathbf{x})\underline{\mathbf{e}}_i(\mathbf{x})$ is expressed as the column vector $\bar{\mathbf{v}}(\mathbf{x}) = {}^t[v^1(\mathbf{x}), \ldots, v^n(\mathbf{x})]$, then the vector field $\nabla \mathbf{v}$ should be expressed as the column vector $d\bar{\mathbf{v}} + \bar{\omega}\bar{\mathbf{v}}$, as indicated by equation (12.9).

(b) Let $(\tilde{\underline{\mathbf{e}}}_1(\mathbf{x}), \ldots, \tilde{\underline{\mathbf{e}}}_n(\mathbf{x}))$ be any other orthonormal frame field on U, related to the original frame field as in equation (12.2). Use equation (12.8) to show that the matrix $\tilde{\bar{\omega}} = [\tilde{\bar{\omega}}_j^i]$ of connection forms associated to the new frame field is given by

$$\tilde{\bar{\omega}} = \bar{A}^{-1} d\bar{A} + \bar{A}^{-1} \bar{\omega} \bar{A}. \tag{12.10}$$

Remark 12.12. Connections on fiber bundles play an important role in theoretical physics, particularly in field theory. In the physics literature, a local section of the orthonormal frame bundle (which amounts to choosing a local trivialization for TM) is called a *gauge*, and a frame transformation of the form (12.2) is called a *gauge transformation.* A connection is called a *gauge field*, and equation (12.10) describes how the gauge field transforms under a gauge transformation.

In addition to the connection forms $(\bar{\omega}^i_j)$, we can associate to an orthonormal frame field $(\underline{\mathbf{e}}_1(\mathbf{x}), \dots, \underline{\mathbf{e}}_n(\mathbf{x}))$ on M the *dual forms* $(\bar{\omega}^1, \dots, \bar{\omega}^n)$. These are defined in essentially the same way as for orthonormal frame fields on $\mathbb{E}^n$:

Definition 12.13. Let M be a Riemannian manifold, and let $(\underline{\mathbf{e}}_1(\mathbf{x}), \dots, \underline{\mathbf{e}}_n(\mathbf{x}))$ be an orthonormal frame field on an open set $U \subset M$. The *dual forms* $(\bar{\omega}^1, \dots, \bar{\omega}^n)$ associated to this frame field are the unique scalar-valued 1-forms on U defined by the property that

$$\bar{\omega}^i(\underline{\mathbf{e}}_j) = \delta^i_j = \begin{cases} 1, & i = j, \\ 0, & i \neq j \end{cases}$$

(cf. (3.7)).

***Exercise 12.14.** (a) Show that the dual forms $(\bar{\omega}^1, \dots, \bar{\omega}^n)$ associated to an orthonormal frame field $(\underline{\mathbf{e}}_1(\mathbf{x}), \dots, \underline{\mathbf{e}}_n(\mathbf{x}))$ form a basis for the 1-forms on U.

(b) Let $(\tilde{\underline{\mathbf{e}}}_1(\mathbf{x}), \dots, \tilde{\underline{\mathbf{e}}}_n(\mathbf{x}))$ be any other orthonormal frame field on U, related to the original frame field as in equation (12.2). Show that the dual forms $(\tilde{\bar{\omega}}^1, \dots, \tilde{\bar{\omega}}^n)$ associated to the new frame field are given by

$$\begin{bmatrix} \tilde{\bar{\omega}}^1 \\ \vdots \\ \tilde{\bar{\omega}}^n \end{bmatrix} = \bar{A}^{-1} \begin{bmatrix} \bar{\omega}^1 \\ \vdots \\ \bar{\omega}^n \end{bmatrix}. \tag{12.11}$$

As you might have guessed from the notation, the dual forms and connection forms $(\bar{\omega}^i, \bar{\omega}^i_j)$ on M associated to an orthonormal frame field $(\underline{\mathbf{e}}_1(\mathbf{x}), \dots, \underline{\mathbf{e}}_n(\mathbf{x}))$ on M are the pullbacks to M of certain 1-forms (ω^i, ω^i_j) on the frame bundle $\mathcal{F}(M)$ via the section

$$\mathbf{f}(\mathbf{x}) = (\mathbf{x}; \underline{\mathbf{e}}_1(\mathbf{x}), \dots, \underline{\mathbf{e}}_n(\mathbf{x}))$$

of $\mathcal{F}(M)$. Fortunately, as the following exercise will show, equations (12.10) and (12.11) tell us precisely how these 1-forms should be defined. (It might be helpful to review §3.3.2 at this point, particularly the derivations of equations (3.3) and (3.4).)

***Exercise 12.15.** Let $(\underline{\mathbf{e}}_1(\mathbf{x}), \dots, \underline{\mathbf{e}}_n(\mathbf{x}))$ be a local orthonormal frame field on an open set $U \subset M$, with associated dual and connection forms $(\bar{\omega}^i, \bar{\omega}^i_j)$. Just as this orthonormal frame field determines a local trivialization of TM (cf. Exercise 12.4), it can also be used to define a local trivialization of the principal bundle $\mathcal{F}(M)$: For any $\mathbf{x} \in U$ and any orthonormal frame $\mathbf{f} = (\mathbf{x}; \mathbf{e}_1, \dots, \mathbf{e}_n)$ based at $\mathbf{x}$, we can write

$$\begin{bmatrix} \mathbf{e}_1 & \cdots & \mathbf{e}_n \end{bmatrix} = \begin{bmatrix} \underline{\mathbf{e}}_1 & \cdots & \underline{\mathbf{e}}_n \end{bmatrix} A \tag{12.12}$$

for some unique matrix $A = [a^i_j] \in SO(n)$. Define a map $\phi : \mathcal{F}(U) \to U \times SO(n)$ by

$$\phi(\mathbf{x}; \mathbf{e}_1, \ldots, \mathbf{e}_n) = (\mathbf{x}, A),$$

where A is the matrix defined by equation (12.12).

(a) Show that the map ϕ defines a diffeomorphism between $\mathcal{F}(U) = \pi^{-1}(U) \subset \mathcal{F}(M)$ and $U \times SO(n)$.

(b) Let $(\tilde{\underline{\mathbf{e}}}_1(\mathbf{x}), \ldots, \tilde{\underline{\mathbf{e}}}_n(\mathbf{x}))$ be any other orthonormal frame field on U, related to the original frame field as in equation (12.2). Show that the analogous local trivialization $\tilde{\phi}$ associated to the orthonormal frame field $(\tilde{\underline{\mathbf{e}}}_1(\mathbf{x}), \ldots, \tilde{\underline{\mathbf{e}}}_n(\mathbf{x}))$ is given by

$$\tilde{\phi}(\mathbf{x}; \mathbf{e}_1, \ldots, \mathbf{e}_n) = \left(\mathbf{x}, \tilde{A}\right) = \left(\mathbf{x}, \left(\bar{A}(\mathbf{x})\right)^{-1} A\right). \tag{12.13}$$

(c) Define 1-forms (ω^i, ω^i_j) on $\mathcal{F}(U)$ by

$$\begin{bmatrix} \omega^1 \\ \vdots \\ \omega^n \end{bmatrix} = A^{-1} \begin{bmatrix} \bar{\omega}^1 \\ \vdots \\ \bar{\omega}^n \end{bmatrix}, \tag{12.14}$$

$$\omega = [\omega^i_j] = A^{-1} dA + A^{-1} \bar{\omega} A.$$

Note the distinction between these 1-forms and those in equations (12.10) and (12.11): In those equations, $\bar{A}$ is an $SO(n)$-valued function on the open set $U \subset M$, whereas in equations (12.14), the entries of A represent local coordinates on $SO(n)$, *independent* of the local coordinates on M. So, while the 1-forms $(\bar{\omega}^i, \bar{\omega}^i_j)$ are 1-forms on M and the $(\bar{\omega}^i_j)$ are linear combinations of the $(\bar{\omega}^i)$, the 1-forms (ω^i, ω^i_j) are linearly independent 1-forms on the orthonormal frame bundle $\mathcal{F}(M)$. Use equations (12.10), (12.11), and (12.13) to show that the 1-forms (12.14) are well-defined, independent of the choice of orthonormal frame field $(\underline{\mathbf{e}}_1(\mathbf{x}), \ldots, \underline{\mathbf{e}}_n(\mathbf{x}))$ used to define the 1-forms $(\bar{\omega}^i, \bar{\omega}^i_j)$ and the local trivialization ϕ of $\mathcal{F}(M)$.

12.3. The Levi-Civita connection

There are many ways of choosing a connection on TM; indeed, given an orthonormal frame field on an open set $U \subset M$, *any* $n \times n$ matrix $\bar{\omega}$ of 1-forms on M can be used to define a connection on TU via the equation (12.8). But some connections have nicer properties than others; the following definition describes two properties that are often considered desirable.

Definition 12.16. An affine connection ∇ on TM is called

(1) *torsion-free* or *symmetric* if for any vector fields $\mathbf{v}, \mathbf{w}$ on M, we have

$$\nabla_{\mathbf{v}}\mathbf{w} - \nabla_{\mathbf{w}}\mathbf{v} = [\mathbf{v}, \mathbf{w}],$$

where $[\mathbf{v}, \mathbf{w}]$ denotes the usual Lie bracket of vector fields (cf. §1.4);

(2) *compatible with the metric* g on M if for any vector fields $\mathbf{v}, \mathbf{w}$ on M, we have

$$d\langle \mathbf{v}, \mathbf{w}\rangle = \langle \nabla\mathbf{v}, \mathbf{w}\rangle + \langle \mathbf{v}, \nabla\mathbf{w}\rangle.$$

(This condition is often written as $\nabla g = 0$.)

A given connection ∇ on TM may have one, both, or neither of these properties. But it turns out that there is exactly one connection on TM that has both:

Theorem 12.17 (Levi-Civita). *Given a Riemannian manifold M, there exists a unique connection ∇ on TM that is both torsion-free and compatible with the metric. This connection is called the* Levi-Civita connection *on TM.*

Remark 12.18. The existence of a canonical "nice" connection is a very important feature of Riemannian geometry, and unless otherwise stated, the Levi-Civita connection is almost always the connection of choice for the tangent bundle of a Riemannian manifold. But in other types of geometry (e.g., equi-affine, projective), there is often no single "best" connection, and then the choice of connection must be stated explicitly. In some contexts, it is even desirable to consider a whole family of connections!

In the following exercises, we will show how to prove Theorem 12.17. The strategy of the proof is to show that if a torsion-free, metric-compatible connection exists, then it must be unique. In the process, an explicit formula for this unique connection is derived, which serves to prove the existence result as well.

To begin the proof, let $(\mathbf{e}_1(\mathbf{x}), \ldots, \mathbf{e}_n(\mathbf{x}))$ be an orthonormal frame field on an open set $U \subset M$, with associated dual forms $(\bar{\omega}^1, \ldots, \bar{\omega}^n)$. (At this point, we will drop the underscore notation on the vector fields $\mathbf{e}_i(\mathbf{x})$ since we no longer need to think of them as "basis" vector fields.) Let ∇ be an affine connection on M, and let $(\bar{\omega}^i_j)$ be the corresponding connection forms associated to the given frame field. Since the dual forms are a basis for the

1-forms on M, we can write the connection forms as

$$\bar{\omega}^i_j = a^i_{jk}\bar{\omega}^k \tag{12.15}$$

for some functions (a^i_{jk}) on U.

***Exercise 12.19.** Suppose that ∇ is compatible with the metric.

(a) Show that the connection forms must satisfy the condition $\bar{\omega}^j_i = -\bar{\omega}^i_j$; i.e., the connection forms are skew-symmetric in their indices. (Hint: Differentiate the equations $\langle \mathbf{e}_i, \mathbf{e}_j \rangle = \delta_{ij}$.)

(b) Conclude that if ∇ is compatible with the metric, then the functions (a^i_{jk}) must satisfy

$$a^i_{jk} = -a^j_{ik}. \tag{12.16}$$

***Exercise 12.20.** Suppose that ∇ is torsion-free, and let $c^k_{ij}(\mathbf{x}) = -c^k_{ji}(\mathbf{x})$ be the functions on U defined by the Lie bracket relations

$$[\mathbf{e}_i, \mathbf{e}_j] = c^k_{ij}\mathbf{e}_k.$$

(a) Show that for each i, j, k,

$$\bar{\omega}^k_j(\mathbf{e}_i) - \bar{\omega}^k_i(\mathbf{e}_j) = c^k_{ij}.$$

(Hint: Apply the torsion-free condition with $\mathbf{v} = \mathbf{e}_i$, $\mathbf{w} = \mathbf{e}_j$.)

(b) Conclude that if ∇ is torsion-free, then the functions (a^i_{jk}) must satisfy

$$a^k_{ji} - a^k_{ij} = c^k_{ij}. \tag{12.17}$$

***Exercise 12.21.** Now, suppose that ∇ is both compatible with the metric and torsion-free.

(a) Use equations (12.16) and (12.17) to show that

$$a^k_{ij} = \frac{1}{2}\left(c^j_{ki} - c^i_{jk} - c^k_{ij}\right). \tag{12.18}$$

(Hint: This is an exercise in index juggling. Start with a^k_{ij} and apply equations (12.16) and (12.17) alternately until you come back to the index arrangement that you started with. And keep in mind that $c^k_{ij} = -c^k_{ji}$!)

(b) Conclude that Theorem 12.17 is true and that the Levi-Civita connection is defined by equation (12.15), with (a^k_{ij}) as in equation (12.18).

From now on, we will assume that ∇ is the Levi-Civita connection on TM.

12.4. The structure equations

Let (ω^i, ω^i_j) be the dual forms and connection forms on $\mathcal{F}(M)$ associated to the Levi-Civita connection on TM. In order to compute geometric invariants associated to M, we first need to compute the structure equations of these 1-forms. The process is analogous to that of §3.3.2, but it is complicated by the fact that we must use the connection ∇ rather than the exterior derivative d to differentiate vector-valued quantities.

As in the case of $\mathbb{E}^n$, we can define projection maps $\mathbf{x} : \mathcal{F}(M) \to M$ and $\mathbf{e}_i : \mathcal{F}(M) \to TM$ in the obvious way: If $\mathbf{f} = (\mathbf{x}; \mathbf{e}_1, \dots, \mathbf{e}_n) \in \mathcal{F}(M)$, then

$$\begin{aligned}\mathbf{x}(\mathbf{f}) &= \mathbf{x} \in M,\\ \mathbf{e}_i(\mathbf{f}) &= (\mathbf{x}, \mathbf{e}_i) \in T_{\mathbf{x}}M.\end{aligned}$$

The differentials of $(\mathbf{x}, \mathbf{e}_i)$ are maps

$$\begin{aligned}d\mathbf{x} &: T_{\mathbf{f}}\mathcal{F}(M) \to T_{\mathbf{x}}M,\\ d\mathbf{e}_i &: T_{\mathbf{f}}\mathcal{F}(M) \to T_{(\mathbf{x},\mathbf{e}_i)}(TM).\end{aligned}$$

The map $d\mathbf{x}$ is exactly analogous to the Euclidean case: The vectors $(\mathbf{e}_1, \dots, \mathbf{e}_n)$ form a basis for $T_{\mathbf{x}}M$ at each point, and the dual forms are defined precisely so that

$$d\mathbf{x} = \mathbf{e}_i\omega^i. \tag{12.19}$$

The maps $(d\mathbf{e}_i)$ are a bit more complicated. For each $\mathbf{f} \in \mathcal{F}(M)$, the image of $(d\mathbf{e}_i)_{\mathbf{f}}$ takes values in the $2n$-dimensional tangent space $T_{(\mathbf{x},\mathbf{e}_i)}(TM)$. The Levi-Civita connection ∇ determines a linear projection operator

$$\pi_2 : T_{(\mathbf{x},\mathbf{e}_i)}(TM) \to T_{(\mathbf{x},\mathbf{e}_i)}(T_{\mathbf{x}}M) \cong T_{\mathbf{x}}M,$$

defined by the condition that

$$(\pi_2 \circ d)(\mathbf{e}_i) = \nabla\mathbf{e}_i = \mathbf{e}_j\omega^j_i \tag{12.20}$$

(cf. Remark 12.9). This equation is the analog of the equation for $d\mathbf{e}_i$ in equations (3.1).

In order to compute the structure equations for the forms (ω^i, ω^i_j), we will first need to differentiate equation (12.19). This requires some care: Since both sides of the equation are 1-forms that take values in TM, we must use the connection ∇ to differentiate them. And in general, it is *not* necessarily true that $\nabla \circ d = 0$, so we cannot directly apply any obvious analog of the identity $d \circ d = 0$. Fortunately, we can get around this problem by considering two different expressions for the TM-valued 1-form $d\mathbf{x}$, as the following two exercises show.

***Exercise 12.22.** Recall that with respect to any local coordinate system $\mathbf{x} = (x^1, \ldots, x^n)$ on an open set $U \subset M$, we can write

$$d\mathbf{x} = \frac{\partial}{\partial x^i}\, dx^i, \tag{12.21}$$

where $\left(\frac{\partial}{\partial x^1}, \ldots, \frac{\partial}{\partial x^n}\right)$ are the coordinate vector fields on U.

(a) Show that the coordinate vector fields $\left(\frac{\partial}{\partial x^1}, \ldots, \frac{\partial}{\partial x^n}\right)$ have pairwise Lie brackets equal to zero; i.e., $\left[\frac{\partial}{\partial x^i}, \frac{\partial}{\partial x^j}\right] = 0$ (cf. Exercise 1.34).

(b) Apply ∇ to equation (12.21) and use an argument similar to that of Exercise 12.11, part (a), to show that

$$\nabla\,(d\mathbf{x}) = \nabla\left(\frac{\partial}{\partial x^i}\right) \wedge dx^i + \frac{\partial}{\partial x^i} d(dx^i).$$

Conclude that

$$\nabla\,(d\mathbf{x}) = \nabla\left(\frac{\partial}{\partial x^i}\right) \wedge dx^i. \tag{12.22}$$

(c) Use the definition (12.7) for the TM-valued 1-form $\nabla\left(\frac{\partial}{\partial x^i}\right)$ to show that

$$\nabla\left(\frac{\partial}{\partial x^i}\right) = \nabla_{\frac{\partial}{\partial x^j}}\left(\frac{\partial}{\partial x^i}\right) dx^j. \tag{12.23}$$

(d) Use equations (12.22) and (12.23) to show that

$$\nabla\,(d\mathbf{x}) = \sum_{i<j}\left(\nabla_{\frac{\partial}{\partial x^i}}\left(\frac{\partial}{\partial x^j}\right) - \nabla_{\frac{\partial}{\partial x^j}}\left(\frac{\partial}{\partial x^i}\right)\right) dx^i \wedge dx^j.$$

(e) Conclude from part (a) and the fact that the Levi-Civita connection is torsion-free that $\nabla\,(d\mathbf{x}) = \mathbf{0}$.

***Exercise 12.23.** Now that we know that $\nabla\,(d\mathbf{x}) = \mathbf{0}$, we can differentiate equation (12.19) to obtain the structure equations for the dual forms.

(a) Apply ∇ to equation (12.19) to show that

$$\nabla \mathbf{e}_i \wedge \omega^i + \mathbf{e}_i\, d\omega^i = \mathbf{0}. \tag{12.24}$$

(b) Apply equation (12.20) (and a bit of index juggling) to equation (12.24) to obtain

$$\mathbf{e}_i\left(d\omega^i + \omega^i_j \wedge \omega^j\right) = \mathbf{0}.$$

(c) Use the linear independence of the vectors $(\mathbf{e}_1, \ldots, \mathbf{e}_n)$ in each tangent space $T_{\mathbf{x}}M$ to conclude that the dual forms $(\omega^1, \ldots, \omega^n)$ satisfy the structure equations

$$d\omega^i = -\omega^i_j \wedge \omega^j. \tag{12.25}$$

***Exercise 12.24.** In this exercise, we will show that the structure equations (12.25) can be used to compute the connection forms directly from the dual forms associated to any orthonormal frame field on M. To this end, let $(\mathbf{e}_1(\mathbf{x}), \ldots, \mathbf{e}_n(\mathbf{x}))$ be an orthonormal frame field on an open set $U \subset M$, with associated dual forms $(\bar{\omega}^1, \ldots, \bar{\omega}^n)$.

(a) Show that if there exist 1-forms $(\bar{\omega}^i_j)$ on U that satisfy the metric compatibility condition $\bar{\omega}^j_i = -\bar{\omega}^i_j$ and the equations

$$d\bar{\omega}^i = -\bar{\omega}^i_j \wedge \bar{\omega}^j, \tag{12.26}$$

then the 1-forms (ω^i, ω^i_j) on $\mathcal{F}(U)$ defined by equations (12.14) satisfy the structure equations (12.25).

(b) Show that if there exist 1-forms $(\bar{\omega}^i_j)$ on U that satisfy the condition $\bar{\omega}^i_j = -\bar{\omega}^j_i$ and equations (12.26), then they are unique. (Hint: Suppose that there are two such sets of 1-forms $(\bar{\omega}^i_j)$, $(\tilde{\bar{\omega}}^i_j)$. Subtract the two equations, and then use Cartan's lemma and the skew-symmetry of the $(\bar{\omega}^i_j)$ to show that $\tilde{\bar{\omega}}^i_j = \bar{\omega}^i_j$.)

(c) In order to show that there exist such 1-forms $(\bar{\omega}^i_j)$, let $c^k_{ij} = -c^k_{ji}$ be the functions on U defined by the conditions

$$d\bar{\omega}^k = -c^k_{ij}\bar{\omega}^i \wedge \bar{\omega}^j. \tag{12.27}$$

Suppose that the desired 1-forms $(\bar{\omega}^i_j)$ exist; then we must have

$$\bar{\omega}^i_j = a^i_{jk}\bar{\omega}^k$$

for some functions (a^i_{jk}) on U. Since $\bar{\omega}^i_j = -\bar{\omega}^j_i$, the functions (a^i_{jk}) must satisfy the conditions

$$a^i_{jk} = -a^j_{ik}.$$

Show that the structure equations (12.26) are equivalent to the equations

$$a^k_{ji} - a^k_{ij} = c^k_{ij}.$$

Observe that these are precisely the same conditions as equations (12.16), (12.17), and conclude that the functions (a^k_{ij}) defined by equation (12.18) produce the desired connection forms $(\bar{\omega}^i_j)$.

Our next task is to compute the structure equations for the connection forms (ω^i_j). Differentiating equation (12.20) as we did for the analogous equation in the Euclidean case turns out to be impractical: We would have to use the connection ∇ to differentiate the equation, and there is no useful identity for $\nabla \circ \nabla$ that would allow us to differentiate the left-hand side effectively. Instead, we will differentiate the structure equations (12.25); these are scalar-valued equations and can be differentiated with the usual exterior derivative d.

***Exercise 12.25.** Differentiate equation (12.25) (and do some index juggling) to show that for $i = 1, \dots, n$,

$$(d\omega^i_j + \omega^i_k \wedge \omega^k_j) \wedge \omega^j = 0. \tag{12.28}$$

Definition 12.26. Given any affine connection ∇ on TM, the matrix-valued 2-form

$$\Omega = [\Omega^i_j] = [d\omega^i_j + \omega^i_k \wedge \omega^k_j] \tag{12.29}$$

on $\mathcal{F}(M)$ is called the *curvature tensor* of ∇. When ∇ is the Levi-Civita connection of a Riemannian manifold M, Ω is also called the *Riemann curvature tensor* of M.

***Exercise 12.27.** Let $(\mathbf{e}_1(\mathbf{x}), \dots, \mathbf{e}_n(\mathbf{x}))$ be an orthonormal frame field on an open set $U \subset M$, with $\bar{\omega} = [\bar{\omega}^i_j]$ the matrix of associated connection forms on U. Let $(\tilde{\mathbf{e}}_1(\mathbf{x}), \dots, \tilde{\mathbf{e}}_n(\mathbf{x}))$ be any other orthonormal frame field on U, related to the original frame field as in equation (12.2). Equation (12.10) shows how the matrix $\bar{\tilde{\omega}} = [\bar{\tilde{\omega}}^i_j]$ of connection forms associated to the new frame field is related to $\bar{\omega}$; in particular, the presence of the exterior derivative $d\bar{A}$ indicates that the connection matrix $\bar{\omega}$ is *not* a tensor on U because the transformation rule (12.10) is not an algebraic function of the components of $\bar{A}$. Use equations (12.10) and (12.29) to show that, by contrast, the pullbacks $\bar{\Omega}$, $\bar{\tilde{\Omega}}$ of Ω to U via the two frame fields are related by the equation

$$\bar{\tilde{\Omega}} = \bar{A}^{-1}\bar{\Omega}\bar{A}. \tag{12.30}$$

This transformation rule indicates that, unlike the matrix of connection forms ω, the curvature tensor Ω is, in fact, a tensor on U.

Now, your instinctive reaction is probably to apply Cartan's lemma to equation (12.28). But Cartan's lemma only applies to equations of the form

$$\phi_j \wedge \omega^j = 0,$$

where $\phi_1, \dots, \phi_n$ are 1-forms (cf. Lemma 2.49), whereas equation (12.28) has the form

$$\Omega^i_j \wedge \omega^j = 0,$$

where $\Omega^i_1, \dots, \Omega^i_n$ are 2-forms. So Cartan's lemma cannot be applied directly, and we will need to explore the implications of equation (12.28) via direct computation.

***Exercise 12.28.** Let $\Omega^i_j = d\omega^i_j + \omega^i_k \wedge \omega^k_j$.

(a) Use the fact that $\omega^i_j = -\omega^j_i$ to show that $\Omega^i_j = -\Omega^j_i$.

(b) Show that we can write

$$\Omega^i_j = \tfrac{1}{2} R^i_{jk\ell}\omega^k \wedge \omega^\ell + S^{ik}_{j\ell m}\omega^\ell_k \wedge \omega^m + T^{ikm}_{j\ell n}\omega^\ell_k \wedge \omega^n_m \tag{12.31}$$

for some functions $(R^i_{jk\ell})$, $(S^{ik}_{j\ell m})$, $(T^{ikm}_{j\ell n})$ on $\mathcal{F}(M)$ that satisfy the symmetry conditions

$$\begin{aligned} R^i_{jk\ell} &= -R^j_{ik\ell} = -R^i_{j\ell k}, \\ S^{ik}_{j\ell m} &= -S^{jk}_{i\ell m} = -S^{i\ell}_{jkm}, \\ T^{ikm}_{j\ell n} = -T^{jkm}_{i\ell n} &= -T^{i\ell m}_{jkn} = -T^{ikn}_{j\ell m} = -T^{imk}_{jn\ell}. \end{aligned} \tag{12.32}$$

(Hint: Ω^i_j is a 2-form on $\mathcal{F}(M)$, and the 1-forms (ω^i, ω^i_j) are a basis for the 1-forms on $\mathcal{F}(M)$.)

(c) Equation (12.28) can now be written as

$$\left(\tfrac{1}{2} R^i_{jk\ell}\,\omega^k \wedge \omega^\ell + S^{ik}_{j\ell m}\,\omega^\ell_k \wedge \omega^m + T^{ikm}_{j\ell n}\,\omega^\ell_k \wedge \omega^n_m\right) \wedge \omega^j = 0.$$

By considering the various types of 3-forms appearing in this expression, show that $T^{ikm}_{j\ell n} = 0$ and that

$$S^{ik}_{j\ell m} = S^{ik}_{m\ell j}, \tag{12.33}$$

$$R^i_{jk\ell} + R^i_{k\ell j} + R^i_{\ell jk} = 0. \tag{12.34}$$

(d) Use equations (12.32) and (12.33) to show that $S^{ik}_{j\ell m} = 0$. (Hint: This is an exercise in index juggling, very similar to that in Exercise 12.21.)

(e) Conclude that the Riemann curvature tensor (cf. Definition 12.26) has the form

$$\Omega = \left[\Omega^i_j\right] = \left[\tfrac{1}{2} R^i_{jk\ell}\,\omega^k \wedge \omega^\ell\right],$$

where the functions $(R^i_{jk\ell})$ satisfy the skew-symmetry relations (12.32) and the condition (12.34). Equivalently, the connection forms (ω^i_j) satisfy the structure equations

$$d\omega^i_j = -\omega^i_k \wedge \omega^k_j + \tfrac{1}{2} R^i_{jk\ell}\,\omega^k \wedge \omega^\ell. \tag{12.35}$$

Equation (12.34) is known as the *first Bianchi identity*, and together with the other symmetries, it implies that the functions $R^i_{jk\ell}$ satisfy the additional relation

$$R^i_{jk\ell} = R^k_{\ell ij}.$$

Remark 12.29. The reason for including the factor of $\frac{1}{2}$ in equations (12.31) and (12.35) is so that the curvature 2-forms (Ω^i_j) may be written as

$$\Omega^i_j = \tfrac{1}{2}R^i_{jk\ell}\,\omega^k \wedge \omega^\ell = \sum_{k<\ell} R^i_{jk\ell}\,\omega^k \wedge \omega^\ell. \tag{12.36}$$

In particular, when $n = 2$, the Riemann curvature tensor has a single nontrivial component R^1_{212}, which is equal to the Gauss curvature K.

Exercise 12.30. The Nash embedding theorem [**Nas56**] states that any n-dimensional Riemannian manifold M can be isometrically embedded into a Euclidean space $\mathbb{E}^{n+m}$ of sufficiently large dimension. The structure equations for the Maurer-Cartan forms on M can then be derived from those on $\mathbb{E}^{n+m}$, as follows.

Suppose that $M \subset \mathbb{E}^{n+m}$ is a smooth submanifold of $\mathbb{E}^{n+m}$. Choose an orthonormal frame field $(\mathbf{e}_1(\mathbf{x}), \ldots, \mathbf{e}_{n+m}(\mathbf{x}))$ for $T_\mathbf{x}\mathbb{E}^{n+m}$ along M so that for each $\mathbf{x} \in M$, the tangent space $T_\mathbf{x}M$ is spanned by the vectors $(\mathbf{e}_1(\mathbf{x}), \ldots, \mathbf{e}_n(\mathbf{x}))$. Let $(\bar{\omega}^\alpha, \bar{\omega}^\alpha_\beta)$, where $1 \leq \alpha, \beta \leq (n+m)$, denote the Maurer-Cartan forms associated to this frame field on M.

(a) Show that

$$\bar{\omega}^{n+1} = \cdots = \bar{\omega}^{n+m} = 0 \tag{12.37}$$

(cf. Proposition 4.18).

(b) Differentiate equations (12.37) and use Cartan's lemma to show that there exist functions (h^a_{ij}) on M, with $1 \leq i \leq n$, $(n+1) \leq a \leq (n+m)$, and $h^a_{ij} = h^a_{ji}$, such that

$$\bar{\omega}^a_i = h^a_{ij}\,\bar{\omega}^j. \tag{12.38}$$

These are the coefficients of the *second fundamental form*

$$\mathrm{II} = h^a_{ij}\mathbf{e}_a \otimes \bar{\omega}^i\bar{\omega}^j$$

of $M \subset \mathbb{E}^{n+m}$ (cf. Exercise 4.27).

(c) We can define a connection ∇ on M in much the same way that we did for $\mathbb{S}^n$ and $\mathbb{H}^n$ (cf. §11.3), as follows: For any vector field $\mathbf{v}$ on M and any vector $\mathbf{w} \in T_\mathbf{x}M$, define the *covariant derivative* of $\mathbf{v}$ with respect to $\mathbf{w}$ to be the vector $\nabla_\mathbf{w}\mathbf{v} \in T_\mathbf{x}M$ given by the orthogonal projection of the Euclidean directional derivative $d\mathbf{v}(\mathbf{w})$ in $\mathbb{E}^{n+m}$ onto the tangent plane $T_\mathbf{x}M \subset T_\mathbf{x}\mathbb{E}^{n+m} \cong \mathbb{E}^{n+m}$. Then let $\nabla\mathbf{v} : TM \to TM$ be the TM-valued 1-form on M defined by

$$\nabla\mathbf{v}(\mathbf{w}) = \nabla_\mathbf{w}\mathbf{v},$$

as in Exercise 12.10. Show that for $1 \leq i \leq n$,

$$\nabla\mathbf{e}_i(\mathbf{x}) = \mathbf{e}_j(\mathbf{x})\,\bar{\omega}^j_i,$$

where the sum ranges over $1 \leq j \leq n$. Conclude that the 1-forms $(\bar{\omega}^i_j)$ with $1 \leq i, j \leq n$ are the connection forms associated to the connection ∇ and the orthonormal frame field $(\mathbf{e}_1(\mathbf{x}), \ldots, \mathbf{e}_n(\mathbf{x}))$ on M. Use the fact that $\bar{\omega}^i_j = -\bar{\omega}^j_i$ and the structure equations for the $(d\bar{\omega}^i)$ on $\mathbb{E}^{n+m}$ to argue that ∇ is, in fact, the Levi-Civita connection on M.

(d) Substitute the result of part (b) into the structure equations

$$d\bar{\omega}^i_j = -\bar{\omega}^i_\alpha \wedge \bar{\omega}^\alpha_j, \qquad 1 \leq i, j \leq n,$$

on $\mathbb{E}^{n+m}$ (where the sum ranges over $1 \leq \alpha \leq (n+m)$), and compare with the structure equations (12.35) to obtain the *Gauss equations*

$$R^i_{jk\ell} = \sum_{a=n+1}^{n+m} \left(h^a_{ik} h^a_{j\ell} - h^a_{i\ell} h^a_{jk}\right). \tag{12.39}$$

These equations are the higher-dimensional analog of the Gauss equation (4.11) for surfaces in $\mathbb{E}^3$.

Similarly, the structure equations for $(d\bar{\omega}^a_i)$ with $1 \leq i \leq n$, $(n+1) \leq a \leq (n+m)$ are the higher-dimensional analog of the *Codazzi equations* (4.12), while the structure equations for $(d\bar{\omega}^a_b)$ with $(n+1) \leq a, b \leq (n+m)\}$ are called the *Ricci equations.* (The Ricci equations have no analog for surfaces in $\mathbb{E}^3$ because they only appear for embeddings of codmension greater than or equal to 2.)

The remainder of this chapter will be devoted to the study of curves and surfaces in a 3-dimensional Riemannian manifold M.

12.5. Moving frames for curves in 3-dimensional Riemannian manifolds

Let M be a 3-dimensional Riemannian manifold, and consider a smooth, parametrized curve $\alpha : I \to M$ that maps some open interval $I \subset \mathbb{R}$ into M. An adapted frame field along α should be a lifting $\tilde{\alpha} : I \to \mathcal{F}(M)$. Any such lifting can be written as

$$\tilde{\alpha}(t) = (\alpha(t); \mathbf{e}_1(t), \mathbf{e}_2(t), \mathbf{e}_3(t)),$$

where for each $t \in I$, $(\mathbf{e}_1(t), \mathbf{e}_2(t), \mathbf{e}_3(t))$ is an oriented, orthonormal basis for the tangent space $T_{\alpha(t)}M$. Such an adapted frame field is usually called an *orthonormal frame field* along α.

The construction of a canonical adapted frame field along α is very similar to that for curves in $\mathbb{S}^3$ and $\mathbb{H}^3$ (cf. §11.4), in that we must use the covariant derivative to differentiate vector fields along α. As in the Euclidean case, we say that α is *regular* if $\alpha'(t) \neq \mathbf{0}$ for every $t \in I$; as usual, we will only consider regular curves, and we will assume that $\alpha = \alpha(s)$ is parametrized by arc length. In addition, we will say that a regular curve α is *nondegenerate* if $\nabla_{\alpha'(s)}\alpha'(s) \neq \mathbf{0}$ for all $s \in I$.

***Exercise 12.31.** A regular curve $\alpha : I \to M$ is called a *geodesic* in M if $\nabla_{\alpha'(t)}\alpha'(t) = \mathbf{0}$ for all $t \in I$, where ∇ denotes the Levi-Civita connection on M (cf. Exercise 11.18). In this exercise, we will show that the geodesics on M may be described in terms of the integral curves of a vector field called the *geodesic spray* on the tangent bundle TM.

(a) Recall (cf. Remark 12.9) that the connection ∇ determines a decomposition

$$T_{(\mathbf{x},\mathbf{v})}(TM) = H_{(\mathbf{x},\mathbf{v})} \oplus V_{(\mathbf{x},\mathbf{v})} \tag{12.40}$$

of each tangent space $T_{(\mathbf{x},\mathbf{v})}(TM)$ into horizontal and vertical tangent spaces, where $V_{(\mathbf{x},\mathbf{v})} = T_{(\mathbf{x},\mathbf{v})}(T_{\mathbf{x}}M) \subset T_{(\mathbf{x},\mathbf{v})}(TM)$ is canonically defined and the complementary subspace $H_{(\mathbf{x},\mathbf{v})}$ is determined by ∇.

Let $\pi : TM \to M$ denote the standard projection operator from TM to M, and let $d\pi : T_{(\mathbf{x},\mathbf{v})}(TM) \to T_{\mathbf{x}}M$ denote its differential at the point $(\mathbf{x}, \mathbf{v}) \in TM$. Show that for each vector $\mathbf{v} \in T_{\mathbf{x}}M$, there exists a unique vector $\hat{\mathbf{v}} \in H_{(\mathbf{x},\mathbf{v})}$ such that $d\pi(\hat{\mathbf{v}}) = \mathbf{v}$.

(b) Any regular curve $\alpha : I \to M$ has a unique lifting to a curve $\alpha' : I \to TM$ defined by the condition that for each $t \in I$, $\alpha'(t) \in T_{\alpha(t)}$ is the tangent vector to α at the point $\alpha(t)$. This curve, in turn, has a unique lifting to a curve $(\alpha')' : I \to T(TM)$ defined analogously: For each $t \in I$, $(\alpha')'(t) \in T_{\alpha'(t)}(TM)$ is the tangent vector to α' at the point $\alpha'(t) \in TM$.

Show that under the decomposition (12.40) of the tangent space $T_{\alpha'(t)}(TM)$, we have

$$(\alpha')'(t) = \widehat{\alpha'(t)} + \nabla_{\alpha'(t)}\alpha'(t),$$

where $\widehat{\alpha'(t)} \in H_{\alpha'(t)}$ and $\nabla_{\alpha'(t)}\alpha'(t) \in V_{\alpha'(t)}$. (Hint: It follows from the discussion in Remark 12.9 that $\nabla_{\alpha'(t)}\alpha'(t)$ is the projection of $(\alpha')'(t)$ onto the vertical tangent space $V_{\alpha'(t)}$, so you only need to show that the projection of $(\alpha')'(t)$ onto the horizontal tangent space $H_{\alpha'(t)}$ is equal to $\widehat{\alpha'(t)}$.)

(c) Conclude from part (b) that α is a geodesic if and only if $(\alpha')'(t) \in H_{\alpha'(t)}$ for all $t \in I$.

(d) A vector field $\mathbf{w}$ on TM is called *horizontal* if $\mathbf{w}(\mathbf{x}, \mathbf{v}) \in H_{(\mathbf{x},\mathbf{v})}$ for all $(\mathbf{x}, \mathbf{v}) \in TM$. The horizontal vector field

$$\mathbf{w}(\mathbf{x}, \mathbf{v}) = \hat{\mathbf{v}}$$

(where $\hat{\mathbf{v}}$ is as in part (a)) is called the *geodesic spray* on TM. Show that a curve $\gamma : I \to TM$ is an integral curve of $\mathbf{w}$ (i.e., $\gamma'(t) = \mathbf{w}(\gamma(t))$ for all $t \in I$) if and only if the curve $\alpha = \pi \circ \gamma : I \to M$ is a geodesic in M.

Remark 12.32. It is not necessary to assume that the curve α in Exercise 12.31 is parametrized by arc length; it turns out that any geodesic is automatically parametrized by a constant multiple of arc length, sometimes called the *speed* of the curve.

***Exercise 12.33.** Now suppose that $\alpha : I \to M$ is a nondegenerate curve, parametrized by arc length. Show that the construction in §11.4 for the Frenet frame may be carried out in exactly the same way and that we have the following analog of the Frenet equations:

$$\begin{aligned}(12.41)\quad &\left[\alpha'(s)\ \nabla_{\mathbf{e}_1(s)}\mathbf{e}_1(s)\ \nabla_{\mathbf{e}_1(s)}\mathbf{e}_2(s)\ \nabla_{\mathbf{e}_1(s)}\mathbf{e}_3(s)\right]\\ &= \left[\alpha(s)\ \mathbf{e}_1(s)\ \mathbf{e}_2(s)\ \mathbf{e}_3(s)\right]\begin{bmatrix} 0 & 0 & 0 & 0 \\ 1 & 0 & -\kappa(s) & 0 \\ 0 & \kappa(s) & 0 & -\tau(s) \\ 0 & 0 & \tau(s) & 0 \end{bmatrix},\end{aligned}$$

where ∇ is the Levi-Civita connection on M and $\kappa, \tau : I \to \mathbb{R}$ are smooth functions along α with

$$\kappa(s) = |\nabla_{\mathbf{e}_1(s)}\mathbf{e}_1(s)| > 0.$$

As in the Euclidean case, the functions $\kappa(s), \tau(s)$ are called the *curvature* and *torsion*, respectively, of α. In general, there is no analog of the result in Exercise 11.20(b); a curve with torsion $\tau(s)$ identically zero is not necessarily contained in any special hypersurface of M. Indeed, even a geodesic need not be contained in any special hypersurface of M; we will explore this question in Exercises 12.41 and 12.42.

12.6. Moving frames for surfaces in 3-dimensional Riemannian manifolds

Now, let U be an open set in $\mathbb{R}^2$, and let $\mathbf{x} : U \to M$ be an immersion whose image is a surface $\Sigma = \mathbf{x}(U)$. Just as for curves, an adapted frame field along Σ is a lifting $\tilde{\mathbf{x}} : U \to \mathcal{F}(M)$ of the form

$$\tilde{\mathbf{x}}(\mathbf{u}) = (\mathbf{x}(\mathbf{u}); \mathbf{e}_1(\mathbf{u}), \mathbf{e}_2(\mathbf{u}), \mathbf{e}_3(\mathbf{u})),$$

where for each $\mathbf{u} \in U$, $(\mathbf{e}_1(\mathbf{u}), \mathbf{e}_2(\mathbf{u}), \mathbf{e}_3(\mathbf{u}))$ is an oriented, orthonormal basis for the tangent space $T_{\mathbf{x}(\mathbf{u})}M$.

***Exercise 12.34.** Convince yourself that the following statements, which we proved for surfaces in $\mathbb{E}^3$ in Chapter 4 and for surfaces in $\mathbb{S}^3$ and $\mathbb{H}^3$ in Chapter 11, remain true for a surface $\Sigma = \mathbf{x}(U) \subset M$:

(a) We can choose an orthonormal frame $(\mathbf{e}_1(\mathbf{u}), \mathbf{e}_2(\mathbf{u}), \mathbf{e}_3(\mathbf{u}))$ for each tangent space $T_{\mathbf{x}(\mathbf{u})}M$ so that that $\mathbf{e}_3(\mathbf{u})$ is orthogonal to $T_{\mathbf{x}(\mathbf{u})}\Sigma$ and $(\mathbf{e}_1(\mathbf{u}), \mathbf{e}_2(\mathbf{u}))$ span $T_{\mathbf{x}(\mathbf{u})}\Sigma$. Any such choice defines a lifting $\tilde{\mathbf{x}} : U \to \mathcal{F}(M)$ by

$$\tilde{\mathbf{x}}(\mathbf{u}) = (\mathbf{x}(\mathbf{u}); \mathbf{e}_1(\mathbf{u}), \mathbf{e}_2(\mathbf{u}), \mathbf{e}_3(\mathbf{u})) .$$

(b) Let $(\bar{\omega}^i, \bar{\omega}^i_j)$ denote the pullbacks of the Maurer-Cartan forms (ω^i, ω^i_j) on $\mathcal{F}(M)$ to U via $\tilde{\mathbf{x}}$. Then we have $\bar{\omega}^3 = 0$, and the 1-forms $(\bar{\omega}^1, \bar{\omega}^2)$ form a basis for the 1-forms on U.

(c) The metric on M naturally induces a metric on $\Sigma = \mathbf{x}(U) \subset M$, given by the *first fundamental form*

$$\mathrm{I} = (\bar{\omega}^1)^2 + (\bar{\omega}^2)^2.$$

(d) Differentiating the equation $\bar{\omega}^3 = 0$ and applying Cartan's lemma implies that there exist functions h_{11}, h_{12}, h_{22} on U such that

$$\begin{bmatrix} \bar{\omega}^3_1 \\ \bar{\omega}^3_2 \end{bmatrix} = \begin{bmatrix} h_{11} & h_{12} \\ h_{12} & h_{22} \end{bmatrix} \begin{bmatrix} \bar{\omega}^1 \\ \bar{\omega}^2 \end{bmatrix}.$$

As for surfaces in $\mathbb{S}^3$ and $\mathbb{H}^3$, the Gauss map of Σ must be defined as a map from Σ to the tangent bundle TM:

Definition 12.35. Let $U \subset \mathbb{R}^2$ be an open set, and let $\mathbf{x} : U \to M$ be an immersion with image $\Sigma = \mathbf{x}(U)$. The *Gauss map* of $\Sigma = \mathbf{x}(U)$ is the map $N : \Sigma \to TM$ defined by

$$N(\mathbf{x}(\mathbf{u})) = \mathbf{e}_3(\mathbf{u}) \in T_{\mathbf{x}(\mathbf{u})}M,$$

where $(\mathbf{e}_1(\mathbf{u}), \mathbf{e}_2(\mathbf{u}), \mathbf{e}_3(\mathbf{u}))$ is any adapted frame field on $\Sigma = \mathbf{x}(U)$.

Definition 12.36. Let $U \subset \mathbb{R}^2$ be an open set, and let $\mathbf{x} : U \to M$ be an immersion. The *second fundamental form* of $\Sigma = \mathbf{x}(U)$ is the quadratic form II on TU defined by

$$\mathrm{II}(\mathbf{v}) = -\langle \nabla_{\mathbf{v}}\mathbf{e}_3, d\mathbf{x}(\mathbf{v})\rangle$$

for $\mathbf{v} \in T_{\mathbf{u}}U$, where $(\mathbf{e}_1(\mathbf{u}), \mathbf{e}_2(\mathbf{u}), \mathbf{e}_3(\mathbf{u}))$ is any adapted frame field on $\Sigma = \mathbf{x}(U)$.

***Exercise 12.37** (Cf. Exercise 11.24)**.** Show that

$$\mathrm{II} = \bar{\omega}^3_1\,\bar{\omega}^1 + \bar{\omega}^3_2\,\bar{\omega}^2 = h_{11}\,(\bar{\omega}^1)^2 + 2h_{12}\,\bar{\omega}^1\,\bar{\omega}^2 + h_{22}\,(\bar{\omega}^2)^2.$$

As for surfaces in $\mathbb{E}^3$, the first and second fundamental forms are *invariants* of the surface. Consequently, if two surfaces have different first and second fundamental forms, then they cannot be equivalent via an isometry of M. However, the converse question is not so straightforward: In general, when there is no group structure on $\mathcal{F}(M)$, there are no analogs of Lemmas 4.2 and 4.12 that tell us when we have a complete set of invariants for a surface, or when two surfaces are equivalent via a symmetry of M. In particular, given arbitrary quadratic forms I and II on an open set $U \subset \mathbb{R}^2$, checking that the structure equations are satisfied is not sufficient to guarantee that there exists an immersion $\mathbf{x} : U \to M$ whose first and second fundamental forms are I and II. The best that we can do is to arbitrarily specify *one* of these quadratic forms, and even then the existence question may be very difficult. For instance, we have the following local isometric embedding theorem, which is due independently to Cartan [**Car27**] and Janet [**Jan26**].

Theorem 12.38 (Cartan-Janet). *Let M be a 3-dimensional real analytic Riemannian manifold; let $U \subset \mathbb{R}^2$ be an open set, and let I be a real analytic metric on U (i.e., a positive definite quadratic form on U with real analytic coefficients). Then every point $\mathbf{u} \in U$ has a neighborhood $V \subset U$ on which there exists an embedding $\mathbf{x} : V \to M$ for which the metric on $\Sigma = \mathbf{x}(V)$ induced from M agrees with the given quadratic form I.*

The proof of this theorem uses exterior differential systems; it may be found in [**BCG$^+$91**]. The assumption of real analyticity is a crucial ingredient of the proof; without this hypothesis, much less is known about the existence of isometric embeddings, even locally. Even the local existence question for embedding a surface with a C^∞ metric into $\mathbb{E}^3$ is not completely solved. (For a comprehensive treatment of the current status of the isometric embedding problem, see [**HH06**].)

For another example of how the geometry of surfaces in a general 3-dimensional Riemannian manifold M differs from the geometry of surfaces in the homogeneous spaces $\mathbb{E}^3$, $\mathbb{S}^3$, and $\mathbb{H}^3$, consider the family of *totally geodesic* surfaces in M, which are defined exactly as in Definition 11.30:

Definition 12.39. A regular surface $\Sigma = \mathbf{x}(U) \subset M$ is called *totally geodesic* if every geodesic in Σ is also a geodesic in M.

As for surfaces in $\mathbb{E}^3$, $\mathbb{S}^3$, and $\mathbb{H}^3$, a surface $\Sigma \subset M$ is totally geodesic if and only if

$$\nabla_{\mathbf{w}} \mathbf{e}_3 = \mathbf{0} \quad \text{for all } \mathbf{w} \in T\Sigma, \tag{12.42}$$

where $\mathbf{e}_3$ is a unit normal vector field along Σ.

As we saw in Remark 11.31 and Exercise 11.32, totally geodesic surfaces in the homogeneous spaces $X = \mathbb{E}^3$, $\mathbb{S}^3$, and $\mathbb{H}^3$ are plentiful: Given any point $\mathbf{x} \in X$ and any 2-dimensional subspace $P \subset T_{\mathbf{x}}X$, there exists a totally geodesic surface Σ in X (a plane in $\mathbb{E}^3$, a great sphere in $\mathbb{S}^3$, or a great hyperboloid in $\mathbb{H}^3$) with $\mathbf{x} \in \Sigma$ and $T_{\mathbf{x}}\Sigma = P$. And for a general 3-dimensional Riemannian manifold M, the results of Exercise 12.31 imply that a similar statement is true for *curves*: Given a point $\mathbf{x} \in M$ and a vector $\mathbf{v} \in T_{\mathbf{x}}M$, there exists a geodesic curve $\alpha : I \to M$ with $\alpha(0) = \mathbf{x}$ and $\alpha'(0) = \mathbf{v}$. However, the existence of totally geodesic surfaces in a general 3-dimensional Riemannian manifold M is a bit more complicated, and the Riemann curvature tensor plays a fundamental role.

In order to understand how the Riemann curvature tensor Ω is related to the existence of totally geodesic surfaces, it will be helpful to understand how its component functions transform under a change of orthonormal frame field. So, let $(\mathbf{e}_1(\mathbf{x}), \ldots, \mathbf{e}_n(\mathbf{x}))$ be an orthonormal frame field on an open set $U \subset M$, and let $(\tilde{\mathbf{e}}_1(\mathbf{x}), \ldots, \tilde{\mathbf{e}}_n(\mathbf{x}))$ be any other orthonormal frame field on U, related to the original frame field as in equation (12.2). From Exercise 12.27, we know that the curvature $\bar{\tilde{\Omega}}$ associated to the transformed frame field is related to the curvature $\bar{\Omega}$ associated to the original frame field by equation (12.30), and the dual forms $(\bar{\tilde{\omega}}^i)$ associated to the transformed frame field are related to the dual forms $(\bar{\omega}^i)$ associated to the original frame field by equation (12.11). The combination of these two transformations makes for a fairly complicated transformation rule for the functions $(\bar{R}^i_{jk\ell})$; fortunately, in dimension 3 it can be simplified considerably.

***Exercise 12.40.** (a) Let $\bar{A} \in SO(3)$, and write $\bar{A}$ as

$$\bar{A} = \begin{bmatrix} \mathbf{a}_1 & \mathbf{a}_2 & \mathbf{a}_3 \end{bmatrix},$$

so that the vectors $(\mathbf{a}_1, \mathbf{a}_2, \mathbf{a}_3)$ in $\mathbb{R}^3$ represent the columns of $\bar{A}$. Show that for any cyclic permutation (i, j, k) of the indices $(1, 2, 3)$, we have

$$\mathbf{a}_i \times \mathbf{a}_j = \mathbf{a}_k.$$

Similarly, if $(\mathbf{a}^1, \mathbf{a}^2, \mathbf{a}^3)$ represent the rows of $\bar{A}$, then

$$\mathbf{a}^i \times \mathbf{a}^j = \mathbf{a}^k.$$

(b) Let $\bar{\Omega}^\dagger$, $\bar{\tilde{\Omega}}^\dagger$ denote the *vectors* of curvature 2-forms

$$\bar{\Omega}^\dagger = \begin{bmatrix} \bar{\Omega}^2_3 \\ \bar{\Omega}^3_1 \\ \bar{\Omega}^1_2 \end{bmatrix}, \qquad \bar{\tilde{\Omega}}^\dagger = \begin{bmatrix} \bar{\tilde{\Omega}}^2_3 \\ \bar{\tilde{\Omega}}^3_1 \\ \bar{\tilde{\Omega}}^1_2 \end{bmatrix}.$$

Use equation (12.30) and part (a) to show that

$$\tilde{\bar{\Omega}}^\dagger = \bar{A}^{-1}\bar{\Omega}^\dagger.$$

(Hint: Recall that for $\bar{A} \in SO(3)$, we have $\bar{A}^{-1} = {}^t\bar{A}$.)

(c) Let $\bar{\omega}^\dagger$, $\tilde{\bar{\omega}}^\dagger$ denote the vectors of 2-forms

$$\bar{\omega}^\dagger = \begin{bmatrix} \bar{\omega}^2 \wedge \bar{\omega}^3 \\ \bar{\omega}^3 \wedge \bar{\omega}^1 \\ \bar{\omega}^1 \wedge \bar{\omega}^2 \end{bmatrix}, \qquad \tilde{\bar{\omega}}^\dagger = \begin{bmatrix} \tilde{\bar{\omega}}^2 \wedge \tilde{\bar{\omega}}^3 \\ \tilde{\bar{\omega}}^3 \wedge \tilde{\bar{\omega}}^1 \\ \tilde{\bar{\omega}}^1 \wedge \tilde{\bar{\omega}}^2 \end{bmatrix}.$$

Use equation (12.11) and part (a) to show that

$$\tilde{\bar{\omega}}^\dagger = \bar{A}^{-1}\bar{\omega}^\dagger.$$

(d) For a 3-dimensional Riemannian manifold M, the Riemann curvature tensor has six nontrivial components, represented by the functions

$$(12.43) \qquad \begin{gathered} R^2_{323},\ R^3_{131},\ R^1_{212}, \\ R^3_{112} = R^1_{231},\ R^1_{223} = R^2_{312},\ R^2_{331} = R^3_{123}. \end{gathered}$$

Let $\bar{R}$, $\tilde{\bar{R}}$ denote the matrices

$$\bar{R} = \begin{bmatrix} \bar{R}^2_{323} & \bar{R}^2_{331} & \bar{R}^2_{321} \\ \bar{R}^3_{123} & \bar{R}^3_{131} & \bar{R}^3_{112} \\ \bar{R}^1_{223} & \bar{R}^1_{231} & \bar{R}^1_{212} \end{bmatrix}, \qquad \tilde{\bar{R}} = \begin{bmatrix} \tilde{\bar{R}}^2_{323} & \tilde{\bar{R}}^2_{331} & \tilde{\bar{R}}^2_{321} \\ \tilde{\bar{R}}^3_{123} & \tilde{\bar{R}}^3_{131} & \tilde{\bar{R}}^3_{112} \\ \tilde{\bar{R}}^1_{223} & \tilde{\bar{R}}^1_{231} & \tilde{\bar{R}}^1_{212} \end{bmatrix}.$$

The symmetries in equation (12.43) imply that these matrices are symmetric, and by definition, we have

$$(12.44) \qquad \bar{\Omega}^\dagger = \bar{R}\,\bar{\omega}^\dagger, \qquad \tilde{\bar{\Omega}}^\dagger = \tilde{\bar{R}}\,\tilde{\bar{\omega}}^\dagger.$$

Use parts (b), (c), and equation (12.44) to show that

$$(12.45) \qquad \tilde{\bar{R}} = \bar{A}^{-1}\bar{R}\bar{A}.$$

Next, we will see how the curvature tensor can impose constraints on the existence of totally geodesic surfaces.

***Exercise 12.41.** Suppose that $\Sigma = \mathbf{x}(U) \subset M$ is a totally geodesic surface in M. Let $(\mathbf{e}_1(\mathbf{u}), \mathbf{e}_2(\mathbf{u}), \mathbf{e}_3(\mathbf{u}))$ be an adapted frame field along Σ, with associated dual and connection forms $(\bar{\omega}^i, \bar{\omega}^i_j)$ on U.

(a) Show that the condition (12.42) is equivalent to the condition that

$$(12.46) \qquad \bar{\omega}^3_1 = \bar{\omega}^3_2 = 0.$$

(b) Differentiate equations (12.46) and show that

$$\bar{R}^3_{112}\,\bar{\omega}^1 \wedge \bar{\omega}^2 = \bar{R}^2_{312}\,\bar{\omega}^1 \wedge \bar{\omega}^2 = 0, \tag{12.47}$$

where $\bar{R}^3_{112}$, $\bar{R}^2_{312}$ denote the pullbacks to U of the corresponding components of the Riemann curvature tensor of M on $\mathcal{F}(M)$ via the frame field $(\mathbf{e}_1(\mathbf{u}), \mathbf{e}_2(\mathbf{u}), \mathbf{e}_3(\mathbf{u}))$. Since we must have $\bar{\omega}^1 \wedge \bar{\omega}^2 \neq 0$, conclude that

$$\bar{R}^3_{112} = \bar{R}^2_{312} = 0. \tag{12.48}$$

(c) The transformation rule (12.45) indicates that R may be regarded as a well-defined, self-adjoint linear transformation on the tangent space $T_{\mathbf{x}}M$ and $\bar{R}$ is its matrix expression with respect to the basis $(\mathbf{e}_1(\mathbf{u}), \mathbf{e}_2(\mathbf{u}), \mathbf{e}_3(\mathbf{u}))$. Recall that any symmetric 3×3 matrix can be diagonalized by a transformation of the form (12.45) with $\bar{A} \in SO(3)$. Conclude from part (b) that, given an orthonormal frame $(\mathbf{e}_1, \mathbf{e}_2, \mathbf{e}_3)$ at a point $\mathbf{x} \in M$, the plane $P \subset T_{\mathbf{x}}M$ spanned by $(\mathbf{e}_1, \mathbf{e}_2)$ cannot be tangent to a totally geodesic surface in M unless $\mathbf{e}_3$ is an eigenvector for R.

Finally, we will consider an example to show that the necessary condition developed in Exercise 12.41(c) is not sufficient to guarantee the existence of a totally geodesic surface in M passing through the point $\mathbf{x} \in M$ and tangent to the plane $P \subset T_{\mathbf{x}}M$; indeed, a given Riemannian 3-manifold M may contain very few totally geodesic surfaces, or even none at all.

***Exercise 12.42** (MAPLE recommended). Let $a > b > c > 0$, and let $M = \mathbb{R}^3$, with metric

$$g = \frac{1}{(ax^2 + by^2 + cz^2 + 1)^2}\left(dx^2 + dy^2 + dz^2\right).$$

(a) Show that the frame field

$$\begin{aligned} \underline{\mathbf{e}}_1(x, y, z) &= (ax^2 + by^2 + cz^2 + 1)\frac{\partial}{\partial x}, \\ \underline{\mathbf{e}}_2(x, y, z) &= (ax^2 + by^2 + cz^2 + 1)\frac{\partial}{\partial y}, \\ \underline{\mathbf{e}}_3(x, y, z) &= (ax^2 + by^2 + cz^2 + 1)\frac{\partial}{\partial z} \end{aligned} \tag{12.49}$$

on M is an orthonormal frame field for g and that its dual forms are

$$\begin{aligned} \bar{\omega}^1 &= \frac{dx}{(ax^2 + by^2 + cz^2 + 1)}, \\ \bar{\omega}^2 &= \frac{dy}{(ax^2 + by^2 + cz^2 + 1)}, \\ \bar{\omega}^3 &= \frac{dz}{(ax^2 + by^2 + cz^2 + 1)}. \end{aligned} \tag{12.50}$$

(b) Show that the Levi-Civita connection forms associated to the dual forms (12.50) are

$$\begin{aligned}\bar{\omega}_3^2 &= \frac{2}{(ax^2+by^2+cz^2+1)}(by\,dz - cz\,dy),\\ \bar{\omega}_1^3 &= \frac{2}{(ax^2+by^2+cz^2+1)}(cz\,dx - ax\,dz),\\ \bar{\omega}_2^1 &= \frac{2}{(ax^2+by^2+cz^2+1)}(ax\,dy - by\,dx).\end{aligned} \tag{12.51}$$

(c) Show that the curvature matrix $\bar{R}$ has the form

$$\bar{R} = \begin{bmatrix} \bar{R}^2_{323} & 0 & 0 \\ 0 & \bar{R}^3_{131} & 0 \\ 0 & 0 & \bar{R}^1_{212} \end{bmatrix},$$

where

$$\begin{aligned}\bar{R}^2_{323} &= 2\left((b+c-2a)ax^2 + (b-c)(cz^2-by^2) + (b+c)\right),\\ \bar{R}^3_{131} &= 2\left((c+a-2b)by^2 + (c-a)(ax^2-cz^2) + (c+a)\right),\\ \bar{R}^1_{212} &= 2\left((a+b-2c)cz^2 + (a-b)(by^2-ax^2) + (a+b)\right).\end{aligned} \tag{12.52}$$

Verify that the eigenvalues $(\bar{R}^2_{323}, \bar{R}^3_{131}, \bar{R}^1_{212})$ of $\bar{R}$ are distinct for all $\mathbf{x} = (x,y,z) \in M$, and conclude that the corresponding eigenvectors of $\bar{R}$ in each tangent space $T_{\mathbf{x}}M$ are precisely the coordinate directions $(\underline{\mathbf{e}}_1, \underline{\mathbf{e}}_2, \underline{\mathbf{e}}_3)$.

(d) Conclude from the result of Exercise 12.41(c) that the only potential tangent planes $P \subset T_{\mathbf{x}}M$ to totally geodesic surfaces in M are the coordinate planes

$$P_1 = \operatorname{span}(\underline{\mathbf{e}}_2, \underline{\mathbf{e}}_3), \qquad P_2 = \operatorname{span}(\underline{\mathbf{e}}_3, \underline{\mathbf{e}}_1), \qquad P_3 = \operatorname{span}(\underline{\mathbf{e}}_1, \underline{\mathbf{e}}_2).$$

It follows that if there exists a totally geodesic surface $\Sigma \subset M$, it must be a level surface for one of the coordinate functions (x,y,z) on M.

(e) Let $U \subset \mathbb{R}^2$, and consider a candidate surface $\Sigma \subset M$ with parametrization $\mathbf{x} : U \to M$ given by

$$\mathbf{x}(u,v) = (u, v, z_0)$$

for some $z_0 \in \mathbb{R}$. The frame field (12.49) is an adapted frame field along Σ. Compute the pullbacks of the connection forms (12.51) to U; in particular, show that

$$\bar{\omega}_1^3 = \frac{2cz_0\,du}{(au^2+bv^2+cz_0^2+1)}, \qquad \bar{\omega}_2^3 = \frac{2cz_0\,dv}{(au^2+bv^2+cz_0^2+1)}.$$

Conclude from the result of Exercise 12.41(a) that Σ is a totally geodesic surface if and only if $z_0 = 0$.

A similar argument shows that the only totally geodesic surfaces in M are contained in the coordinate "planes"

$$\Sigma_1 = \{(0, y, z) \mid y, z \in \mathbb{R}\} \subset M,$$
$$\Sigma_2 = \{(x, 0, z) \mid x, z \in \mathbb{R}\} \subset M,$$
$$\Sigma_3 = \{(x, y, 0) \mid x, y \in \mathbb{R}\} \subset M.$$

Any open submanifold $M_0 \subset M$ that does not intersect these surfaces—for instance, the first octant $M_0 = \{(x, y, z) \in M \mid x, y, z > 0\}$—contains no totally geodesic surfaces whatsoever.

12.7. Maple computations

Exercise 12.30: It's not so simple to tell Maple how to do computations on a manifold of arbitrary dimension, so we'll go through this exercise for the case $n = m = 3$, which represents a 3-dimensional Riemannian manifold M embedded in $\mathbb{E}^6$. (This is the embedding dimension for which the isometric embedding problem is a determined system of PDEs, rather than an overdetermined or underdetermined system, when $n = 3$.)

After loading the `Cartan` and `LinearAlgebra` packages into Maple, we first need to declare the Maurer-Cartan forms on $\mathcal{F}(\mathbb{E}^6)$. This is rather a lot of forms, and it's helpful to use the `seq` command to generate the necessary lists of forms:

```
> Form(seq(omega[i], i=1..6));
  Form(seq(seq(omega[i,j], j=1..6), i=1..6));
```

Tell Maple about the symmetries in the connection forms; for convenience, do this in such a way that the resulting basis includes forms indexed as $(\bar{\omega}_i^a)$ with $1 \leq i \leq 3$, $4 \leq a \leq 6$ rather than the other way around:

```
> for i from 1 to 6 do
    omega[i,i]:= 0;
  end do;

  omega[2,1]:= -omega[1,2];
  omega[3,2]:= -omega[2,3];
  omega[1,3]:= -omega[3,1];

  omega[5,4]:= -omega[4,5];
  omega[6,5]:= -omega[5,6];
  omega[4,6]:= -omega[6,4];

  for a from 4 to 6 do
    for j from 1 to 3 do
```

```
      omega[j,a]:= -omega[a,j];
    end do;
  end do;
```

Tell MAPLE how to differentiate these forms according to the Cartan structure equations (3.8), and be sure to define structure equations only for those connection forms that have not been defined in terms of other forms:

```
> for i from 1 to 6 do
    d(omega[i]):= sum('-omega[i,j] &^ omega[j]', 'j'=1..6);
  end do;

  d(omega[1,2]):= sum('-omega[1,k] &^ omega[k,2]', 'k'=1..6);
  d(omega[2,3]):= sum('-omega[2,k] &^ omega[k,3]', 'k'=1..6);
  d(omega[3,1]):= sum('-omega[3,k] &^ omega[k,1]', 'k'=1..6);

  d(omega[4,5]):= sum('-omega[4,k] &^ omega[k,5]', 'k'=1..6);
  d(omega[5,6]):= sum('-omega[5,k] &^ omega[k,6]', 'k'=1..6);
  d(omega[6,4]):= sum('-omega[6,k] &^ omega[k,4]', 'k'=1..6);

  for a from 4 to 6 do
    for j from 1 to 3 do
      d(omega[a,j]):= sum('-omega[a,k] &^ omega[k,j]',
      'k'=1..6);
    end do;
  end do;
```

Now, consider an adapted orthonormal frame field along M for which the frame vectors $(\mathbf{e}_1(\mathbf{u}), \mathbf{e}_2(\mathbf{u}), \mathbf{e}_3(\mathbf{u}))$ are tangent to M at each point. According to equation (12.37), for such a frame field, we must have $\bar{\omega}^4 = \bar{\omega}^5 = \bar{\omega}^6 = 0$. Set up a substitution for the Maurer-Cartan forms associated to an adapted frame field:

```
> adaptedsub:= [omega[4]=0, omega[5]=0, omega[6]=0];
```

Now differentiate these equations:

```
> Simf(subs(adaptedsub, Simf(d(omega[4]))));
  Simf(subs(adaptedsub, Simf(d(omega[5]))));
  Simf(subs(adaptedsub, Simf(d(omega[6]))));
```

Applying Cartan's lemma tells us that equations (12.38) hold, and we can add these equations to our substitution as follows:

```
> for a from 4 to 6 do
    h[a,2,1]:= h[a,1,2];
    h[a,3,2]:= h[a,2,3];
```

```
    h[a,1,3]:= h[a,3,1];
  end do;
> unassign('i', 'j', 'a');
> adaptedsub:= [op(adaptedsub),
  seq(seq(omega[a,i] = sum(h[a,i,j]*omega[j],
  j=1..3), i=1..3), a=4..6)];
```

(The `unassign` command is needed because one of the previous `for` loops assigned values to these indices that need to be removed before using them in a `seq` command.)

Now look at the structure equations for $d\bar{\omega}^1, d\bar{\omega}^2, d\bar{\omega}^3$:

```
> Simf(subs(adaptedsub, Simf(d(omega[1]))));
```

$$(\omega_2)\ \&\hat{}\ (\omega_{1,2}) - (\omega_3)\ \&\hat{}\ (\omega_{3,1})$$

```
> Simf(subs(adaptedsub, Simf(d(omega[2]))));
```

$$-(\omega_1)\ \&\hat{}\ (\omega_{1,2}) + (\omega_3)\ \&\hat{}\ (\omega_{2,3})$$

```
> Simf(subs(adaptedsub, Simf(d(omega[3]))));
```

$$(\omega_1)\ \&\hat{}\ (\omega_{3,1}) - (\omega_2)\ \&\hat{}\ (\omega_{2,3})$$

It follows from these equations that the 1-forms $(\bar{\omega}^i_j)$ with $1 \leq i, j \leq 3$ are the Levi-Civita connection forms associated to the metric

$$\mathrm{I} = \left(\bar{\omega}^1\right)^2 + \left(\bar{\omega}^2\right)^2 + \left(\bar{\omega}^3\right)^2$$

on M. Therefore, they satisfy the structure equations (12.35), where $(R^i_{jk\ell})$ are the components of the Riemann curvature tensor of M. Now compare equations (12.35) to the structure equations for the $(d\bar{\omega}^i_j)$ as Maurer-Cartan forms on $\mathbb{E}^6$:

```
> zero1:= Simf(subs(adaptedsub, Simf(d(omega[1,2])
  + omega[1,3] &^ omega[3,2] - R[1,2,1,2]*omega[1] &^ omega[2]
  - R[1,2,2,3]*omega[2] &^ omega[3]
  - R[1,2,3,1]*omega[3] &^ omega[1])));
> zero2:= Simf(subs(adaptedsub, Simf(d(omega[2,3])
  + omega[2,1] &^ omega[1,3] - R[2,3,1,2]*omega[1] &^ omega[2]
  - R[2,3,2,3]*omega[2] &^ omega[3]
  - R[2,3,3,1]*omega[3] &^ omega[1])));
> zero3:= Simf(subs(adaptedsub, Simf(d(omega[3,1])
  + omega[3,2] &^ omega[2,1] - R[3,1,1,2]*omega[1] &^ omega[2]
  - R[3,1,2,3]*omega[2] &^ omega[3]
  - R[3,1,3,1]*omega[3] &^ omega[1])));
```

The scalar coefficients in these quantities are equivalent to the Gauss equations (12.39).

Next, look at the structure equations for the connection forms $(\bar{\omega}^a_i)$ with $1 \le i \le 3$, $4 \le a \le 6$:

```
> zero41:= Simf(subs(adaptedsub, Simf(d(omega[4,1]))
  - d(Simf(subs(adaptedsub, omega[4,1])))));
```

The resulting expression has the form

$$\phi_1 \wedge \bar{\omega}^1 + \phi_2 \wedge \bar{\omega}^2 + \phi_3 \wedge \bar{\omega}^3,$$

where ϕ_1, ϕ_2, ϕ_3 are the following 1-forms:

```
> pick(zero41, omega[1]);
```

$$-2\, h_{4,1,2}\, \omega_{1,2} + 2\, h_{4,3,1}\, \omega_{3,1} - h_{5,1,1}\, \omega_{4,5} + h_{6,1,1}\, \omega_{6,4} - d(h_{4,1,1})$$

```
> pick(zero41, omega[2]);
```

$$-\,(h_{4,2,2} - h_{4,1,1})\, \omega_{1,2} - h_{4,3,1}\, \omega_{2,3} + h_{4,2,3}\, \omega_{3,1} \\ - h_{5,1,2}\, \omega_{4,5} + h_{6,1,2}\, \omega_{6,4} - d(h_{4,1,2})$$

```
> pick(zero41, omega[3]);
```

$$-\, h_{4,2,3}\, \omega_{1,2} + h_{4,1,2}\, \omega_{2,3} - (-h_{4,3,3} + h_{4,1,1})\, \omega_{3,1} \\ - h_{5,3,1}\, \omega_{4,5} + h_{6,3,1}\, \omega_{6,4} - d(h_{4,3,1})$$

Applying Cartan's lemma to all 9 structure equations for the $(d\bar{\omega}^a_i)$ shows that there exist functions (h^a_{ijk}), symmetric in their lower indices, such that these equations take the form

$$dh^a_{ij} = \text{(terms involving the functions } (h^b_{k\ell}) \\ \text{and the connection forms } (\bar{\omega}^i_j, \bar{\omega}^a_b)) + h^a_{ijk}\bar{\omega}^k.$$

These are the Codazzi equations; there are 18 of them, one for each of the components (h^a_{ij}) of the second fundamental form.

Remark 12.43. From the PDE perspective, the symmetry of the (h^a_{ijk}) is equivalent to a system of PDEs of the form

$$\frac{\partial h^a_{ij}}{\partial x^k} - \frac{\partial h^a_{ik}}{\partial x^j} = \text{(lower order terms)}.$$

From this point of view, there are 27 Codazzi equations, 24 of which are independent.

Finally, look at the structure equations for the connection forms $(\bar{\omega}^a_b)$ with $4 \le a, b \le 6$:

```
> Simf(subs(adaptedsub, Simf(d(omega[4,5]))));
> Simf(subs(adaptedsub, Simf(d(omega[5,6]))));
> Simf(subs(adaptedsub, Simf(d(omega[6,4]))));
```

These all have the form

$$d\bar{\omega}^a_b = -\bar{\omega}^a_c \wedge \bar{\omega}^c_b + S^a_{bij}\, \bar{\omega}^i \wedge \bar{\omega}^j,$$

where the functions (S^a_{bij}) are quadratic expressions in the functions (h^a_{ij}), similar to those found in the Gauss equations. These are the Ricci equations; the coefficients (S^a_{bij}) represent the components of the curvature tensor for the connection on the normal bundle of M induced from the Euclidean connection on $\mathbb{E}^6$.

EXERCISE 12.42: Since this exercise requires different dual and connection forms with different structure equations from the previous exercise, it's probably best to restart MAPLE and reload the `Cartan` and `LinearAlgebra` packages.

Declare the dual and connection forms on $\mathcal{F}(M)$, and tell Maple about their symmetries:

```
> Form(omega[1], omega[2], omega[3]);
  Form(omega[1,2], omega[3,1], omega[3,2]);
> omega[1,1]:= 0;
  omega[2,2]:= 0;
  omega[3,3]:= 0;
  omega[2,1]:= -omega[1,2];
  omega[1,3]:= -omega[3,1];
  omega[2,3]:= -omega[3,2];
```

Declare constants a, b, c:

```
> Form(a=-1, b=-1, c=-1);
```

Set up a substitution for the dual forms of the given orthonormal frame field on M, as well as the reverse substitution:

```
> dualformssub:= [
  omega[1]  = d(x)/(a*x^2 + b*y^2 + c*z^2 + 1),
  omega[2]  = d(y)/(a*x^2 + b*y^2 + c*z^2 + 1),
  omega[3]  = d(z)/(a*x^2 + b*y^2 + c*z^2 + 1)];
> dualformsbacksub:= makebacksub(dualformssub);
```

In order to compute the connection forms, differentiate the dual forms and compare the result to the structure equations. (It's helpful to go ahead and write the $(\bar{\omega}_j^i)$ as linear combinations of the $(\bar{\omega}^i)$ with undetermined coefficients and then solve for these coefficients.)

```
> connectionformssub:= [
  omega[1,2] = h[1,2,1]*omega[1] + h[1,2,2]*omega[2]
  + h[1,2,3]*omega[3],
  omega[3,1] = h[3,1,1]*omega[1] + h[3,1,2]*omega[2]
  + h[3,1,3]*omega[3],
  omega[3,2] = h[3,2,1]*omega[1] + h[3,2,2]*omega[2]
  + h[3,2,3]*omega[3]];
```

According to the structure equations, the following expressions should be zero:

```
> zero1:= Simf(d(Simf(subs(dualformssub, omega[1])))
  + Simf(subs(dualformssub, Simf(subs(connectionformssub,
  omega[1,2] &^ omega[2] + omega[1,3] &^ omega[3])))));
  zero2:= Simf(d(Simf(subs(dualformssub, omega[2])))
  + Simf(subs(dualformssub, Simf(subs(connectionformssub,
  omega[2,1] &^ omega[1] + omega[2,3] &^ omega[3])))));
  zero3:= Simf(d(Simf(subs(dualformssub, omega[3])))
  + Simf(subs(dualformssub, Simf(subs(connectionformssub,
  omega[3,1] &^ omega[1] + omega[3,2] &^ omega[2])))));
```

$$zero1 := \frac{(2cz - h_{3,1,1})\,(d(x))\,\&\hat{}\,(d(z))}{(ax^2+by^2+cz^2+1)^2} - \frac{(2by + h_{1,2,1})\,(d(y))\,\&\hat{}\,(d(x))}{(ax^2+by^2+cz^2+1)^2} - \frac{(h_{1,2,3} + h_{3,1,2})\,(d(y))\,\&\hat{}\,(d(z))}{(ax^2+by^2+cz^2+1)^2}$$

$$zero2 := \frac{(h_{1,2,3} - h_{3,2,1})\,(d(x))\,\&\hat{}\,(d(z))}{(ax^2+by^2+cz^2+1)^2} + \frac{(2ax - h_{1,2,2})\,(d(y))\,\&\hat{}\,(d(x))}{(ax^2+by^2+cz^2+1)^2} + \frac{(2cz - h_{3,2,2})\,(d(y))\,\&\hat{}\,(d(z))}{(ax^2+by^2+cz^2+1)^2}$$

$$zero3 := -\frac{(2ax + h_{3,1,3})\,(d(x))\,\&\hat{}\,(d(z))}{(ax^2+by^2+cz^2+1)^2} + \frac{(h_{3,1,2} - h_{3,2,1})\,(d(y))\,\&\hat{}\,(d(x))}{(ax^2+by^2+cz^2+1)^2} - \frac{(2by + h_{3,2,3})\,(d(y))\,\&\hat{}\,(d(z))}{(ax^2+by^2+cz^2+1)^2}$$

We can set the scalar coefficients of these forms equal to zero and solve for the (h_{ijk}) to determine the connection forms:

```
> solve({op(ScalarForm(zero1)), op(ScalarForm(zero2)),
  op(ScalarForm(zero3))}, {h[1,2,1], h[1,2,2], h[1,2,3],
  h[3,1,1], h[3,1,2], h[3,1,3], h[3,2,1], h[3,2,2],
  h[3,2,3]});
```

$$\{h_{1,2,1} = -2by, h_{1,2,2} = 2ax, h_{1,2,3} = 0, h_{3,1,1} = 2cz, h_{3,1,2} = 0, h_{3,1,3} = -2ax, h_{3,2,1} = 0, h_{3,2,2} = 2cz, h_{3,2,3} = -2by\}$$

```
> assign(%);
```

So here are the connection forms, expressed in terms of the coordinate 1-forms:

```
> connectionformssub:= Simf(subs(dualformssub,
  Simf(connectionformssub)));
```

$$connectionformssub := \Bigg[$$

$$\omega_{1,2} = -\frac{2by\,d(x)}{ax^2+by^2+cz^2+1} + \frac{2ay\,d(y)}{ax^2+by^2+cz^2+1},$$

$$\omega_{3,1} = \frac{2cz\,d(x)}{ax^2+by^2+cz^2+1} - \frac{2ax\,d(z)}{ax^2+by^2+cz^2+1},$$

$$\omega_{3,2} = \frac{2cz\,d(y)}{ax^2+by^2+cz^2+1} - \frac{2by\,d(z)}{ax^2+by^2+cz^2+1}\Bigg]$$

Now compute the curvature 2-forms, expressed in terms of the dual forms:

```
> Omega[2,3]:= Simf(subs(dualformsbacksub,
  Simf(d(Simf(subs(connectionformssub, omega[2,3])))
  + Simf(subs(connectionformssub,
  omega[2,1] &^ omega[1,3])))));
  Omega[3,1]:= Simf(subs(dualformsbacksub,
  Simf(d(Simf(subs(connectionformssub, omega[3,1])))
  + Simf(subs(connectionformssub,
  omega[3,2] &^ omega[2,1])))));
  Omega[1,2]:= Simf(subs(dualformsbacksub,
  Simf(d(Simf(subs(connectionformssub, omega[1,2])))
  + Simf(subs(connectionformssub,
```

```
omega[1,3] &^ omega[3,2])))));
```

$$\Omega_{2,3} := \left(2cax^2 + 2cby^2 - 2c^2z^2 + 2c + 2bax^2 - 2b^2y^2 + 2bcz^2 + 2b - 4a^2x^2\right)(\omega_2) \ \&\hat{}\ (\omega_3)$$

$$\Omega_{3,1} := \left(-2cax^2 - 2cby^2 + 2c^2z^2 - 2c + 2a^2x^2 - 2aby^2 - 2acz^2 - 2a + 4b^2y^2\right)(\omega_1) \ \&\hat{}\ (\omega_3)$$

$$\Omega_{1,2} := \left(2bax^2 - 2b^2y^2 + 2bcz^2 + 2b - 2a^2x^2 + 2aby^2 + 2acz^2 + 2a - 4c^2z^2\right)(\omega_1) \ \&\hat{}\ (\omega_2)$$

So the matrix $\bar{R}$ is diagonal, with the following diagonal entries:

```
> R[2,3,2,3]:= pick(Omega[2,3], omega[2], omega[3]);
  R[3,1,3,1]:= pick(Omega[3,1], omega[3], omega[1]);
  R[1,2,1,2]:= pick(Omega[1,2], omega[1], omega[2]);
```

Check that the diagonal entries of $\bar{R}$ are distinct:

```
> factor(R[2,3,2,3] - R[3,1,3,1]);
```

$$-2\,(-b+a)\,(ax^2+by^2+cz^2+1)$$

```
> factor(R[3,1,3,1] - R[1,2,1,2]);
```

$$-2\,(-c+b)\,(ax^2+by^2+cz^2+1)$$

```
> factor(R[1,2,1,2] - R[2,3,2,3]);
```

$$2\,(a-c)\,(ax^2+by^2+cz^2+1)$$

It follows that any totally geodesic surface must be a level surface for one of the coordinate functions (x, y, z). Suppose that Σ is contained in the level set $\{(x, y, z) \in M \mid z = z_0\}$ for some constant z_0 and compute the pullbacks of the connection forms to Σ:

```
> Form(z0=-1);
> Simf(subs([x=u, y=v, z=z0], connectionformssub));
```

$$\left[\omega_{1,2} = -\frac{2bv\,d(u)}{au^2+bv^2+cz0^2+1} + \frac{2au\,d(v)}{au^2+bv^2+cz0^2+1},\right.$$

$$\left.\omega_{3,1} = \frac{2cz0\,d(u)}{au^2+bv^2+cz0^2+1},\ \omega_{3,2} = \frac{2cz0\,d(v)}{au^2+bv^2+cz0^2+1}\right]$$

Since a totally geodesic surface must have $\bar{\omega}_1^3 = \bar{\omega}_2^3 = 0$, Σ is not a totally geodesic surface unless $z_0 = 0$.

Bibliography

[AEG06] Juan A. Aledo, José M. Espinar, and José A. Gálvez, *Timelike surfaces in the Lorentz-Minkowski space with prescribed Gaussian curvature and Gauss map*, J. Geom. Phys. **56** (2006), no. 8, 1357–1369.

[AI79] Robert L. Anderson and Nail H. Ibragimov, *Lie-Bäcklund Transformations in Applications*, SIAM Studies in Applied Mathematics, vol. 1, Society for Industrial and Applied Mathematics (SIAM), Philadelphia, PA, 1979.

[AR93] M. A. Akivis and B. A. Rosenfeld, *Élie Cartan (1869–1951)*, Translations of Mathematical Monographs, vol. 123, American Mathematical Society, Providence, RI, 1993, translated from the Russian manuscript by V. V. Goldberg.

[B8̈3] A. V. Bäcklund, *Om ytor med konstant negativ krökning*, Lunds Universitets Årsskrift **19** (1883), 1–48.

[BCG+91] R. L. Bryant, S. S. Chern, R. B. Gardner, H. L. Goldschmidt, and P. A. Griffiths, *Exterior Differential Systems*, Mathematical Sciences Research Institute Publications, vol. 18, Springer-Verlag, New York, 1991.

[BG80] Richard L. Bishop and Samuel I. Goldberg, *Tensor Analysis on Manifolds*, Dover Publications Inc., New York, 1980, corrected reprint of the 1968 original.

[Bia79] Luigi Bianchi, *Ricerche sulle superficie elicoidali e sulle superficie a curvatura costante*, Ann. Scuola Norm. Sup. Pisa Cl. Sci. **2** (1879), 285–341.

[Bla23] Wilhelm Blaschke, *Vorlesungen über Differentialgeometrie, vol. II*, Springer, Berlin, 1923.

[Bla85] ———, *Gesammelte Werke. Band 4*, Thales-Verlag, Essen, 1985, Affine Differentialgeometrie. Differentialgeometrie der Kreis- und Kugelgruppen. [Affine differential geometry. Differential geometry of circle and ball groups], with commentaries by Werner Burau and Udo Simon, edited by Burau, S. S. Chern, K. Leichtweiß, H. R. Müller, L. A. Santaló, Simon and K. Strubecker.

[Cal00] James J. Callahan, *The Geometry of Spacetime: An Introduction to Special and General Relativity*, Undergraduate Texts in Mathematics, Springer-Verlag, New York, 2000.

[Car04] Élie Cartan, *Sur la structure des groupes infinis de transformation*, Ann. Sci. École Norm. Sup. (3) **21** (1904), 153–206.

[Car20] ———, *Sur la déformation projective des surfaces*, Ann. Sci. École Norm. Sup. (3) **37** (1920), 259–356.

[Car27] ———, *Sur la possibilité de plonger un espace riemannien donné dans un espace euclidéen*, Ann. Soc. Polon. Math. **6** (1927), 1–7.

[Car30] ———, *La théorie des groupes finis et continus et l'analysis situs*, Mémorial des sciences mathématiques, no. 42, Gauthier-Villars et Cie, 1930.

[Car35] ———, *La Méthode du Repère Mobile, la Théorie des Groupes Continus, et les Espaces Généralisés*, Exposés de Géométrie, no. 5, Hermann, Paris, 1935.

[Car46] ———, *Leçons sur la Géométrie des Espaces de Riemann*, 2nd ed., Gauthier-Villars, Paris, 1946.

[Car92] ———, *Leçons sur la Géométrie Projective Complexe. La Théorie des Groupes Finis et Continus et la Géométrie Différentielle Traitées Par la Méthode du Repère Mobile. Leçons sur la Théorie des Espaces à Connexion Projective*, Les Grands Classiques Gauthier-Villars. [Gauthier-Villars Great Classics], Éditions Jacques Gabay, Sceaux, 1992, reprint of the 1931, 1937, and 1937 editions.

[CEM+14] Jeanne Clelland, Edward Estrada, Molly May, Jonah Miller, and Sean Peneyra, *A Tale of Two Arc Lengths: Metric Notions for Curves in Surfaces in Equiaffine Space*, Proc. Amer. Math. Soc. **142** (2014), 2543–2558.

[Che42] Shiing-shen Chern, *On integral geometry in Klein spaces*, Ann. of Math. (2) **43** (1942), 178–189.

[Cle99] Jeanne N. Clelland, *Lie groups and the method of moving frames*, 1999, lecture videos from Mathematical Sciences Research Institute Summer Graduate Workshop, http://www.msri.org/publications/video/index2.html.

[Cle12] ———, *Totally quasi-umbilic timelike surfaces in* $\mathbb{R}^{1,2}$, Asian J. Math. **16** (2012), 189–208.

[CT80] Shiing Shen Chern and Chuu Lian Terng, *An analogue of Bäcklund's theorem in affine geometry*, Rocky Mountain J. Math. **10** (1980), no. 1, 105–124.

[CT86] Shiing-shen Chern and Keti Tenenblat, *Pseudo-spherical surfaces and evolution equations*, Stud. Appl. Math. **74** (1986), 55–83.

[Dac08] Bernard Dacorogna, *Introduction to the Calculus of Variations*, 2nd ed., Imperial College Press, London, 2008.

[Dar72a] Gaston Darboux, *Leçons sur la Théorie Générale des Surfaces et les Applications Géométriques du Calcul Infinitésimal. Deuxième partie*, Chelsea Publishing Co., Bronx, N.Y., 1972, Les Congruences et les Équations Linéaires aux Dérivées Partielles. Les Lignes Tracées sur les Surfaces, réimpression de la deuxième édition de 1915.

[Dar72b] ———, *Leçons sur la Théorie Générale des Surfaces et les Applications Géométriques du Calcul Infinitésimal. Première partie*, Chelsea Publishing Co., Bronx, N.Y., 1972, Généralités. Coordonnées curvilignes. Surfaces minima, réimpression de la deuxième édition de 1914.

[Dar72c] ———, *Leçons sur la Théorie Générale des Surfaces et les Applications Géométriques du Calcul Infinitésimal. Quatrième partie*, Chelsea Publishing Co., Bronx, N.Y.,

1972, Déformation Infiniment Petite et Representation Sphérique, réimpression de la première édition de 1896.

[Dar72d] ______, *Leçons sur la Théorie Générale des Surfaces et les Applications Géométriques du Calcul Infinitésimal. Troisième partie*, Chelsea Publishing Co., Bronx, N.Y., 1972, Lignes Géodésiques et Courbure Géodésique. Paramètres Différentiels. Déformation des Surfaces, réimpression de la première édition de 1894.

[dC76] Manfredo P. do Carmo, *Differential Geometry of Curves and Surfaces*, Prentice-Hall Inc., Englewood Cliffs, N.J., 1976, translated from the Portuguese.

[dC92] ______, *Riemannian Geometry*, Mathematics: Theory & Applications, Birkhäuser Boston Inc., Boston, MA, 1992, translated from the second Portuguese edition by Francis Flaherty.

[DFN92] B. A. Dubrovin, A. T. Fomenko, and S. P. Novikov, *Modern Geometry—Methods and Applications. Part I: The Geometry of Surfaces, Transformation Groups, and Fields*, 2nd ed., Graduate Texts in Mathematics, vol. 93, Springer-Verlag, New York, 1992, translated from the Russian by Robert G. Burns.

[Dou31] Jesse Douglas, *Solution of the problem of Plateau*, Trans. Amer. Math. Soc. **33** (1931), no. 1, 263–321.

[Eis60] Luther Pfahler Eisenhart, *A Treatise on the Differential Geometry of Curves and Surfaces*, Dover Publications Inc., New York, 1960.

[Fav57] J. Favard, *Cours de géométrie différentielle locale*, Gauthier-Villars, Paris, 1957.

[FO98] Mark Fels and Peter J. Olver, *Moving coframes. I. A practical algorithm*, Acta Appl. Math. **51** (1998), no. 2, 161–213.

[FO99] ______, *Moving coframes. II. Regularization and theoretical foundations*, Acta Appl. Math. **55** (1999), no. 2, 127–208.

[Fom90] A. T. Fomenko, *The Plateau Problem. Part I*, Studies in the Development of Modern Mathematics, vol. 1, Gordon and Breach Science Publishers, New York, 1990, Historical survey, translated from the Russian.

[Fre47] J. F. Frenet, *Sur les Courbes à Double Courbure*, Ph.D. thesis, Toulouse, 1847.

[Gál09] José A. Gálvez, *Surfaces of constant curvature in 3-dimensional space forms*, Mat. Contemp. **37** (2009), 1–42.

[Gar89] Robert B. Gardner, *The Method of Equivalence and Its Applications*, CBMS-NSF Regional Conference Series in Applied Mathematics, vol. 58, Society for Industrial and Applied Mathematics (SIAM), Philadelphia, PA, 1989.

[Gre78] Mark L. Green, *The moving frame, differential invariants and rigidity theorems for curves in homogeneous spaces*, Duke Math. J. **45** (1978), no. 4, 735–779.

[Gri74] P. Griffiths, *On Cartan's method of Lie groups and moving frames as applied to uniqueness and existence questions in differential geometry*, Duke Math. J. **41** (1974), 775–814.

[Gug77] Heinrich W. Guggenheimer, *Differential Geometry*, Dover Publications, Inc., New York, 1977, corrected reprint of the 1963 edition, Dover Books on Advanced Mathematics.

[HCV52] D. Hilbert and S. Cohn-Vossen, *Geometry and the Imagination*, Chelsea Publishing Company, New York, N.Y., 1952, translated by P. Neményi.

[HH06] Qing Han and Jia-Xing Hong, *Isometric Embedding of Riemannian Manifolds in Euclidean Spaces*, Mathematical Surveys and Monographs, vol. 130, American Mathematical Society, Providence, RI, 2006.

[IL03] Thomas A. Ivey and J. M. Landsberg, *Cartan for Beginners: Differential Geometry via Moving Frames and Exterior Differential Systems*, Graduate Studies in Mathematics, vol. 61, American Mathematical Society, Providence, RI, 2003.

[Jan26] M. Janet, *Sur la possibilité de plonger un espace Riemannien donné dans un espace euclidien*, Ann. Soc. Polon. Math. **5** (1926), 38–43.

[Jen77] Gary R. Jensen, *Higher Order Contact of Submanifolds of Homogeneous Spaces*, Springer-Verlag, Berlin, 1977, Lecture Notes in Mathematics, Vol. 610.

[Kle93a] Felix Klein, *A comparative review of recent researches in geometry*, Bull. Amer. Math. Soc. **2** (1893), no. 10, 215–249.

[Kle93b] ———, *Vergleichende Betrachtungen über neuere geometrische Forschungen*, Math. Ann. **43** (1893), no. 1, 63–100.

[Kob95] Shoshichi Kobayashi, *Transformation Groups in Differential Geometry*, Classics in Mathematics, Springer-Verlag, Berlin, 1995, reprint of the 1972 edition.

[Kog03] Irina Kogan, *Two algorithms for a moving frame construction*, Canad. J. Math. **55** (2003), 266–291.

[Lee13] John M. Lee, *Introduction to Smooth Manifolds*, 2nd ed., Graduate Texts in Mathematics, vol. 218, Springer, New York, 2013.

[Man10] Elizabeth Louise Mansfield, *A Practical Guide to the Invariant Calculus*, Cambridge Monographs on Applied and Computational Mathematics, vol. 26, Cambridge University Press, Cambridge, 2010.

[Mil83] Tilla Klotz Milnor, *Harmonic maps and classical surface theory in Minkowski 3-space*, Trans. Amer. Math. Soc. **280** (1983), no. 1, 161–185.

[Min78] Hermann Minkowski, *Die Grundgleichungen für die elektromagnetischen Vorgänge in bewegten Körpern*, Nachrichten von der Gesellschaft der Wissenschaften zu Göttingen, Mathematisch-Physikalische Klasse (1907/8), 53–111.

[Min89] ———, *Raum und Zeit*, Jahresbericht der Deutschen Mathematiker-Vereinigung (1908/9), 75–88.

[MR05] Sebastián Montiel and Antonio Ros, *Curves and surfaces*, Graduate Studies in Mathematics, vol. 69, American Mathematical Society, Providence, RI; Real Sociedad Matemática Española, Madrid, 2005, translated and updated from the 1998 Spanish edition by the authors.

[Nas56] John Nash, *The imbedding problem for Riemannian manifolds*, Ann. of Math. (2) **63** (1956), 20–63.

[NMS08] Mehdi Nadjafikhah and Ali Mahdipour Sh., *Affine classification of n-curves*, Balkan J. Geom. Appl. **13** (2008), no. 2, 66–73.

[NS94] Katsumi Nomizu and Takeshi Sasaki, *Affine Differential Geometry*, Geometry of affine immersions, Cambridge Tracts in Mathematics, vol. 111, Cambridge University Press, Cambridge, 1994.

[Olv00] Peter Olver, *Applications of Lie Groups to Differential Equations*, 2nd ed., Graduate Texts in Mathematics, Springer-Verlag, New York, 2000.

[Olv10] ———, *Recent advances in the theory and application of Lie pseudo-groups*, XVIII International Fall Workshop on Geometry and Physics, AIP Conf. Proc., vol. 1260, Amer. Inst. Phys., Melville, NY, 2010, pp. 35–63.

[Olv11a] ———, *Lectures on Moving Frames*, Symmetries and Integrability of Difference Equations (Decio Levi, Peter Olver, Zora Thomova, and Pavel Winternitz, eds.), London Mathematical Society Lecture Note Series, vol. 381, Cambridge University Press, Cambridge, 2011, Lectures from the Summer School (Séminaire de Máthematiques Supérieures) held at the Université de Montréal, Montréal, QC, June 8–21, 2008.

[Olv11b] ———, *Recursive moving frames*, Results Math. **60** (2011), 423–452.

[O'N83] Barrett O'Neill, *Semi-Riemannian Geometry*, with applications to relativity, Pure and Applied Mathematics, vol. 103, Academic Press Inc. [Harcourt Brace Jovanovich Publishers], New York, 1983.

[O'N06] ———, *Elementary Differential Geometry*, 2nd ed., Elsevier/Academic Press, Amsterdam, 2006.

[Opr07] John Oprea, *Differential Geometry and Its Applications*, 2nd ed., Classroom Resource Materials Series, Mathematical Association of America, Washington, DC, 2007.

[OT05] V. Ovsienko and S. Tabachnikov, *Projective Differential Geometry Old and New*, Cambridge Tracts in Mathematics, vol. 165, Cambridge University Press, Cambridge, 2005.

[Rad30] Tibor Radó, *On Plateau's problem*, Ann. of Math. (2) **31** (1930), no. 3, 457–469.

[Rey98] E. G. Reyes, *Pseudo-spherical surfaces and integrability of evolution equations*, J. Diff. Eq. **147** (1998), 195–230.

[RS82] C. Rogers and W. F. Shadwick, *Bäcklund Transformations and Their Applications*, Mathematics in Science and Engineering, vol. 161, Academic Press Inc. [Harcourt Brace Jovanovich Publishers], New York, 1982.

[RS02] C. Rogers and W. K. Schief, *Bäcklund and Darboux Transformations*, Geometry and modern applications in soliton theory, Cambridge Texts in Applied Mathematics, Cambridge University Press, Cambridge, 2002.

[Sen31] C. E. Senff, *Theoremata Princioalia e Theoria Curvarum et Su Perficierum*, Dorpat Univ., 1831.

[Ser51] J. Serret, *Mémoire sur quelques formules relatives à la théorie des courbes à double courbure*, J. Math. Pures Appl. **16** (1851), 193–207.

[She99] Mary D. Shepherd, *Line congruences as surfaces in the space of lines*, Differential Geom. Appl. **10** (1999), no. 1, 1–26.

[Spi79] Michael Spivak, *A Comprehensive Introduction to Differential Geometry. 5 Vols.*, 2nd ed., Publish or Perish Inc., Wilmington, DE, 1979.

[Ste07] James Stewart, *Calculus: Early Transcendentals*, 6th ed., Stewart's Calculus Series, Brooks Cole, Belmont, CA, 2007.

[Su83] Bu Chin Su, *Affine Differential Geometry*, Science Press, Beijing, 1983.

[Tsu96] Kazumi Tsukada, *Totally geodesic submanifolds of Riemannian manifolds and curvature-invariant subspaces*, Kodai Math. J. **19** (1996), no. 3, 395–437.

[Wei66] K. Weierstrass, *Über die Flächen deren mittlere Krümmung überall gleich null ist.*, Monatsber. Berliner Akad. (1866), 612–625, 855–856.

[Wil61] T. J. Willmore, *The definition of Lie derivative*, Proc. Edinburgh Math. Soc. (2) **12** (1960/1961), 27–29.

[Wil62] E. J. Wilczynski, *Projective Differential Geometry of Curves and Ruled Surfaces*, Chelsea Publishing Co., New York, 1962.

Index

0-form, 42
1-form, 35
  on $\mathbb{R}^n$, 40–41
  on a manifold, 47

$\mathbb{A}^n$, *see* Equi-affine space
Adapted frame field, 109
  on a surface in $\mathbb{E}^3$, 118
  on a surface in $\mathbb{A}^3$, 178
  on a surface in $\mathbb{P}^3$, 221
  on a timelike surface in $\mathbb{M}^{1,2}$, 150
  equi-affine principal adapted frame field on an elliptic surface in $\mathbb{A}^3$, 186
  null adapted frame field
    on a hyperbolic surface in $\mathbb{A}^3$, 190
    on a hyperbolic surface in $\mathbb{P}^3$, 232
    on a timelike surface in $\mathbb{M}^{1,2}$, 162
  principal adapted frame field
    on a surface in $\mathbb{E}^3$, 123
    on a timelike surface in $\mathbb{M}^{1,2}$, 155
Affine connection, *see* Connection
Affine geometry, 92
Affine Grassmannian, 288, 324
Affine transformation, 93
Arc length, *see* Curve, arc length
Area functional
  on surfaces in $\mathbb{E}^3$, 252
  equi-affine, on surfaces in $\mathbb{A}^3$, 269
Area measure, 324
Associated family of a minimal surface in $\mathbb{E}^3$, 267

Bäcklund transformation
  for Liouville's equation, 302–303
  for pseudospherical surfaces, 290
  for the sine-Gordon equation, 288, 298
Bäcklund's theorem, 287, 290
Bäcklund, Albert, 290
Baker-Jarvis, Duff, xvi
Bartels, Martin, xi
Bianchi, Luigi, 290
Blaschke representation for an elliptic equi-affine minimal surface in $\mathbb{A}^3$, 278
Blaschke, Wilhelm, 178, 274
Bonnet's theorem
  for a surface in $\mathbb{E}^3$
    existence, 127
    uniqueness, 124
  for a surface in $\mathbb{S}^3$ or $\mathbb{H}^3$, 354
  for a timelike surface in $\mathbb{M}^{1,2}$, 155
Bryant, Robert, xv
Bushek, Nathaniel, xv

Canonical isomorphism
  for dual spaces, 36
  for tangent spaces, 16, 50, 76, 339, 367
Carlsen, Brian, xvi
`Cartan` package for MAPLE, xiii, 59–66
  `&^` command, 60
  `d` command, 60
  `Forder` command, 60
  `Form` command, 59

`makebacksub` command, 63
`pick` command, 62
`ScalarForm` command, 63
`Simf` command, 61
`WedgeProduct` command, 60
Cartan structure equations, *see* Structure equations
Cartan's formula for exterior derivative, 48
Cartan's formula for Lie derivative, 58
Cartan's lemma, 55
Cartan, Élie, xi, 70, 223, 233, 383
Cartan-Janet isometric embedding theorem, 383
Catenoid, 128, 260
associated family, 268
conjugate surface, 268
Weierstrass-Enneper representation, 268
Cauchy-Crofton formula, 324, 327
Cauchy-Riemann equations, 264
Chain rule, 24
Chern, Shiing-Shen, 297
Clelland, Richard, xvi
Codazzi equations
for a surface in $\mathbb{E}^3$, 127
for a surface in $\mathbb{S}^3$, 353, 354
for a surface in $\mathbb{H}^3$, 353, 354
for a timelike surface in $\mathbb{M}^{1,2}$, 156, 165
for a submanifold of $\mathbb{E}^{n+m}$, 379
Column vector, *see* Vector, column vector
Commutative diagram, 19
Compatibility equations
for a surface in $\mathbb{E}^3$, 127
for a surface in $\mathbb{S}^3$ or $\mathbb{H}^3$, 353
for a timelike surface in $\mathbb{M}^{1,2}$, 156, 165
for an elliptic surface in $\mathbb{A}^3$, 188
for an elliptic surface in $\mathbb{P}^3$, 229, 247
for a hyperbolic surface in $\mathbb{P}^3$, 235
for a submanifold of $\mathbb{E}^{n+m}$, 379
Complex analytic function, *see* Holomorphic function
Complex structure, 264
Conformal parametrization of a surface, 265
Conformal structure
on a hyperbolic surface in $\mathbb{P}^3$, 232
on an elliptic surface in $\mathbb{P}^3$, 224
Conic section, 177, 212
Conjugate surface of a minimal surface in $\mathbb{E}^3$, 267
Connection, 33
compatibility with a metric, 371
curvature tensor, 376
flat connection on $\mathbb{E}^n$, 366
Levi-Civita, *see* Levi-Civita connection
on a vector bundle, 365
on the tangent bundle, 365–370
horizontal tangent space, 367
vertical tangent space, 366
symmetric, 371
torsion-free, 371
Connection forms
on the orthonormal frame bundle of $\mathbb{E}^n$, 79
on the orthonormal frame bundle of $\mathbb{M}^{1,n}$, 91
on the unimodular frame bundle of $\mathbb{A}^n$, 95
on the projective frame bundle of $\mathbb{P}^n$, 103
for the Levi-Civita connection on $\mathbb{S}^n$ or $\mathbb{H}^n$, 347
determined by a connection, 367, 370
Constant type, 316
Cotangent bundle, 36
Cotangent space, 36
Covariant derivative, 33
for vector fields on $\mathbb{S}^n$ and $\mathbb{H}^n$, 346–347
compatibility with the metric, 347
for vector fields on a submanifold of $\mathbb{E}^{n+m}$, 378
Covector, 37
Covector space, 36
Cruz, Akaxia, xvi
Curvature, *see also* Curve, curvature; Gauss curvature; mean curvature
curvature matrix of a connection matrix, 340
curvature matrix of the connection matrix on $\mathcal{F}(\mathbb{S}^n)$, 342
curvature matrix of the connection matrix on $\mathcal{F}(\mathbb{H}^n)$, 344
curvature tensor of a connection, 376
Curve
in $\mathbb{E}^3$
arc length, 112

binormal vector, 112
complete set of invariants, 115
curvature, 113
Frenet equations, 114
Frenet frame, 112
nondegenerate curve, 112
orthonormal frame field, 111
regular curve, 111
torsion, 113
unit normal vector, 112
unit tangent vector, 111
in $\mathbb{M}^{1,2}$, null curve, 165
in $\mathbb{M}^{1,2}$, timelike curve
Frenet equations, 148
Minkowski curvature, 147
Minkowski torsion, 148
nondegenerate curve, 146
orthonormal frame field, 144
proper time, 144
regular curve, 144
unit normal vector, 146
unit tangent vector, 144
in $\mathbb{A}^2$, 176–178
conic section, 177
equi-affine curvature, 177
in $\mathbb{A}^3$
equi-affine arc length, 174–175
equi-affine curvatures, 176
equi-affine Frenet equations, 176
equi-affine Frenet frame, 175
nondegenerate curve, 172
rational normal curve, 178
unimodular frame field, 172
in $\mathbb{P}^2$
canonical lifting, 205
canonical projective frame field, 205
conic section, 212
nondegenerate curve, 205
projective arc length, 211
projective curvature form, 210
projective frame field, 204
projective Frenet equations, 212
projective parameter, 207
projective parametrization, 207
projective structure, 210
Wilczynski invariants, 206
in $\mathbb{P}^3$
canonical lifting, 215
canonical projective frame field, 215
nondegenerate curve, 215
projective curvature forms, 218
projective frame field, 214
projective Frenet equations, 219
projective parameter, 217
projective parametrization, 217
projective structure, 217
rational normal curve, 220
Wilczynski invariants, 216
in $\mathbb{S}^3$
binormal vector, 350
curvature, 350
Frenet equations, 350
Frenet frame, 350
geodesic, 349
geodesic equation, 349
nondegenerate curve, 349
orthonormal frame field, 348
regular curve, 348
torsion, 350
unit normal vector, 350
in $\mathbb{H}^3$
binormal vector, 350
curvature, 350
Frenet equations, 350
Frenet frame, 350
geodesic, 349
geodesic equation, 349
nondegenerate curve, 349
orthonormal frame field, 348
regular curve, 348
torsion, 350
unit normal vector, 350
in a Riemannian 3-manifold
curvature, 381
Frenet equations, 381
Frenet frame, 381
geodesic, 380
nondegenerate curve, 380
orthonormal frame field, 379
regular curve, 379
torsion, 381

Darboux tangents, 227
Darboux, Jean-Gaston, xi
De Sitter spacetime, 157–158
Derivative
directional, 19, 43, 57, 120, 347, 365
of a map from $\mathbb{R}^m$ to $\mathbb{R}^n$, 16
of a map between manifolds, 23
Diffeomorphism, 25
Differentiable manifold, *see* Manifold

Differential
- of a real-valued function, 35
- of a map from $\mathbb{R}^m$ to $\mathbb{R}^n$, 16
- of a map between manifolds, 24, 49

Differential form
- 0-form, 42
- 1-form, 35
  - on $\mathbb{R}^n$, 40–41
  - on a manifold, 47
- p-form
  - on $\mathbb{R}^n$, 42
  - on a manifold, 47
- algebra of differential forms on $\mathbb{R}^n$, 41, 42
- closed form, 46
- exact form, 46

`DifferentialGeometry` package for MAPLE, xiii

Directional derivative, *see* Derivative, directional

Divergence theorem, 55

Doubly ruled surface, *see* Ruled surface, doubly ruled surface

Dual forms
- on the orthonormal frame bundle of $\mathbb{E}^n$, 79
- on the projective frame bundle of $\mathbb{P}^n$, 103
- associated to an orthonormal frame field, 369

Dual space, 35–36

Dunne, Edward, xv

$\mathbb{E}^n$, *see* Euclidean space

Einstein summation convention, 14–15

Einstein, Albert, 85

Elliptic paraboloid, 272
- Blaschke representation, 280

Elliptic space, 340–342, *see also* Homogeneous space, elliptic space $\mathbb{S}^n$

Elliptic surface
- in $\mathbb{A}^3$, 180–189
- in $\mathbb{P}^3$, 223–232

Embedding, 25

Enneper's surface, 268

Enneper, Alfred, 261

Equi-affine arc length, *see* Curve in $\mathbb{A}^3$, equi-affine arc length

Equi-affine first fundamental form, *see* Surface in $\mathbb{A}^3$, equi-affine first fundamental form

Equi-affine geometry, 92

Equi-affine group $A(n)$, 94
- as a principal bundle over $\mathbb{A}^n$, 95

Equi-affine mean curvature, *see* Surface in $\mathbb{A}^3$, equi-affine mean curvature

Equi-affine minimal surface, *see* Minimal surface, equi-affine, in $\mathbb{A}^3$

Equi-affine normal vector field, *see* Surface in $\mathbb{A}^3$, equi-affine normal vector field

Equi-affine second fundamental form, *see* Surface in $\mathbb{A}^3$, equi-affine second fundamental form

Equi-affine space, 93, *see also* Homogeneous space, equi-affine space $\mathbb{A}^n$
- volume form, 92

Equi-affine sphere
- improper equi-affine sphere, 189
- proper equi-affine sphere, 189

Equi-affine transformation, 93

Equivalence problem, 107

Equivariant, 109

Equivariant moving frame, *see* Moving frame, equivariant moving frame

Estrada, Edward, xvi

Euclidean group $E(n)$, 73
- as a principal bundle over $\mathbb{E}^n$, 75

Euclidean space, 70, *see also* Homogeneous space, Euclidean space $\mathbb{E}^n$

Exterior derivative
- of a real-valued function, 35
- of a p-form on $\mathbb{R}^n$, 43–46
- of a p-form on a manifold, 48–49
- chain rule, 44
- Leibniz rule, 43, 44

Extrinsic curvature of a surface in $\mathbb{S}^3$ or $\mathbb{H}^3$, 353

Fels, Mark, xi

First fundamental form
- of a surface in $\mathbb{E}^3$, 118–120
- of a surface in $\mathbb{S}^3$ or $\mathbb{H}^3$, 352
- of a surface in a Riemannian 3-manifold, 382
- of a timelike surface in $\mathbb{M}^{1,2}$, 150, 163
- equi-affine, of an elliptic surface in $\mathbb{A}^3$, 181
- equi-affine, of a hyperbolic surface in $\mathbb{A}^3$, 190

projective, of an elliptic surface in $\mathbb{P}^3$, 227
Flat connection on $\mathbb{E}^n$, 366
Flat homogeneous space, 339
Flat surface
in $\mathbb{E}^3$, 132–134
in $\mathbb{S}^3$, 355–356
flat torus, 356
in $\mathbb{H}^3$, 356–357
flat cylinder, 356
Frenet, Jean, xi
Frobenius theorem, 46
Fubini-Pick form
of a hyperbolic surface in $\mathbb{A}^3$, 191
of an elliptic surface in $\mathbb{A}^3$, 185
of an elliptic surface in $\mathbb{P}^3$, 226
Fundamental Theorem of Calculus, 54
Fundamental Theorem of Space Curves, 69
existence, 117
uniqueness, 114

$GL(n)$, 28, 29
$\mathfrak{gl}(n)$, 30
Gauge, 368
Gauge field, 368
Gauge transformation, 368
Gauss curvature
of a surface in $\mathbb{E}^3$, 131
of a surface in $\mathbb{S}^3$ or $\mathbb{H}^3$, 353
of a timelike surface in $\mathbb{M}^{1,2}$, 153, 163
Gauss equation
for a surface in $\mathbb{E}^3$, 127
for a surface in $\mathbb{S}^3$, 353, 354
for a surface in $\mathbb{H}^3$, 353, 354
for a timelike surface in $\mathbb{M}^{1,2}$, 156, 165
for a submanifold of $\mathbb{E}^{n+m}$, 379
Gauss map
of a surface in $\mathbb{E}^3$, 121
of a surface in $\mathbb{S}^3$ or $\mathbb{H}^3$, 352
of a surface in a Riemannian 3-manifold, 382
of a timelike surface in $\mathbb{M}^{1,2}$, 151
Gauss, Carl Friedrich, 131
Theorema Egregium, 131
Gelfand, Sergei, xv
General linear group, *see* $GL(n)$
General relativity, 143
Geodesic
in $\mathbb{S}^3$ or $\mathbb{H}^3$, 349
in a Riemannian 3-manifold, 380
Geodesic equation
for curves in $\mathbb{S}^3$ or $\mathbb{H}^3$, 349
for curves in a Riemannian 3-manifold, 380
Geodesic spray, 380–381
Grassmannian, affine, 288, 324
Great hyperboloid in $\mathbb{H}^3$, 351, 355
Great sphere in $\mathbb{S}^3$, 351, 355
Green's theorem, 55
Guggenheimer, Heinrich, xi

$\mathbb{H}^n$, *see* Hyperbolic space
Harmonic function, 264
Helicoid, 261, 268
Helm, Rachel, xvi
Hilbert's theorem, 301–302
Holomorphic function, 263
Homogeneous space, 70, 84, 361
flat homogeneous space, 339
Euclidean space $\mathbb{E}^n$, 70–75
Minkowski space $\mathbb{M}^{1,n}$, 85–92
equi-affine space $\mathbb{A}^n$, 92–96
projective space $\mathbb{P}^n$, 96–103
elliptic space $\mathbb{S}^n$, 340–342
hyperbolic space $\mathbb{H}^n$, 340, 342–344
Horizontal tangent space, 367
Horizontal vector field, 380
Hyperbolic paraboloid, 311, 319
Hyperbolic plane, 301
Hyperbolic space, 340, 342–344, *see also* Homogeneous space, hyperbolic space $\mathbb{H}^n$
Hyperbolic surface
in $\mathbb{A}^3$, 180, 189–191
in $\mathbb{P}^3$, 223, 232–235
Hyperboloid of one sheet, 311

Immersion, 25
Incidence, of a point and a line, 327
Indices
lower index, 9
upper index, 9
in partial derivative operators, 13
Inner product
Euclidean, 70
Minkowski, 86
Integrable system, 288
soliton solution, 288
Interior product, 57
Intrinsic curvature of a surface in $\mathbb{S}^3$ or $\mathbb{H}^3$, 353

Intrinsic invariant, *see* Invariant, intrinsic invariant for surfaces in $\mathbb{E}^3$
Invariant, 107
  for curves in $\mathbb{E}^3$, 69
  for submanifolds of a homogeneous space, 109
  complete set of invariants, 107
    for curves in $\mathbb{E}^3$, 115
  intrinsic invariant for surfaces in $\mathbb{E}^3$, 131
  relative invariant, 226, 315
Isometric embedding, 378–379, 383
  Cartan-Janet theorem, 383
Isotropy group
  of a point in $\mathbb{E}^n$, 73
  of a point in $\mathbb{M}^{1,n}$, 90
  of a point in $\mathbb{A}^n$, 94
  of a point in $\mathbb{P}^n$, 101
  of a point in $\mathbb{S}^n$, 340
  of a point in $\mathbb{H}^n$, 343

Janet, Maurice, 383
Jensen, Andrew, xvi
Joeris, Peter, xvi

Karpel, Joshua, xvi
Kaufman, Bryan, xv
Klein, Felix, 69

Lagrange, Joseph-Louis, 251
Laplace's equation, 356
Left-hook, 57
Levi-Civita connection, 33, 370–372
  on $\mathbb{E}^n$, 366
  on $\mathbb{S}^n$ or $\mathbb{H}^n$, 347
    connection forms, 347
  Riemann curvature tensor, 376–378
Lie algebra, 26–32
Lie bracket, 26
  of vector fields, 27
  on a Lie algebra, 28–29
Lie derivative, 56–59, 258
  Cartan's formula, 58
Lie group, 26–32
  left translation map, 26
  left-invariant vector field, 26–27
  right translation map, 26
Lifting, 109
Light cone, *see* Minkowski space, light cone
Lightlike vector, *see* Minkowski space, lightlike vector
Line congruence, 288–289
  focal surface, 289
  normal congruence, 289
  pseudospherical congruence, 289–290
  surface of reference, 289
Linear fractional transformation, 99
Liouville's equation, 302, 320
  Bäcklund transformation, 302–303
Local coordinates
  on a surface, 4, 5
  on a manifold, 6
Local trivialization
  of a vector bundle, 32
  of a tangent bundle, 364
  of an orthonormal frame bundle, 369
Lorentz group, 89
  proper, orthochronous, 89
Lorentz transformation, 89
  orthochronous, 89
  proper, 89

$\mathbb{M}^{1,n}$, *see* Minkowski space
Mahoney, Michael, xvi
Manifold, 5
  local coordinates, 6
    transition map between, 6
  parametrization, 6
  Riemannian manifold, 362
MAPLE, xiii, 59–66, 103–106, 134–141, 166–169, 191–201, 235–247, 280–286, 303–309, 329–335, 357–360, 388–395
Mapping
  continuous, 15
  differentiable
    from $\mathbb{R}^m$ to $\mathbb{R}^n$, 15
    between manifolds, 18
Mathematical Sciences Research Institute, xvi
Maurer-Cartan equation, *see also* Structure equations
  on a Lie group, 85
  on the Euclidean group $E(n)$, 82
  on the elliptic symmetry group $SO(n+1)$, 342
  on the hyperbolic symmetry group $SO^+(1,n)$, 344
Maurer-Cartan form
  on a Lie group, 85
  on the Euclidean group $E(n)$, 81–82
  on the Poincaré group $M(1,n)$, 91
  on the equi-affine group $A(n)$, 95

on the projective symmetry group $SL(n+1)$, 102
on the elliptic symmetry group $SO(n+1)$, 341
on the hyperbolic symmetry group $SO^+(1,n)$, 344
May, Molly, xvi
Mean curvature
of a surface in $\mathbb{E}^3$, 131
of a surface in $\mathbb{S}^3$ or $\mathbb{H}^3$, 353
of a timelike surface in $\mathbb{M}^{1,2}$, 153, 163
equi-affine, of an elliptic surface in $\mathbb{A}^3$, 185
Measure, 324
area measure, 324
Meromorphic function, 266
Method of moving frames, *see* Moving frame, method of moving frames
Metric, 13–14
Metric structure on a curve in $\mathbb{E}^n$, 209
Miller, Jonah, xvi
Minimal surface
in $\mathbb{E}^3$, 132, 251–268
associated family, 267
catenoid, 128, 260, 268
conjugate surface, 267
Enneper's surface, 268
helicoid, 261, 268
Weierstrass-Enneper representation, 266–267
equi-affine, in $\mathbb{A}^3$, 268–280
Blaschke representation, 278
elliptic paraboloid, 272, 280
Minkowski cross product, 146
Minkowski norm, 88
Minkowski space, 86, *see also* Homogeneous space, Minkowski space $\mathbb{M}^{1,n}$
future-pointing vector, 87
light cone, 87
lightlike vector, 87
Minkowski norm of a vector, 88
null cone, 87
null vector, 87
past-pointing vector, 87
spacelike vector, 87
timelike vector, 87
world line of a particle, 88
Minkowski, Hermann, 85
Moving frame
equivariant moving frame, xi
method of moving frames, 70, 107, 111

Nash embedding theorem, 378
National Science Foundation, xvi
Nondegenerate curve, *see* Curve, nondegenerate
Null adapted frame field
on a timelike surface in $\mathbb{M}^{1,2}$, 162
on a hyperbolic surface in $\mathbb{A}^3$, 190
on a hyperbolic surface in $\mathbb{P}^3$, 232
Null cone, *see* Minkowski space, null cone
Null coordinates on a timelike surface in $\mathbb{M}^{1,2}$, 165
Null curve in $\mathbb{M}^{1,2}$, 165
Null vector, *see* Minkowski space, null vector

$O(1,n)$, 89
$O(n)$, 31
$\mathfrak{o}(n)$, 31
Olver, Peter, xi
Orthogonal group, *see* $O(n)$
Orthonormal basis
for $\mathbb{E}^n$, 72
for $\mathbb{M}^{1,n}$, 87
Orthonormal frame
on $\mathbb{E}^n$, 75
on $\mathbb{M}^{1,n}$, 91
on $\mathbb{S}^n$, 341, 345
on $\mathbb{H}^n$, 343, 345
on a Riemannian manifold, 363
Orthonormal frame bundle
of $\mathbb{E}^n$, 75
of $\mathbb{M}^{1,n}$, 91
of $\mathbb{S}^2$, 34
of $\mathbb{S}^n$, 341, 345
of $\mathbb{H}^n$, 343, 345
of a Riemannian manifold, 363
local trivialization, 369
Orthonormal frame field
on $\mathbb{E}^n$, 83
along a curve in $\mathbb{E}^3$, 111
along a curve in $\mathbb{S}^3$ or $\mathbb{H}^3$, 348
along a curve in a Riemannian 3-manifold, 379
along a timelike curve in $\mathbb{M}^{1,2}$, 144

p-form
on $\mathbb{R}^n$, 42–43
on a manifold, 47

$PGL(m)$, 98
$\mathbb{P}^n$, *see* Projective space
$PSL(m)$, 98
Paraboloid
  elliptic paraboloid, 272
    Blaschke representation, 280
  hyperbolic paraboloid, 311, 319
Parametrization
  of a surface, 4, 5
  of a manifold, 6
  asymptotic, 295
  conformal, 265
  principal, 128, 156, 187
Partial derivative operator
  as a tangent vector, 20
  indices in, 13
Peneyra, Sean, xvi
Pick invariant of an elliptic surface in $\mathbb{A}^3$, 186
Plateau problem, 251
Plateau, Joseph, 251
Poincaré group $M(1, n)$, 90
  as a principal bundle over $\mathbb{M}^{1,n}$, 91
Poincaré lemma, 46
Poincaré-Hopf theorem, 33, 34
Principal adapted frame field
  on a surface in $\mathbb{E}^3$, 123
  on a timelike surface in $\mathbb{M}^{1,2}$, 155
  equi-affine, on an elliptic surface in $\mathbb{A}^3$, 186
Principal bundle, 33–34, 362
  base space, 33
  base-point projection map, 33
  fiber, 33
  section, 33
  total space, 33
  local trivialization, 369
Principal curvatures
  of a surface in $\mathbb{E}^3$, 123
  of a surface in $\mathbb{S}^3$ or $\mathbb{H}^3$, 352
  of a timelike surface in $\mathbb{M}^{1,2}$, 155
  surface in $\mathbb{E}^3$ with constant principal curvatures, 130–131
Principal vectors
  on a surface in $\mathbb{E}^3$, 123
  on a surface in $\mathbb{S}^3$ or $\mathbb{H}^3$, 352
  on a timelike surface in $\mathbb{M}^{1,2}$, 155
Projective arc length, *see* Curve in $\mathbb{P}^2/\mathbb{P}^3$, projective arc length
Projective curvature form, *see* Curve in $\mathbb{P}^2/\mathbb{P}^3$, projective curvature form
Projective first fundamental form, *see* Surface in $\mathbb{P}^3$, projective first fundamental form
Projective frame bundle of $\mathbb{P}^n$, 102
Projective frame field
  along a curve in $\mathbb{P}^2$, 204
    canonical projective frame field, 205
  along a curve in $\mathbb{P}^3$, 214
    canonical projective frame field, 215
Projective frame on $\mathbb{P}^n$, 101
Projective general linear group, 98
Projective parametrization, *see* Curve in $\mathbb{P}^2/\mathbb{P}^3$, projective parametrization
Projective space, 7–9, 96, *see also* Homogeneous space, projective space $\mathbb{P}^n$
  affine coordinates, 97
  homogeneous coordinates, 8
Projective special linear group, 98
Projective sphere, 229–232
Projective structure
  on a curve in $\mathbb{P}^2$, 210
  on a curve in $\mathbb{P}^3$, 217
  on a curve in $\mathbb{P}^n$, 203
Projective transformation, 96, 97
  Schwarzian derivative, 208
Proper time, *see* Curve in $\mathbb{M}^{1,2}$, proper time
Pseudosphere, 287
Pseudospherical line congruence, 289–290
Pseudospherical surface, 287
  1-soliton pseudospherical surface, 301
  asymptotic coordinates, 295
  asymptotic parametrization, 295
Pullback
  for differential forms, 50–53
  for bundles, 108
Push-forward, 50

Quasi-umbilic point on a timelike surface in $\mathbb{M}^{1,2}$, 160

Rational normal curve
  in $\mathbb{A}^3$, 178
  in $\mathbb{P}^3$, 220
Regular curve, *see* Curve, regular
Regular surface, *see* Surface
Relative invariant, 226, 315

Relativity
special relativity, 85, 143
general relativity, 143
Reyes, Enrique, 297
Ricci equations for a submanifold of $\mathbb{E}^{n+m}$, 379
Riemann curvature tensor, 376–378
first Bianchi identity, 377
on a Riemannian 3-manifold, 385
Riemannian manifold, 362
Row vector, *see* Vector, row vector
Ruled surface, 311
doubly ruled surface, 311
0-adapted frame field, 314
1-adapted frame field, 316
2-adapted frame field, 317
classification theorem, 313
hyperbolic paraboloid, 311, 319
hyperboloid of one sheet, 311

$SL(n)$, 30–31
$\mathfrak{sl}(n)$, 30
$SL(n+1)$
as a principal bundle over $\mathbb{P}^n$, 102
as the symmetry group of $\mathbb{P}^n$, 98
S_n, 30
$\mathbb{S}^n$, *see* Elliptic space; Unit sphere
$SO^+(1,n)$, 89
as a principal bundle over $\mathbb{H}^n$, 344
as the symmetry group of $\mathbb{H}^n$, 342
$\mathfrak{so}(1,n)$, 90
$SO(n)$, 31
$SO(n+1)$
as a principal bundle over $\mathbb{S}^n$, 341
as the symmetry group of $\mathbb{S}^n$, 340
Schmidt, Michael, xvi
Schwarzian derivative, 208–209
of a projective transformation, 208
Second fundamental form
of a surface in $\mathbb{E}^3$, 121–122
of a surface in $\mathbb{S}^3$ or $\mathbb{H}^3$, 352
of a surface in a Riemannian 3-manifold, 382
of a timelike surface in $\mathbb{M}^{1,2}$, 151, 163
equi-affine, of an elliptic surface in $\mathbb{A}^3$, 184
equi-affine, of a hyperbolic surface in $\mathbb{A}^3$, 190
of a submanifold of $\mathbb{E}^{n+m}$, 378
Self-adjoint linear operator, 152
Semi-basic forms
on the orthonormal frame bundle of $\mathbb{E}^n$, 79
on the projective frame bundle of $\mathbb{P}^n$, 103
Serret, Joseph, xi
Simple connectivity, 116
Sine-Gordon equation, 288
1-soliton solution, 300
Bäcklund transformation, 288, 298
in characteristic/null coordinates, 296
in space-time coordinates, 296
Skew curvature of a timelike surface in $\mathbb{M}^{1,2}$, 154, 163
Smooth manifold, *see* Manifold
Soliton, 288
1-soliton pseudospherical surface, 301
1-soliton solution of the sine-Gordon equation, 300
Spacelike surface, *see* Surface in $\mathbb{M}^{1,2}$, spacelike surface
Spacelike vector, *see* Minkowski space, spacelike vector
Special affine geometry, *see* Equi-affine geometry
Special linear cross product, 277
Special linear group, *see* $SL(n)$
Special orthogonal group, *see* $SO(n)$
Special relativity, 85, 143
Stokes's theorem, 53–55
Divergence theorem, 55
Fundamental Theorem of Calculus, 54
Green's theorem, 55
Stokes's theorem, multivariable calculus version, 55
Structure equations
on the orthonormal frame bundle of $\mathbb{E}^n$, 80
on the orthonormal frame bundle of $\mathbb{M}^{1,n}$, 91
on the unimodular frame bundle of $\mathbb{A}^n$, 95
on the projective frame bundle of $\mathbb{P}^n$, 102
on the orthonormal frame bundle of $\mathbb{S}^n$, 341
on the orthonormal frame bundle of $\mathbb{H}^n$, 344
on the orthonormal frame bundle of a Riemannian manifold, 374, 377
Submersion, 25

Surface, 3, 5
  parametrization, 4, 5
  local coordinates, 4, 5
    transition map between, 5
  ruled surface, *see* Ruled surface
  doubly ruled surface, *see* Ruled surface, doubly ruled surface
  in $\mathbb{E}^3$
    adapted frame field, 118
    area functional, 252
    Bonnet's theorem, 127
    catenoid, 128, 260, 268
    Codazzi equations, 127
    compatibility equations, 127
    Enneper's surface, 268
    first fundamental form, 118–120
    flat surface, 132–134
    Gauss curvature, 131
    Gauss equation, 127
    Gauss map, 121
    helicoid, 261, 268
    mean curvature, 131
    minimal surface, 132, 251–268
    principal adapted frame field, 123
    principal curvatures, 123
    principal vectors, 123
    pseudosphere, 287
    pseudospherical surface, 287
    second fundamental form, 121–122
    shape operator, 121
    surface with constant principal curvatures, 130–131
    totally umbilic surface, 129
    umbilic point, 124
    variation, 252–255
  in $\mathbb{A}^3$
    0-adapted frame field, 178
  in $\mathbb{A}^3$, elliptic surface, 180–189
    1-adapted frame field, 180
    2-adapted frame field, 183
    compatibility equations, 188
    cubic form, 185
    elliptic paraboloid, 272, 280
    equi-affine area functional, 269
    equi-affine first fundamental form, 181
    equi-affine mean curvature, 185
    equi-affine normal vector field, 183
    equi-affine principal adapted frame field, 186
    equi-affine second fundamental form, 184
    Fubini-Pick form, 185
    improper equi-affine sphere, 189
    minimal surface, 268–280
    Pick invariant, 186
    proper equi-affine sphere, 189
    variation, 269
  in $\mathbb{A}^3$, hyperbolic surface, 180, 189–191
    1-adapted null frame field, 190
    2-adapted null frame field, 190
    equi-affine first fundamental form, 190
    equi-affine second fundamental form, 190
    Fubini-Pick form, 191
    hyperbolic paraboloid, 311, 319
    hyperboloid of one sheet, 311
  in $\mathbb{M}^{1,2}$, spacelike surface, 143
  in $\mathbb{M}^{1,2}$, timelike surface, 143
    adapted frame field, 150
    Codazzi equations, 156, 165
    compatibility equations, 156, 165
    de Sitter spacetime, 157–158
    first fundamental form, 150, 163
    Gauss curvature, 153, 163
    Gauss equation, 156, 165
    Gauss map, 151
    mean curvature, 153, 163
    null adapted frame field, 162
    null coordinates, 165
    principal adapted frame field, 155
    principal curvatures, 155
    principal vectors, 155
    quasi-umbilic point, 160
    second fundamental form, 151, 163
    skew curvature, 154, 163
    totally quasi-umbilic surface, 160, 165–166
    totally umbilic surface, 156–157
    umbilic point, 155
  in $\mathbb{P}^3$
    0-adapted frame field, 221
  in $\mathbb{P}^3$, elliptic surface, 223–232
    1-adapted frame field, 223
    2-adapted frame field, 225
    3-adapted frame field, 226
    4-adapted frame field, 228
    compatibility equations, 229, 247
    conformal structure, 224

cubic form, 226
Darboux tangents, 227
Fubini-Pick form, 226
projective first fundamental form, 227
projective sphere, 229–232
totally umbilic surface, 229–232
umbilic point, 226
in $\mathbb{P}^3$, hyperbolic surface, 223, 232–235
1-adapted null frame field, 232
2-adapted null frame field, 233
3-adapted null frame field, 234
4-adapted null frame field, 234
compatibility equations, 235
conformal structure, 232
in $\mathbb{S}^3$
Bonnet's theorem, 354
Codazzi equations, 353, 354
compatibility equations, 353
extrinsic curvature, 353
first fundamental form, 352
flat surface, 355–356
flat torus, 356
Gauss curvature, 353
Gauss equation, 353, 354
Gauss map, 352
great sphere, 351, 355
intrinsic curvature, 353
mean curvature, 353
principal curvatures, 352
principal vectors, 352
second fundamental form, 352
totally geodesic surface, 355
in $\mathbb{H}^3$
Bonnet's theorem, 354
Codazzi equations, 353, 354
compatibility equations, 353
extrinsic curvature, 353
first fundamental form, 352
flat cylinder, 356
flat surface, 356–357
Gauss curvature, 353
Gauss equation, 353, 354
Gauss map, 352
great hyperboloid, 351, 355
intrinsic curvature, 353
mean curvature, 353
principal curvatures, 352
principal vectors, 352
second fundamental form, 352
totally geodesic surface, 355
in a Riemannian 3-manifold
first fundamental form, 382
Gauss map, 382
second fundamental form, 382
totally geodesic surface, 383–388
Symmetric group, *see* S_n
Symmetric product
of vectors, 39
of 1-forms, 119
Symmetry group
of $\mathbb{E}^n$, 73
of $\mathbb{M}^{1,n}$, 90
of $\mathbb{A}^n$, 94
of $\mathbb{P}^n$, 98
of $\mathbb{S}^n$, 340
of $\mathbb{H}^n$, 342
of a homogeneous space G/H, 84
as a principal bundle over G/H, 85
as the set of frames on G/H, 85

Tangent bundle, 21–23
of a surface, 21–23
of a manifold, 21
base space, 22
total space, 22
fiber, 22
base-point projection map, 23
canonical parametrization, 22
transition map between, 22
local trivialization, 364
Tangent space, 16, 20
tangent plane, 21
Tangent vector, 16, 19
Tenenblat, Keti, 297
Tensor, 9–14
change of basis, 9–10, 12–13
components, 10, 12, 13
metric, 13
rank 1, 10
rank 2, 12
rank k, 38
skew-symmetric, 38–39
symmetric, 38–39
Tensor bundle, 39
Tensor field, 9, 13
rank k, 40
Tensor product, 37–38
symmetric product, 39
wedge product, 39
Theorema Egregium (Gauss), 131

Timelike curve, *see* Curve in $\mathbb{M}^{1,2}$, timelike curve
Timelike surface, *see* Surface in $\mathbb{M}^{1,2}$, timelike surface
Timelike vector, *see* Minkowski space, timelike vector
Totally geodesic surface
  in $\mathbb{S}^3$ or $\mathbb{H}^3$, 355
  in a Riemannian 3-manifold, 383–388
Totally quasi-umbilic timelike surface in $\mathbb{M}^{1,2}$, 160, 165–166
Totally umbilic surface
  in $\mathbb{E}^3$, 129
  in $\mathbb{M}^{1,2}$, timelike surface, 156–157
  in $\mathbb{P}^3$, elliptic surface, 229–232
Transition map
  between local coordinates on a surface, 5
  between local coordinates on a manifold, 6
Transpose notation
  for matrices, 31
  for vectors, 6

Umbilic point
  on a surface in $\mathbb{E}^3$, 124
  on a timelike surface in $\mathbb{M}^{1,2}$, 155
  on an elliptic surface in $\mathbb{P}^3$, 226
Unimodular frame bundle of $\mathbb{A}^n$, 95
Unimodular frame field along a curve in $\mathbb{A}^3$, 172
Unimodular frame on $\mathbb{A}^n$, 94
Unit sphere $\mathbb{S}^n$, 6–7

Variation
  of a surface in $\mathbb{E}^3$, 252–255
    compactly supported, 253
    normal, 253
  of an elliptic surface in $\mathbb{A}^3$, 269
    compactly supported, 269
    normal, 269
Vatuk, Sunita, xv, xvi
Vector
  column vector, 6
  row vector, 6
  tangent vector, 16, 19
  transpose notation for, 6
Vector bundle, 32–33
  base space, 32
  total space, 32
  fiber, 32
  base-point projection map, 32
  rank k, 32
  section, 32–33
    global section, 32
    local section, 32
    zero section, 33
  trivialization
    global trivialization, 32
    local trivialization, 32
Vector field, 24–25
  in local coordinates, 25
  left-invariant vector field on a Lie group, 26–27
  horizontal vector field, 380
Vertical tangent space, 366
Volume form, 92

Wave equation
  in characteristic/null coordinates, 302, 355
  in space-time coordinates, 296
Wedge product
  of vectors, 39
  of 1-forms, 41
Weierstrass, Karl, 261
Weierstrass-Enneper representation for a minimal surface in $\mathbb{E}^3$, 266–267
Wilczynski invariants
  of a curve in $\mathbb{P}^2$, 206
  of a curve in $\mathbb{P}^3$, 216
Wilczynski, Ernest, 206
Wilkens, George, xvi
World line, *see* Minkowski space, world line of a particle

Yu, Yunliang, xiii

Selected Published Titles in This Series

178 **Jeanne N. Clelland,** From Frenet to Cartan: The Method of Moving Frames, 2017
177 **Jacques Sauloy,** Differential Galois Theory through Riemann-Hilbert Correspondence, 2016
176 **Adam Clay and Dale Rolfsen,** Ordered Groups and Topology, 2016
175 **Thomas A. Ivey and Joseph M. Landsberg,** Cartan for Beginners: Differential Geometry via Moving Frames and Exterior Differential Systems, Second Edition, 2016
174 **Alexander Kirillov Jr.,** Quiver Representations and Quiver Varieties, 2016
173 **Lan Wen,** Differentiable Dynamical Systems, 2016
172 **Jinho Baik, Percy Deift, and Toufic Suidan,** Combinatorics and Random Matrix Theory, 2016
171 **Qing Han,** Nonlinear Elliptic Equations of the Second Order, 2016
170 **Donald Yau,** Colored Operads, 2016
169 **András Vasy,** Partial Differential Equations, 2015
168 **Michael Aizenman and Simone Warzel,** Random Operators, 2015
167 **John C. Neu,** Singular Perturbation in the Physical Sciences, 2015
166 **Alberto Torchinsky,** Problems in Real and Functional Analysis, 2015
165 **Joseph J. Rotman,** Advanced Modern Algebra: Third Edition, Part 1, 2015
164 **Terence Tao,** Expansion in Finite Simple Groups of Lie Type, 2015
163 **Gérald Tenenbaum,** Introduction to Analytic and Probabilistic Number Theory, Third Edition, 2015
162 **Firas Rassoul-Agha and Timo Seppäläinen,** A Course on Large Deviations with an Introduction to Gibbs Measures, 2015
161 **Diane Maclagan and Bernd Sturmfels,** Introduction to Tropical Geometry, 2015
160 **Marius Overholt,** A Course in Analytic Number Theory, 2014
159 **John R. Faulkner,** The Role of Nonassociative Algebra in Projective Geometry, 2014
158 **Fritz Colonius and Wolfgang Kliemann,** Dynamical Systems and Linear Algebra, 2014
157 **Gerald Teschl,** Mathematical Methods in Quantum Mechanics: With Applications to Schrödinger Operators, Second Edition, 2014
156 **Markus Haase,** Functional Analysis, 2014
155 **Emmanuel Kowalski,** An Introduction to the Representation Theory of Groups, 2014
154 **Wilhelm Schlag,** A Course in Complex Analysis and Riemann Surfaces, 2014
153 **Terence Tao,** Hilbert's Fifth Problem and Related Topics, 2014
152 **Gábor Székelyhidi,** An Introduction to Extremal Kähler Metrics, 2014
151 **Jennifer Schultens,** Introduction to 3-Manifolds, 2014
150 **Joe Diestel and Angela Spalsbury,** The Joys of Haar Measure, 2013
149 **Daniel W. Stroock,** Mathematics of Probability, 2013
148 **Luis Barreira and Yakov Pesin,** Introduction to Smooth Ergodic Theory, 2013
147 **Xingzhi Zhan,** Matrix Theory, 2013
146 **Aaron N. Siegel,** Combinatorial Game Theory, 2013
145 **Charles A. Weibel,** The K-book, 2013
144 **Shun-Jen Cheng and Weiqiang Wang,** Dualities and Representations of Lie Superalgebras, 2012
143 **Alberto Bressan,** Lecture Notes on Functional Analysis, 2013
142 **Terence Tao,** Higher Order Fourier Analysis, 2012
141 **John B. Conway,** A Course in Abstract Analysis, 2012

For a complete list of titles in this series, visit the
AMS Bookstore at **www.ams.org/bookstore/gsmseries/**.